SCHAUM'S OUTLINE OF

THEORY AND PROBLEMS

of

ENGINEERING MECHANICS
Statics and Dynamics

FIFTH EDITION

•

E. W. NELSON, B.S.M.E., M.Adm.E.

Engineering Supervisor, Retired
Western Electric Company

CHARLES L. BEST, B.S.M.E., M.S., Ph.D.

Emeritus Professor
Lafayette College

W. G. McLEAN, B.S.E.E., Sc.M., Eng.D.

Emeritus Director of Engineering
Lafayette College

•

SCHAUM'S OUTLINE SERIES
McGRAW-HILL

New York   San Francisco   Washington, D.C.   Auckland   Bogotá   Caracas   Lisbon
London   Madrid   Mexico City   Milan   Montreal   New Dehli
San Juan   Singapore   Sydney   Tokyo   Toronto

E. W. NELSON graduated from New York University with a B.S.M.E., and an M.Adm.E. He taught Mechanical Engineering at Lafayette College and later joined the engineering organization of the Western Electric Company (now Lucent Technologies). Retired from Western Electric, he is currently a Fellow of the American Society of Mechanical Engineers. He is a registered Professional Engineer and a member of Tau Beta Pi and Pi Tau Sigma.

CHARLES L. BEST is Emeritus Professor of Engineering at Lafayette College. He holds a B.S. in M.E. from Princeton, an M.S. in Mathematics from Brooklyn Polytechnic Institute, and a Ph.D. in Applied Mechanics from Virginia Polytechnic Institute. He is coauthor of two books on engineering mechanics and coauthor of another book on FORTRAN programming for engineering students. He is a member of Tau Beta Pi.

W. G. McLEAN is Emeritus Director of Engineering at Lafayette College. He holds a B.S.E.E. from Lafayette College, an Sc.M. from Brown University, and an honorary Eng.D. from Lafayette College. Professor McLean is the coauthor of two books on engineering mechanics, is past president of the Pennsylvania Society of Professional Engineers, and is active in the codes and standards committees of the American Society of Mechanical Engineers. He is a registered Professional engineer and a member of Phi Beta Kappa and Tau Beta Pi.

Schaum's Outline of Theory and Problems of
ENGINEERING MECHANICS

3 4 5 6 7 8 9 10 11 12 13 14 15 16 17 18 19 20 PRS PRS 9 0 2 1 0 9

ISBN 0-07-046193-7

Sponsoring Editor: Barbara Gilson
Production Supervisor: Tina Cameron
Editing Supervisor: Maureen B. Walker

Library of Congress Cataloging-in-Publication Data

Nelson, E. W. (Eric William)
    Schaum's outline of theory and problems of engineering mechanics, statics, and dynamics / E.W. Nelson, Charles L. Best, W.G. McLean. — 5th ed.
       p. cm. — (Schaum's outline series)
    McLean's name appears 1st on previous edition.
    Includes index.
    ISBN 0-07-046193-7 (paper)
    1. Mechanics, Applied.   I. Best, Charles L.   II. McLean, W. G. (William G.)   III. Title.   IV. Title: Theory and problems of engineering mechanics, statics, and dynamics.
    TA350.M387   1997
    620.1—DC21   97-24244
                                                                            CIP

McGraw-Hill

A Division of The McGraw·Hill Companies

Preface

This book is designed to supplement standard texts, primarily to assist students of engineering and science in acquiring a more thorough knowledge and proficiency in analytical and applied mechanics. It is based on the authors' conviction that numerous solved problems constitute one of the best means for clarifying and fixing in mind basic principles. While this book will not mesh precisely with any one text, the authors feel that it can be a very valuable adjunct to all.

The previous editions of this book have been very favorably received. This edition incorporates the U.S. Customary units and SI units, as did the third and fourth editions. The units based in the problems are roughly 50 percent U.S. Customary and 50 percent SI; however, the units are not mixed in any one problem. The authors attempt to use the best mathematical tools now available to students at the sophomore level. Thus the vector approach is applied in those chapters where its techniques provide an elegance and simplicity in theory and problems. On the other hand, we have not hesitated to use scalar methods elsewhere, since they provide entirely adequate solutions to most of the problems. Chapter 1 is a complete review of the minimum number of vector definitions and operations necessary for the entire book, and applications of this introductory chapter are made throughout the book. Some computer solutions are given, but most problems can be readily solved using other means.

Chapter topics correspond to material usually covered in standard introductory mechanics courses. Each chapter begins with statements of pertinent definitions, principles, and theorems. The text material is followed by graded sets of solved and supplementary problems. The solved problems serve to illustrate and amplify the theory, present methods of analysis, provide practical examples, and bring into sharp focus those fine points that enable the student to apply the basic principles correctly and confidently. Numerous proofs of theorems and derivations of formulas are included among the solved problems. The many supplementary problems serve as a complete review of the material covered in each chapter.

In the first edition the authors gratefully acknowledged their indebtedness to Paul B. Eaton and J. Warren Gillon. In the second edition the authors received helpful suggestions and criticism from Charles L. Best and John W. McNabb. Also in that edition Larry Freed and Paul Gary checked the solutions to the problems. In the third edition James Schwar assisted us in preparing the computer solutions in Appendix C. For the fourth edition we extend thanks again to James Schwar and to Michael Regan, Jr., for their help with the Appendix C computer solutions. For this fifth edition the authors thank William Best for checking the solutions to the new problems and reviewing the added new material. For typing the manuscripts of the third and fourth editions we are indebted to Elizabeth Bullock.

E. W. NELSON
C. L. BEST
W. G. McLEAN

Contents

Problems and Sections Also Found in the Companion SCHAUM'S ELECTRONIC TUTOR

Some of the problems and sections in this book have software components in the companion *Schaum's Electronic Tutor*. The Mathcad Engine, which "drives" the Electronic Tutor, allows every number, formula, and graph chosen to be completely live and interactive. To identify those items that are available in the Electronic Tutor software, please look for the Mathcad icons, [Mathcad icon], placed adjacent to a numbered item or problem number. A complete list of these Mathcad entries follows below. For more information about the software, including the sample screens, see Appendix D on page 501.

Problem 1.3	Problem 5.6	Problem 9.60	Problem 14.10	Problem 16.131
Problem 1.5	Problem 5.15	Problem 10.11	Problem 14.27	Problem 16.132
Problem 1.10	Problem 5.42	Problem 10.28	Problem 14.50	Problem 16.134
Problem 1.15	Problem 5.57	Problem 10.42	Problem 15.12	Problem 16.156
Problem 1.17	Problem 6.1	Problem 10.95	Problem 15.28	Problem 17.42
Section 2.3	Problem 6.2	Problem 10.96	Problem 15.32	Problem 17.43
Problem 2.3	Problem 6.5	Problem 10.108	Problem 15.51	Problem 17.48
Problem 2.4	Problem 6.9	Problem 11.6	Problem 15.57	Problem 17.56
Problem 2.12	Problem 6.15	Problem 11.8	Problem 15.74	Problem 17.89
Problem 2.13	Problem 6.29	Problem 12.3	Problem 15.76	Problem 17.92
Problem 2.28	Section 7.3	Problem 12.7	Problem 16.8	Problem 18.27
Problem 2.30	Problem 7.14	Problem 12.17	Problem 16.25	Problem 18.40
Problem 3.1	Problem 7.16	Problem 12.18	Problem 16.27	Problem 18.82
Problem 3.4	Problem 7.18	Problem 12.36	Problem 16.31	Problem 18.83
Problem 3.9	Problem 8.3	Problem 12.85	Problem 16.34	Problem 18.84
Problem 3.46	Problem 8.7	Problem 13.4	Problem 16.62	Problem 18.109
Problem 3.50	Problem 9.1	Problem 13.10	Problem 16.70	Problem 19.29
Problem 4.1	Problem 9.2	Problem 13.53	Problem 16.76	Problem 19.32
Problem 4.2	Problem 9.20	Problem 13.57	Problem 16.80	Problem 19.43
Problem 4.4	Problem 9.35	Problem 14.5	Problem 16.125	Problem 19.48
Problem 4.19	Problem 9.47	Problem 14.9	Problem 16.128	Problem 19.50
Problem 5.5				

Chapter 1

Vectors

1.1 DEFINITIONS

Scalar quantities possess only magnitude, e.g., time, volume, energy, mass, density, work. Scalars are added by ordinary algebraic methods, e.g., $2\,s + 7\,s = 9\,s$; $14\,kg - 5\,kg = 9\,kg$.

Vector quantities possess both magnitude and direction,* e.g., force, displacement, velocity, impulse. A vector is represented by an arrow at the given inclination. The head of the arrow indicates the sense, and the length represents the magnitude of the vector. The symbol for a vector is shown in print in boldface type, such as $\mathbf{P}$. The magnitude is represented by $|\mathbf{P}|$ or P.

A *free vector* may be moved anywhere in space provided it maintains the same direction and magnitude.

A *sliding vector* may be applied at any point along its line of action. By the *principle of transmissibility* the external effects of a sliding vector remain the same.

A *bound* or *fixed vector* must remain at the same point of application.

A *unit vector* is a vector one unit in length.

The *negative* of a vector $\mathbf{P}$ is the vector $-\mathbf{P}$ that has the same magnitude and inclination but is of the opposite sense.

The *resultant* of a system of vectors is the least number of vectors that will replace the given system.

1.1 ADDITION OF TWO VECTORS

(*a*) The *parallelogram law* states that the resultant $\mathbf{R}$ of two vectors $\mathbf{P}$ and $\mathbf{Q}$ is the diagonal of the parallelogram for which $\mathbf{P}$ and $\mathbf{Q}$ are adjacent sides. All three vectors $\mathbf{P}, \mathbf{Q}$, and $\mathbf{R}$ are concurrent as shown in Fig. 1-1(*a*). $\mathbf{P}$ and $\mathbf{Q}$ are also called the components of $\mathbf{R}$.

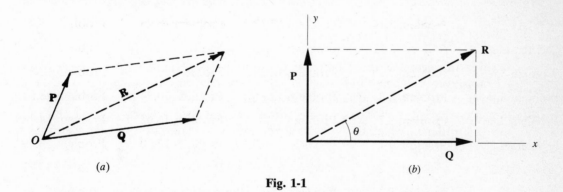

(*a*)   (*b*)

Fig. 1-1

(*b*) If the sides of the parallelogram in Fig. 1-1(*a*) are perpendicular, the vectors $\mathbf{P}$ and $\mathbf{Q}$ are said to be *rectangular* components of the vector $\mathbf{R}$. The rectangular components are illustrated in Fig. 1-1(*b*). The magnitude of the rectangular components is given by

$$\mathbf{Q} = \mathbf{R} \cos \theta$$

and

$$\mathbf{P} = \mathbf{R} \cos (90° - \theta) = \mathbf{R} \sin \theta$$

---

* Direction is understood to include both the inclination (angle) that the line of action makes with a given reference line and the sense of the vector along the line of action.

1

(*c*)   *Triangle law.* Place the tail end of either vector at the head end of the other. The resultant is drawn from the tail end of the first vector to the head end of the other. The triangle law follows from the parallelogram law because opposite sides of the paralleloram are free vectors as shown in Fig. 1-2.

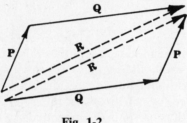

Fig. 1-2

(*d*)   Vector addition is commutative; i.e., $\mathbf{P} + \mathbf{Q} = \mathbf{Q} + \mathbf{P}$.

1.3   SUBTRACTION OF A VECTOR

Subtraction of a vector is accomplished by adding the negative of the vector; i.e.,

$$\mathbf{P} - \mathbf{Q} = \mathbf{P} + (-\mathbf{Q})$$

Note also that

$$-(\mathbf{P} + \mathbf{Q}) = -\mathbf{P} - \mathbf{Q}$$

1.4   ZERO VECTOR

A *zero vector* is obtained when a vector is subtracted from itself; i.e., $\mathbf{P} - \mathbf{P} = 0$. This is also called a *null vector*.

1.5   COMPOSITION OF VECTORS

Composition of vectors is the process of determining the resultant of a system of vectors. A vector polygon is drawn placing the tail end of each vector in turn at the head end of the preceding vector as shown in Fig. 1-3. The resultant is drawn from the tail end of the first vector to the head end (terminus) of the last vector. As will be shown later, not all vector systems reduce to a single vector. Since the order in which the vectors are drawn is immaterial, it can be seen that for three given vectors $\mathbf{P}$, $\mathbf{Q}$, and $\mathbf{S}$,

$$\mathbf{R} = \mathbf{P} + \mathbf{Q} + \mathbf{S} = (\mathbf{P} + \mathbf{Q}) + \mathbf{S}$$
$$= \mathbf{P} + (\mathbf{Q} + \mathbf{S}) = (\mathbf{P} + \mathbf{S}) + \mathbf{Q}$$

The above equation may be extended to any number of vectors.

1.6   MULTIPLICATION OF VECTORS BY SCALARS

(*a*)   The product of vector $\mathbf{P}$ and scalar m is a vector $m\mathbf{P}$ whose magnitude is $|m|$ times as great as the magnitude of $\mathbf{P}$ and that is similarly or oppositely directed to $\mathbf{P}$, depending on whether m is positive or negative.

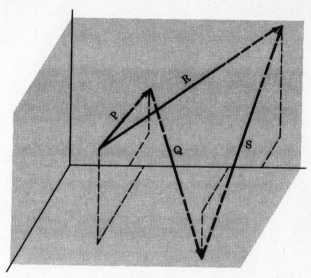

Fig. 1-3

(*b*) Other operations with scalars *m* and *n* are

$$(m + n)\mathbf{P} = m\mathbf{P} + n\mathbf{P}$$

$$m(\mathbf{P} + \mathbf{Q}) = m\mathbf{P} + m\mathbf{Q}$$

$$m(n\mathbf{P}) = n(m\mathbf{P}) = (mn)\mathbf{P}$$

1.7 ORTHOGONAL TRIAD OF UNIT VECTORS

An *orthogonal triad* of unit vectors **i**, **j**, and **k** is formed by drawing unit vectors along the *x*, *y*, and *z* axes respectively. A right-handed set of axes is shown in Fig. 1–4.

A vector **P** is written as

$$\mathbf{P} = P_x\mathbf{i} + P_y\mathbf{j} + P_z\mathbf{k}$$

where $P_x\mathbf{i}$, $P_y\mathbf{j}$, and $P_z\mathbf{k}$ are the vector components of **P** along the *x*, *y*, and *z* axes respectively as shown in Fig. 1-5.

Note that $P_x = P \cos \theta_x$, $P_y = P \cos \theta_y$, and $P_z = P \cos \theta_z$.

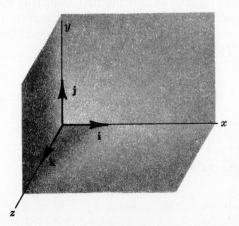

Fig. 1-4

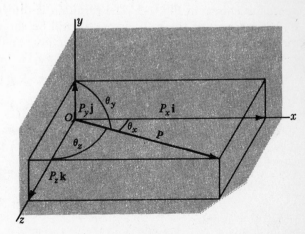

Fig. 1-5

1.8 POSITION VECTOR

The *position vector* **r** of a point (x, y, z) in space is written

$$\mathbf{r} = x\mathbf{i} + y\mathbf{j} + z\mathbf{k}$$

where $r = \sqrt{x^2 + y^2 + z^2}$. See Fig. 1–6.

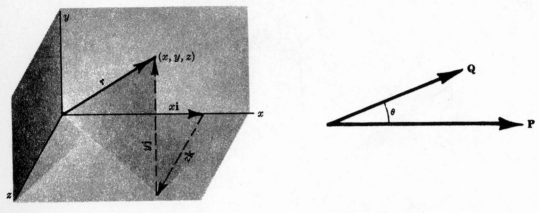

Fig. 1-6   Fig. 1-7

1.9 DOT OR SCALAR PRODUCT

The *dot* or *scalar product* of two vectors **P** and **Q**, written **P** · **Q**, is a scalar quantity and is defined as the product of the magnitudes of the two vectors and the cosine of their included angle θ (see Fig. 1-7). Thus

$$\mathbf{P} \cdot \mathbf{Q} = PQ \cos \theta$$

The following laws hold for dot products, where m is a scalar:

$$\mathbf{P} \cdot \mathbf{Q} = \mathbf{Q} \cdot \mathbf{P}$$

$$\mathbf{P} \cdot (\mathbf{Q} + \mathbf{S}) = \mathbf{P} \cdot \mathbf{Q} + \mathbf{P} \cdot \mathbf{S}$$

$$(\mathbf{P} + \mathbf{Q}) \cdot (\mathbf{S} + \mathbf{T}) = \mathbf{P} \cdot (\mathbf{S} + \mathbf{T}) + \mathbf{Q} \cdot (\mathbf{S} + \mathbf{T}) = \mathbf{P} \cdot \mathbf{S} + \mathbf{P} \cdot \mathbf{T} + \mathbf{Q} \cdot \mathbf{S} + \mathbf{Q} \cdot \mathbf{T}$$

$$m(\mathbf{P} \cdot \mathbf{Q}) = (m\mathbf{P}) \cdot \mathbf{Q} = \mathbf{P} \cdot (m\mathbf{Q})$$

Since **i**, **j**, and **k** are orthogonal,

$$\mathbf{i} \cdot \mathbf{j} = \mathbf{i} \cdot \mathbf{k} = \mathbf{j} \cdot \mathbf{k} = (1)(1) \cos 90° = 0$$

$$\mathbf{i} \cdot \mathbf{i} = \mathbf{j} \cdot \mathbf{j} = \mathbf{k} \cdot \mathbf{k} = (1)(1) \cos 0° = 1$$

Also, if $\mathbf{P} = P_x\mathbf{i} + P_y\mathbf{j} + P_x\mathbf{k}$ and $\mathbf{Q} = Q_x\mathbf{i} + Q_y\mathbf{j} + Q_z\mathbf{k}$ then

$$\mathbf{P} \cdot \mathbf{Q} = P_xQ_x + P_yQ_y + P_zQ_z$$

$$\mathbf{P} \cdot \mathbf{P} = P^2 = P_x^2 + P_y^2 + P_z^2$$

The magnitudes of the vector components of **P** along the rectangular axes can be written

$$P_x = \mathbf{P} \cdot \mathbf{i} \qquad P_y = \mathbf{P} \cdot \mathbf{j} \qquad P_z = \mathbf{P} \cdot \mathbf{k}$$

since, for example,

$$\mathbf{P} \cdot \mathbf{i} = (P_x\mathbf{i} + P_y\mathbf{j} + P_z\mathbf{k}) \cdot \mathbf{i} = P_x + 0 + 0 = P_x$$

Similarly, the magnitude of the vector component of **P** along any line L can be written $\mathbf{P} \cdot \mathbf{e}_L$, where $\mathbf{e}_L$ is the unit vector along the line L. (Some authors use **u** as unit vector.) Figure 1-8 shows a plane through the tail end A of vector **P** and a plane through the head B, both planes being perpendicular to line L. The planes intersect line L at points C and D. The vector **CD** is the component of **P** along L, and its magnitude equals $\mathbf{P} \cdot \mathbf{e}_L = Pe_L \cos \theta$.

Applications of these principles can be found in Problems 1.15 and 1.16.

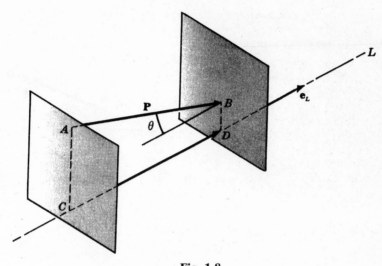

Fig. 1-8

1.10 THE CROSS OR VECTOR PRODUCT

The *cross* or *vector product* of two vectors **P** and **Q**, written $\mathbf{P} \times \mathbf{Q}$, is a vector **R** whose magnitude is the product of the magnitudes of the two vectors and the sine of their included angle. The vector $\mathbf{R} = \mathbf{P} \times \mathbf{Q}$ is normal to the plane of **P** and **Q** and points in the direction of advance of a right-handed screw when turned in the direction from **P** to **Q** through the smaller included angle θ. Thus if **e** is the unit vector that gives the direction of $\mathbf{R} = \mathbf{P} \times \mathbf{Q}$, the cross product can be written

$$\mathbf{R} = \mathbf{P} \times \mathbf{Q} = (PQ \sin \theta)\mathbf{e} \qquad 0 \leq \theta \leq 180°$$

Figure 1-9 indicates that $\mathbf{P} \times \mathbf{Q} = -\mathbf{Q} \times \mathbf{P}$ (not commutative).

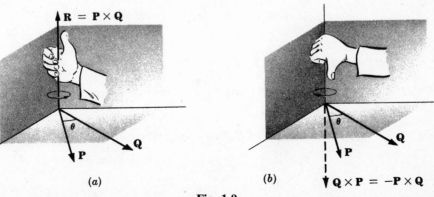

Fig. 1-9

The following laws hold for cross products, where m is a scalar:

$$\mathbf{P} \times (\mathbf{Q} + \mathbf{S}) = \mathbf{P} \times \mathbf{Q} + \mathbf{P} \times \mathbf{S}$$

$$(\mathbf{P} + \mathbf{Q}) \times (\mathbf{S} + \mathbf{T}) = \mathbf{P} \times (\mathbf{S} + \mathbf{T}) + \mathbf{Q} \times (\mathbf{S} + \mathbf{T})$$

$$= \mathbf{P} \times \mathbf{S} + \mathbf{P} \times \mathbf{T} + \mathbf{Q} \times \mathbf{S} + \mathbf{Q} \times \mathbf{T}$$

$$m(\mathbf{P} \times \mathbf{Q}) = (m\mathbf{P}) \times \mathbf{Q} = \mathbf{P} \times (m\mathbf{Q})$$

Since $\mathbf{i}$, $\mathbf{j}$, and $\mathbf{k}$ are orthogonal,

$$\mathbf{i} \times \mathbf{i} = \mathbf{j} \times \mathbf{j} = \mathbf{k} \times \mathbf{k} = 0$$

$$\mathbf{i} \times \mathbf{j} = \mathbf{k} \qquad \mathbf{j} \times \mathbf{k} = \mathbf{i} \qquad \mathbf{k} \times \mathbf{i} = \mathbf{j}$$

Also, if $\mathbf{P} = P_x\mathbf{i} + P_y\mathbf{j} + P_z\mathbf{k}$ and $\mathbf{Q} = Q_x\mathbf{i} + Q_y\mathbf{j} + Q_z\mathbf{k}$ then

$$\mathbf{P} \times \mathbf{Q} = (P_yQ_z - P_zQ_y)\mathbf{i} + (P_zQ_x - P_xQ_z)\mathbf{j} + (P_xQ_y - P_yQ_x)\mathbf{k} = \begin{vmatrix} \mathbf{i} & \mathbf{j} & \mathbf{k} \\ P_x & P_y & P_z \\ Q_x & Q_y & Q_z \end{vmatrix}$$

For proof of this cross-product determination see Problem 1.12.

1.11 VECTOR CALCULUS

(a) *Differentiation* of a vector $\mathbf{P}$ that varies with respect to a scalar quantity such as time t is performed as follows.

Let $\mathbf{P} = \mathbf{P}(t)$; that is, $\mathbf{P}$ is a function of time t. A change $\Delta\mathbf{P}$ in $\mathbf{P}$ as time changes from t to $(t + \Delta t)$ is

$$\Delta\mathbf{P} = \mathbf{P}(t + \Delta t) - \mathbf{P}(t)$$

Then
$$\frac{d\mathbf{P}}{dt} = \lim_{\Delta t \to 0} \frac{\Delta\mathbf{P}}{\Delta t} = \lim_{\Delta t \to 0} \frac{\mathbf{P}(t + \Delta t) - \mathbf{P}(t)}{\Delta t}$$

If $\mathbf{P}(t) = P_x\mathbf{i} + P_y\mathbf{j} + P_z\mathbf{k}$, where P_x, P_y, and P_z are functions of time t, we have

$$\frac{d\mathbf{P}}{dt} = \lim_{\Delta t \to 0} \frac{(P_x + \Delta P_x)\mathbf{i} + (P_y + \Delta P_y)\mathbf{j} + (P_z + \Delta P_z)\mathbf{k} - P_x\mathbf{i} - P_y\mathbf{j} - P_z\mathbf{k}}{\Delta t}$$

$$= \lim_{\Delta t \to 0} \frac{\Delta P_x\mathbf{i} + \Delta P_y\mathbf{j} + \Delta P_z\mathbf{k}}{\Delta t} = \frac{dP_x}{dt}\mathbf{i} + \frac{dP_y}{dt}\mathbf{j} + \frac{dP_z}{dt}\mathbf{k}$$

The following operations are valid:

$$\frac{d}{dt}(\mathbf{P} + \mathbf{Q}) = \frac{d\mathbf{P}}{dt} + \frac{d\mathbf{Q}}{dt}$$

$$\frac{d}{dt}(\mathbf{P} \cdot \mathbf{Q}) = \frac{d\mathbf{P}}{dt} \cdot \mathbf{Q} + \mathbf{P} \cdot \frac{d\mathbf{Q}}{dt}$$

$$\frac{d}{dt}(\mathbf{P} \times \mathbf{Q}) = \frac{d\mathbf{P}}{dt} \times \mathbf{Q} + \mathbf{P} \times \frac{d\mathbf{Q}}{dt}$$

$$\frac{d}{dt}(\psi\mathbf{P}) = \psi\frac{d\mathbf{P}}{dt} + \frac{d\psi}{dt}\mathbf{P} \qquad \text{where } \psi \text{ is a scalar function of } t$$

(b) *Integration* of a vector **P** that varies with respect to a scalar quantity such as time t is performed as follows. Let $\mathbf{P} = \mathbf{P}(t)$; that is, **P** is a function of time t. Then

$$\int_{t_0}^{t_1} \mathbf{P}(t)\, dt = \int_{t_0}^{t_1} (P_x\mathbf{i} + P_y\mathbf{j} + P_z\mathbf{k})\, dt$$

$$= \mathbf{i}\int_{t_0}^{t_1} P_x\, dt + \mathbf{j}\int_{t_0}^{t_1} P_y\, dt + \mathbf{k}\int_{t_0}^{t_1} P_z\, dt$$

1.12 DIMENSIONS AND UNITS

In the study of mechanics, the characteristics of a body and its motion can be described in terms of a set of fundamental quantities called dimensions. In the United States, engineers have been accustomed to a gravitational system using the dimensions of force, length, and time. Most countries throughout the world use an absolute system in which the selected dimensions are mass, length, and time. There is a growing trend to use this second system in the United States.

Both systems derive from Newton's second law of motion, which is often written as

$$\mathbf{R} = M\mathbf{a}$$

where **R** is the resultant of all forces acting on a particle, **a** is the acceleration of the particle, and M is the constant of proportionality called the mass.

U.S. Customary System

In this engineering system, the unit of length is the foot (ft), the unit of time is the second (s), and the unit of force is the pound (lb). A mass M falling freely near the earth's surface is pulled toward the earth's centre by a force W with an acceleration of gravity g. The force W is the weight measured in pounds and the acceleration g is in ft/s². Hence Newton's second law becomes, in scalar form,

$$W = Mg$$

The value of the acceleration of gravity g varies with the observer's location on the surface of the earth. In this book the value of 32.2 ft/s² will be used. An object that weighs 1 lb at or near the earth's surface will have a free-fall acceleration g of 32.2 ft/s². The above equation yields

$$M = \frac{W}{g} = \frac{1\ \text{lb}}{32.2\ \text{ft/s}^2} = \frac{1}{32.2}\ \frac{\text{lb s}^2}{\text{ft}} = \frac{1}{32.2}\ \text{slug}$$

In solving statics problems, the mass is not mentioned. It is important to realize that the mass in slugs is a constant for a given body. On the surface of the moon, this same given mass will have acting on it a force of gravity approximately one-sixth of that on the earth.

The International System (SI)

In the International System (SI),* the unit of mass is the kilogram (kg), the unit of length is the meter (m), and the unit of time is the second (s). The unit of force is the newton (N) and is defined

---

* SI is the acronym for Système International d'Unités (modernized international metric system).

as the force that will accelerate a mass of one kilogram one meter per second squared (m/s²). Thus

$$1 \text{ N} = (1 \text{ kg})(1 \text{ m/s}^2) = 1 \text{ kg} \cdot \text{m/s}^2 = 1 \text{ kg} \cdot \text{m} \cdot \text{s}^{-2}$$

A mass of 1 kg falling freely near the surface of the earth has an acceleration of gravity g that varies from place to place. In this book we shall assume an average value of 9.80 m/s². Thus the force of gravity acting on a 1-kg mass becomes

$$W = Mg = (1 \text{ kg})(9.80 \text{ m/s}^2) = 9.80 \text{ kg} \cdot \text{m/s}^2 = 9.80 \text{ N}$$

Of course, problems in statics involve forces; but, in a problem, a mass given in kilograms is not a force. The gravitational force acting on the mass must be used. In all work involving mass, the student must remember to multiply the mass in kilograms by 9.80 m/s² to obtain the gravitational force in newtons. A 5-kg mass has a gravitational force of $5 \times 9.8 = 49$ N acting on it.

The student should further note that, in SI, the millimeter (mm) is the standard linear dimension unit for engineering drawings. Therefore, all engineering drawing dimensions must be in millimeters (1 mm = 10^{-3} m). Further, a space should be left between the number and unit symbol; e.g., 2.85 mm, not 2.85mm. When using five or more figures, space them in groups of three starting at the decimal point as 12 832.325. Do not use commas in SI. A number with four figures can be written without the space unless it is in a column of quantities involving five or more figures.

Tables of SI units, SI prefixes, and conversion factors for the modern metric system (SI) are included in Appendix A. In this text about 50 percent of the problems are in U.S. Customary units and 50 percent in SI units.

Solved Problems

1.1.   In a plane, add 120-lb force at 30° and a −100-lb force at 90° using the parallelogram method. Refer to Fig. 1-10(a).

SOLUTION

Draw a sketch of the problem, not necessarily to scale. The negative sign indicates that the 100-lb force acts along the 90° line downward toward the origin. This is equivalent to a positive 100-lb force along the 270° line, according to the principle of transmissibility.

As in Fig. 1-10(b), place the tail ends of the two vectors at a common point and draw the vectors to a suitable scale. Complete the parallelogram. The resulting R measures to the chosen scale 111 lb. By protractor, it is at an angle with the x axis of $\theta_x = 339°$.

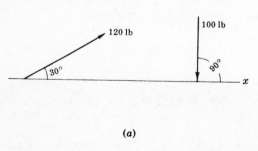

(a)

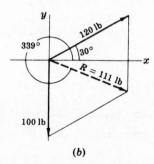

(b)

Fig. 1-10

Consider the triangle, one side of which is the y axis, in Fig. 1-10(b). The sides of this triangle are R, 100, and 200. The angle between the 100 and 120 sides is 60°. Applying the Law of Cosines,

$$R^2 = \overline{120}^2 + \overline{100}^2 - 2(120)(100) \cos 60° \qquad R = 111 \text{ lb}$$

Now applying the Law of Sines,

$$\frac{120}{\sin \alpha} = \frac{111}{\sin 60} \qquad \alpha = 69°$$

The angle of 60° added to 270° yields the measured angle of 339°.

1.2. Use the triangle law for Problem 1.1. See Fig. 1-11.

SOLUTION

It is immaterial which vector is chosen first. Take the 120-lb force. To the head of this vector attach the tail end of the 100-lb force. Draw the resultant from the tail end of the 120-lb force to the head end of the 100-lb force. When measured to the chosen scale and the direction determined, the results are the same as in Problem 1.1.

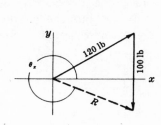

Fig. 1-11

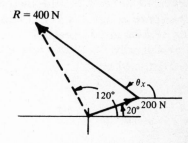

Fig. 1-12

1.3. The resultant of two forces in a plane is 400 N at 120°. One of the forces is 200 N at 20°. Determine the missing force. See Fig. 1-12.

SOLUTION

Select a point through which to draw the resultant and the given force to a convenient scale.

Draw the line connecting the head ends of the given force and the resultant. Place a head on the end of this line near the resultant. This line represents the missing force. When measured to scale, the desired force is 477 N with $\theta_x = 144°$.

This result is also obtained analytically by the laws of trigonometry. The angle between **R** and the 200-N force is 100°, and hence, by the Law of Cosines, the unknown force F is

$$F^2 = \overline{400}^2 + \overline{200}^2 - 2(400)(200) \cos 100° \qquad F = 477 \text{ N}$$

Call the angle between **F** and the 200-N force α. Then, by the Law of Sines,

$$\frac{477}{\sin 100} = \frac{400}{\sin \alpha} \qquad \alpha = 55.7° \qquad \theta_x = 144°$$

1.4. In a plane, subtract 130 N, 60° from 280 N, 320°. See Fig. 1-13.

SOLUTION

To the 280-N, 320° force add the negative of the 130-N, 60° force, obtaining a resultant force of 330 N, 297°. All angles are measured with respect to the x axis.

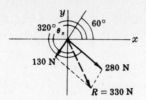

Fig. 1-13

1.5. Determine the resultant of the following coplanar system of forces: 26 lb, 10°; 39 lb, 114°; 63 lb, 183°; 57 lb, 261°. See Fig. 1–14.

SOLUTION

Applying the polygon method, place the tail of each vector in turn at the head of the preceding vector. See Fig. 1-14(*a*).

The resultant vector is the force drawn from the tail of the first vector to the head of the last vector. Measured to scale, $R = 65$ lb with $\theta_x = 197°$.

This problem can also be solved analytically using the idea of rectangular components. Resolve each force in Fig. 1–14(*b*) into x and y rectangular components. Since all the x components are collinear, they can be added algebraically, as can the y components. Now, if the x components and y components are added, the two sums form the x and y components of the resultant. Thus

$$R_x = 26 \cos 10° + 39 \cos 114° + 63 \cos 183° + 57 \cos 261° = -62.1$$

$$R_y = 26 \sin 10° + 39 \sin 114° + 63 \sin 183° + 57 \sin 261° = -19.5$$

$$R = \sqrt{(-62.1)^2 + (-19.5)^2} \qquad R = 65 \text{ lb}$$

$$\tan \theta_x = \frac{-19.5}{-62.1} \qquad \theta_x = 17° \qquad \theta = 180° + 17° = 197°$$

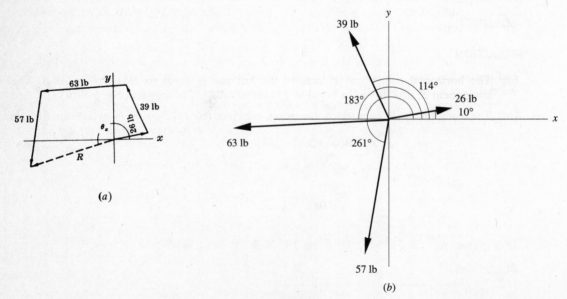

(*a*)

(*b*)

Fig. 1-14

1.6. In Fig. 1-15 the rectangular component of the force **F** is 10 lb in the direction of *OH*. The force **F** acts at 60° to the positive *x* axis. What is the magnitude of the force?

SOLUTION

The component of **F** in the direction of *OH* is $F \cos \theta$. Hence, $F \cos 15° = 10$ or $F = 10.35$ lb

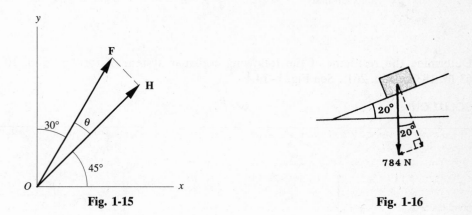

Fig. 1-15   Fig. 1-16

1.7. An 80-kg person is standing on a board inclined 20° with the horizontal. What is the gravitational component (*a*) normal to the board and (*b*) parallel to the board? See Fig. 1–16.

SOLUTION

(*a*) The normal component is at an angle of 20° with the gravitational force vector, which has a magnitude of 80 (9.8) = 784 N. To scale, the normal component measures 740 N. By trigonometry, the normal component is $784 \cos 20° = 737$ N.

(*b*) To scale, the parallel component is 270 N. By trigonometry, it is $784 \cos 70° = 268$ N.

1.8. A force *P* of 235 N acts at an angle of 60° with the horizontal on a block resting on a 22° inclined plane. Determine algebraically (*a*) the horizontal and vertical components of *P* and (*b*) the components of *P* perpendicular to and along the plane. Refer to Fig. 1-17(*a*).

SOLUTION

(*a*) The horizontal component P_h acts to the left and is equal to $235 \cos 60° = 118$ N. The vertical component P_v acts up and is equal to $235 \sin 60° = 204$ N as shown in Fig. 1-17(*b*).

(*b*) The component $P_{\parallel}$ parallel to the plane $= 235 \cos (60° - 22°) = 185$ N acting up the plane. The component $P_{\perp}$ normal to the plane $= 235 \sin 38° = 145$ N as shown in Fig. 1-17(*c*).

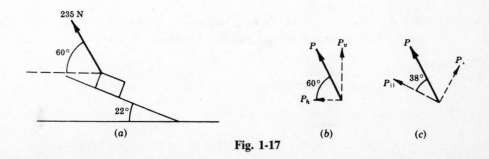

(*a*)   (*b*)   (*c*)

Fig. 1-17

1.9. The three forces shown in Fig. 1-18 produce a resultant force of 20 lb acting upward along the *y* axis. Determine the magnitudes of **F** and **P**.

SOLUTION

For the resultant to be a force of 20 lb upward along the *y* axis, $R_x = 0$ and $R_y = 20$ lb. As the sum of the *x* components must be equal to the *x* component of the resultant $R_x = P \cos 30° - 90 \cos 40° = 0$, from which $P = 79.6$ lb. Similarly, $R_y = P \sin 30° + 90 \sin 40° - F = 20$ and $F = 77.7$ lb.

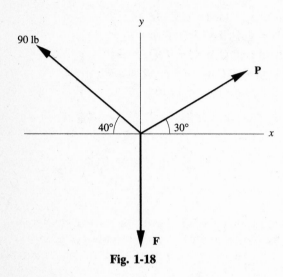

Fig. 1-18

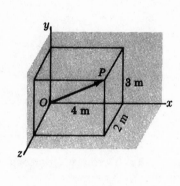

Fig. 1-19

Mathcad

1.10. Refer to Fig. 1-19. The *x*, *y*, and *z* edges of a rectangular parallelepiped are 4, 3, and 2 m respectively. If the diagonal *OP* drawn from the origin represents a 50-N force, determine the *x*, *y*, and *z* components of the force. Express the force as a vector in terms of the unit vectors **i**, **j**, and **k**.

SOLUTION

Let θ_x, θ_y, θ_z represent respectively the angles between the diagonal *OP* and the *x*, *y*, *z* axes. Then

$$P_x = P \cos \theta_x \qquad P_y = P \cos \theta_y \qquad P_z = P \cos \theta_z$$

Length of $OP = \sqrt{4^2 + 3^2 + 2^2} = 5.38$ m. Hence,

$$\cos \theta_x = \frac{4}{5.38} \qquad \cos \theta_y = \frac{3}{5.38} \qquad \cos \theta_z = \frac{2}{5.38}$$

Since each component in the sketch is in the positive direction of the axis along which it acts,

$$P_x = 50 \cos \theta_x = 37.2 \text{ N} \qquad P_y = 50 \cos \theta_y = 27.9 \text{ N} \qquad P_z = 50 \cos \theta_z = 18.6 \text{ N}$$

The vector $\mathbf{P} = P_x\mathbf{i} + P_y\mathbf{j} + P_z\mathbf{k} = 37.2\mathbf{i} + 27.9\mathbf{j} + 18.6\mathbf{k}$ N.

1.11. Determine the *x*, *y*, and *z* components of a 100-N force passing from the origin through point $(2, -4, 1)$. Express the vector in terms of the unit vectors **i**, **j**, and **k**.

SOLUTION

The direction cosines of the force line are

$$\cos \theta_x = \frac{2}{\sqrt{(2)^2 + (-4)^2 + (1)^2}} = 0.437 \qquad \cos \theta_y = \frac{-4}{\sqrt{21}} = -0.873 \qquad \cos \theta_z = 0.281$$

Hence $P_x = 43.7$ N, $P_y = -87.3$ N, $P_z = 21.8$ N; and the vector $\mathbf{P} = 43.7\mathbf{i} - 87.3\mathbf{j} + 21.8\mathbf{k}$ N.

1.12. Show that the cross product of two vectors **P** and **Q** can be written as

$$\mathbf{P} \times \mathbf{Q} = \begin{vmatrix} \mathbf{i} & \mathbf{j} & \mathbf{k} \\ P_x & P_y & P_z \\ Q_x & Q_y & Q_z \end{vmatrix}$$

SOLUTION

Write the given vectors in component form and expand the cross product to obtain

$$\mathbf{P} \times \mathbf{Q} = (P_x\mathbf{i} + P_y\mathbf{j} + P_z\mathbf{k}) \times (Q_x\mathbf{i} + Q_y\mathbf{j} + Q_z\mathbf{k})$$
$$= (P_xQ_x)\mathbf{i} \times \mathbf{i} + (P_xQ_y)\mathbf{i} \times \mathbf{j} + (P_xQ_z)\mathbf{i} \times \mathbf{k}$$
$$+ (P_yQ_x)\mathbf{j} \times \mathbf{i} + (P_yQ_y)\mathbf{j} \times \mathbf{j} + (P_yQ_z)\mathbf{j} \times \mathbf{k}$$
$$+ (P_zQ_x)\mathbf{k} \times \mathbf{i} + (P_zQ_y)\mathbf{k} \times \mathbf{j} + (P_zQ_z)\mathbf{k} \times \mathbf{k}$$

But $\mathbf{i} \times \mathbf{i} = \mathbf{j} \times \mathbf{j} = \mathbf{k} \times \mathbf{k} = 0$; and $\mathbf{i} \times \mathbf{j} = \mathbf{k}$ and $\mathbf{j} \times \mathbf{i} = -\mathbf{k}$, etc. Hence

$$\mathbf{P} \times \mathbf{Q} = (P_xQ_y)\mathbf{k} - (P_xQ_z)\mathbf{j} - (P_yQ_x)\mathbf{k} + (P_yQ_z)\mathbf{i} + (P_zQ_x)\mathbf{j} - (P_zQ_y)\mathbf{i}$$

These terms can be grouped as

$$\mathbf{P} \times \mathbf{Q} = (P_yQ_z - P_zQ_y)\mathbf{i} + (P_zQ_x - P_xQ_z)\mathbf{j} + (P_xQ_y - P_yQ_x)\mathbf{k}$$

or in determinant form as

$$\mathbf{P} \times \mathbf{Q} = \begin{vmatrix} \mathbf{i} & \mathbf{j} & \mathbf{k} \\ P_x & P_y & P_z \\ Q_x & Q_y & Q_z \end{vmatrix}$$

Be careful to observe that the scalar components of the first vector **P** in the cross product are written in the middle row of the determinant.

1.13. A force $\mathbf{F} = 2.63\mathbf{i} + 4.28\mathbf{j} - 5.92\mathbf{k}$ N acts through the origin. What is the magnitude of this force and what angles does it make with the x, y, and z axes?

SOLUTION

$$F = \sqrt{(2.63)^2 + (4.28)^2 + (-5.92)^2} = 7.75 \text{ N}$$

$$\cos\theta_x = +\frac{2.63}{7.75} \qquad \theta_x = 70.2°$$
$$\cos\theta_y = +\frac{4.28}{7.75} \qquad \theta_y = 56.3°$$
$$\cos\theta_z = -\frac{5.92}{7.75} \qquad \theta_z = 139.8°$$

1.14. Find the dot product of $\mathbf{P} = 4.82\mathbf{i} - 2.33\mathbf{j} + 5.47\mathbf{k}$ N and $\mathbf{Q} = -2.81\mathbf{i} - 6.09\mathbf{j} + 1.12\mathbf{k}$ m.

SOLUTION

$$\mathbf{P} \cdot \mathbf{Q} = P_xQ_x + P_yQ_y + P_zQ_z = (4.82)(-2.81) + (-2.33)(-6.09) + (5.47)(1.12) = 6.72 \text{ N} \cdot \text{m}$$

1.15. Determine the unit vector $\mathbf{e}_L$ for a line L that originates at point $(2, 3, 0)$ and passes through point $(-2, 4, 6)$. Next determine the projection of the vector $\mathbf{P} = 2\mathbf{i} + 3\mathbf{j} - \mathbf{k}$ along the line L.

SOLUTION

The line L changes from $+2$ to -2 in the x direction, or a change of -4. The change in the y direction is $4 - 3 = +1$. The change in the z direction is $6 = 0 = +6$. The unit vector is

$$\mathbf{e}_L = \frac{-4}{\sqrt{(-4)^2 + (+1)^2 + (+6)^2}}\mathbf{i} + \frac{1}{\sqrt{53}}\mathbf{j} + \frac{6}{\sqrt{53}}\mathbf{k} = -0.549\mathbf{i} + 0.137\mathbf{j} + 0.823\mathbf{k}$$

The projection of $\mathbf{P}$ is thus

$$\mathbf{P} \cdot \mathbf{e}_L = 2(-0.549) + 3(0.137) - 1(0.823) = -1.41$$

1.16. Determine the projection of the force $\mathbf{P} = 10\mathbf{i} - 8\mathbf{j} + 14\mathbf{k}$ lb on the directed line L which originates at point $(2, -5, 3)$ and passes through point $(5, 2, -4)$.

SOLUTION

The unit vector along L is

$$\mathbf{e}_L = \frac{5-2}{\sqrt{(5-2)^2 + [2-(-5)]^2 + (-4-3)^3}}\mathbf{i} + \frac{2-(-5)}{\sqrt{107}}\mathbf{j} + \frac{-4-3}{\sqrt{107}}\mathbf{k}$$

$$= 0.290\mathbf{i} + 0.677\mathbf{j} - 0.677\mathbf{k}$$

The projection of $\mathbf{P}$ on L is

$$\mathbf{P} \cdot \mathbf{e}_L = (10\mathbf{i} - 8\mathbf{j} + 14\mathbf{k}) \cdot (0.29\mathbf{i} + 0.677\mathbf{j} - 0.677\mathbf{k})$$

$$= 2.90 - 5.42 - 9.48 = -12.0 \text{ lb}$$

The minus sign indicates that the projection is directed opposite to the direction of L.

1.17. Find the cross product of $\mathbf{P} = 2.85\mathbf{i} + 4.67\mathbf{j} - 8.09\mathbf{k}$ ft and $\mathbf{Q} = 28.3\mathbf{i} + 44.6\mathbf{j} + 53.3\mathbf{k}$ lb.

SOLUTION

$$\mathbf{P} \times \mathbf{Q} = \begin{vmatrix} \mathbf{i} & \mathbf{j} & \mathbf{k} \\ P_x & P_y & P_z \\ Q_x & Q_y & Q_z \end{vmatrix} = \begin{vmatrix} \mathbf{i} & \mathbf{j} & \mathbf{k} \\ 2.85 & 4.67 & -8.09 \\ 28.3 & 44.6 & 53.3 \end{vmatrix}$$

$$= \mathbf{i}[(4.67)(53.3) - (44.6)(-8.09)] - \mathbf{j}[(2.85)(53.3) - (28.3)(-8.09)]$$

$$+ \mathbf{k}[(2.85)(44.6) - (28.3)(4.67)]$$

$$= \mathbf{i}(249 + 361) - \mathbf{j}(152 + 229) + \mathbf{k}(127 - 132) = 610\mathbf{i} - 381\mathbf{j} - 5\mathbf{k} \text{ lb-ft}$$

1.18. Determine the time derivative of the position vector $\mathbf{r} = x\mathbf{i} + 6y^2\mathbf{j} - 3z\mathbf{k}$, where $\mathbf{i}, \mathbf{j}, \mathbf{k}$ are fixed vectors.

SOLUTION

The time derivative is

$$\frac{d\mathbf{r}}{dt} = \frac{dx}{dt}\mathbf{i} + 12y\frac{dy}{dt}\mathbf{j} - 3\frac{dz}{dt}\mathbf{k}$$

1.19. Determine the time integral from time $t_1 = 1$ s to time $t_2 = 3$ s of the velocity vector

$$\mathbf{v} = t^2\mathbf{i} + 2t\mathbf{j} - \mathbf{k} \text{ ft/s}$$

where $\mathbf{i}, \mathbf{j}$, and $\mathbf{k}$ are fixed vectors.

SOLUTION

$$\int_1^3 (t^2\mathbf{i} + 2t\mathbf{j} - \mathbf{k})\, dt = \mathbf{i}\int_1^3 t^2\, dt + \mathbf{j}\int_1^3 2t\, dt - \mathbf{k}\int_1^3 dt = 8.67\mathbf{i} + 8.00\mathbf{j} - 2.00\mathbf{k}$$

Supplementary Problems

1.20. Determine the resultant of the coplanar forces 100 N, 0° and 200 N, 90°.   *Ans.*   224 N, $\theta_x = 64°$

1.21. Determine the resultant of the coplanar forces 32 N, 20° and 64 N, 190°.   *Ans.*   33.0 N, $\theta_x = 180°$

1.22. Find the resultant of the coplanar forces 80 N, −30° and 60 N, 60°.   *Ans.*   100 N, $\theta_x = 6.87°$

1.23. Find the resultant of the concurrent coplanar forces 120 N, 78° and 70 N, 293°.
Ans.   74.7 N, $\theta_x = 45.2°$

1.24. The resultant of two coplanar forces is 18 oz at 30°. If one of the forces is 28 oz at 0°, determine the other.   *Ans.*   15.3 oz, 144°

1.25. The resultant of two coplanar forces is 36 N at 45°. If one of the forces is 24 N at 0°, find the other force.
Ans.   25.5 N, 87°

1.26. The resultant of two coplanar forces is 50 N at 143°. One of the forces is 120 N at 238°. Determine the missing force.   *Ans.*   134 N, $\theta_x = 79.6°$

1.27. The resultant of two forces, one in the positive *x* direction and the other in the positive *y* direction, is 100 lb at 50° counterclockwise from the positive *x*-direction. What are the two forces?
Ans. $R_x = 64.3$ lb, $R_y = 76.6$ lb

1.28. A force of 120 N has a rectangular component of 84 N acting along a line making an angle of 20° counterclockwise from the positive *x* axis. What angle does the 120-N force make with the positive *x* axis?
Ans.   65.6°

1.29. Determine the resultant of the coplanar forces: 6 oz, 38°; 12 oz, 73°; 18 oz, 67°; 24 oz, 131°.
Ans.   50.0 oz, $\theta_x = 91°$

1.30. Determine the resultant of the coplanar forces: 20 lb, 0°; 20 lb, 30°; 20 lb, 60°; 20 lb, 90°; 20 lb, 120°; 20 lb, 150°.   *Ans.*   77.2 lb, $\theta_x = 75°$

1.31. Determine the single force that will replace the following coplanar forces: 120 N, 30°; 200 N, 110°; 340 N, 180°; 170 N, 240°; 80 N, 300°.   *Ans.* 351 N, 175°

1.31. Determine the single force that will replace the following coplanar forces: 120 N, 30°; 200 N, 110°; 340 N, 180°; 170 N, 240°; 80 N, 300°.   *Ans.*   351 N, 175°

1.32. Find the single force to replace the following coplanar forces: 150 N, 78°; 320 N, 143°; 485 N, 249°; 98 N, 305°; 251 N, 84°.   *Ans.*   321 N, 171°

1.33. A sled is being pulled by a force of 25 lb exerted in a rope inclined 30° with the horizontal. What is the effective component of the force pulling the sled? What is the component tending to lift the sled vertically?   *Ans.* $P_h = 21.7$ lb, $P_v = 12.5$ lb

1.34. Determine the resultant of the following coplanar forces: 15 N, 30°; 55 N, 80°; 90 N, 210°; and 130 N, 260°.   *Ans.*   136 N, $\theta_x = 235°$

1.35. A car is traveling at a constant speed in a tunnel, up a 1-percent grade. If the car and passenger weigh 3100 lb, what tractive force must the engine supply to just overcome the component of the gravitational force on the car along the bottom of the tunnel?   *Ans.*   31 lb

1.36. A telephone pole is supported by a guy wire that exerts a pull of 200 lb on the top of the pole. If the angle between the wire and the pole is 50°, what are the horizontal and vertical components of the pull on the pole?   *Ans.* $P_h = 153$ lb, $P_v = 129$ lb

1.37. A boat is being towed through a canal by a horizontal cable that makes an angle of 10° with the shore. If the pull on the cable is 200 N, find the force tending to move the boat along the canal. *Ans.*   197 N

1.38. Express in terms of the unit vectors **i**, **j**, and **k** the force of 200 N that starts at the point $(2, 5, -3)$ and passes through the point $(-3, 2, 1)$.   *Ans.* $\mathbf{F} = -141\mathbf{i} - 84.9\mathbf{j} + 113\mathbf{k}$ N

1.39. Determine the resultant of the three forces $\mathbf{F}_1 = 2.0\mathbf{i} + 3.3\mathbf{j} - 2.6\mathbf{k}$ lb, $\mathbf{F}_2 = -\mathbf{i} + 5.2\mathbf{j} - 2.9\mathbf{k}$ lb, and $F_3 = 8.3\mathbf{i} - 6.6\mathbf{j} + 5.8\mathbf{k}$ lb, which are concurrent at the point $(2, 2, -5)$.
Ans. $\mathbf{R} = 9.3\mathbf{i} + 1.9\mathbf{j} + 0.3\mathbf{k}$ lb at $(2, 2, -5)$

1.40. The pulley shown in Fig. 1-20 is free to ride on the supporting guide wire. If the pulley supports a 160-lb weight, what is the tension in the wire?   *Ans.* $T = 234$ lb

Fig. 1-20   Fig. 1-21

1.41. Two cables support a 500-lb weight as shown in Fig. 1-21. Determine the tension in each cable.
Ans. $T_{AB} = 433$ lb, $T_{BC} = 250$ lb

1.42. What horizontal force P is required to hold the 10-lb weight W in the position shown in Fig. 1-22?
Ans. $P = 3.25$ lb

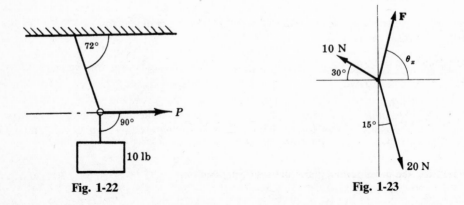

Fig. 1-22   Fig. 1-23

1.43. A charged particle is at rest under the action of three other charged particles. The forces exerted by two of the particles are shown in Fig. 1-23. Determine the magnitude and direction of the third force.
Ans. $F = 14.7$ N, $\theta_x = 76.8°$

1.44. Determine the resultant of the coplanar forces 200 N, 0° and 400 N, 90°.
Ans. Since each force in Problem 1.20 has been multiplied by the scalar 2, the magnitude of the resultant in this problem should be double that of Problem 1.20. The angle should be the same.

1.45. What vector must be added to the vector $\mathbf{F} = 30\,\text{N}$, 60° to yield the zero vector?
Ans. 30 N, $\theta_x = 240°$

1.46. At time $t = 2\,\text{s}$, a point moving on a curve has coordinates $(3, -5, 2)$. At time $t = 3\,\text{s}$, the coordinates of the point are $(1, -2, 0)$. What is the change in the position vector?   *Ans.* $\Delta\mathbf{r} = -2\mathbf{i} + 3\mathbf{j} - 2\mathbf{k}$

1.47. Determine the dot product of $\mathbf{P} = 4\mathbf{i} + 2\mathbf{j} - \mathbf{k}$ and $\mathbf{Q} = -3\mathbf{i} + 6\mathbf{j} - 2\mathbf{k}$.   *Ans.* $+2$

1.48. Find the dot product of $\mathbf{P} = 2.12\mathbf{i} + 8.15\mathbf{j} - 4.28\mathbf{k}\,\text{N}$ and $\mathbf{Q} = 6.29\mathbf{i} - 8.93\mathbf{j} - 10.5\mathbf{k}\,\text{m}$.
Ans. $-14.5\,\text{N}\cdot\text{m}$

1.49. Determine the cross product of the vectors in Problem 1.47.   *Ans.* $\mathbf{P} \times \mathbf{Q} = 2\mathbf{i} + 11\mathbf{j} + 30\mathbf{k}$

1.50. Determine the cross product of $\mathbf{P} = 2.12\mathbf{i} + 8.15\mathbf{j} - 4.28\mathbf{k}$ and $\mathbf{Q} = 2.29\mathbf{i} - 8.93\mathbf{j} - 10.5\mathbf{k}$.
Ans. $-124\mathbf{i} + 12.5\mathbf{j} - 37.6\mathbf{k}$

1.51. Determine the derivative with respect to time of $\mathbf{P} = x\mathbf{i} + 2y\mathbf{i} - z^2\mathbf{k}$.
Ans. $\dfrac{d\mathbf{P}}{dt} = \dfrac{dx}{dt}\mathbf{i} + 2\dfrac{dy}{dt}\mathbf{j} - 2z\dfrac{dz}{dt}\mathbf{k}$

1.52. If $\mathbf{P} = 2t\mathbf{i} + 3t^2\mathbf{j} - t\mathbf{k}$ and $\mathbf{Q} = t\mathbf{i} + t^2\mathbf{j} + t^3\mathbf{k}$, show that

$$\frac{d}{dt}(\mathbf{P} \cdot \mathbf{Q}) = 4t + 8t^3$$

Check the result by using

$$\frac{d\mathbf{P}}{dt} \cdot \mathbf{Q} + \mathbf{P} \cdot \frac{d\mathbf{Q}}{dt} = \frac{d}{dt}(\mathbf{P} \cdot \mathbf{Q})$$

1.53. In the preceding problem show that

$$\frac{d}{dt}(\mathbf{P} \times \mathbf{Q}) = (15t^4 + 3t^2)\mathbf{i} - (8t^3 + 2t)\mathbf{j} - 3t^2\mathbf{k}$$

Check the result by using

$$\frac{d\mathbf{P}}{dt} \times \mathbf{Q} + \mathbf{P} \times \frac{d\mathbf{Q}}{dt} = \frac{d}{dt}(\mathbf{P} \times \mathbf{Q})$$

1.54. Determine the dot product for the following vectors:

	P	**Q**	*Ans.*
(a)	$3\mathbf{i} - 2\mathbf{j} + 8\mathbf{k}$	$-\mathbf{i} - 2\mathbf{j} - 3\mathbf{k}$	-23
(b)	$0.86\mathbf{i} + 0.29\mathbf{j} - 0.37\mathbf{k}$	$1.29\mathbf{i} - 8.26\mathbf{j} + 4.0\mathbf{k}$	-2.77
(c)	$a\mathbf{i} + b\mathbf{j} - c\mathbf{k}$	$d\mathbf{i} - e\mathbf{j} + f\mathbf{k}$	$ad - be - cf$

1.55. Determine the cross products for the following vectors:

	P	**Q**	*Ans.*
(a)	$3\mathbf{i} - 2\mathbf{j} + 8\mathbf{k}$	$-\mathbf{i} - 2\mathbf{j} - 3\mathbf{k}$	$22\mathbf{i} + \mathbf{j} - 8\mathbf{k}$
(b)	$0.86\mathbf{i} + 0.29\mathbf{j} - 0.37\mathbf{k}$	$1.29\mathbf{i} - 8.26\mathbf{j} + 4.0\mathbf{k}$	$-1.90\mathbf{i} - 3.92\mathbf{j} - 7.48\mathbf{k}$
(c)	$a\mathbf{i} + b\mathbf{j} - c\mathbf{k}$	$d\mathbf{i} - e\mathbf{j} + f\mathbf{k}$	$(bf - ec)\mathbf{i} - (af + cd)\mathbf{j} - (ae + bd)\mathbf{k}$

1.56. Determine the component of the vector $\mathbf{Q} = 10\mathbf{i} - 20\mathbf{j} - 20\mathbf{k}$ along a line drawn from point $(2, 3, -2)$ through the point $(1, 0, 5)$.   *Ans.* -11.72

1.57. Determine the component of the vector $\mathbf{P} = 1.52\mathbf{i} - 2.63\mathbf{j} + 0.83\mathbf{k}$ on the line that originates at the point $(2, 3, -2)$ and passes through the point $(1, 0, 5)$.   *Ans.* $P_L = +1.59$

1.58. Given the vectors $\mathbf{P} = \mathbf{i} + P_y\mathbf{j} - 3\mathbf{k}$ and $\mathbf{Q} = 4\mathbf{i} + 3\mathbf{j}$, determine the value of P_y so that the cross product of the two vectors will be $9\mathbf{i} - 12\mathbf{j}$.   *Ans.* $P_y = 0.75$

1.59. Given the vectors $\mathbf{P} = \mathbf{i} - 3\mathbf{j} + P_z\mathbf{k}$ and $\mathbf{Q} = 4\mathbf{i} - \mathbf{k}$, determine the value of P_z so that the dot product of the two vectors will be 14.   *Ans.* $P_z = -10$

1.60. Express the vectors shown in Fig. 1-24 in $\mathbf{i}$, $\mathbf{j}$, and $\mathbf{k}$ notation.
   Ans.   (a) $\mathbf{P} = -223\mathbf{i} + 306\mathbf{j} - 129\mathbf{k}$; (b) $\mathbf{Q} = +75\mathbf{i} + 50\mathbf{j} - 43.3\mathbf{k}$; (c) $\mathbf{S} = +144\mathbf{i} + 129\mathbf{j} + 52.4\mathbf{k}$

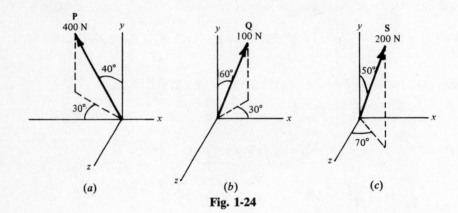

Fig. 1-24

Chapter 2

Operations with Forces

2.1 THE MOMENT OF A FORCE

The moment $\mathbf{M}$ of a force $\mathbf{F}$ with respect to a point O is the cross product $\mathbf{M} = \mathbf{r} \times \mathbf{F}$, where $\mathbf{r}$ is the position vector relative to point O of any point P on the action line of force $\mathbf{F}$. Physically, $\mathbf{M}$ represents the tendency of the force $\mathbf{F}$ to rotate the body (on which it acts) about an axis that passes through O and is perpendicular to the plane containing the force $\mathbf{F}$ and the position vector $\mathbf{r}$.

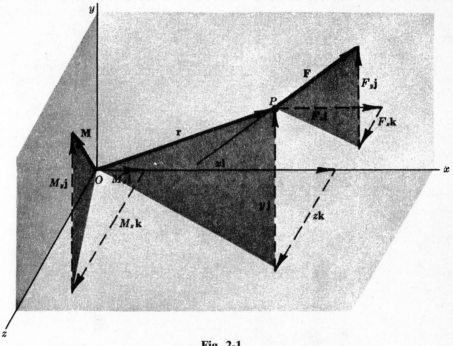

Fig. 2-1

If a set of x, y, and z axes is drawn through O as shown in Fig. 2-1,

$$\mathbf{r} = x\mathbf{i} + y\mathbf{j} + z\mathbf{k} \qquad \mathbf{F} = F_x\mathbf{i} + F_y\mathbf{j} + F_z\mathbf{k} \qquad \mathbf{M} = M_x\mathbf{i} + M_y\mathbf{j} + M_z\mathbf{k}$$

and, by definition,

$$\mathbf{M} = \mathbf{r} \times \mathbf{F} = \begin{vmatrix} \mathbf{i} & \mathbf{j} & \mathbf{k} \\ x & y & z \\ F_x & F_y & F_z \end{vmatrix}$$

Expanding the determinant,

$$\mathbf{M} = \mathbf{i}(F_z y - F_y z) + \mathbf{j}(F_x z - F_z x) + \mathbf{k}(F_y x - F_x y)$$

Comparing this expression for $\mathbf{M}$ with the one listed above, it can be seen that

$$M_x = F_z y - F_y z \qquad M_y = F_x z - F_z x \qquad M_z = F_y x - F_x y$$

19

The scalar quantities M_x, M_y, and M_z are the magnitudes of the respective moments of the force **F** about the x, y, and z axes through O. See Problems 2.3 and 2.4.

Note that M_x can be obtained by the dot product of the moment **M** and the unit vector **i** along the x axis. Thus,

$$\mathbf{M} \cdot \mathbf{i} = (M_x\mathbf{i} + M_y\mathbf{j} + M_z\mathbf{k}) \cdot \mathbf{i} = M_x(1) + M_y(0) + M_z(0) = M_x$$

Similarly, the magnitude of the moment of **F** about any axis L through O is the scalar component of **M** on L. It can be obtained by the dot product of **M** and unit vector $\mathbf{e}_L$ along the line L. Thus,

$$M_L = \mathbf{M} \cdot \mathbf{e}_L$$

2.2 A COUPLE

A *couple* consists of two forces equal in magnitude and parallel, but oppositely directed.

2.3 THE MOMENT OF A COUPLE

The moment **C** of a couple with respect to any point O is the sum of the moments with respect to O of the two forces that constitute the couple.

The moment **C** of the couple shown in Fig. 2-2 is

$$\mathbf{C} = \sum \mathbf{M}_O = \mathbf{r}_1 \times \mathbf{F} + \mathbf{r}_2 \times (-\mathbf{F}) = (\mathbf{r}_1 - \mathbf{r}_2) \times \mathbf{F} = \mathbf{a} \times \mathbf{F}$$

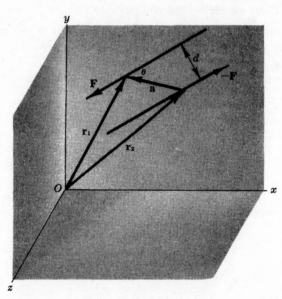

Fig. 2-2

Thus **C** is a vector perpendicular to the plane containing the two forces (**a** is in the same plane). By definition of the cross product, the magnitude of **C** is $|\mathbf{a} \times \mathbf{F}| = a\mathbf{F}\sin\theta$. Since d, the perpendicular distance between the two forces of the couple, is equal to $a\sin\theta$, the magnitude of **C** is $C = Fd$.

Couples obey the laws of vectors. Any couple **C** can be written $\mathbf{C} = C_x\mathbf{i} + C_y\mathbf{j} + C_z\mathbf{k}$, where C_x, C_y, and C_z are the magnitudes of the components.

Note that point O is any point; hence, the moment of a couple is independent of the choice of point O.

2.4 REPLACING A SINGLE FORCE

A single force **F** acting at point P may be replaced by (a) an equal and similarly directed force acting through any point O and (b) a couple $\mathbf{C} = \mathbf{r} \times \mathbf{F}$, where **r** is the vector from O to P. See Problems 2.11 and 2.12.

2.5 COPLANAR FORCE SYSTEMS

Coplanar force systems occur in many problems of mechanics. The following scalar treatment is useful in dealing with these two-dimensional problems.

1. The *moment* M_O of a force about a point O in a plane containing the force is the scalar moment of the force about an axis through the point and perpendicular to the plane. As such, the moment is the product of (a) the force and (b) the perpendiculat distance from the point to the line of action of the force. It is customary to assign a positive sign to the moment if the force tends to turn in a counterclockwise direction about the point. See Problem 2.1.

2. *Varignon's theorem* states that the moment of a force about any point is equal to the algebraic sum of the moments of the components of the force about that point. See Problem 2.2.

3. The moment of a couple will not be changed if (a) the couple is rotated or translated in its plane, (b) the couple is transferred to a parallel plane, or (c) the size of its forces is changed, provided the moment arm is also altered to keep the magnitude of the moment the same.

4. A couple and a single force in the same plane or parallel planes may be combined into one force of the same magnitude and sense as the given force and parallel to it. See Problem 2.9.

5. Conversely, a single force as indicated above may be replaced by (a) an equal and similarly directed force acting through any point and (b) a couple lying in the same plane as the single force and the chosen point. See Problem 2.11.

2.6 NOTES

In some of the solved problems vector equations are used, but in other problems the equivalent scalar equations are used. In figures, vectors are identified by their magnitudes when the directions are obvious.

Also note that in the U.S. Customary System the units for moments are pound-feet (lb-ft). In SI the units for moments are newton-meters (N · m).

Solved Problems

2.1. Determine the moment of the 20-lb force about the point O. See Fig. 2-3.

SOLUTION

Drop the perpendicular OD from O to the action line of the 20-lb force. Its length to scale is 4.33 ft. The moment of the force about O (actually about an axis through O perpendicular to the xy plane) is therefore $-(20 \times 4.33) = -86.6$ lb-ft.

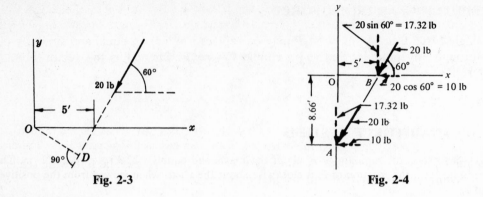

Fig. 2-3   **Fig. 2-4**

The minus sign is used because the direction of rotation viewed from the positive end of the z axis (not shown) is clockwise.

2.2.   Solve Problem 2.1 using Varignon's theorem. See Fig. 2-4.

SOLUTION

In using this theorem the 20-lb force is replaced with its rectangular components parallel to the x and y aces and acting at *any* convenient point along the line of action.

If the point B is chosen on the x axis, then it should be apparent that the x component has no moment about O. The moment of the 20-lb force about O is then only the moment of the y component about O, or $-(17.32 \times 5) = -86.6$ lb-ft.

If the point A on the y axis is chosen, then the y component has no moment about O. The moment of the 20-lb force about O is then only the moment of the x component about O, or $-(10 \times 8.66) = -86.6$ lb-ft.

2.3.   A 100-N force is directed along the line drawn from the point whose x, y, z coordinates are $(2, 0, 4)$ m to the point whose coordinates are $(5, 1, 1)$ m. What are the moments of this force about the x, y, and z axes?

SOLUTION

In Fig. 2-5, assume the scale is such that the 100-N force is measured by the diagonal of the parallelepiped whose sides are parallel to the axes. The sides represent to the same scale the components of the force.

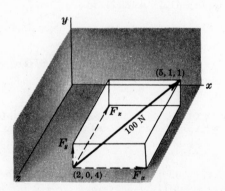

Fig. 2-5

The x side is $5 - 2 = 3$ m long; the y side is $1 - 0 = 1$ m long, and the z side is $1 - 4 = -3$ m long. This means that the component F_z is directed toward the back or negative direction of the z axis.

$$F_x = \frac{\text{length of } x \text{ side}}{\text{length of diagonal}} \times 100\,\text{N} = \frac{3}{\sqrt{3^2 + 1^2 + 3^2}} \times 100 = \frac{3}{\sqrt{19}} \times 100 = 68.7\,\text{N}$$

Similarly, $$F_y = \frac{1}{\sqrt{19}} \times 100 = 22.9\,\text{N} \qquad F_z = \frac{-3}{\sqrt{19}} \times 100 = -68.7\,\text{N}$$

To find the moment of the 100-N force about the x axis, determine the moments of its components about the x axis. By inspection the only component that has such a moment is F_y. Therefore M_x for the 100-N force is the moment of F_y about the x axis and equals $-22.9 \times 4 = -91.6\,\text{N} \cdot \text{m}$. The minus sign indicates that the rotation of F_y is clockwise about the x axis when viewed from the positive end of the x axis.

In finding the moment about the y axis, note that F_y is parallel to the y axis and has no moment about it. Now, however, both F_z and F_x must be considered. It is better to determine the sign of the moment by inspection rather than by writing signs for the component and its arm. Accordingly, $M_y = +(68.7 \times 2) + (68.7 \times 4) = +412\,\text{N} \cdot \text{m}$.

By similar reasoning using F_y only (since F_z is parallel to the z axis and F_x intersects it), $M_z = +(22.9 \times 2) = +45.8\,\text{N} \cdot \text{m}$.

Be sure to affix signs for moments and to understand the significance thereof.

Mathcad

2.4.   Repeat Problem 2.3 using the cross-product definition of moment.

SOLUTION

From Problem 2.3, $\mathbf{F} = 68.7\mathbf{i} + 22.9\mathbf{j} - 68.7\mathbf{k}$.

The vector $\mathbf{r}$ is the position vector of any point on the action line of $\mathbf{F}$ with respect to the origin. If we use point $(2, 0, 4)$, $\mathbf{r} = 2\mathbf{i} + 0\mathbf{j} + 4\mathbf{k}$. Then

$$\mathbf{M} = \mathbf{r} \times \mathbf{F} = \begin{vmatrix} \mathbf{i} & \mathbf{j} & \mathbf{k} \\ 2 & 0 & 4 \\ 68.7 & 22.9 & -68.7 \end{vmatrix}$$

$$= \mathbf{i}[0 - 4(22.9)] - \mathbf{j}[2(-68.7) - 4(68.7)] + \mathbf{k}[2(22.9) - 0]$$

$$= -91.6\mathbf{i} + 412\mathbf{j} + 45.8\mathbf{k}\,\text{N} \cdot \text{m}$$

Next, using point $(5, 1, 1)$ on the action line of $\mathbf{F}$, $\mathbf{r} = 5\mathbf{i} + \mathbf{j} + \mathbf{k}$. Then

$$\mathbf{M} = \begin{vmatrix} \mathbf{i} & \mathbf{j} & \mathbf{k} \\ 5 & 1 & 1 \\ 68.7 & 22.9 & -68.7 \end{vmatrix}$$

$$= \mathbf{i}[-1(68.7) - 22.9(1)] - \mathbf{j}[5(-68.7) - 1(68.7)] + \mathbf{k}[5(22.9) - 68.7(1)]$$

$$= -91.6\mathbf{i} + 412\mathbf{j} + 45.8\mathbf{k}\,\text{N} \cdot \text{m}$$

The scalar moments about the x, y, and z axes are the coefficients of the unit vectors, $\mathbf{i}$, $\mathbf{j}$, and $\mathbf{k}$.

2.5.   Determine the moment of the force $\mathbf{F} = 2\mathbf{i} + 3\mathbf{j} - \mathbf{k}$ lb acting through the point $(3, 1, 1)$ with respect to the line passing from $(2, 5, -2)$ through $(3, -1, 1)$. The coordinates are in feet.

SOLUTION

The moment arm $\mathbf{r}$ may be found by using the vector from either point on the line to the point on the force. From $(2, 5, -2)$ the vector $\mathbf{r} = \mathbf{i} - 4\mathbf{j} + 3\mathbf{k}$. The moment $\mathbf{M}$ about the chosen point is

$$\mathbf{M} = \mathbf{r} \times \mathbf{F} = \begin{vmatrix} \mathbf{i} & \mathbf{j} & \mathbf{k} \\ 1 & -4 & 3 \\ 2 & 3 & -1 \end{vmatrix} = -5\mathbf{i} + 7\mathbf{j} + 11\mathbf{k}$$

Now
$$e_L = \frac{[(3-2)\mathbf{i} + (-1-5)\mathbf{j} + (1+2)\mathbf{k}]}{\sqrt{(1)^2 + (-6)^2 + (3)^2}} = \frac{\mathbf{i} - 6\mathbf{j} + 3\mathbf{k}}{\sqrt{46}}$$

Hence the moment of $\mathbf{F}$ about the line is

$$\mathbf{M}_L = \mathbf{M} \cdot e_L = (-5\mathbf{i} + 7\mathbf{j} + 11\mathbf{k}) \cdot \frac{\mathbf{i} - 6\mathbf{j} + 3\mathbf{k}}{\sqrt{46}} = \frac{-5 - 42 + 33}{\sqrt{46}} = \frac{-14}{\sqrt{46}} = -2.06 \text{ lb-ft}$$

If the moment arm is chosen from the point $(3, -1, 1)$, the arm is $\mathbf{r} = 2\mathbf{j}$. The moment $\mathbf{M}$ is

$$\mathbf{M} = \mathbf{r} \times \mathbf{F} = \begin{vmatrix} \mathbf{i} & \mathbf{j} & \mathbf{k} \\ 0 & 2 & 0 \\ 2 & 3 & -1 \end{vmatrix} = -2\mathbf{i} - 4\mathbf{k}$$

Hence the moment of $\mathbf{M}$ about the line is

$$\mathbf{M} \cdot e_L = (-2\mathbf{i} + 0\mathbf{j} - 4\mathbf{k}) \cdot \frac{\mathbf{i} - 6\mathbf{j} + 3\mathbf{k}}{\sqrt{46}} = \frac{-2 - 12}{\sqrt{46}} = \frac{-14}{\sqrt{46}} = -2.06 \text{ lb-ft}$$

2.6 Determine the moment of a force $\mathbf{P}$ whose rectangular components are $P_x = 22$ N, $P_y = 23$ N, $P_z = 7$ N and acting at a point $(1, -1, -2)$. Take the moment about a line from the origin through point $(3, -1, 0)$. Coordinates are in meters.

SOLUTION
$$\mathbf{P} = 22\mathbf{i} + 23\mathbf{j} + 7\mathbf{k} \text{ N}$$

The moment arm, $\mathbf{r} = (1 - 0)\mathbf{i} + (-1 - 0)\mathbf{j} + (-2 - 0)\mathbf{k} = \mathbf{i} - \mathbf{j} - 2\mathbf{k}$ m

$$\mathbf{M} = \mathbf{r} \times \mathbf{F} = \begin{vmatrix} \mathbf{i} & \mathbf{j} & \mathbf{k} \\ 1 & -1 & -2 \\ 22 & 23 & 7 \end{vmatrix} = 39\mathbf{i} - 51\mathbf{j} + 45\mathbf{k} \text{ N} \cdot \text{m}$$

2.7. A couple of moment $+60$ lb-ft acts in the plane of the paper. Indicate this couple with (a) 10-lb forces and (b) 30-lb forces.

SOLUTION

In (a) the moment arm must be 6 ft, while in (b) it is 2 ft.

The direction of rotation must be counterclockwise. The parallel forces may be drawn at any angle, as shown in Fig. 2-6.

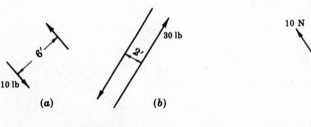

Fig. 2-6   Fig. 2-7

2.8. Combine couple $C_1 = +20$ N $\cdot$ m with couple $C_2 = -50$ N $\cdot$ m, both in the same plane. See Fig. 2-7.

SOLUTION

To combine graphically, show both couples with forces of the same magnitude, say 10 N, and drawn in such a way that two of the forces, one from each couple, are collinear but oppositely directed.

It is evident that the collinear forces cancel, leaving two 10-N forces with an arm of 3 m. The resultant couple is -30 N $\cdot$ m, a result which can also be obtained by algebraic addition.

2.9.   Replace a couple of moment -100 N $\cdot$ m and a vertical force of 50 N, acting at the origin, by a single force. Where does the single force act?

SOLUTION

In Fig. 2-8 the couple is represented by two equal and opposite forces of 50 N at a perpendicular distance of 2 m. One force of the couple is aligned with the given 50-N force at the origin. These two forces cancel leaving the single, upward force of 50 N acting 2 m to the left of the origin.

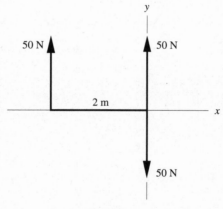

Fig. 2-8

2.10.   Combine a force of 30 N, 60° with a $+50$ N $\cdot$ m couple in the same plane. See Fig. 2-9.

SOLUTION

A couple as such cannot be reduced to a simpler tystem, but it can be combined with another force.

Draw the given couple with 30-N forces and in such a way that one of its forces is collinear with the given single 30-N force but oppositely directed.

By inspection the collinear forces cancel, leaving only a single force of 30 N parallel to and in the dame direction as the original force but at a distance 1.67 m from it.

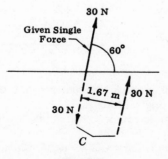

Fig. 2-9

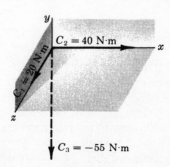

Fig. 2-10

2.11. As shown in Fig. 2-10, a couple C_1 of 20 N · m acts in the xy plane, a couple C_2 of 40 N · m acts in the yz plane, and a couple C_3 of -55 N · m acts in the xz plane. Determine the resultant couple.

SOLUTION

The couple C_1 is positive and acting in the xy plane. When viewed from the positive end of the z axis, it tends to turn in a counterclockwise direction about the z axis. By the *right-hand rule,* it is represented by a vector along the z axis drawn toward the positive end. Using this type of reasoning, all three couples are drawn in the figure. Adding vectorially,

$$C = \sqrt{C_1^2 + C_2^2 + C_3^2} = \sqrt{(20)^2 + (40)^2 + (-55)^2} = 70.9 \text{ N} \cdot \text{m}$$

$$\cos \phi_x = \frac{C_2}{C} = +0.564 \qquad \cos \phi_y = \frac{C_3}{C} = -0.777 \qquad \cos \phi_z = \frac{C_1}{C} = +0.282$$

These are the direction cosines of the couple **C**. The couple acts in a plane perpendicular to this vector.

The couple **C** may be written in vector notation,

$$\mathbf{C} = +40\mathbf{i} - 55\mathbf{j} + 20\mathbf{k} \text{ N} \cdot \text{m}$$

from which the value of **C** is derived as above.

2.12. A 2-in-diameter pipe is subjected to a force of 25 lb applied vertically downward to the horizontal rod at an arm of 14 in. Replace the 25-lb force with (1) a force at the end of the pipe which causes bending and (2) a couple that twists the shaft, placing it in torsion. What are the moments of the force and the couple? See Fig. 2-11(a).

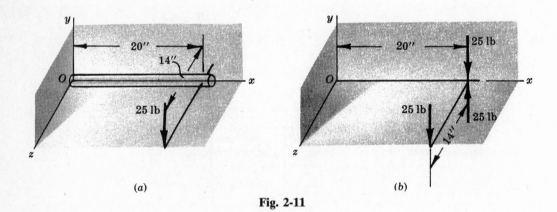

(a) (b)

Fig. 2-11

SOLUTION

Place two vertical 25-lb forces oppositely directed through the center of the pipe as shown in Fig. 2-11(b). The three forces are still equivalent to the original force.

The upward force combines with the original to form a couple $C = 25 \times 14 = 350$ lb-in. This couple tends to twist the pipe counterclockwise when viewed from the right.

The other 25-lb force down on the pipe causes a bending moment $M = -25 \times 20 = -500$ lb-in about the z axis.

2.13. Solve Problem 2.12 by determining the moment of the 25-lb force about the point O.

SOLUTION

The position vector of the point of application of the 25-lb force with respect to the origin is $\mathbf{r} = 20\mathbf{i} + 14\mathbf{k}$. The force $\mathbf{F} = -25\mathbf{j}$. Thus the moment of the 25-lb force with respect to the origin is

$$\mathbf{M} = \mathbf{r} \times \mathbf{F} = \begin{vmatrix} \mathbf{i} & \mathbf{j} & \mathbf{k} \\ 20 & 0 & 14 \\ 0 & -25 & 0 \end{vmatrix} = \mathbf{i}[0 - 14(-25)] - \mathbf{j}[0 - 0] + \mathbf{k}[20(-25) - 0] = 350\mathbf{i} - 500\mathbf{k} \text{ lb-in}$$

This agrees with the results of Problem 2.12.

2.14. The crane in Fig. 2–12 is on level ground. The x axis is through the contact points of the rearmost wheels with the ground, the y axis is parallel to the front-to-back centerline, and the z axis is vertical as shown. The bed (platform) of the crane is 3 ft above the ground. For practical purposes the pivot point of the bottom of the boom can be considered in the bed of the crane and 6 ft from the center of the cab. The center of the cab is on the centerline 15 ft forward (to the left) of the rearmost axle. The 50-ft boom makes an angle of 60° with the bed of the crane in a vertical plane, and the cab and boom are swiveled 45° horizontally from the fore and aft centerline of the truck bed. The distance between the contact points of the rear wheels is considered to be 8 ft. Determine the turning moment of the 4000-lb load about the x axis.

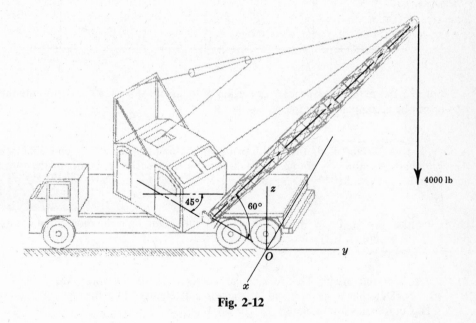

Fig. 2-12

SOLUTION

Relative to the origin O of the axes, the coordinates of the cab center are $(-4, -15, +3)$. The coordinates of the bottom of the boom are $(-4 + 6 \sin 45°, -15 + 6 \cos 45°, +3)$ or $(+0.24, -10.8, +3)$. The coordinates of the top of the boom are $(+0.24 + 50 \cos 60° \sin 45°, -10.8 + 50 \cos 60° \cos 45°, +3 + 50 \sin 60°)$ or $(+17.9, +6.91, +46.3)$.

The moment of the 4000-lb weight about O is

$$\mathbf{M} = \mathbf{r} \times \mathbf{F} = \begin{vmatrix} \mathbf{i} & \mathbf{j} & \mathbf{k} \\ 17.9 & 6.91 & 46.3 \\ 0 & 0 & -4000 \end{vmatrix}$$

The scalar coefficient of the $\mathbf{i}$ term is the moment about the x axis. Hence $M_x = -27{,}600$ lb-ft. Thus the moment is clockwise about the x axis when viewed from the front.

Supplementary Problems

2.15. In each case find the moment of the force $\mathbf{F}$ about the origin. Use Varignon's theorem.

Magnitude of $\mathbf{F}$	Angle of $\mathbf{F}$ with horizontal	Coordinates of point of application of $\mathbf{F}$	*Answer*
20 lb	30°	(5, −4) ft	+119 lb-ft
64 lb	140°	(−3, 4) ft	+72.9 lb-ft
15 lb	337°	(8, −2) ft	−19.3 lb-ft
8 oz	45°	(6, 1) in	+28.3 oz-in
4 N	90°	(0, −20) m	0
96 N	60°	(4, 2) m	+236 N · m

2.16. In Problem 2.15 use the cross-product definition of moment ($\mathbf{M} = \mathbf{r} \times \mathbf{F}$) to determine the moment. Each answer will be accompanied by the unit vector $\mathbf{k}$.

2.17. A 50-N force is directed along the line drawn from the point whose x, y, z coordinates are $(8, 2, 3)$ m to the point whose coordinates are $(2, -6, 5)$ m. What are the scalar moments of the force about the x, y, and z axes?   *Ans.* $M_x = +137$ N · m, $M_y = -167$ N · m, $M_z = -255$ N · m

2.18. Given the force $\mathbf{P} = 32.4\mathbf{i} - 29.3\mathbf{j} + 9.9\mathbf{k}$ lb acting at the origin. Find the moment about a line through the points $(0, -1, 3)$ and $(3, 1, 1)$. Coordinates are in feet.   *Ans.* $M = -88.2$ lb-ft

2.19. A force acts at the origin. The rectangular components of the force are $P_x = 68.7$ N, $P_y = 22.9$ N, $P = -68.7$ N. Determine the moment of the force $\mathbf{P}$ about a line through the points $(1, 0, -1)$ and $(4, 4, -1)$. Coordinates are in meters.   *Ans.* $M = -13.7$ N · m

2.20. Combine $C_1 = +20$ lb-ft, $C_2 = -80$ lb-ft, and $C_3 = -18$ lb-ft, all acting in the same plane
Ans. $C = -78$ lb-ft acting in the same or parallel plane

2.21. Replace a vertical force of 270 lb acting down at the origin by a vertical force of 270 lb acting at $x = -5$ in and a couple.   *Ans.* $C = -1350$ lb-in

2.22. Determine the resultant vector of the three couples $+16 \, \text{N} \cdot \text{m}$, $-45 \, \text{N} \cdot \text{m}$, $+120 \, \text{N} \cdot \text{m}$ acting respectively in the *xy, yz,* and *xz* planes.
Ans. $C = +129 \, \text{N} \cdot \text{m}$, $\cos \theta_x = -0.349$, $\cos \theta_y = 0.931$, $\cos \theta_z = 0.124$

2.23. Add the couple $\mathbf{C} = 30\mathbf{i} - 20\mathbf{j} + 35\mathbf{k} \, \text{N} \cdot \text{m}$ to the resultant couple in Problem 2.22.
Ans. $\mathbf{C} = -15\mathbf{i} + 100\mathbf{j} + 51\mathbf{k} \, \text{N} \cdot \text{m}$

2.24. The 24-lb forces applied at the corners *A* and *B* of the parallelepiped shown in Fig. 2-13 act along *AE* and *BF* respectively. Show that the given couple may be replaced by a set of vertical forces consisting of a 16-lb force up at point *C* and a 16-lb force down at point *D*.

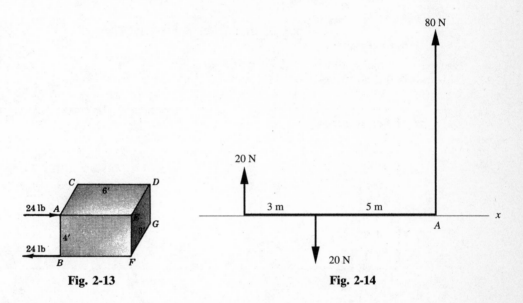

Fig. 2-13   Fig. 2-14

2.25. Replace the set of three parallel vertical forces, shown in Fig. 2-14, by a single force. What is the magnitude, sense, and location of the single force?   *Ans.*   80 N, vertically up, 0.75 m left of *A*

2.26. A horizontal bar 8 m long is acted upon by a downward vertical 12-N force at the right end as shown in Fig. 2-15. Show that this is equivalent to a 12-N vertical force acting down at the left end *and* a clockwise couple of $96 \, \text{N} \cdot \text{m}$.

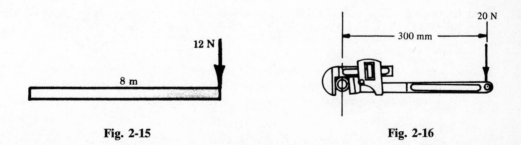

Fig. 2-15   Fig. 2-16

2.27. A wrench in the horizontal position is locked around a pipe at its left end. A vertical force of 20 N is going to be applied at the right end at an effective arm of 300 mm. Show that this will be equivalent to a vertical downward force of 20 N acting through the pipe center *and* a clockwise couple of $6 \, \text{N} \cdot \text{m}$. Refer to Fig. 2-16.

2.28. Reduce the system of forces in the belts shown in Fig. 2-17 to a single force at O and a couple. Forces are either vertical or horizontal.   *Ans.*   78.3 lb, $\theta_x = 296.5°$, $C = 0$

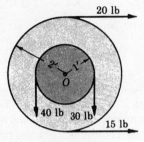

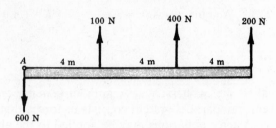

Fig. 2-17   **Fig. 2-18**

2.29. Reduce the system of forces acting on the beam shown in Fig. 2-18 into a force at A and a couple.
Ans. $R = 100$ N up at A, $C = 6000$ N $\cdot$ m

2.30. Referring to Fig. 2-19, reduce the system of forces and couples to the simplest system using point A.
Ans. $R_x = +48.1$ lb, $R_y = -3.9$ lb, $C = +36.2$ lb-ft

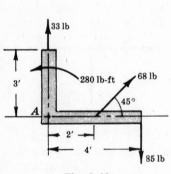

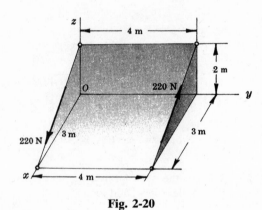

Fig. 2-19   **Fig. 2-20**

2.31. Determine the moments of the two forces about the x, y, and z axes shown in Fig. 2-20.
Ans. $\mathbf{M} = 488\mathbf{i} + 732\mathbf{k}$ N $\cdot$ m or $M_x = +488$ N $\cdot$ m, $M_y = 0$, $M_z = +732$ N $\cdot$ m

Chapter 3

Resultants of Coplanar Force Systems

3.1 COPLANAR FORCES

Coplanar forces lie in one plane. A concurrent system consists of forces that intersect at a point called the concurrence. A parallel system consists of forces that intersect at infinity. A nonconcurrent, nonparallel system consists of forces that are not all concurrent and not all parallel.

Vector equations may be applied to the above systems to determine resultants, but the following derived scalar equations will be more useful for a given system.

3.2 CONCURRENT SYSTEM

The resultant **R** may be (*a*) a single force through the concurrence or (*b*) zero. Algebraically,

$$R = \sqrt{\left(\sum F_x\right)^2 + \left(\sum F_y\right)^2} \qquad \text{and} \qquad \tan \theta_x = \frac{\sum F_y}{\sum F_x}$$

where $\sum F_x, \sum F_y$ = algebraic sums of the x and y components, respectively, of the forces of the system

θ_x = angle that the resultant **R** makes with the x axis.

3.3 PARALLEL SYSTEM

The resultant may be (*a*) a single force **R** parallel to the system, (*b*) a couple in the plane of the system or in a parallel plane, or (*c*) zero. Algebraically,

$$R = \sum F \qquad \text{and} \qquad R\bar{a} = \sum M_O$$

where $\sum F$ = algebraic sum of the forces of the system

O = any moment center in the plane

$\bar{a}$ = perpendicular distance from the moment center O to the resultant R

$R\bar{a}$ = moment of R with respect to O

$\sum M_O$ = algebraic sum of the moments of the forces of the system with respect to O

If $\sum F$ is not zero, apply the equation $R\bar{a} = \sum M_O$ to determine $\bar{a}$ and hence the action line of R. If $\sum F = 0$, the resultant couple, if there is one, has a magnitude $\sum M_O$.

3.4 NONCONCURRENT, NONPARALLEL SYSTEM

The resultant may be (*a*) a single force **R**, (*b*) a couple in the plane of the system or in a parallel plane, or (*c*) zero. Algebraically,

$$R = \sqrt{\left(\sum F_x\right)^2 + \left(\sum F_y\right)^2} \qquad \text{and} \qquad \tan \theta_x = \frac{\sum F_y}{\sum F_x}$$

where $\sum F_x, \sum F_y$ = algebraic sums of x and y components, respectively, of forces of system

θ_x = angle that the resultant **R** makes with x axis

To determine the action line of the resultant force, employ the equation

$$R\bar{a} = \sum M_O$$

where O = any moment center in the plane
               $\bar{a}$ = perpendicular distance from moment center O to the resultant R
               $R\bar{a}$ = moment of R with respect to O
               $\Sigma\, M_O$ = algebraic sum of the moments of the forces of the system with respect to O

Note that even if $R = 0$, a couple may exist with magnitude equal to $\Sigma\, M_O$.

3.5 RESULTANTS OF DISTRIBUTED FORCE SYSTEMS

A distributed force system is one in which forces cannot be represented by individual force vectors acting at specific points in space; they must be represented by an infinite number of vectors, each of which is a function of the point at which it acts. Consider the coplanar (parallel) distributed force system shown in Fig. 3-1. In the U.S. Customary System the units for $f(x)$ would be, for example, lb/ft. In the International System, the units might be N/m. The resultant R of the force system and its location can be found by integration. Thus

$$R = \int_A^B f(x)\, dx \qquad \text{and} \qquad Rd = \int_A^B xf(x)\, dx$$

Problems 3.13 through 3.15 are specific examples.

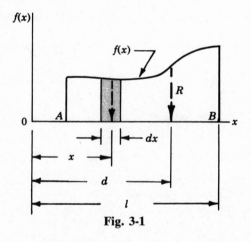

Fig. 3-1

Solved Problems

3.1.   Determine the resultant of the concurrent force system shown in Fig. 3-2.

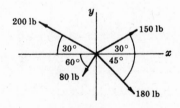

Fig. 3-2

SOLUTION

Find the x and y components of each of the four given forces. Add the x components algebraically to determine ΣF_x. Find ΣF_y for the y components. A tabular form may present the information more clearly.

Force	$\cos \theta_x$	$\sin \theta_x$	F_x	F_y
150	+0.866	+0.500	+129.9	+75.0
200	−0.866	+0.500	−173.2	+100.0
80	−0.500	−0.866	−40.0	−69.2
180	+0.707	−0.707	+127.3	−127.3

$\Sigma F_x = +44.0$, $\Sigma F_y = -21.5$, and $R = \sqrt{(\Sigma F_x)^2 + (\Sigma F_y)^2} = \sqrt{(44.0)^2 + (-21.5)^2} = 49.0$ lb.

$$\tan \theta_x = \frac{\Sigma F_y}{\Sigma F_x} = \frac{-21.5}{+44.0} = -0.489$$

from which
$$\theta_x = 360° - 26° = 334°$$

3.2. Determine the resultant of the force system shown in Fig. 3.3. Note that the slope of the action line of each force is indicated in the figure.

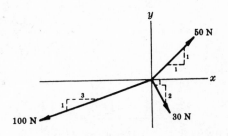

Fig. 3-3

SOLUTION

Force	F_x	F_y
50	$+50 \times \dfrac{1}{\sqrt{1^2 \times 1^2}}$	$+50 \times \dfrac{1}{\sqrt{2}}$
100	$-100 \times \dfrac{3}{\sqrt{1^2 + 3^2}}$	$-100 \times \dfrac{1}{\sqrt{10}}$
30	$+30 \times \dfrac{1}{\sqrt{1^2 + 2^2}}$	$-30 \times \dfrac{2}{\sqrt{5}}$

$\Sigma F_x = -46.1$, $\Sigma F_y = -23.0$, and $R = \sqrt{(-46.1)^2 + (-23.0)^2} = 51.6$ N, with $\theta_x = 207°$.

3.3. Find the resultant of the coplanar, concurrent force system of Fig. 3-4.

SOLUTION

$$\Sigma F_x = 70 - 100 \cos 30° - 125 \sin 10° = -38.3 \text{ lb}$$
$$\Sigma F_y = 125 \cos 10° - 100 \sin 30° = 73.1 \text{ lb}$$
$$R = \sqrt{\left(\Sigma F_x\right)^2 + \left(\Sigma F_y\right)^2} = \sqrt{(-38.3)^2 + (73.1)^2} = 82.5 \text{ lb}$$
$$\tan \theta_x = \frac{\Sigma F_y}{\Sigma F_x} = \frac{73.1}{-38.3} = -1.91 \qquad \theta_x = 62°$$

from which $\theta = 180° - 62° = 118°$ to the nearest degree. The resultant is shown as a dashed vector in Fig. 3-4.

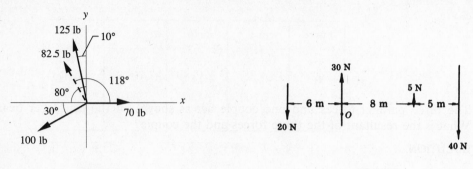

Fig. 3-4   Fig. 3-5

3.4.   Determine the resultant of the parallel system of Fig. 3-5.

SOLUTION

In Fig. 3-5 the action lines of the forces are vertical as shown.

$$R = -20 + 30 + 5 - 40 = -25\,\text{N} \qquad \text{(i.e., down)}$$

To determine the action line of this 25-N force, choose any moment center O. Since the moment of a force about a point on its own action line is zero, it is advisable but not necessary to choose O on one of the given forces. Let O be on the 30-N force.

$$\sum M_O = +(20 \times 6) + (30 \times 0) + (5 \times 8) - (40 \times 13) = -360\,\text{N}\cdot\text{m}$$

Then the moment of R must equal $-360\,\text{N}\cdot\text{m}$. This means that R, which is down $(-)$, must be placed to the right of O because only then will its moment be clockwise $(-)$.
Apply $R\bar{a} = \sum M_O$ to obtain

$$\bar{a} = \frac{360\,\text{N}\cdot\text{m}}{25\,\text{N}} = 14.4\,\text{m} \qquad \text{to the right of } O$$

Note the determination of $\bar{a}$ without regard to the signs of R or $\sum M_O$ but by using reasoning.

3.5.   Determine the resultant of the parallel force system in Fig. 3.6.

SOLUTION

$R = -100 + 200 - 200 + 400 - 300 = 0$. This means that the resultant is not a single force. Next find $\sum M_O$. Choose O, as shown, on the 100-lb force.

$$\sum M_O = +(100 \times 0) + (200 \times 2) - (200 \times 5) + (400 \times 9) - (300 \times 11) = -300\,\text{lb-ft}$$

The resultant is therefore a couple $C = -300$ lb-ft, which can be represented in the plane of the paper in accordance with the laws governing couples.

Fig. 3-6   Fig. 3-7

3.6.   Determine the resultant of the horizontal force system acting on the bar shown in Fig. 3-7.

SOLUTION

$R = \sum F_h = +20 + 20 - 40 = 0$. This means that the resultant is not a single force, but it may be a couple.

$$\sum M_O = -(20 \times 3) + (20 \times 3) = 0$$

Hence in this system the resultant force is zero and the resultant couple is also zero.

3.7.   The three parallel forces and one couple act as shown on the cantilever beam of Fig. 3-8. What is the resultant of the three forces and the couple?

SOLUTION

$$R = \sum F = 500 - 400 - 200 = -100 \text{ N}$$

$$\sum M_O = 2 \times 500 - 4 \times 400 - 6 \times 200 + 1500 = -300 \text{ N} \cdot \text{m}$$

$$R\bar{a} = \sum M_O \qquad \bar{a} = \frac{-300}{-100} = 3 \text{ m}$$

For the resultant, which is downward, to yield a negative moment the resultant must be to the right of point O. The resultant and its location is shown as a dashed vector in Fig. 3-8.

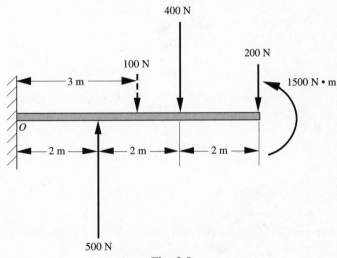

Fig. 3-8

3.8.   Determine the resultant of the coplanar, nonconcurrent force system shown in Fig. 3-9.

SOLUTION

$$\sum F_x = 50 - 100 \cos 45° = -20.7 \text{ lb}$$

$$\sum F_y = 50 - 100 \sin 45° = -20.7 \text{ lb}$$

$$R = \sqrt{(-20.7)^2 + (-20.7)^2} = 29.3 \text{ lb} \qquad \theta = \tan^{-1} \frac{\sum F_y}{\sum F_x} = 45°$$

$$\sum M_O = 5 \times 50 - 4 \times 50 = 50 \text{ lb-ft}$$

$$R\bar{a} = \sum M_O = 50 \qquad \bar{a} = 50/29.3 = 1.71 \text{ ft}$$

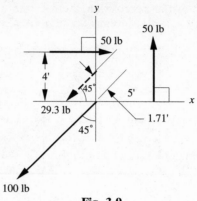

Fig. 3-9

R is directed downward to the left, and hence must be above the origin to produce a positive (+) moment.

3.9. Determine the resultant of the nonconcurrent, nonparallel system shown in Fig. 3-10(*a*). Assume that the coordinates are in meters.

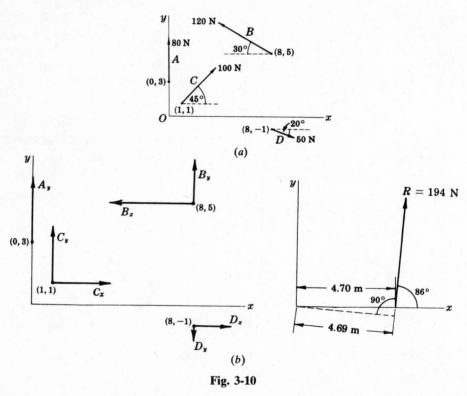

Fig. 3-10

SOLUTION

For convenience the forces are lettered *A*, *B*, *C*, *D*. The simplest method of attack is to use a tabular form listing *x* and *y* components for each force and also the moment of each component about some moment center—for this example *O*. The forces are now replaced with their components at the same points in the action lines as indicated in Fig. 3-10(*b*). It may be convenient at times to use components at a different point in the action line than that which is given, e.g., force *C* acting at 45° has an action line that passes through the origin *O*. The moment about *O* is easily seen to be zero in this case. However, the components used in this example will be shown acting horizontally and vertically through the given points of application.

The following table is useful in compiling the necessary information. Be sure to place the proper sign before each component and to determine the moment signs by inspection.

Force	$\cos \theta_x$	$\sin \theta_x$	F_x	F_y	Moment of F_x about O	Moment of F_y about O	M_O
A	0	+1	0	+80.0	0	0	0
B	−0·866	+0.500	−103.9	+60.0	+519.5	+480.0	+999.5
C	+0.707	+0.707	+70.7	+70.7	−70.7	+70.7	0
D	+0.940	−0.342	+47.0	−17.1	+47.0	−136.8	−89.8

$$\sum F_x = +13.8 \text{ N} \qquad \sum F_y = +193.6 \text{ N} \qquad \sum M_O = +910 \text{ N} \cdot \text{m}$$

$$R = \sqrt{\left(\sum F_x\right)^2 + \left(\sum F_y\right)^2} = \sqrt{(+13.8)^2 + (+193.6)^2} = 194 \text{ N} \qquad \theta_x = \tan^{-1}\frac{+193.6}{+13.8} = 86°$$

To find the moment arm of the resultant, divide 910 by 194 to obtain 4.69 m.

Since R acts upward and slightly to the right, it must be placed as shown because $\sum M_O$ is positive; i.e., R must have a counterclockwise moment.

Another method to locate the action line of the resultant is to determine its intercept with, say, the x axis. If the components of the resultant are drawn through the intercept on the x axis, the x component would have no moment about O. The moment would be determined solely by the y component and would be equal to the product of the y component and the x distance to the intercept (x coordinate of the intercept).

$$\bar{x} = \frac{\sum M_O}{\sum F_y} = \frac{910}{193.6} = 4.70 \text{ m}$$

Draw the resultant as shown with the intercept +4.70 m to the right because $\sum F_y$ is positive and $\sum M_O$ is positive.

3.10. Determine the resultant of the force system shown in Fig. 3-11(a). Assume that the coordinates are in feet.

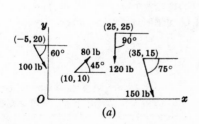

(a)

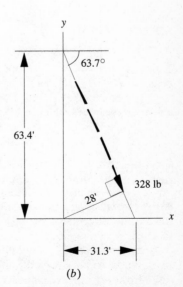

(b)

Fig. 3-11

SOLUTION

The summation of components can be written directly as an alternative to the tabular form of Prob. 3-9:

$$\sum F_x = 100 \cos 60° + 80 \cos 45° + 150 \cos 75° = 145.4 \text{ lb}$$

$$\sum F_y = -100 \sin 60° + 80 \sin 45° - 120 - 150 \sin 75° = -294.9 \text{ lb}$$

$$R = \sqrt{(145.4)^2 + (-294.9)^2} = 328 \text{ lb} \qquad \theta = \tan^{-1}\frac{-294.9}{145.4} = 63.7°$$

Summing the moments of the components about O:

$$\sum M_O = -(20)100 \cos 60° + (5)100 \sin 60° - (10)80 \cos 45°$$
$$+ (10)80 \sin 45° - (25)120 - (15)150 \cos 75° - (35)150 \sin 75° = -9220 \text{ lb-ft}$$

To determine the x intercept of the resultant, use

$$\left(\sum F_y\right)\bar{x} = \sum M_O \qquad \bar{x} = 31.3 \text{ ft}$$

To determine the y intercept of the resultant, use

$$\left(\sum F_x\right)\bar{y} = \sum M_O \qquad \bar{y} = 63.4 \text{ ft}$$

The resultant is shown in Fig. 3-11(b).

3.11. Determine the resultant of the four forces tangent to the circle of radius 3 ft shown in Fig. 3-12(a). What will be its location with respect to the center of the circle?

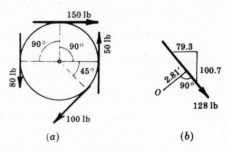

(a)   (b)

Fig. 3-12

SOLUTION

Note that the horizontal and vertical components of the 100-lb force are both -70.7 lb. Hence $\sum F_h = +150 - 70.7 = +79.3$ lb, i.e., to the right; and $\sum F_v = +50 - 80 - 70.7 = -100.7$ lb, i.e., down. The resultant $R = \sqrt{(\sum F_h)^2 + (\sum F_v)^2} = 128$ lb.

The moment of R with respect to O is $R \times a$, and this equals the sum of the moments of all the given forces about O. Hence $128a = +50 \times 3 - 150 \times 3 + 80 \times 3 - 100 \times 3 = -360$. The resultant is shown, in Fig. 3-12(b), at a distance of 2.81 ft from the center O of the circle, causing a negative moment.

3.12. Find the resultant of the force system acting on the thin gusset plate of Fig. 3-13. Locate the resultant by giving the x intercept.

SOLUTION

$$\sum F_x = 150 \cos 45° + 200 - 225 \cos 30° = 111.2 \text{ lb}$$
$$\sum F_y = 150 \sin 45° - 225 \sin 30° + 200 = 193.6 \text{ lb}$$
$$R = \sqrt{(111.2)^2 + (193.6)^2} = 223 \text{ lb} \qquad \theta = \tan^{-1}\frac{193.6}{111.2} = 60°$$
$$\sum M_O = (3)200 - (15)150 \cos 45° + (12)150 \sin 45° - (6)200 + 900 + (6)225 \cos 30° - (3)225 \sin 30°$$
$$= 813 \text{ lb-ft}$$

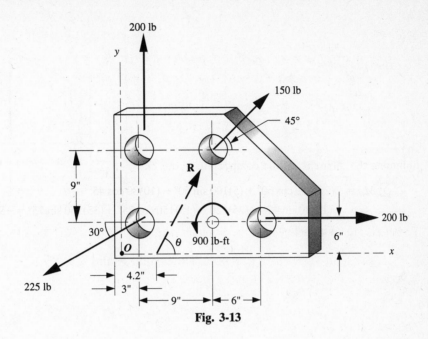

Fig. 3-13

To locate the resultant by its x intercept, use

$$\left(\sum F_y\right)\bar{x} = \sum M_O \qquad \bar{x} = \frac{813}{194} = 4.2 \text{ in}$$

3.13. In Fig. 3-14, the load of 20 lb/ft is uniformly distributed over the beam of length 6 ft. Determine R and d.

SOLUTION

$$R = \int_0^6 20\, dx = 120 \text{ lb} \qquad Rd = \int_0^6 x(20)\, dx = 360 \text{ lb-ft} \qquad d = \frac{360}{120} = 3 \text{ ft}$$

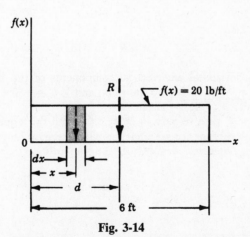

Fig. 3-14

3.14. In Fig. 3-15, the load is triangular in shape. The height of the diagram at distance x from the point O is by proportion equal to $(x/9)30$ N/m. Determine R and d.

SOLUTION

$$R = \int_0^9 \frac{x}{9}(30)\, dx = 135 \text{ N} \qquad Rd = \int_0^9 x\left[\frac{x}{9}(30)\right] dx = 810 \text{ N} \cdot \text{m} \qquad d = \frac{810}{135} = 6 \text{ m}$$

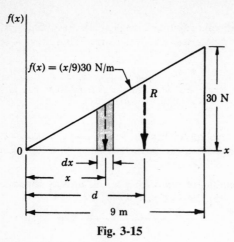

Fig. 3-15

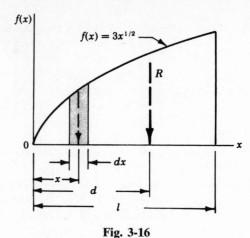

Fig. 3-16

3.15. In Fig. 3-16, the load varies as a parabola. Determine R and d.

SOLUTION

$$R = \int_0^l 3x^{1/2}\, dx = 2x^{3/2}\Big]_0^l = 2l^{3/2} \qquad Rd = \int_0^l x(3x^{1/2})\, dx = \tfrac{6}{5}l^{5/2} \qquad d = \frac{\tfrac{6}{5}l^{5/2}}{2l^{3/2}} = 0.6l$$

Supplementary Problems

3.16. Two forces of 200 N and 300 N pull in a horizontal plane on a vertical post. If the angle between them is 85°, what is their resultant? What angle does it make with the 200-N force? Solve both graphically and algebraically.   *Ans.* $R = 375\,\text{N}$, $\theta = 53°$

In Problems 3.17 through 3.20, find the resultant of each concurrent force system. The angle that each force makes with the x axis (measured counterclockwise) is given. Forces are in pounds.

3.17. Force   85   126   65   223
      θ_x 38°   142°   169°   295°   *Ans.* $R = 59.8\,\text{lb}$, $\theta_x = 268°$

3.18. Force   22   13   19   8
      θ_x 135°   220°   270°   358°   *Ans.* $R = 21.3\,\text{lb}$, $\theta_x = 214°$

3.19. Force   1250   1830   855   2300
      θ_x 62°   125°   340°   196°   *Ans.* $R = 2520\,\text{lb}$, $\theta_x = 138°$

3.20. Force   285   860   673   495   241
      θ_x 270°   180°   45°   330°   100°   *Ans.* $R = 181\,\text{lb}$, $\theta_x = 89°$

3.21. The 100-lb resultant of four forces together with three of those four forces is shown in Fig. 3-17. Determine the fourth force.   *Ans.* $F = 203\,\text{lb}$, $\theta_x = 49°$

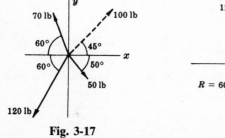

Fig. 3-17

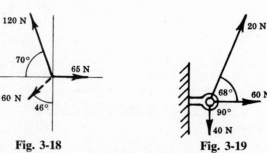

Fig. 3-18   Fig. 3-19

3.22. Three coplanar forces of 80 N each pull on a small ring (diameter negligible). Assuming that their lines of action make equal angles with each other (120°), determine the resultant. This system is said to be in equilibrium.   *Ans.   R = 0*

3.23. The resultant of three forces is 60 N as shown in Fig. 3-18. Two of the three forces are also shown as 120 N and 65 N. Determine the third force.   *Ans.   169 N, $\theta_x = 246°$*

3.24. Three wires exert the tensions indicated on the eyebolt in Fig. 3-19. Assuming a concurrent system, determine the force in a single wire that will replace the three wires.   *Ans.   T = 70.8 N, $\theta_x = 343°$*

3.25. Determine the resultant of the three forces originating at point $(3, -3)$ and passing through the points indicated: 126 N through $(8, 6)$, 183 N through $(2, -5)$, 269 N through $(-6, 3)$.
Ans.   R = 263 N, $\theta_x = 159°$ through $(3, -3)$

3.26. Determine the values of P and Q in Fig. 3-20 so that the resultant of the three coplanar forces is 100 lb at 20° to the x axis.   *Ans.   P = 240 lb; Q = 161 lb*

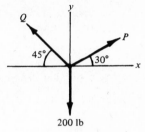

Fig. 3-20

Find the resultant in each of Problems 3.27 through 3.29. The forces are horizontal and expressed in pounds. The y distances to the action lines are expressed in feet.

3.27.   Force   +50   +20   −10
              y +3   −5   +6   *Ans.   R = +60 lb, $\bar{y} = -0.167$ ft*

3.28.   Force   +800   −300   +1000   −600
              y −6   −5   −4   0   *Ans.   R = +900 lb, $\bar{y} = -8.11$ ft*

3.29.   Force   +160   −220   +80   −180   +160
              y +3   −7   −3   +10   0   *Ans.   C = +20 lb-ft*

3.30. See Fig. 3-21. Find the resultant of the three loads shown acting on the beam.
Ans.   R = 38 T down, distance from left support = 8.37 ft

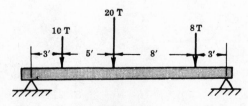

Fig. 3-21

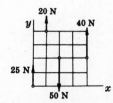

Fig. 3-22

3.31. Determine the resultant of the four forces shown in Fig. 3-22. The side of each small square is 1 m.
      Ans. $R = +35\,\text{N},\ \bar{x} = 2.99\,\text{m}$

3.32. Six weights of 30, 20, 40, 25, 10, and 35 lb hang in one plane from a horizontal support at distances 2, 3, 5, 7, 10, and 12 ft, respectively, from a wall. What single force would replace these six?
      Ans. $R = -160\,\text{lb}$, 6.34 ft from wall

3.33. Three forces act on the beam, two of which are shown in Fig. 3-23 together with the resultant of all three. What is the third force?   *Ans.* $F = 20\,\text{T}$ down, $\bar{x} = 10\,\text{ft}$ from left support

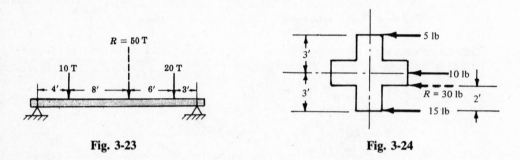

Fig. 3-23   Fig. 3-24

3.34. Determine the resultant of the coplanar parallel system shown in Fig. 3-24.
      Ans.   30 lb to left, 2 ft above bottom

3.35. Determine the resultant of the coplanar parallel system shown in Fig. 3-25.
      Ans. $R = 100\,\text{N}$ down, 55 mm up along plane from A

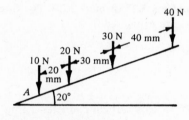

Fig. 3-25

In Problems 3.36 through 3.38, determine the resultant of the nonconcurrent, nonparallel systems of forces. F is in newtons, and the coordinates are in meters.

3.36.

F	20	30	50	10
θ_x	45°	120°	190°	270°

Coordinates of point of application $(1, 3)$ $(4, -5)$ $(5, 2)$ $(-2, -4)$
Ans. $R = 54.7\,\text{N},\ \theta_x = 157°$, x intercept $= 3.52\,\text{m}$

3.37.

F	50	100	200	90
θ_x	90°	150°	30°	45°

Coordinates of point of application $(2, 2)$ $(4, 6)$ $(3, -2)$ $(7, 2)$
Ans. $R = 303\,\text{N},\ \theta_x = 60.3°$, x intercept $= 6.77\,\text{m}$

3.38.

F	2	4	5	8
θ_x	45°	290°	183°	347°

Coordinates of point of application   (0, 5)   (4, 3)   (9, −4)   (2, −6)
Ans. $R = 7.12$ N, $\theta_x = 322°$, *x* intercept = 1.20 m

3.39.   Determine completely the resultant of the five forces shown in Fig. 3-26. The forces are in ounces, and the squares are 1 in by 1 in   *Ans.* $C = -268$ oz-in

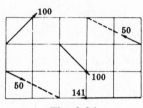

Fig. 3-26

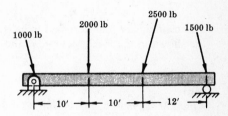

Fig. 3-27

3.40.   Determine completely the resultant of the four forces shown in Fig. 3-27. Each force makes a 15° angle with the vertical, except the 2000-lb force, which is vertical.
Ans. $R = 6830$ lb down, horizontal distance from hinge = 16.8 ft

3.41.   A thin steel plate is subjected to the three forces shown in Fig. 3-28. What single force would have an equivalent effect on the plate?
Ans. $R = 18.7$ N, $\theta_x = 285°$, bottom intercept = 4.23 m to left of *O*

3.42.   Determine the resultant of the forces acting on the bell crank shown in Fig. 3-29.
Ans. $R = 247$ lb, $\theta_x = 259°$, horizontal intercept = −6.3 in from *O*

3.43.   Determine the resultant of the three forces acting on the pulley shown in Fig. 3-30.
Ans. $R = 742$ N, $\theta_x = 357°$, *R* cuts vertical diameter 2.27 m above *O*

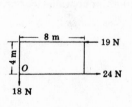

Fig. 3-28

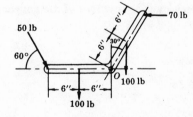

Fig. 3-29

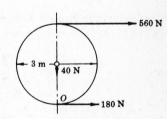

Fig. 3-30

3.44.   Solve for the resultant of the six loads on the truss shown in Fig. 3-31. The loads are given in kips (K) (1 kip = 1000 lb). Three loads are vertical. The wind loads are perpendicular to the side. The truss is symmetrical.
Ans. $R = 10.7$ k, $\theta_x = 281°$, *R* cuts lower member of truss +15.3 ft from left support

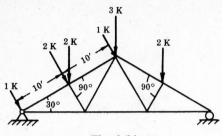

Fig. 3-31

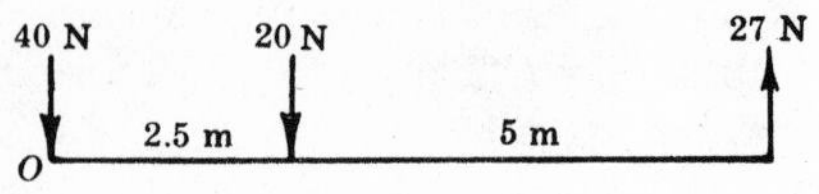

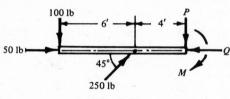

Fig. 3-32

Mathcad

3.45. The resultant of four vertical forces is a couple of 300 N · m counterclockwise. Three of the four forces are shown in Fig. 3-32. Determine the fourth.   *Ans.*   33 N up at 4.46 m to right of O

3.46. Determine M, P, and Q in Fig. 3-33 so that the resultant of the coplanar nonconcurrent force system is zero.
Ans. $M = 293$ lb-ft, $P = 76.7$ lb, $Q = 227$ lb

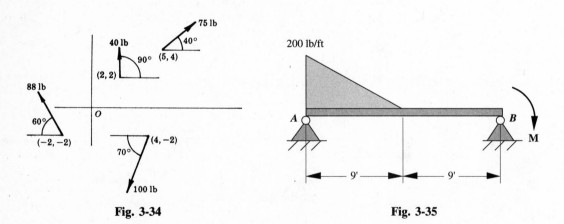

Fig. 3-33

3.47. Determine the resultant of the forces shown in Fig. 3-34. The coordinates are in feet.
Ans.   73.4 lb, $\theta_x = 107°$, x intercept = 8.38 ft to left of O

Fig. 3-34

Fig. 3-35

3.48. What is the maximum value of the moment M such that the beam in Fig. 3-35 does not leave the support at A?
Ans. $M = 13,500$ lb-ft

3.49. Determine the resultant of the force system shown in Fig. 3-36. Coordinates are in millimeters.
Ans. $\mathbf{R} = 24.7\mathbf{i} + 12.9\mathbf{j}$ N, x intercept is 24.5 mm

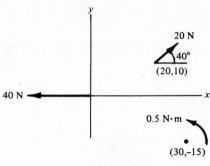

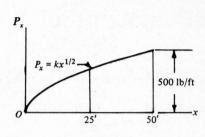

Fig. 3-36   **Fig. 3-37**

3.50. A 50-ft airplane wing is subjected to a test load that varies parabolically from 0 to 500 lb/ft. Refer to Fig. 3-37 and determine k, the resultant load and its location.   *Ans.* $R = 16,700$ lb, $\bar{x} = 30$ ft, $k = 70.7$

3.51. Do Prob. 3.50 if the test load is the first 90° of a sinusoidal cycle that varies from zero to 500 lb/ft in 50 ft. *Hint:* $P_x = k \sin(\pi x/100)$.
Ans. $k = 500$, $R = 15,900$ lb, $x = 31.9$ ft

Chapter 4

Resultants of Noncoplanar Force Systems

4.1 NONCOPLANAR FORCE SYSTEMS

Noncoplanar force systems are categorized as follows. A concurrent system consists of forces that intersect at a point called the concurrence. A parallel system consists of forces that intersect at infinity. The most general system is called nonconcurrent, nonparallel (or skew), and, as the name implies, the forces are not all concurrent and not all parallel.

4.2 RESULTANTS OF A NONCOPLANAR FORCE SYSTEM

The resultant of a noncoplanar force system is a force $\mathbf{R}$ and a couple $\mathbf{C}$, where $\mathbf{R} = \sum \mathbf{F}$, the vector sum of all forces of the system, and $\mathbf{C} = \sum \mathbf{M}$, the vector sum of the moments (relative to a selected base point) of all the forces of the system. The value of $\mathbf{R}$ is independent of the choice of the base point, but the value of $\mathbf{C}$ depends on the base point. For any force system it is possible to select a unique base point so that the vector $\mathbf{C}$ representing the couple is parallel to $\mathbf{R}$. This special combinaiton is called a *wrench* or *screw*.

The vector equations in the preceding paragraph may be applied directly to noncoplanar systems to determine the resultant, or the following derived scalar equations may be used.

4.3 CONCURRENT SYSTEM

The resultant $\mathbf{R}$ may be (*a*) a single force through the concurrence or (*b*) zero. Algebraically,

$$R = \sqrt{\left(\sum F_x \right)^2 + \left(\sum F_y \right)^2 + \left(\sum F_z \right)^2}$$

with direction cosines

$$\cos \theta_x = \frac{\sum F_x}{R} \qquad \cos \theta_y = \frac{\sum F_y}{R} \qquad \cos \theta_z = \frac{\sum F_z}{R}$$

where $\sum F_x, \sum F_y, \sum F_z$ = algebraic sums of the *x*, *y*, and *z* components, respectively, of the forces of the system

$\theta_x, \theta_y, \theta_z$ = angles that the resultant $\mathbf{R}$ makes with the *x*, *y*, and *z* axes, respectively.

4.4 PARALLEL SYSTEM

The resultant may be (*a*) a single force $\mathbf{R}$ parallel to the system, (*b*) a couple, or (*c*) zero. Assume the *y* axis is parallel to the system. Then, algebraically,

$$R = \sum F \qquad R\bar{x} = \sum M_z \qquad R\bar{z} = \sum M_x$$

where $\sum F$ = algebraic sum of the forces of the system

$\bar{x}$ = perpendicular distance from the *yz* plane to the resultant

$\bar{z}$ = perpendicular distance from the xy plane to the resultant

$\sum M_x$, $\sum M_z$ = algebraic sums of the moments of the forces of the system about the x and z axes, respectively.

If $\sum F = 0$, the resultant couple **C**, if there is one, will be determined by the following equation:

$$C = \sqrt{\left(\sum M_x\right)^2 + \left(\sum M_z\right)^2} \qquad \text{with} \qquad \tan \phi = \frac{\sum M_z}{\sum M_x}$$

where ϕ = angle that the vector representing the resultant couple makes with the x axis.

4.5 NONCONCURRENT, NONPARALLEL SYSTEM

As already indicated, the resultant is a force and a couple where the couple varies with the choice of a base point. In the following discussion a set of x, y, and z axes is placed with their origin at the base point.

Replace each force of the given system by the following setup: (1) an equal parallel force but acting through any chosen origin and (2) a couple acting in the plane containing the given force and the origin.

The magnitude of the resultant **R** of the concurrent system at the origin is given by the equation

$$R = \sqrt{\left(\sum F_x\right)^2 + \left(\sum F_y\right)^2 + \left(\sum F_z\right)^2}$$

with direction cosines

$$\cos \theta_x = \frac{\sum F_x}{R} \qquad \cos \theta_y = \frac{\sum F_y}{R} \qquad \cos \theta_z = \frac{\sum F_z}{R}$$

where the above quantities have the same meaning as listed in Section 4.3.

The magnitude of the resultant couple **C** is given by

$$C = \sqrt{\left(\sum M_x\right)^2 + \left(\sum M_y\right)^2 + \left(\sum M_z\right)^2}$$

with direction cosines

$$\cos \phi_x = \frac{\sum M_x}{C} \qquad \cos \phi_y = \frac{\sum M_y}{C} \qquad \cos \phi_z = \frac{\sum M_z}{C}$$

where $\sum M_x$, $\sum M_y$, $\sum M_z$ = algebraic sums of the moments of the forces of the system about the x, y, and z axes respectively

ϕ_x, ϕ_y, ϕ_z = angles which the vector representing the couple **C** makes with the x, y, and z axes respectively.

Solved Problems

In the following problems equivalent scalar equations are used when more convenient than the vector equations. Similarly, in the diagrams the forces are indicated by their magnitudes if the directions are clearly indicated.

4.1.   Forces of 20, 15, 30, and 50 lb are concurrent at the origin and are directed through the points whose coordinates are (2, 1, 6), (4, −2, 5), (−3, −2, 1), and (5, 1, −2), respectively. Determine the resultant of the system.

SOLUTION

F	Coordinates	$\cos \theta_x$	$\cos \theta_y$	$\cos \theta_z$	F_x	F_y	F_z
20	(2, 1, 6)	$\dfrac{+2}{\sqrt{41}} = +0.313$	$\dfrac{+1}{\sqrt{41}} = +0.156$	$\dfrac{+6}{\sqrt{41}} = +0.938$	+6.26	+3.12	+18.8
15	(4, −2, 5)	$\dfrac{+4}{\sqrt{45}} = +0.597$	$\dfrac{-2}{\sqrt{45}} = -0.298$	$\dfrac{+5}{\sqrt{45}} = +0.745$	+8.96	−4.47	+11.2
30	(−3, −2, 1)	$\dfrac{-3}{\sqrt{14}} = -0.803$	$\dfrac{-2}{\sqrt{14}} = -0.535$	$\dfrac{+1}{\sqrt{14}} = +0.268$	−24.1	−16.1	+8.04
50	(5, 1, −2)	$\dfrac{+5}{\sqrt{30}} = +0.912$	$\dfrac{+1}{\sqrt{30}} = +0.183$	$\dfrac{-2}{\sqrt{30}} = -0.365$	+45.6	+9.15	−18.3

      The denominator in each case is determined by taking the square root of the sum of the squares of x, y, and z intervals of differences. For the 30-lb force this is $\sqrt{(-3)^2 + (-2)^2 + (1)^2} = \sqrt{14}$.

      F_x is the product of F and $\cos \theta_x$. Watch signs. It is advisable to write signs before inserting any values.

      $\sum F_x = +6.26 + 8.96 - 24.1 + 45.6 = +36.7$. Similarly, $\sum F_y = -8.30$ and $\sum F_z = +19.7$. Then

$$R = \sqrt{\left(\sum F_x\right)^2 + \left(\sum F_y\right)^2 + \left(\sum F_z\right)^2} = 42.5 \text{ lb}$$

$$\cos \theta_x = \frac{\sum F_x}{R} = \frac{+36.7}{42.5} = +0.864 \qquad \theta_x = 30.2°$$

$$\cos \theta_y = \frac{\sum F_y}{R} = \frac{-8.30}{42.5} = -0.192 \qquad \theta_y = 79.0°$$

$$\cos \theta_z = \frac{\sum F_z}{R} = \frac{+19.7}{42.5} = +0.463 \qquad \theta_z = 62.4°$$

      The negative value of $\cos \theta_y$ signifies that the resultant has a negative component in the y direction. This is illustrated in Fig. 4-1.

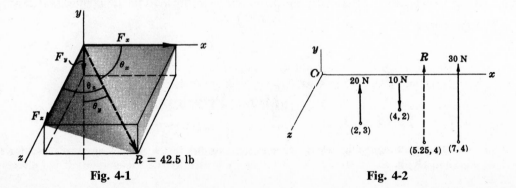

Fig. 4-1   Fig. 4-2

4.2. Three forces of $+20$ N, -10 N, and $+30$ N are shown in Fig. 4-2. The y axis is chosen parallel to the action lines of the forces. These lines pierce the xz plane at the points whose x and z coordinates in meters are respectively $(2, 3)$, $(4, 2)$, and $(7, 4)$. Locate the resultant.

SOLUTION

$$R = \sum F = +20 - 10 + 30 = +40 \text{ N}$$

To determine the x coordinate of the resultant (i.e., of the point where the action line of the resultant pierces the xz plane), use the projected system in the xy plane as shown in Fig. 4-3. Apply the equation $R\bar{x} = \sum M_z$:

$$\sum M_z = \sum M_O = +(20 \times 2) - (10 \times 4) + (30 \times 7)$$
$$= +210 \text{ N} \cdot \text{m}$$

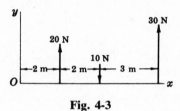

Fig. 4-3

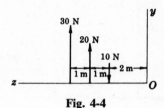

Fig. 4-4

The x coordinate must be such that a force of $+40$ N (acting up) will have a positive or counterclockwise moment. Therefore R must be to the right of O.

$$\bar{x} = +\frac{210}{40} = +5.25 \text{ m}$$

Be sure to determine the sign by inspection as indicated in the previous paragraph, and not by combining the signs of the moment and the force.

Figure 4-4 shows the projection of the system into the yz plane.

$$\sum M_x = \sum M_O = -(30 \times 4) - (20 \times 3) + (10 \times 2) = -160 \text{ N} \cdot \text{m}$$

The z coordinate must be such that a force of 40 N (acting up) will have a negative or clockwise moment of 160 N $\cdot$ m. Therefore R must be to the left of O. In this case the z coordinate is positive when it is to the left of O (refer to the space diagram).

$$\bar{z} = +\frac{160}{40} = +4.00 \text{ m}$$

The problem may now be summarized by saying that the resultant is a 40-N force acting up. Its action line is parallel to the y axis and pierces the xz plane at the point whose x, z coordinates are $(+5.25, +4.00)$ m. This is shown in Fig. 4-2.

4.3. Find the resultant of the system of forces shown in Fig. 4-5. The coordinates are in meters.

SOLUTION

$$R = \sum F = +100 + 50 - 150 = 0$$

This indicates that the resultant is not a single force. It may, however, be a couple.

Next find $\sum M_x$ and $\sum M_z$ as in the preceding problem.

$$\sum M_x = -(100 \times 2) + (50 \times 2) + (150 \times 3) = +350 \text{ N} \cdot \text{m}$$
$$\sum M_z = +(100 \times 2) + (50 \times 4) - (150 \times 8) = -800 \text{ N} \cdot \text{m}$$

Since $\sum F = 0$, $\sum M_x$ and $\sum M_z$ represent couples in the yz and xy planes respectively. These are shown in Fig. 4-6.

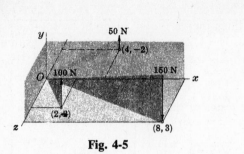

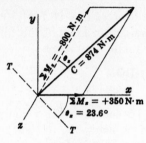

Fig. 4-5   Fig. 4-6

The two vectors representing the couples are shown combined into a resultant couple **C** with magnitude

$$C = \sqrt{\left(\sum M_x\right)^2 + \left(\sum M_z\right)^2} = 874\,\text{N}\cdot\text{m}$$

The vector **C** in the xz plane makes an angle θ_z with the z axis as shown in figure, where $\theta_z = \theta_x$.

According to the convention about couples, the resultant couple acts in a plane perpendicular to the vector **C** representing it. In the figure this could be in a plane containing the y axis with trace TT in the xz plane.

This trace makes an angle with the x axis of

$$\theta_x = \tan^{-1}\frac{\sum M_x}{\sum M_z} = \tan^{-1}\frac{350}{800} = 23.6°$$

4.4.   Determine the resultant of the nonconcurrent, nonparallel system of forces shown in Fig. 4-7.

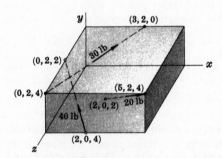

Fig. 4-7

SOLUTION

In working this example, replace each force by an equal force parallel to it through the origin and a couple. The forces at the origin are resolved into x, y, z components using direction cosines. Thus the 40-lb force has direction cosines that are determined by the differences in the coordinates of the two given points on its line of action.

The x difference is $0 - 2 = -2$; the y difference is $2 - 0 = +2$; the z difference is $2 - 4 = -2$. The cosine of the angle the 40-lb force makes with the x axis is

$$\cos\theta_x = \frac{-2}{\sqrt{(-2)^2 + (+2)^2 + (-2)^2}} = \frac{-2}{\sqrt{12}}$$

Similarly, $\cos\theta_y = +2/\sqrt{12}$ and $\cos\theta_z = -2/\sqrt{12}$.

The results are most conveniently set down in tabular form

F	$\cos \theta_x$	$\cos \theta_y$	$\cos \theta_z$	F_x	F_y	F_z
40	$\dfrac{-2}{\sqrt{12}}$	$\dfrac{+2}{\sqrt{12}}$	$\dfrac{-2}{\sqrt{12}}$	-23.1	$+23.1$	-23.1
30	$\dfrac{+3}{\sqrt{25}}$	0	$\dfrac{-4}{\sqrt{25}}$	$+18.0$	0	-24.0
20	$\dfrac{+3}{\sqrt{17}}$	$\dfrac{+2}{\sqrt{17}}$	$\dfrac{+2}{\sqrt{17}}$	$+14.6$	$+9.71$	$+9.71$

$$\sum F_x = +9.5 \qquad \sum F_y = +32.8 \qquad \sum F_z = -37.4$$

From the above table determine the resultant of this system of transferred forces concurrent at the origin.

$$R = \sqrt{\left(\sum F_x\right)^2 + \left(\sum F_y\right)^2 + \left(\sum F_z\right)^2} = \sqrt{(+9.5)^2 + (+32.8)^2 + (-37.4)^2} = 50.8 \text{ lb}$$

$$\cos \theta_x = \frac{\sum F_x}{R} = \frac{+9.5}{50.8} = +0.187 \qquad \cos \theta_y = \frac{\sum F_y}{R} = +0.645 \qquad \cos \theta_z = \frac{\sum F_z}{R} = -0.737$$

The resultant of the transferred forces is shown graphically in Fig. 4-8.

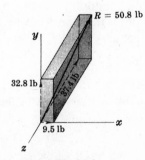

Fig. 4-8

The foregoing did not use the couple associated with each transferred force. Determine the moments of each of the three forces about the three axes to esgablish the magnitude and direction of these couples. Referring to Fig. 4-7, consider the 40-lb force acting at the point $(2, 0, 4)$. Its moment about the x axis is the algebraic sum of the moments of its three components about the x axis. However, its only component possessing a moment about the x axis is the y component. The moment of the 40-lb force about the x axis is therefore $-(23.1 \times 4) = -92.4$ lb-ft. In finding the moment about the y axis (i.e., M_y), consider the moments of both the x and z components. The moment of the x component about the y axis is $-(23.1 \times 4) = -92.4$ lb-ft. The moment of the z component about the y axis is $+(23.1 \times 2) = +46.2$ lb-ft. Therefore M_y is equal to $-92.4 + 46.2 = -46.2$ lb-ft. The moment of the 40-lb force about the z axis is the same as the moment of its y component about the z axis. Hence $M_z = +(23.1 \times 2) = +46.2$ lb-ft.

A tabular solution is given for the moments of the forces.

F	F_x	F_y	F_z	M_x	M_y	M_z
40	-23.1	$+23.1$	-23.1	-92.4	-46.2	$+46.2$
30	$+18.0$	0	-24.0	-48.0	$+72.0$	-36.0
20	$+14.6$	$+9.71$	$+9.71$	-19.4	$+9.8$	$+19.4$

$$\sum M_x = -159.8 \qquad \sum M_y = +35.6 \qquad \sum M_z = +29.6$$

The magnitude of the resultant couple is

$$C = \sqrt{\left(\sum M_x\right)^2 + \left(\sum M_y\right)^2 + \left(\sum M_z\right)^2} = \sqrt{(-159.8)^2 + (+35.6)^2 + (+29.6)^2} = 166 \text{ lb-ft}$$

with direction cosines

$$\cos \phi_x = \frac{\sum M_x}{C} = \frac{-159.8}{166} = -0.963 \qquad \cos \phi_y = \frac{\sum M_y}{C} = +0.214 \qquad \cos \phi_z = \frac{\sum M_z}{C} = +0.178$$

The vector $\mathbf{C}$ is shown in Fig. 4-9. The resultant couple acts in a plnae perpendicular to the vector $\mathbf{C}$. The resultant of the system is the combination of the force $\mathbf{R}$ and the couple $\mathbf{C}$.

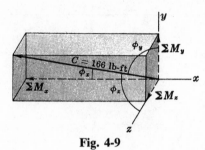

Fig. 4-9

4.5. Find the general force-couple resultant of the noncoplanar, nonconcurrent force system consisting of a 150-lb force along the line from $(2, 0, 0)$ through $(0, 0, 1)$, a 90-lb force along the line from $(0, -2, -1)$ through $(-1, 0, -1)$, and a 160 lb-ft couple lying in the xy plane. Distances are in feet. Rather the moments to the origin.

SOLUTION

Express each force in $\mathbf{i}, \mathbf{j}, \mathbf{k}$ notation and find the moment of each force relative to the origin.

$$\mathbf{F}_1 = 150 \frac{(0-2)\mathbf{i} + (0-0)\mathbf{j} + (1-0)\mathbf{k}}{\sqrt{(-2)^2 + (0)^2 + (1)^2}} = -134\mathbf{i} + 67.1 \mathbf{k} \text{ lb}$$

$$\mathbf{F}_2 = 90 \frac{(-1)\mathbf{i} + (2)\mathbf{j} + (0)\mathbf{k}}{\sqrt{(-1)^2 + (2)^2 + (0)^2}} = -40.25\mathbf{i} + 80.5\mathbf{j} \text{ lb}$$

Then

$$\mathbf{R} = \mathbf{F}_1 + \mathbf{F}_2 = -174\mathbf{i} + 80.5\mathbf{j} + 67.1\mathbf{k} \text{ lb}$$

Next determine $\mathbf{C}_1$ and $\mathbf{C}_2$.

$$\mathbf{C}_1 = \mathbf{r}_1 \times \mathbf{F}_1 = \begin{vmatrix} \mathbf{i} & \mathbf{j} & \mathbf{k} \\ 2 & 0 & 0 \\ -134 & 0 & 67.1 \end{vmatrix} = -134\mathbf{j} \text{ lb-ft}$$

$$\mathbf{C}_2 = \mathbf{r}_2 \times \mathbf{F}_2 = \begin{vmatrix} \mathbf{i} & \mathbf{j} & \mathbf{k} \\ 0 & -2 & -1 \\ -40.25 & 80.5 & 0 \end{vmatrix} = 80.5\mathbf{i} + 40.25\mathbf{j} - 80.5\mathbf{k} \text{ lb-ft}$$

The given couple in the xy plane can be written as $\mathbf{C}_3 = 160\mathbf{k}$ lb-ft. The total resultant couple is

$$\mathbf{C} = \mathbf{C}_1 + \mathbf{C}_2 + \mathbf{C}_3 = 80.5\mathbf{i} - 93.8\mathbf{j} + 79.5\mathbf{k} \text{ lb-ft}$$

Supplementary Problems

Determine the resultants in Problems 4.6 through 4.9, which involve concurrent systems of forces. The forces are in pounds and the coordinates of the points on the lines of action are in feet. All forces in each problem are concurrent at the origin. (Negative answers involve negative components.)

4.6. $\quad$ F $\qquad$ 100 $\qquad$ 200 $\qquad$ 500 $\qquad$ 300

Coordinates $\quad$ $(1,1,1)$ $\quad$ $(2,3,1)$ $\quad$ $(-2,-3,4)$ $\quad$ $(-1,1,-2)$

Ans. $\quad$ $R = 286$ lb, $\theta_x = -60°$, $\theta_y = 78°$, $\theta_z = 33°$

4.7. $\quad$ F $\qquad$ 5 $\qquad$ 2 $\qquad$ 3 $\qquad$ 4 $\qquad$ 8

Coordinates $\quad$ $(2,2,3)$ $\quad$ $(5,1,-2)$ $\quad$ $(-3,-4,5)$ $\quad$ $(2,1,-4)$ $\quad$ $(5,2,3)$

Ans. $\quad$ $R = 13.3$ lb, $\theta_x = 33°$, $\theta_y = 70°$, $\theta_z = 66°$

4.8. $\quad$ F $\qquad$ 1000 $\qquad$ 1500 $\qquad$ 1800

Coordinates $\quad$ $(-5,2,1)$ $\quad$ $(6,-3,-2)$ $\quad$ $(-2,-1,-1)$

Ans. $\quad$ $R = 1780$ lb, $\theta_x = -52°$, $\theta_y = -55°$, $\theta_z = -57°$

4.9. $\quad$ F $\qquad$ 40 $\qquad$ 80 $\qquad$ 30 $\qquad$ 20

Coordinates $\quad$ $(6,5,4)$ $\quad$ $(1,-3,-2)$ $\quad$ $(8,10,-7)$ $\quad$ $(-10,-9,-10)$

Ans. $\quad$ $R = 80.1$ lb, $\theta_x = 49°$, $\theta_y = -67°$, $\theta_z = -51°$

4.10. Determine the resultant of the three forces shown in Fig. 4-10. Note that their lines of action lie in the three coordinate planes and pass through the origin.

Ans. $\quad$ $R = 19.7$ lb, $\theta_x = 43°$, $\theta_y = 56°$, $\theta_z = 66°$

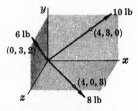

Fig. 4-10

In each of Problems 4.11 through 4.14, find the resultant and the coordinates of the intersection of its line of action with the xz plane. The given forces are in newtons and parallel to the y axis, and the coordinates of the intersection of each action line with the xz plane are in meters.

4.11. $\quad$ F $\qquad$ 100 $\qquad$ 150 $\qquad$ 200 $\qquad$ 300

(x, z) $\quad$ $(3, -2)$ $\quad$ $(1, 6)$ $\quad$ $(2, -3)$ $\quad$ $(-1, -1)$

Ans. $\quad$ $R = 750$ N, $\bar{x} = 0.733$ m, $\bar{z} = -0.267$ m

4.12. $\quad$ F $\qquad$ -25 $\qquad$ 18 $\qquad$ -12 $\qquad$ -30 $\qquad$ 36

(x, z) $\quad$ $(1, 2)$ $\quad$ $(2, -1)$ $\quad$ $(0, 0)$ $\quad$ $(-6, -2)$ $\quad$ $(3, 2)$

Ans. $\quad$ $R = -13$ N, $\bar{x} = -23.0$ m, $\bar{z} = -4.92$ m

4.13. $\quad$ F $\qquad$ 3 $\qquad$ -4 $\qquad$ 2 $\qquad$ -5

(x, z) $\quad$ $(2, 5)$ $\quad$ $(1, -5)$ $\quad$ $(3, 3)$ $\quad$ $(-4, -4)$

Ans. $\quad$ $R = -4$ N, $\bar{x} = -7.00$ m, $\bar{z} = -15.3$ m

4.14. $\quad$ F $\qquad$ $+10$ $\qquad$ $+20$ $\qquad$ -30

(x, z) $\quad$ $(1, 1)$ $\quad$ $(2, -5)$ $\quad$ $(3, -4)$

Ans. $\quad$ $C_x = -30$ N $\cdot$ m, $C_z = -40$ N $\cdot$ m

4.15. Five weights of 20, 15, 12, 6, and 10 lb rest on a table with the coordinates $(0.5, 15°)$, $(1.5, 90°)$, $(0.8, 185°)$, $(0.7, 262°)$, and $(1.2, 340°)$, respectively, where the first number in parentheses represents the radial distance in feet from the center of the table and the angle is measured counterclockwise from a reference radius as the table is viewed from above. Determine the resultant weight.

Ans. $\quad$ $R = 63$ lb down, $r = 0.3$, $\theta = 56°$

4.16.   Find the resultant of the system shown in Fig. 4-11. The forces are in pounds and the distances in feet.
Ans. $R = 40.3$ lb, $\cos \theta_x = +0.594$, $\cos \theta_y = +0.673$, $\cos \theta_z = -0.428$ at origin, $C = 251$ lb-ft, $\cos \phi_x = -0.660$, $\cos \phi_y = +0.633$, $\cos \phi_z = +0.396$

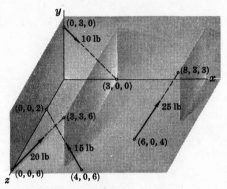

Fig. 4-11

4.17.   In Problem 4.4 find the couple **C** by determining the sum of the moments of the three forces about the origin using vector cross products. The reader should be aware that in finding the moment of a force about a moment center, the position vector **r** may be drawn from the moment center to *any* point on the action line of the force. The reader should prove this by using more than one point on the action line of one of the forces.

4.18.   Determine the resultant of the three forces shown in Fig. 4-12. Coordinates are in meters. Use the origin as the base point.
Ans. $\mathbf{R} = 3530\mathbf{i} + 267\mathbf{j} + 1200\mathbf{k}$ N through the origin, and $\mathbf{C} = -3200\mathbf{i} + 4810\mathbf{j} - 534\,\mathbf{k}$ N · m

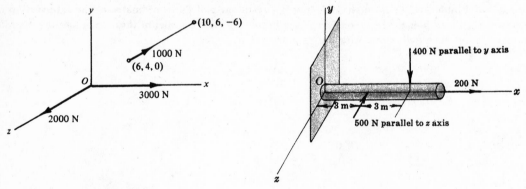

Fig. 4-12   **Fig. 4-13**

4.19.   Replace the three forces shown in Fig. 4-13 by a resultant force **R** through *O* and a couple.
Ans. $\mathbf{R} = 200\mathbf{i} - 400\mathbf{j} - 500\mathbf{k}$ N
          or $R = 671$ N with $\cos \theta_x = 0.298$, $\cos \theta_y = -0.597$, $\cos \theta_z = -0.745$
          $\mathbf{C} = 1500\mathbf{j} - 2400\mathbf{k}$ N · m
          or $C = 2830$ N · m with $\cos \phi_x = 0$, $\cos \phi_y = 0.530$, $\cos \phi_z = -0.848$

4.20.   Given the two forces $\mathbf{F}_1 = 20\mathbf{i} - 10\mathbf{j} + 60\mathbf{k}$ lb acting at $(0, -1, +1)$ and $\mathbf{F}_2 = 30\mathbf{i} + 20\mathbf{j} - 40\mathbf{k}$ lb acting at $(-1, -1, -1)$, and a couple of moment -80 lb-ft in the *xy* plane, determine the resultant force-couple system. Coordinates are in feet.
Ans. $\mathbf{R} = 50\mathbf{i} + 10\mathbf{j} + 20\mathbf{k}$ lb, $\mathbf{C} = 10\mathbf{i} - 50\mathbf{j} - 50\mathbf{k}$ lb-ft

4.21. Replace the three forces shown in Fig. 4-14 by a resultant force at A and a couple.

   Ans. $\mathbf{R} = 80\mathbf{i} - 100\mathbf{j} - 50\mathbf{k}$ lb, $\mathbf{C} = 100\mathbf{i} + 100\mathbf{j} - 200\mathbf{k}$ lb-ft

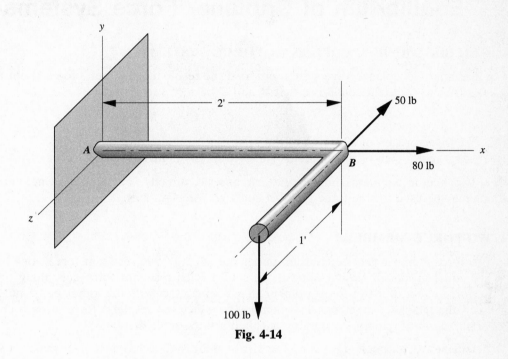

Fig. 4-14

4.22. Repeat Prob. 4.21 with the resultant force acting at B rather than A.

   Ans. $\mathbf{R} = 80\mathbf{i} - 100\mathbf{j} - 50\mathbf{k}$ lb, $\mathbf{C} = 100\mathbf{i}$ lb-ft

Chapter 5

Equilibrium of Coplanar Force Systems

5.1 EQUILIBRIUM OF A COPLANAR FORCE SYSTEM

Equilibrium of a coplanar force system occurs if the resultant is neither a force **R** nor a couple **C**. The necessary and sufficient conditions that **R** and **C** be zero vectors are

$$\mathbf{R} = \sum \mathbf{F} = 0 \quad \text{and} \quad \mathbf{C} = \sum \mathbf{M} = 0$$

where $\sum \mathbf{F}$ = vector sum of all forces of the system
   $\sum \mathbf{M}$ = vector sum of the moments (relative to any point) of all the forces of the system.

The two vector equations above may be applied directly, or the following derived scalar equations may be used for the three types of coplanar force systems.

5.2 TWO-FORCE MEMBERS

A two-force member is in equilibrium under the effect of two resultant forces, one at each end, which are equal in magnitude but opposite in sense. Each resultant force acts along the member. Thus the force effect of a two-force member that is in contact with any other body must act in the direction of the two-force member. Any member on which the resultant force at each end does not act in the direction of (along) the number is called a three-force member.

5.3 CONCURRENT SYSTEM

Any of the following sets of equations ensures equilibrium of a concurrent system; i.e., the resultant is zero. Assume concurrency at origin.

Set	Equations of Equilibrium	Remarks
A	(1) $\sum F_x = 0$	$\sum F_x$ = algebraic sum of the x components of the forces of the system.
	(2) $\sum F_y = 0$	$\sum F_y$ = algebraic sum of the y components of the forces of the system.
B	(1) $\sum F_x = 0$ (2) $\sum M_A = 0$	$\sum M_A$ = algebraic sum of the moments of the forces of the system about A, which may be chosen any place in the plane except on the y axis
C	(1) $\sum M_A = 0$ (2) $\sum M_B = 0$	$\sum M_A$ and $\sum M_B$ = algebraic sums of the moments of the forces of the system about A and B, which may be chosen any place in the plane provided A, B, and the origin do not lie on the same straight line

If only three nonparallel forces act in a plane on a body in equilibrium, these three forces must be concurrent.

5.4 PARALLEL SYSTEM

Any of the following sets of equations ensures equilibrium of a parallel system; i.e., the resultant is neither a force nor a couple.

Set	Equations	Remarks
A	(1) $\sum F = 0$	$\sum F$ = algebraic sum of the forces of the system parallel to the action lines of the forces
	(2) $\sum M_A = 0$	$\sum M_A$ = algebraic sum of the moments of the forces of the system about any point A in the plane
B	(1) $\sum M_A = 0$ (2) $\sum M_B = 0$	$\sum M_A$ and $\sum M_B$ = algebraic sums of the moments of the forces of the system about A and B, which may be chosen any place in the plane provided the line joining A and B is not parallel to the forces of the system.

5.5 NONCONCURRENT, NONPARALLEL SYSTEM

Any of the following sets of equations ensures equilibrium of a nonconcurrent, nonparallel system; i.e., the resultant is neither a force nor a couple.

Set	Equations	Remarks
A	(1) $\sum F_x = 0$	$\sum F_x$ = algebraic sum of x components of the forces
	(2) $\sum F_y = 0$	$\sum F_y$ = algebraic sum of y components of the forces
	(3) $\sum M_A = 0$	$\sum M_A$ = algebraic sum of the moments of the forces of the system about any point A in the plane
B	(1) $\sum F_x = 0$	$\sum F_x$ = algebraic sum of x components of the forces
	(2) $\sum M_A = 0$ (3) $\sum M_B = 0$	$\sum M_A$ and $\sum M_B$ = algebraic sums of the moments of the forces of the system about any two points A and B in the plane, provided that the line joining A and B is not perpendicular to the x-axis
C	(1) $\sum M_A = 0$ (2) $\sum M_B = 0$ (3) $\sum M_C = 0$	$\sum M_A, \sum M_B$, and $\sum M_C$ = algebraic sums of the moments of the forces of the system about any three points A, B, and C in the plane, provided that A, B, and C do not lie on the same straight line

5.6 REMARKS—FREE-BODY DIAGRAMS

In the solution of the problems, the following comments may be of assistance.

1. Draw free-body diagrams. The force system being analyzed will be holding a body or system of bodies in equilibrium. A free-body diagram is a sketch of the body (bodies) showing all external forces acting on the body (bodies). These include (a) all active forces, such as applied loads and gravity forces, and (b) all reactive forces. The latter forces are supplied by the ground, walls, pins, rollers, cables, or other means. A roller or knife-edge support means that the reaction there is shown perpendicular to the member. A pin connection means the reaction can be at any angle—it is represented by a force at an unknown angle or by using components of the pin reaction, e.g., A_x and A_y in a plane.

2. Note further that if the angle the reaction makes is known, the sense is then assumed along the reaction line. A positive sign in the result indicates that proper sense was assumed. A negative sign indicates the reactive force has the opposite sense to that assumed.

3. It may not be necessary to use all three equations of a set to obtain a solution. The proper choice of a moment center, for example, may yield an equation containing only one unknown.

4. The x and y axes in the above equations need not necessarily be chosen horizontally and vertically, respectively. Actually, if a system is in equilibrium, the algebraic sum of the scalar components of the system along any axis must be zero.

5. In the diagrams, a force is identified by its magnitude if the direction and sense are readily apparent.

6. The force in a spring is equal to the product of its spring constant k and its deformation from the unstressed position. In U.S. Customary units, k is in lb/in; thus $F = [k(\text{lb/in})][x(\text{in})] = kx(\text{lb})$. In SI Units, k is in N/m or N/mm; thus $F = [k(\text{N/m})][x(\text{m})] = kx(\text{N})$.

Solved Problems

5.1. Figure 5-1(a) shows a 25-lb lamp supported by two cables AB and AC. Find the tension in each cable.

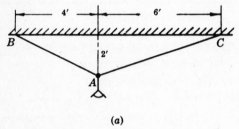
(a)

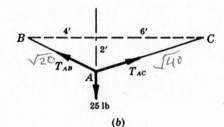
(b)

Fig. 5-1

SOLUTION

A free-body diagram of the knot at A is shown in Fig. 5-1(b) with the 25-lb force (weight of lamp) acting vertically down and the tensions in AC and AB.

Using Set A of the equations for a *concurrent system,* we have

$$\sum F_x = 0 = +T_{AC}\frac{6}{\sqrt{40}} - T_{AB}\frac{4}{\sqrt{20}} \tag{1}$$

$$\sum F_y = 0 = T_{AC}\frac{2}{\sqrt{40}} + T_{AB}\frac{2}{\sqrt{20}} - 25 \tag{2}$$

There are two equations in two unknowns. The problem is therefore statically determinate; i.e., it can be solved.

From equation (1), $T_{AC} = \frac{2}{3}\sqrt{2}\,T_{AB} = 0.942T_{AB}$. Substuting into (2),

$$0.942T_{AB}\frac{2}{\sqrt{40}} + T_{AB}\frac{2}{\sqrt{20}} - 25 = 0$$

from which $T_{AB} = 33.6\,\text{lb}$ and $T_{AC} = 0.942T_{AB} = 31.7\,\text{lb}$.

The solution could be obtained by using Sets B or C for a *concurrent system.* By choosing a moment center on one of the unknown forces, an equation is obtained that yields one unknown. Suppose, for example, point B is chosen as a moment center. Then

$$\sum M_B = 0 = -25 \times 4 + T_{AC} \times \frac{2}{\sqrt{40}} \times 4 + T_{AC} \times \frac{6}{\sqrt{40}} \times 2 \qquad \text{or} \qquad T_{AC} = 31.7\,\text{lb}$$

The moment of the force T_{AC} is equal to the moment of its components taken about point B. Another moment center, say at C, will yield an equation involving only the unknown T_{AB}.

5.2. Determine the horizontal force P necessary to push the 100-lb roller in Fig. 5-2(a) over the 2-in obstruction. The roller is smooth.

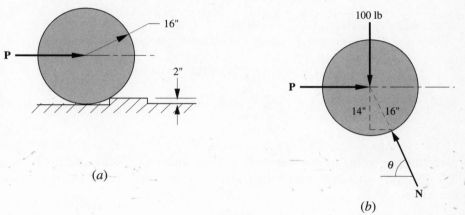

(a)

(b)

Fig. 5-2

SOLUTION

The free-body diagram is shown in Fig. 5-2(b). At the instant the roller starts to pass over the obstruction the reaction at the floor goes to zero. The reaction moves to N at the corner of the obstruction, and is normal to the surface of the roller. The angle $\theta = \sin^{-1}(14/16) = 61°$. The equations of equilibrium are

$$\sum F_v = N \sin 61° - 100 = 0 \qquad N = 114.3 \text{ lb}$$

$$\sum F_h = P - N \cos 61° = 0 \qquad P = 55.4 \text{ lb}$$

5.3. A boom 20 m long supports a load of 1200 kg as shown in Fig. 5-3(a). The cable BC is horizontal and 10 m long. Solve for the forces in the cable and the boom.

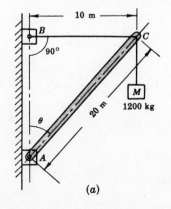

(a)

(b)

Fig. 5-3

SOLUTION

Both the cable and the boom are examples of two-force members. Thus the force F_1 in the cable is along the cable and the force F_2 in the boom is along the boom. By inspection, the arrows are placed as shown in Fig. 5-3(b), indicating tension in the cable and compression in the boom.

$$AB = \sqrt{(20)^2 - (10)^2} = 17.3 \text{ m} \qquad \cos \theta = \frac{17.3}{20} = 0.866$$

By taking moments about A, only one unknown enters into the equation.

$$\sum M_A = 0 = +(F_1 \times AB) - (11\,760 \times 10) \qquad 0 = F_1(17.3) - 117\,600 \qquad F_1 = 6800 \text{ N}$$

Summing forces vertically, $\sum F_v = 0 = -11\,760 + F_2 \cos \theta,\ 0 = -11\,760 + 0.866F_2,\ F_2 = 13\,600 \text{ N}.$

5.4. Solve for the forces in members AB and BC under the actions of the horizontal and vertical 1000-lb forces shown in Fig. 5-4(a). AB is at an angle θ and BC is at an angle β with the horizontal.

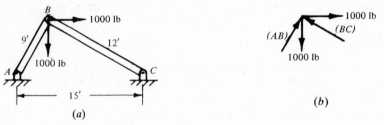

$$(a) \hspace{8em} (b)$$

Fig. 5-4

SOLUTION

Draw the free-body diagram in Fig. 5-4(b), assuming that (AB) and (BC) are compressive forces. The equations of equilibrium are

$$\sum F_x = 0 = (AB) \cos \theta - (BC) \cos \beta + 1000 \tag{1}$$

$$\sum F_y = 0 = (AB) \sin \theta + (BC) \sin \beta - 1000 \tag{2}$$

$$\cos \theta = 9/15 = \tfrac{3}{5} \qquad \cos \beta = 12/15 = \tfrac{4}{5} \qquad \sin \theta = \tfrac{4}{5} \qquad \sin \beta = \tfrac{3}{5}$$

Substitution yields

$$\tfrac{3}{5}(AB) - \tfrac{4}{5}(BC) + 1000 = 0$$

$$\tfrac{4}{5}(AB) + \tfrac{3}{5}(BC) - 1000 = 0$$

Thus,
$$(AB) = 200 \text{ lb compression}$$

$$(BC) = 1400 \text{ lb compression}$$

5.5. In Fig. 5-5(a) the bar AB weighs 10 lb/ft and is supported by a cable AC and a pin at B. Determine the reaction at B and the tension in the cable.

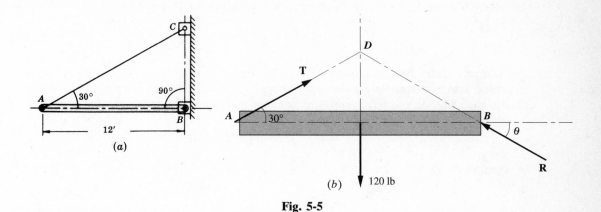

Fig. 5-5

SOLUTION

The tension in the cable is along the cable (two-force member). The boom AB is not a two-force member, since forces act at three points. In the free-body diagram of Fig. 5-5(b) the force at B is shown with unknown magnitude and direction. Because there are three non-parallel forces acting on the boom, these three forces must be concurrent at a point, in this case the point D. From trigonometry $\theta = 30°$. The equilibrium equations yield

$$F_x = T \cos 30° - R \cos 30° = 0$$
$$F_y = T \sin 30° + R \sin 30° - 120 = 0$$

From which $T = R = 120$ lb.

5.6. A and B, weighing 40 lb and 30 lb, respectively, rest on smooth planes as shown in Fig. 5-6. They are connected by a weightless cord passing over a frictionless pulley. Determine the angle θ and the tension in the cord for equilibrium.

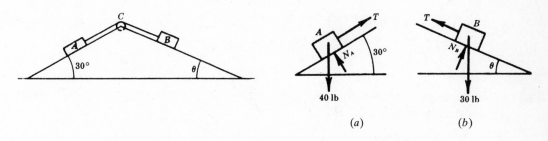

Fig. 5-6   Fig. 5-7

SOLUTION

Draw free-body diagrams of Fig. 5-7(a) and (b).

There are three unknowns in Fig. 5-7(b): T, N_B, and θ. Since only two equations are available, the system seems statically indeterminate as it stands. However, Fig. 5-7(a) contains only two unknowns including T which also occurs in Fig. 5-7(b), thereby making the system determinate when T has been found.

Summing forces parallel to the 30° plane, the equation of equilibrium obtained is

$$\sum F_{\parallel} = 0 = +T - 40 \sin 30° \qquad \text{or} \qquad T = 20 \, \text{lb}$$

Returning to Fig. 5-7(b) and summing forces parallel to the plane, we have

$$\sum F_{\parallel} = 0 = +T - 30 \sin \theta \qquad \sin \theta = \frac{T}{30} = \frac{2}{3} \qquad \theta = 41.8°$$

5.7. A uniform slender beam of mass M has its center of gravity as shown in Fig. 5-8(a). The corner on which it rests is a knife-edge; hence the reaction N is perpendicular to the beam. The vertical wall on the left is smooth. What is the value of the angle θ for equilibrium?

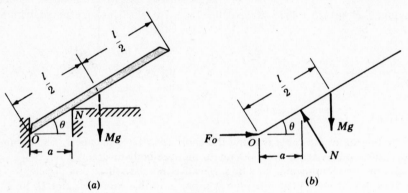

(a)   (b)

Fig. 5-8

SOLUTION

The free-body diagram is shown in Fig. 5-8(b).
Summing moments about O,

$$\sum M_O = 0 = +N \frac{a}{\cos \theta} - Mg(\tfrac{1}{2}l \cos \theta)$$

Summing forces vertically,

$$\sum F_v = 0 = N \cos \theta - Mg \qquad \text{or} \qquad N = \frac{Mg}{\cos \theta}$$

Substitute $N = Mg/\cos \theta$ into the first equation to obtain

$$\frac{Mga}{\cos^2 \theta} - \frac{Mgl \cos \theta}{2} = 0 \qquad \text{or} \qquad \cos \theta = \sqrt[3]{\frac{2a}{l}}$$

5.8. A beam considered weightless is loaded with concentrated loads as shown in Fig. 5-9(a). Determine the reactions at A and B.

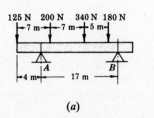

(a)

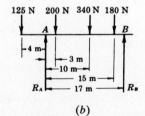

(b)

Fig. 5-9

SOLUTION

To find beam reactions R_A and R_B shown in Fig. 5-9(b), it is advisable to take moments about A and then about B. In this way each equation yields only one unknown. Each reaction is thus found independently of the other. Then the summation of the forces should equal zero, providing an excellent check. Many readers may prefer to determine one reaction by a moment equation and then determine the other by the sum of the forces. This is, of course, a correct procedure. But the authors prefer the two summations of moments, thereby reserving the summation of forces as a check equation. Using this procedure, the two equations are

$$\sum M_A = 0 = +(125 \times 4) - (200 \times 3) - (340 \times 10) - (180 \times 15) + R_B \times 17 \qquad \text{or} \qquad R_B = 365 \text{ N} \quad (1)$$

$$\sum M_B = 0 = +(125 \times 21) - R_A \times 17 + (200 \times 14) + (340 \times 7) + (180 \times 2) \qquad \text{or} \qquad R_A = 480 \text{ N} \quad (2)$$

Checking, $\sum F = -125 + 480 - 200 - 340 - 180 + 365 = 0$. This sum should, within the limits of accuracy, equal zero. Since it does, the beam reactions are correct.

5.9. Determine the reactions for the beam with both concentrated and distributed loads, as shown in Fig. 5-10.

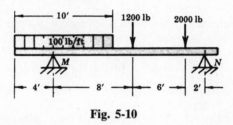

Fig. 5-10

SOLUTION

Note that the distributed load of 100 lb/ft is replaced in the free body diagram (Fig. 5-11) with a concentrated load of 1000 lb at its midpoint. This is permissible only in determining beam reactions.

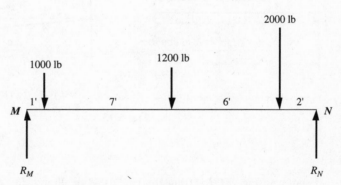

Fig. 5-11

The equilibrium equations are

$$\sum M_M = 16R_N - 1000 - (8)(1200) - (14)(2000) = 0, \qquad R_N = 2410 \text{ lb}$$

$$\sum M_N = -16R_M + (15)(1000) + (8)(1200) + (2)(2000) = 0, \qquad R_M = 1790 \text{ lb}$$

5.10. Determine the force P required to hold a mass of 10 kg in equilibrium utilizing the system of pulleys shown in Fig. 5–12(a). Assume that all pulleys are the same size.

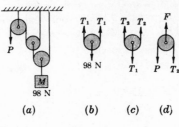

Fig. 5-12

SOLUTION

Figure 5-12(b) is a free-body diagram of the lowest pulley. The gravitational force of 10×9.8 N acts down. Acting up is the pull in the rope on each side of the pulley. Since the rope is continuous and since frictionless pulleys are assumed, the tension in the rope leaving one side is the same as the tension in the rope entering the other side. The tension T_1 is therefore equal to 49 N, since a vertical summation of forces yields $\sum F_v = 0 = +2T_1 - 98$.

Next draw a free-body diagram of the middle pulley, Fig. 5-12(c). From the reasoning just explained, the tension in the rope around this pulley is T_2. Summing forces vertically, the equation obtained is $\sum F_v = 0 = +2T_2 - T_1 = 2T_2 - 49$, whence $T_2 = 24.5$ N.

Finally, draw a free-body diagram of the top pulley, Fig. 5-12(d). Since the rope is continuous, $P = T_2$, or $P = 24.5$ N.

5.11. The rigid beam in Fig. 5-13(a) is supported by a pin at A and springs at B and C. If each spring constant is 20 lb/in, determine the reaction at A and the force in each spring. The uniform load is 20 lb/ft.

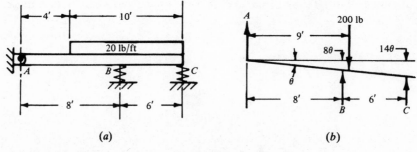

Fig. 5-13

SOLUTION

Assume, as shown in Fig. 5-13(b), that the angle θ is small and thus the spring deflections in feet are 8θ and 14θ. The force at B is then $8\theta \times 20 \times 12 = 1920\theta$ lb. The force at C is similarly $14\theta \times 240 = 3360\theta$ lb. To solve for θ, use the sum of the moments about A.

$$\sum M_A = 0 = -200 \times 9 + 1920\theta \times 8 + 3360\theta \times 14 \qquad \text{or} \qquad 0.0288 \text{ rad}$$

$$\text{Force } B = 1920\theta = 55.3 \text{ lb}$$

$$\text{Force } C = 3360\theta = 96.8 \text{ lb}$$

To find the reaction at A, use a vertical summation of forces as follows:

$$\sum F = 0 = A + 55.3 + 96.8 - 200 \quad \text{or} \quad A = 47.9 \text{ lb up}$$

5.12. A cantilever beam 3.8 m long with a mass of 10 kg/m carries a concentrated load of 1000 N at its free end. The other end of the beam is inserted into a wall 0.8 m thick. What are the reactions on the beam at A and B? Refer to Fig. 5-14(a).

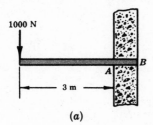

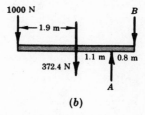

(a)   (b)

Fig. 5-14

SOLUTION

Assume the beam bends so that the wall pushes up at A and down at B on the beam. Draw the free-body diagram showing at the midpoint the gravitational force 372.4 N (3.8 m × 10 kg/m × 9.8 m/s^2). See Fig. 5-14(b).

To determine reaction A, sum moments about B to obtain

$$\sum M_B = 0 = +1000 \times 3.8 + 372.4 \times 1.9 - 0.8A \qquad A = 5630 \text{ N}$$

To determine reaction B, sum moments about A to obtain

$$\sum M_A = 0 = +1000 \times 3.0 + 372.4 \times 1.1 - 0.8B \qquad B = 4260 \text{ N}$$

5.13. Blocks A and B weigh 400 and 200 lb, respectively. They rest on a 30° inclined plane and are attached to the post which is held perpendicular to the plane by force P parallel to the plane [see Fig. 5–15(a)]. Assume that all surfaces are smooth and that the cords are parallel to the plane. Determine the value of P.

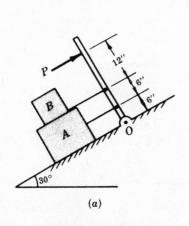

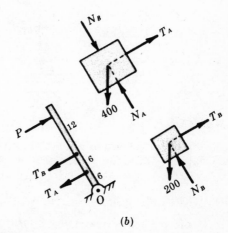

(a)   (b)

Fig. 5-15

SOLUTION

Draw free-body diagrams of A, B, and the post as shown in Fig. 5–15(b). It is seen by inspection that T_A and T_B may be found by a summation of forces parallel to the plane. Thus, $T_A = 400 \sin 30° = 200$ lb. Similarly, $T_B = 100$ lb.

In the free-body diagram of the post, sum moments about O to obtain $-P \times 24 + T_B \times 12 + T_A \times 6 = 0$. Substitute the value of T_A and T_B to obtain $P = 100$ lb.

5.14. A vertical force F of 50 N is applied to the bell crank at point A as shown in Fig. 5-16(a). A force P applied at B prevents rotation of the crank about point O. Determine the force P and the bearing reaction R at O.

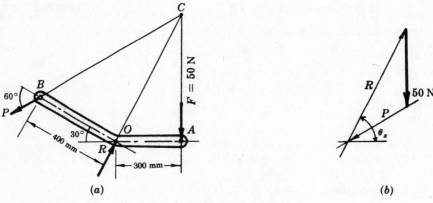

Fig. 5-16

SOLUTION

Since only three forces act on the bell crank, which is in equilibrium, they must be concurrent. This means that the reaction R must pass through point C where F and P intersect.

The force triangle is shown in Fig. 5-16(b). The reaction R is parallel to the line OC. Measurement of the force triangle yields $P = 43.0$ N, and $R = 80.8$ N with $\theta_x = 62.5°$.

5.15. Determine the reactions on the beam loaded as shown in Fig. 5-17(a). The loads are in kilonewtons (kN). Neglect the thickness and mass of the beam.

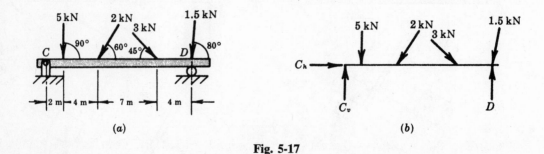

Fig. 5-17

SOLUTION

In the free-body diagram, Fig. 5-17(b), the horizontal and vertical components C_h and C_v of the pin reaction at C are assumed positive. the roller support at D is normal to the beam, as shown.

Summing forces horizontally yields an equation with C_h as the only unknown.

$$\sum F_h = 0 = C_h - 2 \cos 60° + 3 \cos 45° - 1.5 \cos 80° \qquad C_h = -0.86 \text{ kN}$$

This means that C_h acts to the left instead of as assumed.

To find D, take moments about C; this will yield an equation involving only D:

$$\sum M_C = 0 = -5 \times 2 - (2 \sin 60°) \times 6 - (3 \sin 45°) \times 13 - (1.5 \sin 80°) \times 17 + D \times 17$$

Note that the moment of each given force about C is equal to the moment of only its vertical component about C because each horizontal component has a line of action passing through C. Solving this moment equation for D, $D = 4.3$ kN.

To find C_v, the summation of moments about D is convenient.

$$\sum M_D = 0 = -C_v \times 17 + 5 \times 15 + 2 \sin 60° \times 11 + 3 \sin 45° \times 4 \qquad C_v = 6.03 \text{ kN}$$

A check on the values for C_v and D may be obtained by summing the forces vertically, because this equation has not been used yet. This summation, $\sum F_v$, should equal zero when the values of C_v and D are inserted.

$$\sum F_v = 6.03 - 5 - 2 \times 0.866 - 3 \times 0.707 - 1.5 + 4.30 = -0.02$$

Since this is within the limits of accuracy of the problem, the values check closely enough.

5.16. For the beam shown in Fig. 5-18(a) determine the reactions at A and B.

SOLUTION

The free-body diagram is shown in Fig. 5-18(b).
The equilibrium equations are

$$\sum M_A = 4R - (1)200 - (3)200 + 500 = 0 \qquad R_B = 75 \text{ N}$$
$$\sum M_B = -4R + (3)200 + (1)200 + 500 = 0 \qquad R_A = 325 \text{ N}$$

5.17.

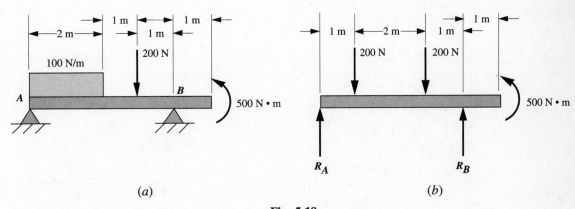

(a)   (b)

Fig. 5-18

Determine the tension in the cable AB that holds a post BC from sliding. Figure 5-19(a) shows the essential data. The weight of the post is 18 lb. Assume that all surfaces are smooth.

SOLUTION

Figure 5-19(b) is the free-body diagram. Note that R_D is normal to the post and R_B is normal to the floor, because frictionless surfaces are assumed.

One procedure is to make moments about B to find R_D, and then to sum the forces horizontally to find T. The following equations ensue.

$$\sum M_B = 0 = -18(7\cos 60°) + R_D \frac{10}{\cos 30°} \qquad \sum F_h = 0 = T - R_D \cos 30° = T - 5.46(0.866)$$

from which $R_D = 5.46$ lb, $T = 4.72$ lb.

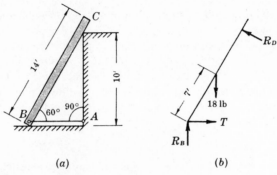

(a)   (b)

Fig. 5-19

5.18. Determine the following forces for the A-frame shown in Fig. 5-20(a): (1) the floor reactions at A and E, (2) the pin reactions at C on CE, (3) the pin reactions at B on AC. The floor is assumed to be smooth. Neglect the weight of the members.

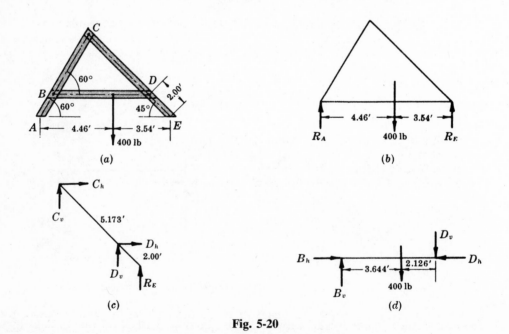

(a)   (b)

(c)   (d)

Fig. 5-20

SOLUTION

To determine the floor reactions at A and E, consider the entire frame as a solid free body as shown in Fig. 5-20(b). The manner in which the frame distributes the 400-lb load within itself has no bearing in determining the external reactions at A and E. Taking moments about A and E, the following solutions result:

$$\sum M_A = 0 = R_E \times 8.00 - 400 \times (8.00 - 3.54) \qquad R_E = 223 \text{ lb}$$

$$\sum M_E = 0 = -R_A \times 8.00 + 400 \times 3.54 \qquad R_A = 177 \text{ lb}$$

As a check, the vertical summation of forces does equal zero.

In working parts (2) and (3) of the problem, draw a free-body diagram of the member CE in Fig. 5-20(c). Assume that the reactions of the pins *on CE* are as shown. There are four unknowns in the figure with only three equatons available. Another free-body diagram must now be drawn involving some of the same unknowns. Draw a free-body diagram of member BD in Fig. 5-20(d) showing the pin reactions acting at D on BD opposite in direction to those assumed in Fig. 5-20(c).

The 2.126' dimension in Fig. 5-20(d) is obtained by subtracting the horizontal projection of the 2.00' dimension from 3.54'. By similar reasoning, the other dimensions are obtained.

The following equations may be written for Fig. 5-20(c):

$$\sum F_h = 0 = +C_h + D_h \tag{1}$$

$$\sum F_v = 0 = +C_v + D_v + 223 \tag{2}$$

$$\sum M_D = 0 = +223 \times 2 \cos 45° - C_v \times 5.173 \cos 45° - C_h \times 5.173 \sin 45° \tag{3}$$

For Fig. 5-20(d) write the following equations:

$$\sum M_B = 0 = -400 \times 3.644 - D_v \times 5.770 \tag{4}$$

$$\sum M_D = 0 = -B_v \times 5.770 + 400 \times 2.126 \tag{5}$$

$$\sum F_h = 0 = +B_h - D_h \tag{6}$$

In solving the above equations, look for equations each with only one unknown.

From (4),
$$D_v = \frac{-400 \times 3.644}{5.770} = -252.6 \text{ lb}$$

From (5),
$$B_v = \frac{400 \times 2.126}{5.770} = 147.4 \text{ lb}$$

Substitute $D_v = -252.6$ lb in equation (2) to obtain $C_v = -223 - (-252.6) = 29.6$ lb.
Substitute $C_v = 29.6$ lb in equation (3) to obtain $C_h = 56.6$ lb.
From (6) and (1), $B_h = +D_h = +(-C_h) = +(-56.6) = -56.6$ lb.
To summarize the results, with particular emphasis on signs:

1. Floor reactions.

$$R_A = 177 \text{ lb up} \qquad R_E = 223 \text{ lb up}$$

2. Pin reactions at C on CE. These were assumed to act on CE in positive directions.

$$C_h = 56.6 \text{ lb to right} \qquad C_v = 29.6 \text{ lb up}$$

3. Pin reactions at B on AC. In Fig. 5-20(d) the pin reactions at B were shown acting on BD. Therefore they act in the opposite direction on AC. In the solution for the reaction on BD it was found that: $B_h = -56.6$ lb, i.e., to left; $B_v = 147.4$ lb, i.e., up. therefore, *on AC* the pin reactions are

$$B_h = 56.6 \text{ lb to right} \qquad B_v = 147.4 \text{ lb down}$$

Note: A computer solution to Problem 5.18 is available in Appendix C.

5.19. A cylinder 1 m in diameter and of 10-kg mass is lodged between cross pieces that make an angle of 60° with each other as shown in Fig. 5-21(a). Determine the tension in the horizontal rope DE, assuming a smooth floor.

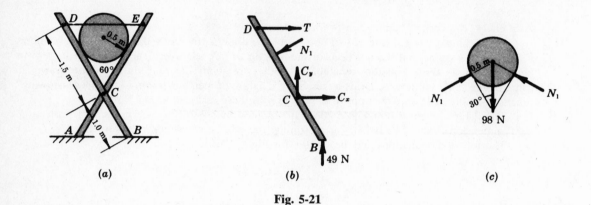

Fig. 5-21

SOLUTION

Treat the entire structure as a free-body diagram. Because of symmetry, it is apparent that $A = B = 49$ N vertically upward (gravitational force is $10 \times 9.8 = 98$ N).

Next draw a free-body diagram of the arm DB showing the rope tension T and the reactions C_x and C_y of pin C [Fig. 5-21(b)]. The reaction N of the cylinder is perpendicular to the arm.

If N_1 were known then a summation of moments about C would yield the tension T. But N_1 can be found by drawing a free-body diagram of the cylinder [Fig. 5-21(c)]. From the geometry involved, N_1 extends through the center of the cylinder. Hence a vertical summation of forces yields

$$\sum F_v = 0 = 2N_1 \sin 30° - 98 \qquad N_1 = 98 \text{ N}$$

Also note that the perpendicular distance from N_1 to C is $0.5/(\tan 30°) = 0.866$ m.

Return to the free-body diagram of the arm BD and sum moments about C to obtain

$$\sum M_c = 0 = -T \times 1.5 \cos 30° + 98 \times 0.866 + 49 \times 1 \sin 30° \qquad T = 84.2 \text{ N}$$

Supplementary Problems

5.20. A weight of 100 lb is suspended by a rope from a ceiling. A horizontal force pulls the weight until the rope makes an angle of 70° with the ceiling. Find the horizintal force H and the tension T in the rope. Use both algebraic and graphical methods.   *Ans.* $H = 36.4$ lb, $T = 106$ lb

5.21. A rubber band has an unstretched length of 8 in. It is pulled until its length is 10 in, as shown in Fig. 5-22. The horizontal force P is 6 oz. What is the tension in the band?   *Ans.* $T = 5$ oz

5.22. As shown in Fig. 5-23, a small piece of round rod is welded to the pinch bar at A to serve as a fulcrum. A force P equal to 1200 N is required to lift the left side of a box B. To lift the right side of the box, the same pinch bar is used, and P is then 1000 N. What is the mass of the box?
Ans. $M = 2240$ kg

5.23.   Refer to Fig. 5-24. The body A weighs 32.8 lb and rests on a smooth surface. The body B weighs 14.3 lb. Determine the tensions in S_1 and S_2 and the normal reaction of the horizontal surface on A.
Ans. $S_1 = 12.4$ lb, $S_2 = 14.3$ lb, $N = 25.7$ lb

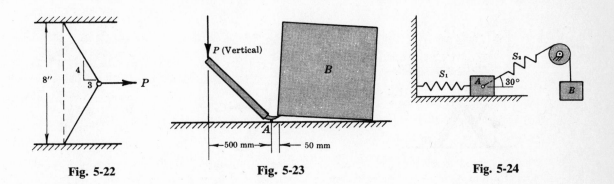

Fig. 5-22   Fig. 5-23   Fig. 5-24

5.24.   Two guy wires are fastened to an anchor bolt in a foundation as shown in Fig. 5-25. What pull does the bolt exert on the foundation?   Ans. $P = 1030$ lbm $\theta_x = 135°$

5.25.   Three concurrent forces have magnitudes of 40, 60, and 50 N, respectively. Determine the angles among them that will produce a state of equilibrium.   Ans.   97°, 138°, 125°

5.26.   A pulley to which is attached W_1 rides on a wire that is attached to a support at the left and that passes over a pulley on the right to weight W (see Fig. 5-26). The horizontal distance between the left support and the pulley (neglect the dimensions of the pulleys) is L. Express the sag d at the center in terms of W, W_1, and L.   Ans. $d = \frac{1}{2}L/\sqrt{(2W/W_1)^2 - 1}$

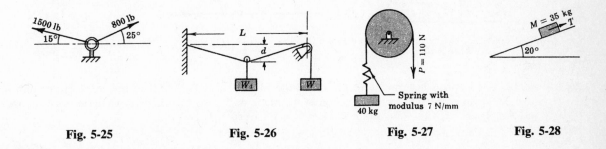

Fig. 5-25   Fig. 5-26   Fig. 5-27   Fig. 5-28

5.27.   What is the normal reaction between the mass that rests on the ground, as shown in Fig. 5-27, and the ground? The pulley is assumed massless and in frictionless bearings.   Ans. $N = 282$ N

5.28.   Refer to Fig. 5-28. What force T parallel to the smooth plane is necessary to hold the 35-kg mass M in equilibrium?   Ans. $T = 117.6$ N

5.29.   Refer to Fig, 5-29. A weight of 80 lb is suspended from a weightless bar AB, which is supported by a cable CB and a pin at A. Determine the tension in the cable and the pin reaction at A on the bar AB.
Ans. $T = 197$ lb, $A_x = 180$ lb, $A_y = 0$ lb

5.30.   Refer to Fig. 5-30. In the figure, three spheres each with 2-kg mass and each 350 mm in diameter rest in a box 760 mm wide. Find (a) the reaction of B on A, (b) the reaction of the wall on C, and (c) the reaction of the floor on B.
Ans.   (a) 12.1 N along the line joining their centers, (b) 7.09 N to left, (c) 29.4 N up

5.31. In Fig. 5-31, a weight of 500 lb is supported by rigid members AB and BC pinned as shown. By drawing a free-body diagram of the pin at B, determine the forces F_1 and F_2 in members AB and BC, respectively.   *Ans.* $F_1 = 433$ lb C, $F_2 = 250$ lb C

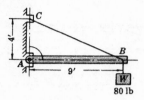

Fig. 5-29

Fig. 5-30

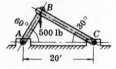

Fig. 5-31

5.32. What horizontal force through the center is necessary to start a 20-kg wheel of 1-m diameter over a block 150 mm high? At the moment motion impends, the force between the wheel and the ground is zero. Note also that the reaction of the block on the wheel must pass through the center of the wheel. *Ans.* $F = 200$ N

5.33. The roller shown in Fig. 5-32 weighs 339 lb. What force T is necessary to start the roller over the block A?   *Ans.* $T = 403$ lb

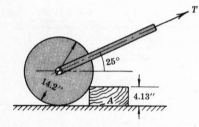

Fig. 5-32

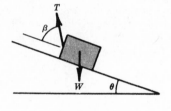

Fig. 5-33

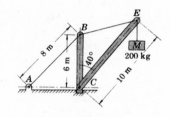

Fig. 5-34

5.34. In Fig. 5-33, express in terms of θ, β, and W the force T necessary to hold the weight in equilibrium. Also derive an expression for the reaction of the plane on W. No friction is assumed between the weight and the plane.   *Ans.* $T = W \sin \theta / \cos \beta$, $N = W \cos (\theta + \beta)/\cos \beta$

5.35. The derrick shown in Fig. 5-34 supports a mass M of 200 kg. Using a free-body diagram of pin E, find the forces in cable BE and boom CE.   *Ans.* $BE = 2450$ N tension, $CE = 3360$ N compression

5.36. If the structure in Problem 5.31 rested at A and C on the smooth floor but a cable joined pins A and C, determine the tension in the cable AC,   *Ans.* $T = 217$ lb

5.37. A 50-kg smooth cylinder is at rest in a box that has smooth walls at right angles to each other. If the box is tipped up to a 45° angle, what is the reaction of the bottom of the box on the cylinder?   *Ans.*   693 N

5.38. A 12-ft beam is simply supported at its ends. The beam supports a uniform load of 200 lb/ft throughout its length and an applied clockwise couple of 2000 lb-ft at its center. Determine the reactions at the ends of the beam.   *Ans.* $R_L = 1033$ lb, $R_R = 1367$ lb

5.39. Determine the reactions on the beam loaded as shown in Fig. 5-35. The uniformly distributed load is 300 kg/m. Neglect the mass of the beam.   *Ans.* $R_A = 3540$ N, $R_B = 930$ N

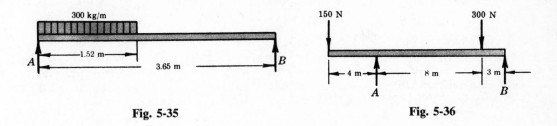

| Fig. 5-35 | Fig. 5-36 |

5.40. Determine beam reactions in Fig. 5-36. Consider only the two concentrated loads.
Ans. $R_A = 286$ N, $R_B = 164$ N

5.41. The bar A shown in Fig. 5-37 weighs 30 lb/ft and is 12 ft long. The left end is inserted into a wall 14 in thick. The 2-ft diameter pulley weighs 40 lb. The tension T in the rope is 80 lb. Determine the reactions at points B and C on the bar, which fits loosely in the wall.
Ans. $R_C = 3910$ lb up, $R_B = 3350$ lb down.

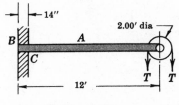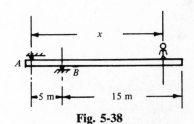

| Fig. 5-37 | Fig. 5-38 |

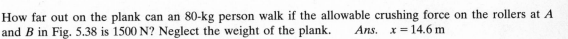

5.42. How far out on the plank can an 80-kg person walk if the allowable crushing force on the rollers at A and B in Fig. 5-38 is 1500 N? Neglect the weight of the plank.   *Ans.* $x = 14.6$ m

5.43. In Fig. 5-39, what force P is required to raise the mass M of 90 kg at constant speed?
Ans. $P = 441$ N

5.44. In Fig. 5-40, what force P is required to hold a weight of 600 lb in equilibrium?   *Ans.* $P = 200$ lb

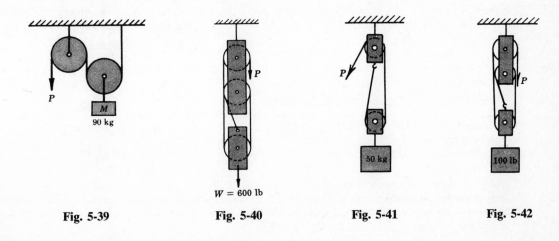

| Fig. 5-39 | Fig. 5-40 | Fig. 5-41 | Fig. 5-42 |

5.45. The upper block in Fig. 5-41 is suspended from a fixed support. The rope is secured to the lower end of the casing of the upper block and then passes around the sheave in the lower block. Then it passes around the sheave in the upper block and is acted upon by a force P. Show that for equilibrium the force P is 245 N when a 50-kg mass is suspended from the bottom of the casing of the lower block.

5.46. The upper block in Fig. 5-42 contains two sheaves, and the lower block has one sheave. The rope is attached to the upper end of the casing of the lower block and then passes around one sheave in the upper block. It then returns to the sheave in the lower block, and finally passes around the second sheave in the upper block, where a force P is exerted upon it. Show that for equilibrium the force P is 33.3 lb when a 100-lb weight is suspended from the bottom of the casing of the lower block.

5.47. In Fig, 5-43 a system of levers is shown supporting a load of 80 N. Determine the reactions at A and B on the lever.   *Ans.* $R_A = -71.1$ N, $R_B = +124$ N

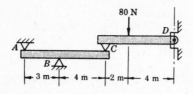

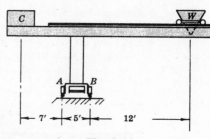

Fig. 5-43   Fig. 5-44

5.48. The wheels of a traveling crane move on tracks at A and B as shown in Fig. 5-44. The weight of the crane is 10 tons, and the center of weight is 3 ft to the right of A. The counterweight C is 4 tons, with its center of weight 7 ft to the left of A. What maximum weight W, 12 ft to the right of B, may be carried without tipping?   *Ans.* $W = 5.67$ tons

5.49. A person whose mass is 70 kg, represented by M, holds the 25-kg mass as shown in Fig. 5-45. The pulley is assumed frictionless. The platform on which the person is standing is suspended by two ropes at A and two ropes at B. What is the tension in one of the ropes at A?   *Ans.* $A = 147$ N

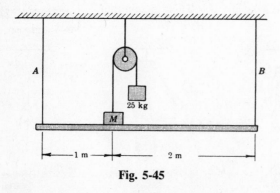

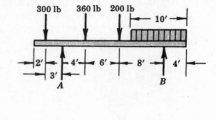

Fig. 5-45   Fig. 5-46

5.50. Figure 5-46 shows an overhanging beam that weighs 32 lb/ft. A uniformly distributed load of 200 lb/ft is shown together with three concentrated loads. Determine beam reactions. Consider the weight of the beam.   *Ans.* $R_A = 1280$ lb, $R_B = 2440$ lb

5.51. A differential chain hoise is shown in Fig. 5-47. The top unit connected to the support consists of two grooved pulleys keyed together but of diameters d_1 and d_2. The lower pulley is of diameter $\frac{1}{2}(d_1 + d_2)$. The weight is attached to the lower pulley. A continuous chain passes over the pulleys as shown. Assume no tension in the slack side (that part of the chain shown to the right of the smaller pulley at the top). What force P is necessary to just start the weight W in the upward direction?
Ans. $P = \frac{1}{2}W(d_2 - d_1)/d_2$

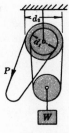

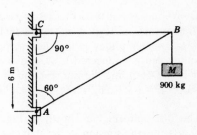

Fig. 5-47   **Fig. 5-48**

5.52. In Fig. 5-48, AB is a rigid rod and CB is a cable. If M is 900 kg, what is the pin reaction at A on the rod AB? What is the tension in the cable?   *Ans.* $A_h = 15.3$ kN, $A_v = 8.82$ kN, $T = 15.3$ kN

5.53. A horizontal force F of 5 lb is applied to the hammer shown in Fig. 5–49. Assuming the hammer pivots about point A, what force is exerted on the vertical nail that is being pulled from the horizontal floor?   *Ans.* $P = 15.8$ lb

5.54. Determine the force P to maintain the bell crank shown in Fig. 5-50 in equilibrium. Neglect friction at the pivot point O.   *Ans.* $P = 52.2$ N

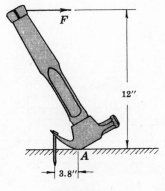

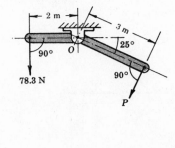

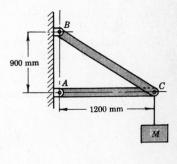

Fig. 5-49   **Fig. 5-50**   **Fig. 5-51**

5.55. The 450-kg mass M is attached to a pin at C as shown in Fig. 5-51. Determine the forces acting in members AC and BC.   *Ans.* $AC = 5880$ N, $BC = 7350$ N

5.56. In Fig. 5-52 the forces are shown acting on the beam divided into equal intervals. The loads are in kN. Determine the reactions at A and B.   *Ans.* $A_h = 1.65$ kN, $A_v = 4.60$ kN, $B = 8.71$ kN

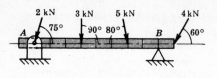

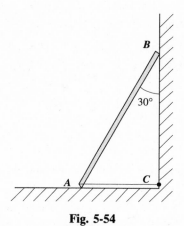

Fig. 5-52   **Fig. 5-53**

Mathcad

5.57. A beam *ED* is loaded as shown in Fig. 5-53. The beam is pin-connected to the wall at *E*. At *D* an 8-in-diameter pulley is attached through frictionless bearings to the beam. A rope passes around the pulley and is connected to *A* and *B* vertically above the extremities of the horizontal diameter of the pulley. Determine the pin reaction at *E* and the tension in the rope.
Ans. $E = 160\,\text{lb}$, $\theta_x = 51°$, $T = 74.4\,\text{lb}$

5.58. The 8-ft uniform bar shown in Fig. 5-54 weighs 40 lb. The floor and the vertical wall are smooth. Determine the tension in the string *AC*.   *Ans.* $T = 11.55\,\text{lb}$

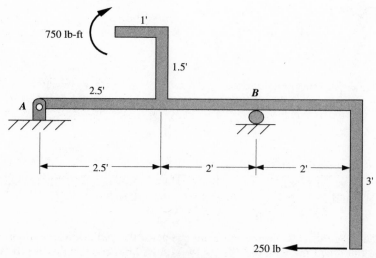

Fig. 5-54

5.59. Determine the reactions at *A* and *B* for the bracket shown in Fig. 5-55.
Ans. $B = 333\,\text{lb}$, $A_y = -333\,\text{lb}$, $A_x = 250\,\text{lb}$

Fig. 5-55

5.60. Refer to Fig. 5-56. Determine the tension in cable *BC*. Neglect the weight of *AB*.
   Ans. $T = 1000 \, \text{lb}$

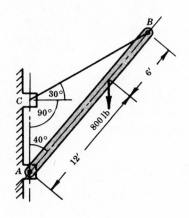

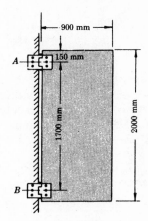

Fig. 5-56   **Fig. 5-57**

5.61. Refer to Fig. 5-57. A uniform door with 18-kg mass is hinged as shown at *A* and *B*. Determine hinge reactions at *A* and *B* on the door. Assume that the vertical components of the reactions at *A* and *B* are equal.   *Ans.* $A = 99.8 \, \text{N}$, $\theta_x = 118°$; $B = 99.8 \, \text{N}$, $\theta_x = 62.1°$

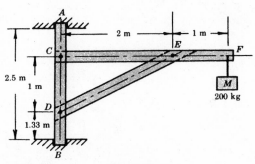

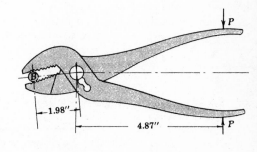

Fig. 5-58   **Fig. 5-59**

5.62. Refer to Fig. 5-58. The frame illustrated is used to support a mass of 200 kg at *F*. Determine (*a*) the pin reaction of *E* on *DE*, (*b*) the pin reaction of *C* on *CF*, and (*c*) the floor reaction at *B* on *AB*.
   Ans.   (*a*) $E = 6570 \, \text{N}$, $\theta_x = 207°$; (*b*) $C = 5960 \, \text{N}$, $\theta_x = 189°$; (*c*) $B_x = 2350 \, \text{N}$, $B_y = 1960 \, \text{N}$

5.63. The bolt *B* is held by a gripping force of 10 lb perpendicular to the jaws of the pliers as shown in Fig. 5-59. What forces *P* must be applied perpendicular to the handles to supply the gripping force?
   Ans. $P = 4.07 \, \text{lb}$

5.64. Referring to Fig. 5-60, determine the components of the pin reaction at *A* on the frame. The member *AB* is horizontal and the member *DBC* is vertical. All pin connections are frictionless.
   Ans. $A_x = 2730 \, \text{N}$ left, $A_y = 250 \, \text{N}$ up

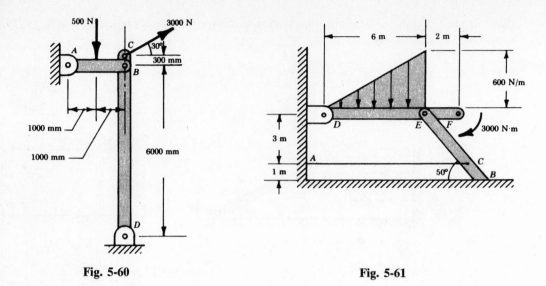

Fig. 5-60   Fig. 5-61

5.65. Figure 5-61 illustrates a frame carrying a distributed load on 6 m of the horizontal member *DEF* of total length 8 m. A moment of 3000 N · m is applied at the end of *DEF*. Determine the tension in the wire *AC*, which is horizontal.   *Ans.* $T = 1900$ N

5.66. The frame in Fig. 5-62 consists of the vertical member *GFHCB* and the horizontal member *CDE*, to which are attached by means of frictionless pins the two pulleys fhown. Each pulley is 400 mm in diameter. The 50-kg mass is held in equilibrium by a cord, which passes over the pulleys and which is parallel along part of its length to the two-force member *FD*. The cord *AB* is needed to hold the entire frame in equilibrium. Determine the tension *T* in *AB* and the magnitude of the pin reaction at *C* on *CDE*.   *Ans.* $T = 490$ N, $C = 1100$ N

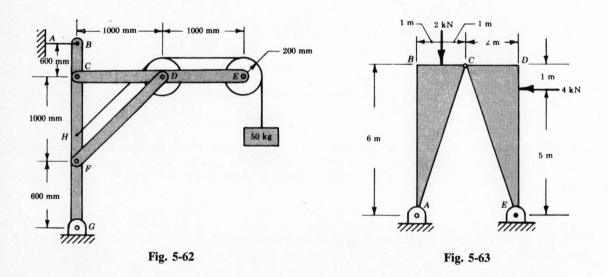

Fig. 5-62   Fig. 5-63

5.67. The two triangular thin plates in Fig. 5-63 have vertical sides and horizontal tops as shown. They are pinned together at *C* by a frictionless pin. The loads shown are either vertical or horizontal. Determine the magnitude of the pin reaction at *C*.   *Ans.* $C = 4.86$ kN

5.68. In the structure shown in Fig. 5-64, determine the magnitude of pin reaction at B on the horizontal member BD. The smooth surface on which the structure rests is horizontal.
Ans. $B = 79.4 \, \text{kN}$

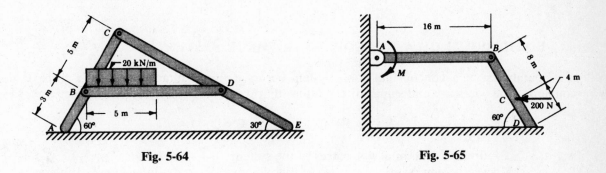

Fig. 5-64   Fig. 5-65

5.69. The horizontal 200-N force in Fig. 5-65 is applied to the sloping member BCD, whose bottom rests on a smooth horizontal plane. Its upper end is pinned at B to the horizontal member AB. What couple M must be applied to the member AB to hold the system in equilibrium? What is the magnitude of the pin reaction at B *Ans.* $M = 3700 \, \text{N} \cdot \text{m}$, $B = 306 \, \text{N}$

5.70. Determine the reactions on a horizontal beam loaded as shown in Fig. 5-66. The system is nonconcurrent, nonparallel.   *Ans.* $A_x = -0.44 \, \text{k}$, $A_y = 2.98 \, \text{k}$, $B = 6.27 \, \text{k}$

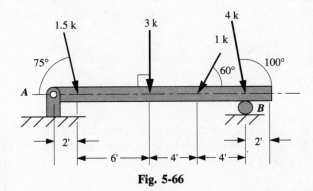

Fig. 5-66

Chapter 6

Equilibrium of Noncoplanar Force Systems

6.1 EQUILIBRIUM OF A NONCOPLANAR FORCE SYSTEM

Equilibrium of a noncoplanar force system occurs if the resultant is neither a force **R** nor a couple **C**. The necessary and sufficient conditions that **R** and **C** be zero vectors are

$$\mathbf{R} = \sum \mathbf{F} = 0 \quad \text{and} \quad \mathbf{C} = \sum \mathbf{M} = 0$$

where $\sum \mathbf{F} = 0$ vector sum of all the forces of the system
$\sum \mathbf{M} =$ vector sum of the moments (relative to any point) of all the forces of the system.

The two vector equations above may be applied directly or, in the simpler problems, the following derived scalar equations may be used for the three types of noncoplanar systems.

6.2 CONCURRENT SYSTEM

The following set of equations ensures equilibrium of a concurrent, noncoplanar system of forces:

$$\sum F_x = 0 \tag{1}$$

$$\sum F_y = 0 \tag{2}$$

$$\sum F_z = 0 \tag{3}$$

where $\sum F_x, \sum F_y, \sum F_z =$ algebraic sums of the x, y, and z components, respectively, of the forces of the system.

$\sum M = 0$ may be used as an alternative to one of the above equations. For example, if it replaces equation (3) then $\sum M$ must be the algebraic sum of the moments of the forces of the system about an axis neither parallel to nor intersecting the z axis.

6.3 PARALLEL SYSTEM

The following set of equations ensures equilibrium of a parallel, noncoplanar system of forces:

$$\sum F_y = 0 \tag{1}$$

$$\sum M_x = 0 \tag{2}$$

$$\sum M_z = 0 \tag{3}$$

where $\sum F_y =$ algebraic sum of the forces of the system along the y axis, which is chosen parallel to the system

$\sum M_x, \sum M_z$ = algebraic sums of the moments of the forces of the system about the x and z axes, respectively.

6.4 NONCONCURRENT, NONPARALLEL SYSTEM

The following six equations are necessary and sufficient conditions for equilibrium of the most general force system in three-dimensional space:

$$\sum F_x = 0 \tag{1}$$

$$\sum F_y = 0 \tag{2}$$

$$\sum F_z = 0 \tag{3}$$

$$\sum M_x = 0 \tag{4}$$

$$\sum M_y = 0 \tag{5}$$

$$\sum M_z = 0 \tag{6}$$

where $\sum F_x, \sum F_y, \sum F_z$ = algebraic sums of the x, y, and z components, respectively, of the forces of the system
$\sum M_x, \sum M_y, \sum M_z$ = algebraic sums of the moments of the forces of the system about the x, y, and z axes, respectively.

All systems previously studied are special cases of this system. Not all of the six equations are necessary in these special cases.

Solved Problems

6.1. In Fig. 6-1, a pole 30 ft high is shown supporting a wire in the xy plane. The wire exerts a force of 150 lb on the top of the pole at an angle of 10° below the horizontal. Two guy wires are affixed as shown. Determine the tension in each guy wire and the compression in the pole.

SOLUTION

Since the pole is subjected only to end loads, it is a two-force member carrying an axial compressive load P. For the concurrent system shown in the free-body diagram (see Fig. 6-2),

$$\sum F_z = 0 = +A \cos 60° \sin 30° - B \cos 60° \sin 30° \quad \text{or} \quad A = B$$

and

$$\sum F_x = 0 = +150 \cos 10° - B \cos 60° \cos 30° - A \cos 60° \cos 30°$$

Substituting A for B and solving, we obtain $A = 171$ lb T.
To determine P, sum the forces vertically along the y axis.

$$\sum F_y = 0 = P - 150 \sin 10° - 2A \sin 60° \quad \text{or} \quad P = 322 \text{ lb } C$$

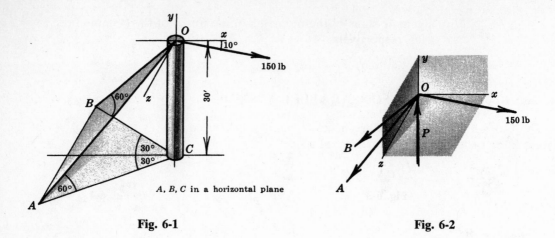

Fig. 6-1   Fig. 6-2

6.2.   Rework Problem 6.1 using $\sum M_C = 0$.

SOLUTION

The pole is in equilibrium under the action of the following four forces as shown in Fig. 6-3.

$$\mathbf{A} = (-A \cos 60° \cos 30°)\mathbf{i} + (-A \sin 60°)\mathbf{j} + (A \cos 60° \sin 30°)\mathbf{k}$$

$$= -0.433A\mathbf{i} - 0.866A\mathbf{j} + 0.25A\mathbf{k} \tag{1}$$

$$\mathbf{B} = (-B \cos 60° \cos 30°)\mathbf{i} + (-B \sin 60°)\mathbf{j} + (-B \cos 60° \sin 30°)\mathbf{k}$$

$$= -0.433B\mathbf{i} - 0.866B\mathbf{j} - 0.25B\mathbf{k} \tag{2}$$

$$\mathbf{C} = C_x\mathbf{i} + C_y\mathbf{j} + C_z\mathbf{k} \tag{3}$$

The 150-lb force is $+(150 \cos 10°)\mathbf{i} + (150 \sin 10°)\mathbf{j} = +149\mathbf{i} - 25.9\mathbf{j}$ $\qquad$ (4)

The position vector of O relative to C is $\mathbf{r} = 30\mathbf{j}$. Then, using $\sum \mathbf{M}_C = 0$.

$$\sum (r \times F) = A \begin{vmatrix} \mathbf{i} & \mathbf{j} & \mathbf{k} \\ 0 & 30 & 0 \\ -0.433 & -0.866 & +0.25 \end{vmatrix} + B \begin{vmatrix} \mathbf{i} & \mathbf{j} & \mathbf{k} \\ 0 & 30 & 0 \\ -0.433 & -0.866 & -0.25 \end{vmatrix} + \begin{vmatrix} \mathbf{i} & \mathbf{j} & \mathbf{k} \\ 0 & 30 & 0 \\ +149 & -25.9 & 0 \end{vmatrix} = 0$$

Expanding the determinants and combining,

$$(7.5A - 7.5B)\mathbf{i} + (0)\mathbf{j} + (13A + 13B - 4470)\mathbf{k} = 0$$

or $7.5A - 7.5B = 0$ and $13A + 13B - 4470 = 0$, from which $A = 171$ lb T, $B = 171$ lb T.
Summing forces in the y direction yields C_y, which is also the compression in the pole:

$$\sum F_y = C_y - 0.866A - 0.866B - 25.9 = 0 \qquad \text{or} \qquad C_y = 322 \text{ lb}$$

The sum of the moments about O indicates $C_x = C_z = 0$.

6.3.   A mass of 6.12 kg is supported by the three wires as shown in Fig. 6-4. AB and AC are in the xz plane. Determine tensions T_1, T_2, and T_3.

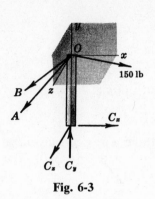

Fig. 6-3

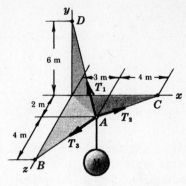

Fig. 6-4

SOLUTION

$$AD = \sqrt{3^2 + 2^2 + 6^2} = 7 \qquad AC = \sqrt{4^2 + 2^2} = 4.47 \qquad AB = 5$$

First sum the forces in the y direction because this equation involves only one unknown, T_1.

$$\sum F_y = 0 = -6.12 \times 9.8 + \tfrac{6}{7}T_1 \qquad T_1 = 70 \text{ N}$$

Now sum forces in the x and z directions:

$$\sum F_x = 0 = T_2\frac{4}{4.47} - T_3\frac{3}{5} - 70\frac{3}{7}$$

$$\sum F_z = 0 = T_3\frac{4}{5} - T_2\frac{2}{4.47} - 70\frac{2}{7}$$

Multiply the $\sum F_z$ equation by 2 and add to the $\sum F_x$ equation to obtain $T_3 = 70$ N. Substitute this value into the $\sum F_x$ or $\sum F_z$ equation to obtain $T_2 = 80.5$ N.

6.4. Solve Problem 6.3 by expressing each force in terms of its **i**, **j**, and **k** components.

SOLUTION

Point A is in equilibrium under the action of the following four concurrent forces:

$$\mathbf{T}_1 = -\tfrac{3}{7}T_1\mathbf{i} + \tfrac{6}{7}T_1\mathbf{j} + \tfrac{2}{7}T_1\mathbf{k} \qquad\qquad (1)$$

$$\mathbf{T}_2 = +\frac{4}{4.47}T_2\mathbf{i} + 0 - \frac{2}{4.47}T_2\mathbf{k} \qquad\qquad (2)$$

$$\mathbf{T}_3 = -\tfrac{3}{5}T_3\mathbf{i} + 0 + \tfrac{4}{5}T_3\mathbf{k} \qquad\qquad (3)$$

$$\mathbf{W} = -6.12 \times 9.8\mathbf{j} = -60\mathbf{j} \qquad\qquad (4)$$

Then $\qquad\qquad \sum \mathbf{F}_y = 0 = \tfrac{6}{7}T_1\mathbf{j} - 60\mathbf{j} \qquad \text{or} \qquad T_1 = 70 \text{ N magnitude}$

In vector form, $\mathbf{T}_1 = -30\mathbf{i} + 60\mathbf{j} - 20\mathbf{k}$.

Also,

$$\sum \mathbf{F}_x = 0 = -\frac{3}{7} T_1 \mathbf{i} + \frac{4}{4.47} T_2 \mathbf{i} - \frac{3}{5} T_3 \mathbf{i}$$

$$\sum \mathbf{F}_z = 0 = -\frac{2}{7} T_1 \mathbf{k} - \frac{2}{4.47} T_2 \mathbf{k} + \frac{4}{5} T_3 \mathbf{k}$$

or $(4/4.47)T_2 - (3/5)T_3 = (3/7)T_1 = 30$ and $-(2/4.47)T_2 + (4/5)T_3 = (2/7)T_1 = 20$, whose simultaneous solution is $T_2 = 80.5$ N and $T_3 = 70.0$ N.

6.5. The 80-kg mass in Fig. 6-5(a) is supported by three wires concurrent at D (2, 0, −1). The wires are attached to points A (1, 3, 0), B (3, 3, −4), and C (4, 3, 0). The coordinates are in meters. Determine the tension in the wire attached to C.

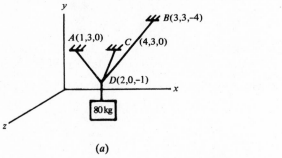

(a)   (b)

Fig. 6-5

SOLUTION

The tension in the wire attached from D to C is most easily found by setting the sum of the moments about the line AB equal to zero. In this equation the moments of the forces in DA and DB will be zero because these forces intersect AB. The only forces with moments about AB will be the force in DC and gravity force (80×9.8 N in the negative $\mathbf{j}$ direction).

The free-body diagram in Fig. 6-5(b) shows all the forces acting at point D. The tension in DC will be written

$$\mathbf{C} = C\frac{(4-2)\mathbf{i} + (3-0)\mathbf{j} + [0-(-1)]\mathbf{k}}{\sqrt{(2)^2 + (3)^2 + (+1)^2}} = C\frac{2\mathbf{i} + 3\mathbf{j} + \mathbf{k}}{\sqrt{14}}$$

To find the sum of the moments about the line AB, we first need the unit vector along the line. Hence,

$$\mathbf{e}_{AB} = \frac{(3-1)\mathbf{i} + (3-3)\mathbf{j} + (-4-0)\mathbf{k}}{\sqrt{20}} = \frac{\mathbf{i} - 2\mathbf{k}}{\sqrt{5}}$$

The moments of the tension C and the gravity force can be found about any point on the line AB. Let us choose point A. The position vector for both forces as chosen here will be from A to D. Hence,

$$\mathbf{r}_{AD} = (2-1)\mathbf{i} + (0-3)\mathbf{j} + (-1-0)\mathbf{k} = \mathbf{i} - 3\mathbf{j} - \mathbf{k}$$

The sum of the moments of the two forces will be

$$\sum \mathbf{r} \times \mathbf{F} = \begin{vmatrix} \mathbf{i} & \mathbf{j} & \mathbf{k} \\ +1 & -3 & -1 \\ 0 & -784 & 0 \end{vmatrix} + \frac{C}{\sqrt{14}} \begin{vmatrix} \mathbf{i} & \mathbf{j} & \mathbf{k} \\ +1 & -3 & -1 \\ +2 & +3 & +1 \end{vmatrix} = -1 + 784\mathbf{i} - 784\mathbf{k} + \frac{C}{\sqrt{14}}(0\mathbf{i} - 3\mathbf{j} + 9\mathbf{k})$$

Finally, we get

$$\mathbf{e}_{AB} \cdot \sum \mathbf{r} \times \mathbf{F} = 0$$

or $\dfrac{\mathbf{i} - 2\mathbf{k}}{\sqrt{5}}(-1 \times 784\mathbf{i} - 784\mathbf{k}) + \dfrac{\mathbf{i} - 2\mathbf{k}}{\sqrt{5}}\left(\dfrac{C}{\sqrt{14}}\right)(-3\mathbf{j} + 9\mathbf{k}) = -1 \times 784\sqrt{5} + \dfrac{2 \times 784}{\sqrt{5}} - \dfrac{18C}{\sqrt{5} \times \sqrt{14}} = 0$

This yields $C = 163$ N.

The reader may wish to use point B on AB as the center for moments. This will yield the same result.

6.6. A 200-lb weight is being lifted out of a 4-ft-diameter hole. Three ropes attached to the weight are held by three people equally spaced around the rim of the hole. What pull is being exerted in each rope when the weight is 4 ft from the top? Assume that (a) each person exerts the same pull, (b) the weight is centered, and (c) each person is holding the rope so that it just clears the rim.

SOLUTION

The vertical components of the three forces must equal the 200-lb weight; that is, $3T \cos \theta = 200$ lb, where θ is the angle between a rope and the vertical through the weight. Then

$$\theta = \tan^{-1} \tfrac{2}{4} = 26.6° \quad \text{and} \quad T = 74.5 \, \text{lb}$$

6.7. The system shown in Fig. 6-6 is subjected to a horizontal load P of 100 N lying in the xy plane. Determine the force in each leg.

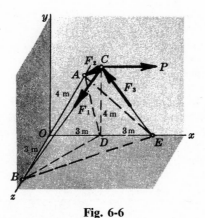

Fig. 6-6

SOLUTION

Calculations yield $CE = 5$, $BC = \sqrt{34}$, and $AC = \sqrt{41}$ m.

Assume that the *three noncoplanar forces* F_1, F_2, F_3 are in the directions shown.

Sum forces parallel to the z axis to obtain a relationship between F_1 and F_2. Take moments about the line AB to determine F_3. Then sum forces parallel to the x axis to obtain another relationship between F_1 and F_2. These equations are

$$\sum F_z = 0 = +\frac{3}{\sqrt{34}} F_1 - \frac{4}{\sqrt{41}} F_2 \tag{1}$$

$$\sum M_z = 0 = +\tfrac{4}{5} F_3 \times 6 - 100 \times 4 \tag{2}$$

$$\sum F_x = 0 = +100 - \frac{3}{5} F_3 - \frac{3}{\sqrt{34}} F_1 - \frac{3}{\sqrt{41}} F_2 \tag{3}$$

The results are $F_1 = 55.6$ tension, $F_2 = 45.7$ N tension, $F_3 = 83.3$ N compression.

6.8. Solve Problem 6.7 using vector notation.

SOLUTION

The four forces may be expressed for the assumed directions as follows:

$$\mathbf{F}_1 = -\frac{3}{\sqrt{34}}F_1\mathbf{i} - \frac{4}{\sqrt{34}}F_1\mathbf{j} + \frac{3}{\sqrt{34}}F_1\mathbf{k}$$

$$\mathbf{F}_2 = -\frac{3}{\sqrt{41}}F_2\mathbf{i} - \frac{4}{\sqrt{41}}F_2\mathbf{j} - \frac{4}{\sqrt{41}}F_2\mathbf{k}$$

$$\mathbf{F}_3 = -\frac{3}{5}F_3\mathbf{i} + \frac{4}{5}F_3\mathbf{j} + 0$$

$$\mathbf{P} = 100\mathbf{i}$$

Since the system is in equilibrium, the sums of the $\mathbf{i}$, $\mathbf{j}$, and $\mathbf{k}$ components must each equal zero:

$$-\frac{3}{\sqrt{34}}F_1 - \frac{3}{\sqrt{41}}F_2 - \frac{3}{5}F_3 + 100 = 0$$

$$-\frac{4}{\sqrt{34}}F_1 - \frac{4}{\sqrt{41}}F_2 + \frac{4}{5}F_3 = 0$$

$$+\frac{3}{\sqrt{34}}F_1 - \frac{4}{\sqrt{41}}F_2 = 0$$

The simultaneous solution of these three equations is $F_1 = 55.6\,\text{N}$ (tension as assumed), $F_2 = 45.7\,\text{N}$ (tension as assumed), $F_3 = 83.3\,\text{N}$ (compression as assumed).

Note: A computer solution to Problem 6.8 is available in Appendix C.

6.9. A table 600 mm by 600 mm is mounted on three legs. Four loads are applied as shown in Fig. 6-7. Determine the three reactions. Since three equations are available for a parallel system, only three supports are necessary.

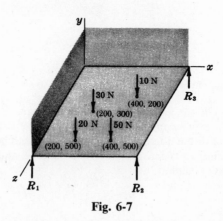

Fig. 6-7

SOLUTION

Applying the three equations of the parallel system, the following equations result:

$$\sum F_y = 0 = R_1 + R_2 + R_3 - 20 - 30 - 10 - 50 \tag{1}$$

$$\sum M_x = 0 = -R_1 \times 600 - R_2 \times 600 + 20 \times 500 + 30 \times 300 + 50 \times 500 + 10 \times 200 \tag{2}$$

$$\sum M_z = 0 = +R_2 \times 600 + R_3 \times 600 - 20 \times 200 - 50 \times 400 - 10 \times 400 - 30 \times 200 \tag{3}$$

Upon simplification, these become

$$R_1 + R_2 + R_3 = 110 \qquad (1')$$

$$R_1 + R_2 = 76.7 \qquad (2')$$

$$R_2 + R_3 = 56.7 \qquad (3')$$

Substitute $R_1 + R_2 = 76.7$ into equation $(1')$ to obtain $76.7 + R_3 = 110$, or $R_3 = 33.3$ N.
Substitute $R_2 + R_3 = 56.7$ into equation $(1')$ to obtain $R_1 + 56.7 = 110$, or $R_1 = 53.3$ N.
Finally, from equation $(1')$, $R_2 = 110 - R_1 - R_3 = 23.4$ N.

Note: Another method of solving is to sum moments about edges R_1R_2 and R_2R_3 to obtain R_3 and R_1, respectively.

6.10. A crankshaft is subjected to pulls F_1 and F_3 parallel to the z axis, and F_2 and F_4 parallel to the y axis. See Fig. 6-8. What are the bearing reactions at A and B if the pulls are each equal to F?

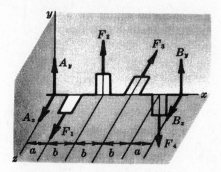

Fig. 6-8

SOLUTION

The bearing reactions are assumed to act in the positive directions of the y and z axes. The equations of the equilibrium are

$$\sum M_z = 0 = +F_2(a + b) - F_4(a + 3b) + B_y(2a + 3b) \qquad (1)$$

$$\sum M_y = 0 = -F_1(a) + F_3(a + 2b) - B_z(2a + 3b) \qquad (2)$$

$$\sum M_{B_z} = 0 = -A_y(2a + 3b) - F_2(a + 2b) + F_4(a) \qquad (3)$$

$$\sum M_{B_y} = 0 = +A_z(2a + 3b) + F_1(a + 3b) - F_3(a + b) \qquad (4)$$

In an actual engine the Fs would not be equal, and each of the above equations would be solved for the unknown it contains. If the Fs are assumed equal then

$$B_y = \frac{2b}{2a + 3b} F \qquad B_z = \frac{2b}{2a + 3b} F \qquad A_y = \frac{-2b}{2a + 3b} F \qquad A_z = \frac{-2b}{2a + 3b} F$$

The minus signs indicate that the components A_z and A_y actually act to the rear and down, respectively. The total reaction at B is parallel to the total reaction at A; it is equal to it in magnitude but opposite in direction. The two form a couple as one would expect, because F_1, F_3, and F_2, F_4 form couples when the forces are assumed to be of equal magnitude.

6.11. Assume that an automobile door weighing 60 lb is of rectangular shape 3 ft wide by 4 ft high with its center of gravity at the geometric center (see Fig. 6-9). The door is opened 45°. A wind load of 50 lb is applied perpendicular ot the door and is assumed concentrated at the geometric center. A door handle is 28 in from the bottom and 3 in from the right edge. What force P, applied in a horizontal plane at the handle but at an angle of 20° with the perpendicular to the door, is necessary to keep the door open? What are the components of the hinge reactions at A and B? Choose the x axis along the door of the automobile. Assume that the lower hinge B carries all vertical loads, that is, $A_y = 0$.

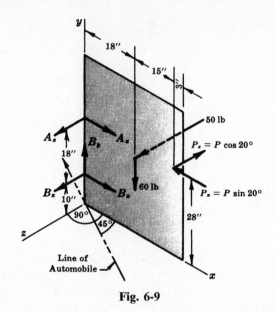

Fig. 6-9

SOLUTION

In Fig. 6-9 the components (assumed positive) of each hinge reaction are shown. Note that the two components of the force P are shown, one perpendicular to the door and one parallel to the door.

Taking moments about the y axis yields an equation with only one unknown P_z, from which P can be found.

Summing forces along the z axis yields an equation in the unknowns A_z and B_z. Taking moments about the x axis produces another equation involving A_z and B_z. Solve simultaneously.

Taking moments about the z axis and summing forces along the x axis yields two equations in A_x and B_x.

Summing forces vertically yields an equation involving B_y.

The above paragraphs indicate the type of analysis that can be made before any equations are written. The equations follow:

$$\sum M_y = 0 = -50 \times 18 + P_z \times 33 \tag{1}$$

$$\sum F_z = 0 = A_z + B_z + 50 - P_z \tag{2}$$

$$\sum M_x = 0 = +A_z \times 28 + B_z \times 10 + 50 \times 24 - P_z \times 28 \tag{3}$$

$$\sum M_z = 0 = -B_x + P_x \times 28 - 60 \times 18 - A_x \times 28 \tag{4}$$

$$\sum F_x = 0 = A_x + B_x - P_x \tag{5}$$

$$\sum F_y = 0 = B_Y - 60 \tag{6}$$

From equation (1), $P_z = (50 \times 18)/33 = 27.3$ lb. But $P \cos 20° = P_z = 27.3$ lb; then $P = 29.1$ lb. Substitute $P_z = 27.3$ lb into equations (2) and (3) and regroup terms as follows:

$$A_z + B_z = -50 + 27.3 \tag{2'}$$

$$28A_z + 10B_z = -1200 + 765 \tag{3'}$$

Solve equations (2') and (3') simultaneously to obtain $A_z = -11.6$ lb, $B_z = -11.1$ lb. Next, substituting in equations (4) and (5), we obtain

$$-10B_x + 28(29.1 \times 0.342) - 28A_x = 1080 \tag{4'}$$

$$A_x + B_x - 29.1 \times 0.342 = 0 \tag{5'}$$

Solve equations (4') and (5') simultaneously to obtain $A_x = -50$ lb, $B_x = +60$ lb. From equation (6), $B_y = 60$ lb.

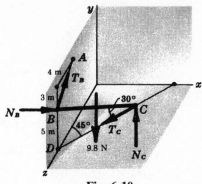

Fig. 6-10

6.12. A homogeneous rod BC 10 m long and having a mass of 1 kg rests against a smooth wall at B and on a smooth floor at C (see Fig. 6-10). Determine the tensions in AB and DC, which are cords to hold the rod in equilibrium. Note in Fig. 6-10 that BD is perpendicular to the z axis, and that AB is in the yz plane.

SOLUTION

Add the wall and floor normal reactions N_B and N_C to complete the free-body diagram of the rod. The following equations of equilibrium apply:

$$\sum F_x = 0 = N_B - T_C \cos 45° \tag{1}$$

$$\sum F_y = 0 = T_B \times \frac{3}{5} - 9.8 + N_C \tag{2}$$

$$\sum M_z = 0 = -9.8 \times 5 \cos 30° \cos 45° + N_C \times 9.8 \cos 30° \cos 45° - N_B \times 5 \tag{3}$$

The three equations contain four unknowns and appear at first glance impossible to solve. However, for a stable position the sum of the forces perpendicular to the plane BCD must be zero. There are only two forces (N_B and T_B) that have components perpendicular to this plane. Hence, $T_B \times \frac{4}{5} \times \cos 45° = N_B \cos 45°$, or $T_B \times \frac{4}{5} = N_B$.

Substitute this value into equation (2) and obtain $N_C = 9.8 - \frac{3}{4}N_B$. This, substituted into equation (3), yields $N_B = 3.03$ N. Then $T_B = \frac{5}{4} \times 3.03$ N $= 3.79$ N.

From equation (1), $T_C = 3.03/0.707 = 4.29$ N.

6.13. Two views of a windlass are shown in Fig. 6-11(a). The bearings are frictionless. What force P perpendicular to the crank is necessary to hold a 200-lb weight in the position shown? What are the bearing reactions at A and B?

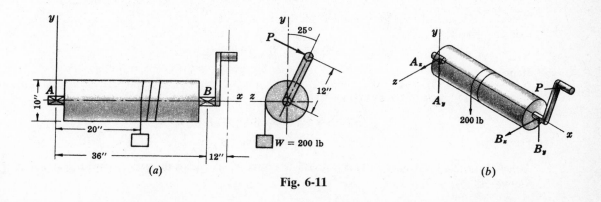

(a)   (b)

Fig. 6-11

SOLUTION

A free-body diagram is drawn in Fig. 6-11(*b*) showing all forces acting on the windlass. Since no forces act along the axis of the windlass, no *x* components of the bearing reactions are shown.

Sum moments about the *x* axis: $\sum M_x = 0 = -P \times 12 + 200 \times 5$. Hence, $P = 83.3$ lb.

To determine the bearing reactions, the following four equations may be used:

$$\sum M_y = 0 = -B_z \times 36 + P \cos 25° \times 48 \tag{1}$$

$$\sum M_z = 0 = -200 \times 20 + B_y \times 36 - P \sin 25° \times 48 \tag{2}$$

$$\sum F_y = 0 = A_y - 200 + B_y - P \sin 25° \tag{3}$$

$$\sum F_z = 0 = A_z + B_z - P \cos 25° \tag{4}$$

Note that the force P (83.3 lb) is resolved into its components $P \cos 25°$ and $P \sin 25°$ along the z and y axes, respectively. The moment, for example, of P about the y axis is then only the moment of its z component about the y axis, because the y component is parallel to the y axis and therefore has no moment about it.

From equation (*1*), $B_z = 101$ lb; from equation (*2*), $B_y = 158$ lb.

Substituting for B_y, B_z, and P in equations (*3*) and (*4*), we obtain $A_y = 77.2$ lb and $A_z = -25.5$ lb.

6.14. Determine the bearing reactions at *A* and *B* on the horizontal shaft shown in Fig. 6-12(*a*). The pulleys are integral with the shaft. The loads on the larger pulley are horizontal, while the loads on the smaller pulley are vertical.

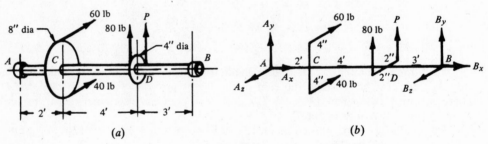

(*a*)   (*b*)

Fig. 6-12

SOLUTION

Figure 6-12(*b*) shows the free-body diagram, with axes chosen conveniently. To determine force *P*, sum moments about the *x* axis, thus obtaining

$$P \times 2 - 80 \times 2 + 40 \times 4 - 60 \times 4 = 0 \qquad \text{or} \qquad P = 120 \text{ lb}$$

Because no external loads are applied in the *x* direction, we can say $A_x = B_x = 0$. The sum of the moments about A_z yields

$$80 \times 6 + 120 \times 6 + B_y \times 9 = 0$$

Hence, $$B_y = -133.3 \text{ lb}$$

The sum of the moments about B_z yields

$$-80 \times 3 - 120 \times 3 - A_y \times 9 = 0$$

From this, $$A_y = -66.7 \text{ lb}$$

As a check, the sum of A_y and B_y is 200 lb down, which equals the two upward forces acting on the smaller pulley.

To determine B_z, use the sum of the moments about A_y. Hence,

$$-60 \times 2 - 40 \times 2 + B_z \times 9 = 0$$

From this, $B_z = +22.2$ lb forward

To determine A_z, use the sum of the moments about B_y. Hence,

$$-60 \times 7 + 40 \times 7 - A_z \times 9 = 0 \quad \text{or} \quad A_z = 77.8 \text{ lb forward}$$

As a check, note that the sum of A_z and B_z is 100 lb forward, which equals the sum of the two back forces acting on the larger pulley.

6.15. The beam EF in Fig. 6-13 weighs 10 lb/ft and carries the 138-lb weight at its end. It is supported by a ball-and-socket joint at E and the cables AB and CD. Determine the tensions in AB and CD. Find the joint reactions at E.

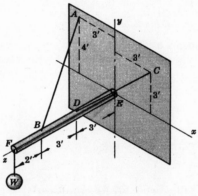

Fig. 6-13

SOLUTION

Select x, y, and z axes as shown in Fig. 6-13. For equilibrium of the beam EF, choose $\sum \mathbf{M}_E = 0$ and $\sum \mathbf{F} = 0$.

The forces acting on the system are as follows:

(1) The weight W of 138 lb acts vertically down and may be written $-138\mathbf{j}$.

(2) The weight of the beam is 8 times 10 lb and may be written $-80\mathbf{j}$.

(3) The ball-and-socket reaction is $E_x\mathbf{i} + E_y\mathbf{j} + E_z\mathbf{k}$.

(4) The tension in AB may be written as $A_x\mathbf{i} + A_y\mathbf{j} + A_z\mathbf{k}$, where

$$A_x = A \cos \theta_x = -\frac{3}{\sqrt{3^2 + 4^2 + 6^2}}A = -\frac{3}{\sqrt{61}}A = -0.384A$$

$$A_y = A \cos \theta_y = +\frac{4}{\sqrt{61}}A = +0.512A$$

$$A_z = A \cos \theta_z = -\frac{6}{\sqrt{61}}A = -0.768A$$

The sign of each component is defined once we assume tension in AB, which must then pull on the beam EF in the direction from B to A, which is in the negative x direction, the positive y direction, and the negative z direction.

(5)   The tension in CD may be written $C_x\mathbf{i} + C_y\mathbf{j} + C_z\mathbf{k}$, where

$$C_x = C \cos \theta_x = +\frac{3}{\sqrt{3^2 + 3^2 + 3^2}} C = +\frac{3}{\sqrt{27}} C = +0.577C$$

$$C_y = C \cos \theta_y = +\frac{3}{\sqrt{27}} C = +0.577C$$

$$C_z = C \cos \theta_z = -\frac{3}{\sqrt{27}} C = -0.577C$$

It is advisable to summarize the five forces and the points on their lines of action to which the position vectors from point E will be drawn:

(1') $-138\mathbf{j}$ at $(0, 0, 8)$

(2') $-80\mathbf{j}$ at $(0, 0, 4)$

(3') $E_x\mathbf{i} + E_y\mathbf{j} + E_z\mathbf{k}$ at $(0, 0, 0)$

(4') $-0.384A\mathbf{i} + 0.512A\mathbf{j} - 0.768A\mathbf{k}$ at $(0, 0, 6)$

(5') $+0.577C\mathbf{i} + 0.577C\mathbf{j} - 0.577C\mathbf{k}$ at $(0, 0, 3)$

The sum of the moments of these five forces about E will be set equal to zero. Using each position vector from point $(0, 0, 0)$ to the point on the force vector listed above, we have

$$\begin{vmatrix} \mathbf{i} & \mathbf{j} & \mathbf{k} \\ 0 & 0 & 8 \\ 0 & -138 & \mathbf{0} \end{vmatrix} + \begin{vmatrix} \mathbf{i} & \mathbf{j} & \mathbf{k} \\ 0 & 0 & 4 \\ 0 & -80 & \mathbf{0} \end{vmatrix} + \begin{vmatrix} \mathbf{i} & \mathbf{j} & \mathbf{k} \\ 0 & 0 & 0 \\ E_x & E_y & E_z \end{vmatrix} + \begin{vmatrix} \mathbf{i} & \mathbf{j} & \mathbf{k} \\ 0 & 0 & 6 \\ -0.384A & +0.512A & -0.768A \end{vmatrix}$$

$$+ \begin{vmatrix} \mathbf{i} & \mathbf{j} & \mathbf{k} \\ 0 & 0 & 3 \\ 0.577C & 0.577C & -0.577C \end{vmatrix} = 0$$

or $8(138)\mathbf{i} + 4(80)\mathbf{i} + [-6(0.512A)\mathbf{i} - 6(0.384A)\mathbf{j}] + [-3(0.577C)\mathbf{i} + 3(0.577C)\mathbf{j}] = 0$

Equating the coefficients of $\mathbf{i}$ to zero, and then the coefficients of $\mathbf{j}$ to zero.

$$1424 - 3.072A - 1.731C = 0 \qquad \text{and} \qquad -2.304A + 1.731C = 0$$

from which $A = 265$ lb and $C = 353$ lb. Hence,

$$\mathbf{A} = -0.384(265)\mathbf{i} + 0.512(265)\mathbf{j} - 0.768(265)\mathbf{k} = -102\mathbf{i} + 136\mathbf{j} - 204\mathbf{k}$$
$$\mathbf{C} = 0.577(353)\mathbf{i} + 0.577(353)\mathbf{j} - 0.577(353)\mathbf{k} = 204\mathbf{i} + 204\mathbf{j} - 204\mathbf{k}$$

To find the pin reaction, equate successively to zero the coefficients of the $\mathbf{i}$, $\mathbf{j}$, and $\mathbf{k}$ terms in the $\sum \mathbf{F} = 0$ equation:

$$E_x - 102 + 204 = 0 \qquad -138 - 80 + E_y + 136 + 204 = 0 \qquad E_z - 204 - 204 = 0$$

from which $E_x = -102$ lb, $E_y = -122$ lb, $E_z = +408$ lb.

Supplementary Problems

6.16.   In Fig. 6-14, a mass of 30 kg is supported by a compression member CD and two tension members AC and BC. CD makes an angle of 40° with the wall. A, B, and C are in a horizontal plane. $AE = EB = 1000$ mm. Find the forces in AC, BC, and CD.
Ans. $AC = BC = 143$ N T, $CD = 384$ N C.

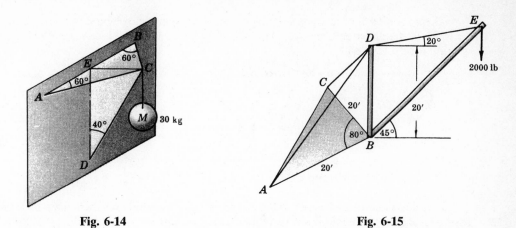

Fig. 6-14   **Fig. 6-15**

6.17. In Fig. 6-15, the crane consists of a boom *BE*, column *BD* (vertical), and three cables *AD, CD, DE*. *A, B,* and *C* are in a horizontal plane. *AC* is bisected by the plane containing *BD, BE,* and *DE*. Determine the forces in *AD, CD,* and *BD*.

    Hint: First consider the coplanar, concurrent system at *E* to obtain the force in *DE*. Then consider the noncoplanar, concurrent system at *D*.   *Ans.* $AD = CD = 2910$ lb *T*, $BD = 2970$ lb *C*.

6.18. In Fig. 6-16, a horizontal pull of 400 N is shown acting on the top of post *DB*. The post is held in equilibrium by the two guy wires *AD* and *CD*. *A, B,* and *C* are on level ground. Find the forces in *AD* and *CD*.   *Ans.* $AD = 366$ N *T*, $CD = 293$ N *T*.

6.19. A video camera with a mass of 2 kg rests on a tripod with legs equally spaced and each making an angle of 18° with the vertical. Assume that the system of forces is concurrent at a point 1200 mm above level ground and determine the forces in each leg.   *Ans.* $C = 6.87$ N

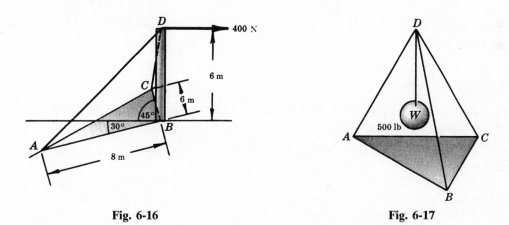

Fig. 6-16   **Fig. 6-17**

6.20. A 500-lb weight is hung on a rope in a tripod with legs of equal length, as shown in Fig. 6-17. Each leg makes an angle of 30° with the rope. *A, B,* and *C* are in the horizontal plane and form an equilateral triangle. Determine the force in each leg.   *Ans.* $AD = BD = CD = 192$ lb *C*

6.21. The circular table 1800 mm in diameter, shown in Fig. 6-18, supports a load of 400 N located on a diameter through the support R_1 and 300 mm from the center on the opposite side from R_1. R_1, R_2, and R_3 are equally spaced. Determine their magnitudes.
Ans. $R_1 = 44$ N, $R_2 = 178$ N, $R_3 = 178$ N

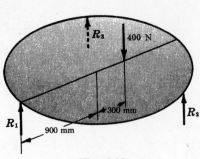

Fig. 6-18

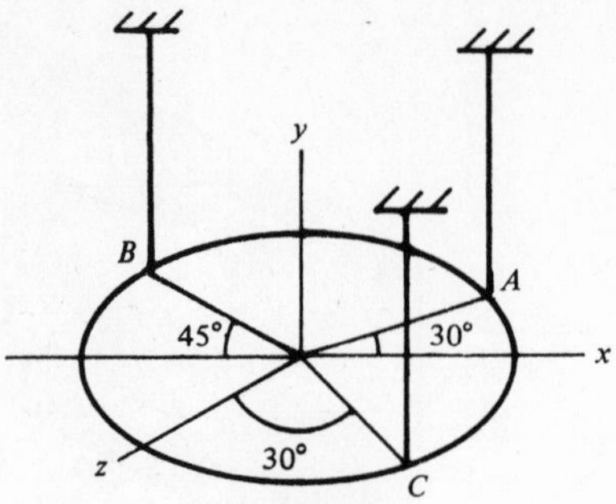

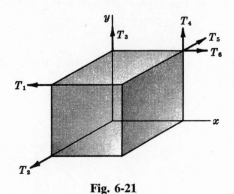

Fig. 6-19

6.22. Refer to Fig. 6-19. If the maximum allowable strength for each cable is 3500 lb, determine the permissible weight of the homogeneous circular plate of radius 5 ft.   *Ans.*   7800 lb

6.23. The triangular plate in Fig. 6-20 carries a 140-N load 1200 mm from the left vertex on the bisector of that vertex angle. T_1, T_2, and T_3 are the tensile loads in three vertical supporting wires. What are their values?   *Ans.* $T_1 = 41.0$ N, $T_2 = T_3 = 49.5$ N

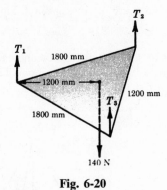

Fig. 6-20

Fig. 6-21

6.24. A uniform cube of weight W is supported by six strings attached to corners, as shown in Fig. 6-21. Each string is perpendicular to a face, i.e., a continuation of an edge of the cube. Determine the tension in each string to hold the cube in equilibrium.   *Ans* $T = W/2$

6.25. See Fig. 6-22. Assume that a motor weighing 500 lb has its center of gravity five-eighths of the overall length from the front on the longitudinal centerline. If the base is 22 in wide and 34 in long, what is the magnitude of the reactions of the supports, assuming one at each front corner and one at the middle in the rear of the motor? A block diagram is shown in the figure.   *Ans* $R_F = 94$ lb, and $R_R = 312$ lb

6.26. See Fig. 6-23. A vertical shaft weighing 40 lb carries two pulleys at B and D weighing 12 and 9 lb, respectively. The pulley at B has a 16-in diameter, and the one at D has a 12-in diameter. The 15- and 60-lb pulls are parallel to the x axis. The 20-lb pull and P are parallel to the z axis. The bearing at C and the step bearing at A are to be considered frictionless. Determine the force P and the reactions at A and C.   *Ans.* $P = 80$ lb, $A_x = 50$ lb, $A_y = 61$ lb, $A_z = -33$ lb, $C_x = 25$ lb, $C_z = 133$ lb, $C_y = 0$

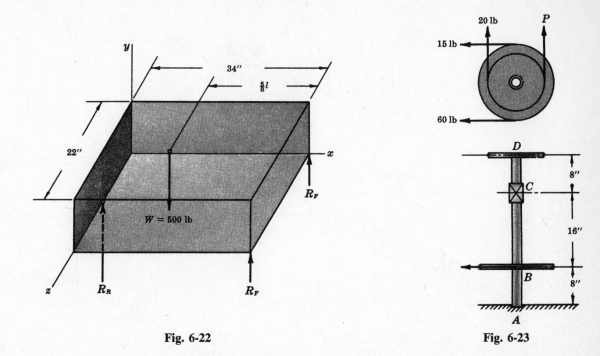

Fig. 6-22   **Fig. 6-23**

6.27. In the simple crane shown in Fig. 6-24, CH is vertical, GD is horizontal, and AC and BC are guy wires. Points A and B are equidistant from the plane which contains CH, DG, and EF. Weight $W = 4000$ lb. Determine the tension in AC and the reactions at H.

Ans. $AC = 2590$ lb T, $H_x = 2800$ lb, $H_y = 8040$ lb, $H_z = 0$

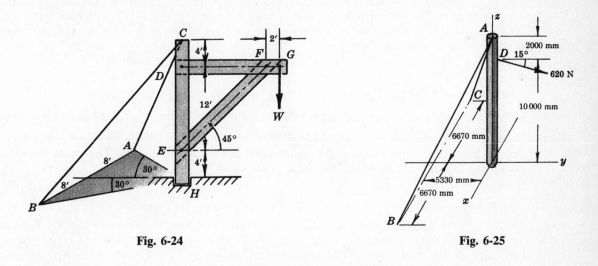

Fig. 6-24   **Fig. 6-25**

6.28. A vertical pole is subjected to a pull of 620 N in the yz plane and 15° below the horizontal. The guy wires AB and AC are attached to supports in the xy plane. The pole rests in a socket. See Fig. 6-25. What is the tension in each cable?   *Ans.* $T_{AB} = T_{AC} = 689$ N

6.29. The homogeneous trapdoor weighs 96 lb. What rope tension T is needed to hold the door in the 26° position shown in Fig. 6-26? What are the hinge reactions at A and B? Assume that the pulley D is on the vertical z axis.

Ans. $T = 52.4$ lb, $A_x + B_x = -28.8$ lb, $A_y = +30.0$ lb, $A_z = +30.7$ lb, $B_y = -4.3$ lb, $B_z = +29.9$ lb

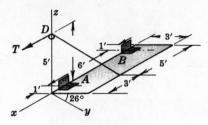

Fig. 6-26

6.30. The boom *EF* shown in Fig. 6-27 may be considered massless. It is supported by the cables *AB* and *CD* and a socket at *E*. Determine the tensions in the two cables and the pin reactions at *E*.
Ans. Tension in $AB = 9020$ N, tension in $CD = 5590$ N, $E_x = -1600$ N, $E_y = 8600$ N, $E_z = -3630$ N

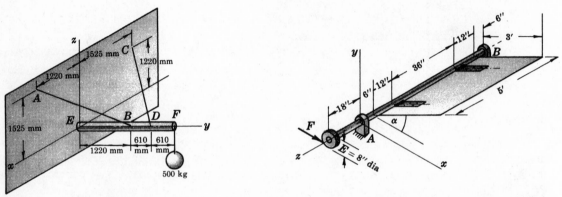

Fig. 6-27   **Fig. 6-28**

6.31. The door shown in Fig. 6-28 is attached to straps that are welded to the rod *EB*. The rod *EB* is carried in bearings at *A* and *B*, and carries a gear *E* at its end. A pinion (not shown) exerts a horizontal force *F* at the bottom of gear *E*. Assuming that the homogeneous door weighs 30 lb, determine the force *F* and the bearing reactions when angle α is 58°.
Ans. $F = 71.5$ lb, $A_x = -89.5$ lb, $A_y = 15$ lb, $A_z = 0$, $B_x = 17.9$ lb, $B_y = 15$ ·lb, $B_z = 0$

6.32. Repeat Prob. 6.31 with $\alpha = 32°$.
Ans. $F = 114.5$ lb, $A_x = -143.1$ lb, $A_y = 15$ lb, $A_z = 0$, $B_x = 28.6$ lb, $B_y = 15$ lb, $B_z = 0$

6.33. In Prob. 6.30, if the maximum strength of either cable is 8000 N, what is the maximum allowable mass at *F*?   *Ans.* $M = 443$ kg

Chapter 7

Trusses and Cables

7.1 TRUSSES AND CABLES

These are examples of coplanar force systems in equilibrium (see Chapter 5).

7.2 TRUSSES

Assumptions

1. Each truss is assumed to be composed of rigid members all lying in one plane. This means that coplanar force systems are involved.
2. The weights of the members are neglected, because they are small in comparison with the loads.
3. Forces are transmitted from one member to another through smooth pins fitting perfectly in the members. These members, which are called two-force members, will be either in tension (T) or compression (C).

Solution by the Method of Joints

To use this techinque, draw a free-body diagram of any pin in the truss, with the proviso that no more than two unknown forces act on that pin. This limitation is imposed because the system of forces is a concurrent one, for which, of course, only two equations are available for a solution. Proceed from one pin to another until all unknowns have been determined.

Solution by the Method of Sections

In the method of joints as just explained, forces in various members are determined by using free-body diagrams of the pins. In the method of sections, a section of the truss is taken as a free-body diagram. This involves cutting through a number of members, including those members whose forces are unknown, in order to isolate one part of the truss. The forces in the members cut act as external forces helping to hold that part of the truss in equilibrium. Since the system is nonconcurrent, nonparallel, three equations are available. Therefore, in any one sectioning no more than three unknown forces can be found. Be sure to isolate the free body completely and at the same time have no more than three unknown forces.

7.3 CABLES

Parabolic

The cable is loaded with w force units per *horizontal* unit of length, e.g., lb/ft or N/m. It assumes a parabolic curve, as shown in Fig. 7-1, which illustrates such a suspension from supports on the same level. Temperature variations, which change the tension, are neglected.

The following equations apply to this coplanar system:

$$d = \frac{wa^2}{8H} \tag{1}$$

$$T = \frac{1}{2}wa\sqrt{1 + \frac{a^2}{16d^2}} \tag{2}$$

$$l = a\left[1 + \frac{8}{3}\left(\frac{d}{a}\right)^2 - \frac{32}{5}\left(\frac{d}{a}\right)^4 + \frac{256}{7}\left(\frac{d}{a}\right)^6 + \cdots\right] \tag{3}$$

where d = sag in feet or meters
w = load in lb/ft or N/m
a = span in feet or meters
H = tension at midpoint in pounds or newtons
T = tension at supports in pounds or newtons
l = length of cable in feet or meters

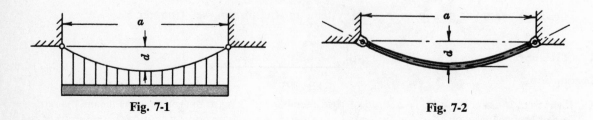

Fig. 7-1   Fig. 7-2

Catenary

This cable carries a load in lb/ft or N/m *along the cable,* rather than horizontally as in the parabolic case. It assumes a catenary curve as shown in Fig. 7-2, which illustrates such a suspension from supports at the same level.

To solve this type of problem, neglecting temperature changes, let

T = tension at distance x from midpoint
s = length along cable from midpoint to point where tension is T
w = load in lb/ft or N/m along cable, e.g., gravitational pull per foot or meter
a = span in feet or meters
d = sag in feet or meters
l = total length in feet or meters
H = tension at midpoint in pounds or newtons
T_{max} = tension at support in pounds or netwons

Referring to the free-body diagram of a portion of the cable to the right of center (Fig. 7-3), note that the x axis is at a distance c below the center of the cable. This simplifies the derivation.

The following equations apply to the catenary. Note that T becomes T_{max} when $x = a/2$ and $y = c + d$.

$$c = \frac{H}{w} \tag{1}$$

$$y = c \cosh \frac{x}{c} \quad \text{and} \quad c + d = c \cosh \frac{a}{2c} \tag{2}$$

$$T = wy \quad \text{and} \quad T_{max} = w(c + d) \tag{3}$$

$$s = c \sinh \frac{x}{c} \quad \text{and} \quad \frac{l}{2} = c \sinh \frac{a}{2c} \tag{4}$$

$$y^2 = c^2 + s^2 \quad \text{and} \quad (c + d)^2 = c^2 + \frac{l^2}{4} \tag{5}$$

The following types of problems involving cables of w force units per unit of length may present themselves for solution:

(a) given the span and sag, that is, a and d;

(b) given the span and length, that is, a and l;

(c) given the sag and length, that is, d and l.

In case (a) solve equation (2) by trial to obtain c. Then equation (3) yields T_{max}, and equation (4) or (5) determines l.

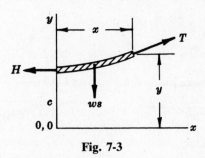

Fig. 7-3

In case (b) solve equation (4) by trial to obtain c. Then equation (5) yields d, and equation (3) T_{max}.

In case (c) solve equation (5) for c. Then T_{max} may be obtained from equation (3). To find a, solve either equation (2) or (4).

Solved Problems

7.1. The simple triangle truss in Fig. 7-4(a) supports two loads as shown. Determine reactions and the forces in each member.

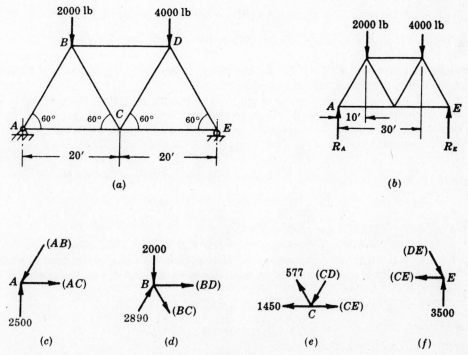

Fig. 7-4

SOLUTION

Figure 7-4(*b*) is a free-body diagram of the entire truss from which to determine R_A and R_E. Since the two loads are vertical, only one component of the pin reaction at *A* is shown.

$$\sum M_A = 0 = R_E \times 40 - 4000 \times 30 - 2000 \times 10 \qquad R_E = 3500\,\text{lb} \tag{1}$$

$$\sum M_E = 0 = -R_A \times 40 + 2000 \times 30 + 4000 + 10 \qquad R_A = 2500\,\text{lb} \tag{2}$$

Of course a vertical summation of the two given forces and the two reactions just determined equals zero, thereby checking the results.

Figure 7-4(*c*) is a free-body diagram of pin *A*. The 2500-lb reaction is drawn up. The only force that can have a downward component to balance R_A is the force in the member *AB*. This is shown acting toward the pin, which means that the member *AB* is in compression. Since the force (*AB*) acts to the left as well as down, some force must act to the right to balance it. Therefore force (*AC*) is shown to the right pulling on the pin. The pin pulls to the left on the member *AC*, which means that (*AC*) is a tensile force.

Writing the equations of the concurrent system of Fig. 7-4(*c*),

$$\sum F_h = 0 = +(AC) - (AB)\cos 60° \tag{3}$$

$$\sum F_x = 0 = +2500 - (AB)\sin 60° \tag{4}$$

Solving, $(AB) = +2500/0.866 = +2890\,\text{lb}$, $(AC) = (AB)\cos 60° = +1450\,\text{lb}$. The plus signs indicate that the directions chosen are correct. Hence, $(AB) = 2890\,\text{lb}\,C$, $(AC) = 1450\,\text{lb}\,T$.

Next draw a free-body diagram for pin *B*. See Fig. 7-4(*d*). Some might have chosen pin *C*, but there are three unknown forces there: (*BC*), (*CD*), and (*CE*). In this figure, the member *AB* is in compression and must be shown pushing on the pin. The 2000-lb load is shown acting directly down on the pin. The directions of forces (*BD*) and (*BC*) are unknown. Instead of spending time trying to decide the direction of each of these, assume they are in tension. A plus sign in the result indicates that tension is correct, whereas a minus sign indicates compression. The equations for this system are

$$\sum F_h = 0 = (BD) + 2890\cos 60° + (BC)\cos 60° \tag{5}$$

$$\sum F_v = 0 = 2890\sin 60° - 2000 - (BC)\sin 60° \tag{6}$$

Solving equation (6), $(BC) = 577\,\text{lb}\,T$. Substituting in equation (5), $(BD) = -1730\,\text{lb}$. Since the sign is minus, this member is actually in compression.

Next draw a free-body diagram of pin *C*, as shown in Fig. 7-4(*e*). The two known values (*AC*) and (*BC*) are inserted. Since (*BC*) has a component acting vertically up. (*CD*) must be shown as compression. If this is not clear, assume it in tension and a minus sign will result, indicating compression. The equations are:

$$\sum F_h = 0 = (CE) - 1450 - 577\cos 60° - (CD)\cos 60° \tag{7}$$

$$\sum F_v = 0 = +577\sin 60° - (C)\sin 60° \tag{8}$$

Solving, $(CD) = 577\,\text{lb}\,C$ and $(CE) = 2020\,\text{lb}\,T$.

The next free-body diagram may be at either *D* or *E* to determine the last force (*DE*). Figure 7-4(*f*) shows the free-body diagram for pin *E*. Note that force (*CE*) is inserted as an unknown. This is done deliberately to provide a check on this value as obtained at pin *C*. The equations are

$$\sum F_x = 0 = 3500 - (DE)\sin 60° \tag{9}$$

$$\sum F_h = 0 = (DE)\cos 60° - (CE) \tag{10}$$

Solving, $(DE) = 4030\,\text{lb}\,C$ and $(CE) = 2020\,\text{lb}\,T$.

7.2. Determine the forces in *FH*, *HG*, *IG*, and *IK* in the truss shown diagrammatically in Fig. 7-5. Each load is 2 kilonewtons (kN). All triangles are equilateral with sides of 4 m.

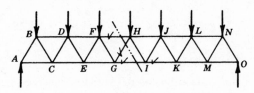

Fig. 7-5

SOLUTION

As a first step, check the members in which the forces are to be found. Cut through as many members as possible, but through no more than three in which the forces are unknown. The first cut should be through *FH*, *HG*, and *GI*. A free-body diagram of either the left or right portion may now be drawn. Choose the one involving as few external forces as possible—the left part in this case. Draw a free-body diagram of this portion as shown in Fig. 7-6. It is usually wise to assume the forces in the members as tension, realizing that a minus sign in the result indicates compression. An arrow pointing away from the free body means that the member pulls on the body and is therefore in tension.

The left reaction of 7 kN is determined by inspection of the *entire* truss, which is symmetrical and symmetrically loaded.

Any three equations of equilibrium may be applied to the free-body diagram. The summation of moments about *G* yields an equation with only one unknown force (*FH*). A summation of moments about *H* (external to the figure) involves one unknown force (*GI*), since members *FH* and *HG* intersect in *H*. Finally, a vertical summation of forces will result in a solution for force (*HG*). Using this procedure, the results are

$$\sum M_G = 0 = -(FH) \times 2 \tan 60° - 7 \times 12 + 2 \times 10 + 2 \times 6 + 2 \times 2 \qquad (FH) = -13.9 \text{ kN } C \qquad (1)$$

$$\sum M_H = 0 = +(GI) \times 2 \tan 30° - 7 \times 14 + 2 \times 12 + 2 \times 8 + 2 \times 4 \qquad (GI) = 14.4 \text{ kN } T \qquad (2)$$

$$\sum F_v = 0 = +7 - 2 - 2 - 2 + (HG) \sin 60° \qquad (HG) = -1.15 \text{ kN } C \qquad (3)$$

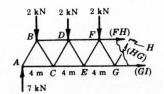

Fig. 7-6

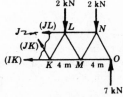

Fig. 7-7

As a check on this particular free-body diagram and its solution, sum the forces horizontally—an equation not used in the solution—to determine whether or not the result is zero:

$$\sum F_h = -13.9 + 14.4 - 1.15 \cos 60° = 0 \qquad (4)$$

To determine the force in member *IK*, make a cut as shown in Fig. 7-7. Take moments about point *J*, yielding

$$\sum M_J = 0 = -(IK) \times 2 \tan 60° - 2 \times 4 - 2 \times 8 + 7 \times 10 \qquad (IK) = 13.3 \text{ kN } T \qquad (5)$$

7.3. The maximum allowable force (tensile or compressive) in members *DC*, *DF*, or *EF* in the pin-connected truss shown in Fig. 7-8 is known to be 40 kips (40,000 lb). Determine the maximum permissible load *P*.

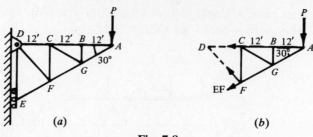

Fig. 7-8

SOLUTION

Using the method of sections, draw the free-body diagram shown in Fig. 7-8(b).

The sum of the moments about A has only one force DF in the equation; hence $DF = 0$.

Next use the sum of the moments about D to find EF. Note that DC and DF intersect at D, and hence their moments equal zero. Also visualize EF moved along its line of action to A; its moment is then the moment of only its vertical component. Thus,

$$\sum M_D = 0 = -36P - (EF \sin 30°)36 \quad \text{or} \quad EF = -2P$$

The sum of the moments about F will yield DC as follows (note $CF = 24 \tan 30° = 13.84$ ft):

$$\sum M_F = 0 = DC \times 13.84 - 24P \quad \text{or} \quad DC = 1.73P$$

To find the maximum P, set $EF = 40,000 = 2P$. Thus,

$$P = 20,000 \text{ lb}$$

7.4. Determine the forces in members BD, CD, and CE of the Fink truss shown in Fig. 7.9.

SOLUTION

Use the method of sections to solve this problem. First determine the vertical reaction at A and the pin reaction at G by treating the entire truss as a free body as shown in Fig. 7-10.

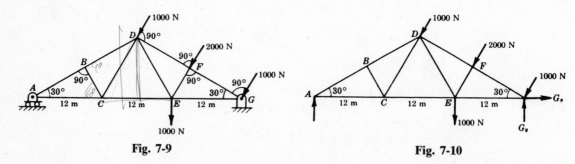

Fig. 7-9   Fig. 7-10

The distance $FG = 12 \cos 30° = 10.4$ m, and distance $DG = 18/(\cos 30°) = 20.8$ m. To find force A, sum moments about G to obtain $\sum M_G = 0 = +1000 \times 12 + 2000 \times 10.4 + 1000 \times 20.8 - 36A$. Hence, $A = 1490$ N.

To determine G_x, sum forces horizontally to obtain $\sum F_h = 0 = G_x - 4000 \sin 30°$. Hence, $G_x = 2000$ N. A vertical summation yields $G_y = 2970$ N.

To determine the forces in the members, select the section as shown in Fig. 7-11. The left portion is chosen, since only one known force A is acting together with the unknowns.

Sum moments about C to determine (BD). Sum moments about D to determine (CE). To obtain (CD), sum the forces vertically. These equations in the order given are as follows:

$$\sum M_C = 0 = -(BD) \times 6 - 1490 \times 12 \tag{1}$$

$$\sum M_D = 0 = -1490 \times 18 + (CE) \times 10.4 \tag{2}$$

$$\sum F_v = 0 = +1490 + (BD) \cos 60° + (CD) \cos \theta \tag{3}$$

Note that $\tan \theta = 6/10.4$; hence, $\cos \theta = 0.866$. From (1), $(BD) = -2980$ N, i.e., compression. From (2), $(CE) = +2580$ N, i.e., tension as assumed.

Substitute (BD) as a negative quantity into equation (3) to obtain $(CD) = 0$.

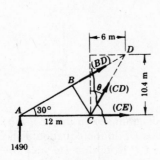

Fig. 7-11

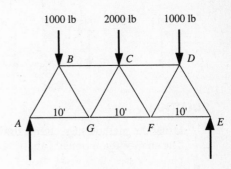

Fig. 7-12

7.5.   Determine the force in each member of the truss shown in Fig. 7-12. The triangles are all equilateral.

SOLUTION

By symmetry, the reactions on the truss at A and E are 2000 lb each. The free-body diagrams of pins A, B and G are shown in Figs 7-13, 14 and 15 respectively. All unknown forces are drawn in the free-body diagrams as tension. When the internal forces are calculated, a plus force means tension and a minus force means compression. From Fig. 7-13,

$$\sum F_y = AB \sin 60° + 2000 = 0 \qquad AB = -2309 \text{ lb } C$$

$$\sum F_x = AG + AB \cos 60° = 0 \qquad AG = +1155 \text{ lb } T$$

From Fig. 7-14,

$$\sum F_y = -AB \cos 30° - 1000 - BG \cos 30° = 0 \qquad BG = +1155 \text{ lb } T$$

$$\sum F_x = -AB \sin 30° + BG \sin 30° + BC = 0 \qquad BC = -1732 \text{ lb } C$$

From Fig. 7-15,

$$\sum F_y = BG \sin 60° + GC \sin 60° = 0 \qquad GC = -1155 \text{ lb } C$$

$$\sum F_x = -AG - BG \cos 60° + GC \cos 60° + GF = 0 \qquad GF = +2309 \text{ lb } T,$$

By the symmetry of the structure and loading,

$$DE = AB = -2309 \text{ lb } C$$

$$FE = AG + 1155 \text{ lb } T$$

$$DF = BG = +1155 \text{ lb } T$$

$$CD = BC = -1732 \text{ lb } C$$

$$CF = CG = -1155 \text{ lb } C$$

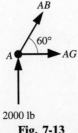

Fig. 7-13

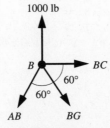

Fig. 7-14

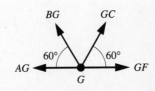

Fig. 7-15

7.6. Determine the forces in the members BD and CD of the truss in Fig. 7-16. All triangles are equilateral.

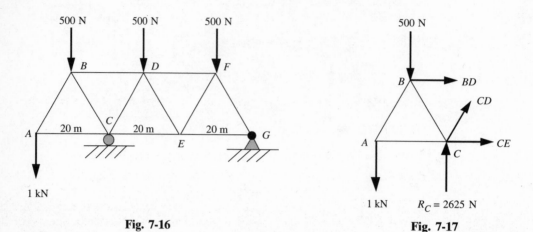

Fig. 7-16   **Fig. 7-17**

SOLUTION

It is convenient in this case to isolate a section of the truss (Fig. 7-17) as pointed out in Section 7.2, and solve it as a free-body diagram. Because, in general, a section of the truss results in a nonconcurrent, nonparallel force system, the maximum number of unknown forces cannot exceed three. In the current problem, rather than solving pin A, then pin B, then pin C in order to find the forces in BD and CD, Fig. 7-17 shows a free-body diagram in which the forces in BD and CD can be found directly once the reaction at C is known.

Using the entire truss of Fig. 7-16 as a free body and writing moments about G, the resulting equation of equilibrium becomes

$$\sum M_G = -40R_C + (60)(1000) + (50)(500) + (30)(500) + (10)(500) = 0$$
$$R_C = 2625 \text{ N}$$

Returning to Fig. 7-17, the equations of equilibrium can be written

$$\sum M_C = (20)(1000) + (10)(500) - (20 \sin 60°)(BD) = 0$$
$$BD = 1440 \text{ N } T$$
$$\sum F_y = 2625 - 1000 - 500 + CD \sin 60° = 0$$
$$CD = -1300 \text{ N } C$$

This so-called "method of sections" allows the determination of the required forces without finding the intermediate values that the "method of joints" would require.

7.7. Each cable of a suspension bridge carries a horizontal load of 800 lb ft. If the span is 600 ft and the sag is 40 ft, determine the tension at either end of the cable and at the midpoint. What is the length of the cable? This is an example of a parabolic cable.

SOLUTION

The tension at either end is the same.

$$T = \frac{1}{2}wa\sqrt{1 + \frac{a^2}{16d^2}} = \frac{1}{2} \times 800 \times 600 \sqrt{1 + \frac{(600)^2}{16(40)^2}} = 932,000 \text{ lb}$$

$$H = \frac{wa^2}{8d} = \frac{800 \times (600)^2}{8 \times 40} = 900,000 \text{ lb}$$

$$l = a\left[1 + \frac{8}{3}\left(\frac{d}{a}\right)^2 - \frac{32}{5}\left(\frac{d}{a}\right)^4 + \frac{256}{7}\left(\frac{d}{a}\right)^6 - \cdots\right]$$

Usually, two or three terms of a series that converges rapidly are sufficient for the accuracy desired. It is well to check as shown below.

$$\frac{d}{a} = \frac{40}{600} = 0.0667 \qquad \left(\frac{d}{a}\right)^2 = 0.0045 \qquad \left(\frac{d}{a}\right)^4 = 0.00002$$

Using two terms, $l = 600[1 + \frac{8}{3}(0.0045)] = 607 \text{ ft}$

Using three terms, $l = 600[1 + \frac{8}{3}(0.0045) - \frac{32}{5}(0.00002)] = 607 \text{ ft}$

If four terms are used, the value will increase slightly, but the magnitude of each term beyond the second is negligible.

7.8.   Solve for the desired tensions in Problem 7.7 without recourse to formulas.

SOLUTION

A free-body diagram of the right half of the cable (see Fig. 7-18) contains the horizontal tension H at the low point, the maximum T at the support, and the load $800(300) = 240,000$ lb. Since only three forces act on the cable, they must be concurrent as shown. Summing the forces horizontally and then vertically,

$$\sum F_h = 0 = T_{\max} \cos \theta - H \tag{1}$$

$$\sum F_v = 0 = T_{\max} \sin \theta - 240,000 \tag{2}$$

from which (1) $T_{\max} \cos \theta = H$ and (2) $T_{\max} \sin \theta = 240,000$. Dividing (2) by (1), $\tan \theta = 240,000/H$. But $\tan \theta = 40/150$; hence $40/150 = 240,000/H$ and $H = 900,000$ lb.

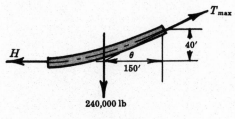

Fig. 7-18

Squaring equations (1) and (2) and adding,

$$T_{\max}^2(\sin^2 \theta + \cos^2 \theta) = (240,000)^2 + H^2 \qquad T_{\max}^2 = (240,000)^2 + (900,000)^2 \qquad T_{\max} = 932,000 \text{ lb}$$

7.9.   A cable suspends a mass of 3 kg/m between two supports not at the same level, as shown in Fig. 7-19. Determine the maximum tension.

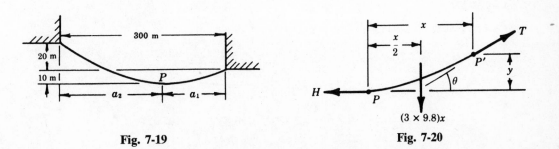

Fig. 7-19   **Fig. 7-20**

SOLUTION

In Fig. 7-20, a free-body diagram is shown of a piece of the cable to the right of the lowest point P, whose location is unknown. Let H be the unknown tension in the cable at point P, and let T be the tension at a point P' a distance x to the right of P. The gravitational force on the cable for this distance x is $(3 \times 9.8)x$ and acts at a distance $x/2$ from P. Summing forces horizontally and vertically,

$$\sum F_x = 0 = T \cos \theta - H \qquad \text{or} \qquad T \cos \theta = H \qquad\qquad (1)$$

$$\sum F_y = 0 = T \sin \theta - (3 \times 9.8)x \qquad \text{or} \qquad T \sin \theta = (3 \times 9.8)x \qquad\qquad (2)$$

Dividing (2) by (1), $\tan \theta = (3 \times 9.8)x/H$. But from the free-body diagram, $\tan \theta = y/(x/2) = 2y/x$; hence $(3 \times 9.8)x/H = 2y/x$ or $Hy = (3 \times 9.8)x^2/2$. Since $x = a_1$ when $y = 10$ m, $10H = (3 \times 9.8)a_1^2/2 = 14.7a_1^2$. Similarly, for the section to the left of P, $30H = 14.7a_2^2$. Now

$$a_1 + a_2 = 300 \qquad \sqrt{\frac{10H}{14.7}} + \sqrt{\frac{30H}{14.7}} = 300 \qquad H = 17.72 \text{ kN}$$

and

$$a_1 = \sqrt{\frac{10H}{14.7}} = 110 \text{ m} \qquad a_2 = \sqrt{\frac{30H}{14.7}} = 190 \text{ m}$$

Squaring equations (1) and (2) and adding, $T^2 = 864x^2 + H^2$. Since the maximum tension occurs at the left support, where $x = -190$ m, $T_{\max}^2 = 864(-190)^2 + (17,720)^2$ or $T_{\max} = 18.58$ kN.

7.10. A horizontal load of 2000 N/m is carried by a TV cable suspended between supports at the same level and 20 m apart. The maximum permissible tension is 140 kN. Determine the required length l of the cable and its sag d.

SOLUTION

$$T = \frac{1}{2}wa\sqrt{1 + \frac{a^2}{16d^2}} \qquad \text{or} \qquad 140\,000 = \frac{1}{2} \times 2000 \times 20 \sqrt{1 + \frac{(20)^2}{16d^2}}$$

Hence

$$d = 0.72 \text{ m}$$

Use

$$l = a\left[1 + \frac{8}{3}\left(\frac{d}{a}\right)^2 - \frac{32}{5}\left(\frac{d}{a}\right)^4 + \frac{256}{7}\left(\frac{d}{a}\right)^6 - \cdots\right]$$

Since $d/a = 0.72/20 = 0.036$, use only the second power to obtain $l = 20.07$ m.

7.11. A wire weighing 10 oz/ft is supported on frictionless pulleys as shown in Fig. 7-21. The load P is 500 lb. The distance between the centers of the pulleys is 80 ft. Determine the sag. Assume the curve to be parabolic and neglect the diameter of the pulleys.

SOLUTION

The tension at the pulleys is the weight P, 500 lb. Then

$$500 = \frac{1}{2} \times \frac{10}{16} \times 80 \sqrt{1 + \frac{(80)^2}{16d^2}} \qquad \text{or} \qquad d = 1 \text{ ft}$$

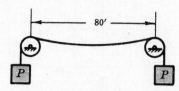

Fig. 7-21

7.12. A wire weighing 0.518 lb/ft is suspended between two towers at the same level and 500 ft apart. If the sag is 50 ft, what is the maximum tension in the wire and what should its minimum length be? This is an example of a catenary.

SOLUTION

Since a and d are given, this is an example of case (a) under the catenary where equation (2) yields

$$c + d = c \cosh \frac{a}{2c} \qquad \text{or} \qquad c + 50 = c \cosh \frac{500}{2c}$$

A graphical solution is perhaps easier than substitution of values. Plot the line $c + 50$ against c, and the curve $c \cosh 500/2c$ against c. The value of c at which these two curves intersect is the solution. The values for plotting are shown in the table below. The extra points at 150 and 250 were chosen to ensure that the curves do not meet at a lower value of c.

c	$c + 50$	$\dfrac{500}{2c}$	$\cosh \dfrac{500}{2c}$	$c \cosh \dfrac{500}{2c}$
0	50	∞	∞	—
100	150	2.5	6.1323	613.2
200	250	1.25	1.8884	377.7
300	350	0.833	1.3678	410.3
400	450	0.625	1.2018	480.7
500	550	0.500	1.1276	563.8
600	650	0.417	1.0882	652.9
700	750	0.357	1.0644	745.1
800	850	0.313	1.0494	839.5
150	200	1.667	2.7427	411.4
250	300	1.000	1.5431	385.8

The curves are plotted against c, and intersect at approximately $c = 650$. See Fig. 7-22.

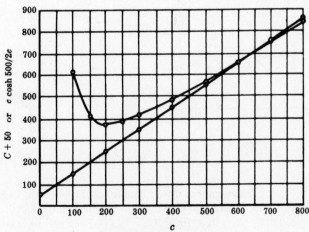

Fig. 7-22

For greater accuracy an enlarged plot could be made of that portion of the graph where the curves cross. However, try $c = 650$ in the equation to determine a more accurate balance:

$$c + 50 = 650 + 50 = 700$$

But

$$650 \cosh \frac{500}{2 \times 650} = 698.8$$

Now try $c = 640$ to obtain $640 + 50$ or 690:

$$640 \cosh \frac{500}{2 \times 640} = 689.5$$

A check on $c = 635$ indicates that 635 is sufficiently accurate.

Using equation (3), $T_{max} = w(c + d) = 0.518(635 + 50) = 355$ lb

Using equation (5), $(c + d)^2 = c^2 + \frac{1}{4}l^2$ $(635 + 50)^2 = (635)^2 + \frac{1}{4}l^2$ $l = 514$ ft

Note: The computer solution in Appendix C shows the power of such a technique.

7.13.   A cable that has a mass of 0.6 kg/m and is 240 m long is to be suspended with a sag of 24 m. Determine the maximum tension and maximum span.

SOLUTION

This is an example of case (c), in Section 7.3, "Catenary."

Using equation (5), $(c + d)^2 = c^2 + \frac{1}{4}l^2$ $(c + 24)^2 = c^2 + \frac{1}{4}(240)^2$ $c = 288$

Using equation (3), $T_{max} = w(c + d) = 0.6 \times 9.8(288 + 24) = 1835$ N

Using equation (4), $\frac{1}{2}l = c \sinh \frac{a}{2c}$ $\frac{1}{2}(240) = 288 \sinh \frac{a}{576}$ $a = 234$ m

Supplementary Problems

7.14.   The Howe truss supports the three loads shown in Fig. 7-23. Determine the forces in *AB*, *BD*, *CD*, and *EF* by the method of joints.
Ans. $AB = 25.0$ kN C, $BD = 15.0$ kN C, $CD = 12.5$ kN C, $EF = 22.5$ kN T

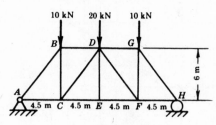

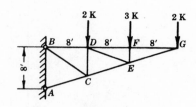

Fig. 7-23   Fig. 7-24

7.15.   Determine the forces in all members of the cantilever truss shown diagrammatically in Fig. 7-24. Loads are in kips. Solution may be started with pin *G*. Use method of joints.
Ans. $BD = 10,500$ lb T, $BC = 4200$ lb T, $AC = 14,800$ lb C, $DF = 6000$ lb T, $DE = 4730$ lb T,
$CE = 11,100$ lb C, $FG = 6000$ lb T, $EG = 6330$ lb C, $CD = 3490$ lb C, $EF = 3000$ lb C

7.16. In the truss shown in Fig. 7-25, determine the forces in members *AC* and *BD* by the method of joints.
Ans.   *AC* = 35.3 kN *C*, *BD* = 47.9 kN *T*

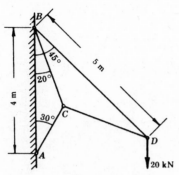

Fig. 7-25

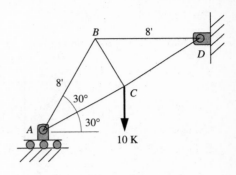

Fig. 7-26

7.17. Determine the forces in all of the members of the truss shown in Fig. 7-26.
Ans.   *AB* = *BD* = 8.66 K *C*, *AC* = 5 K *T*, *CD* = 10 K *T*, *CB* = 8.66 K *T*

7.18. The Fink truss is subjected to the loading shown in Fig. 7-27. Determine the forces in all members by the method of joints.
Ans.   *AB* = 10.04 K *T*, *AC* = 9.87 K *C*, *CD* = 8.87 K *C*, *BD* = 3.72 K *T*, *BE* = 6.33 K *T*, *DF* = 8.30 K *C*,
FG = 9.30 K *C*, *EG* = 8.06 K *T*

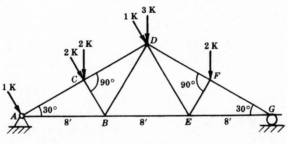

Fig. 7-27

7.19. Determine forces in *AB*, *AC*, *EG*, and *FG* in Fig. 7-28.
Ans.   *AB* = 1.73 kN *T*, *AC* = 0.866 kN *C*, *EG* = 2.89 kN *C*, *FG* = 5.77 kN *T*

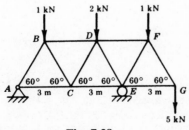

Fig. 7-28

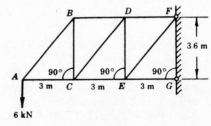

Fig. 7-29

7.20. Determine the forces in *AB* and *CD* in the cantilever truss using the method of joints. See Fig. 7-29.
Ans. *AB* = 7.81 kN *T*, *CD* = 7.81 kN *T*

7.21. Determine the forces in *AC* and *AB* in the Howe truss using the method of joints. See Fig. 7-30.
Ans. *AC* = 4.3 K *C*, *AB* = 2.8 K *T*

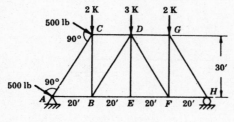

Fig. 7-30

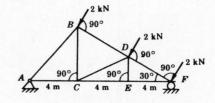

Fig. 7-31

7.22. Find the forces in members *AB* and *CD* in Fig. 7-31.   *Ans.*   *AB* = 3.06 kN *C*, *CD* = 2.31 kN *C*

Solve Problems 7.23 through 7.26 by the method of sections for the forces in the members checked in the individual figures.

7.23. Figure 7-32.   *Ans.*   *DF* = 3.46 kN *C*, *DE* = 0, *CE* = 3.46 kN *T*

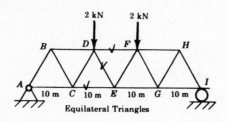

Fig. 7-32

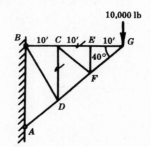

Fig. 7-33

7.24. Figure 7-33.   *Ans.*   *CE* = 11.9 K *T*, *CD* = 0

7.25. Figure 7-34.   *Ans.*   *DF* = 833 N *C*, *DE* = 322 N *T*

7.26. Figure 7-35.   *Ans.*   *BD* = 5000 lb *C*, *EF* = 0

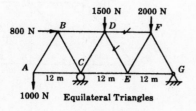

Fig. 7-34

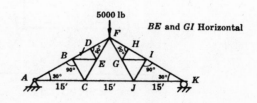

Fig. 7-35

7.27. Solve for the forces in kips in the members *CE* and *DF* in the truss of Fig. 7-36. All triangles are equilateral. Use the method of sections.   *Ans.*   *CE* = 4.04 K *T*, *DF* = 4.04 K *C*

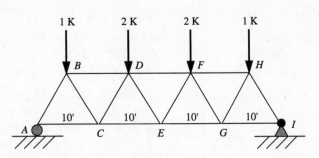

Fig. 7-36

7.28. Solve Prob. 7.15 for members *DF* and *CE* only. Use method of sections.

7.29. Solve Prob. 7.25 by the method of joints.

7.30. Determine the forces in all members in Fig. 7-37.
Ans.   *AB* = 4.00 kN *C*, *AD* = 2.00 kN *C*, *BD* = 4.00 kN *T*, *BC* = 3.46 kN *C*, *CE* = 3.46 kN *C*,
CD = 4.00 kN *C*, *DE* = 2.00 kN *T*

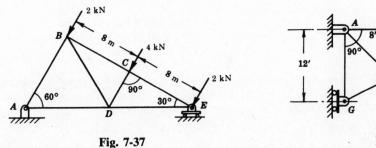

Fig. 7-37   **Fig. 7-38**

7.31. A car is located on the span *CD* as shown in Fig. 7-38. It impresses a load of 4000 lb, which is equally distributed on all four wheels. Since there are two identical trusses to support the rails, the load is 1000 lb at *C* and 1000 lb at *D*. What are the forces in the members of the truss?
Ans.   *AB* = 3.0 K *T*, *AF* = 0.47 K *T*, *AG* = 1.67 K *T*, *GF* = 3.72 K *C*, *BF* = 0.5 K *C*, *BC* = 2.0 K *T*,
BE = 1.12 K *T*, *FE* = 3.35 K *C*, *CE* = 1.0 K *C*, *CD* = 2.0 K *T*, *ED* = 2.24 K *C*

7.32. In Fig. 7-39, the 1800-kN load is applied horizontally to the truss shown with a pin support at *A* and a roller support at *E*. Determine the forces in *AB* and *CE*.   *Ans.*   *CE* = 4070 *N* T, *AB* = 1300 N *C*

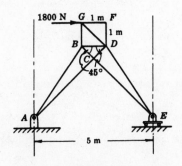

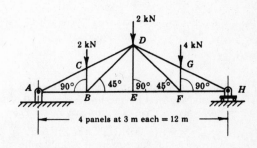

Fig. 7-39   **Fig. 7-40**

7.33. Determine the forces in all members of the truss subjected to the three loads shown in Fig. 7-40.
　Ans.　$AC = 7.83$ kN C, $AB = 7.00$ kN T, $BC = 2.00$ kN C, $CD = 7.83$ kN C, $BD = 2.83$ kN T
　　　　$DE = 0$, $DG = 10.1$ kN C, $DF = 5.67$ kN T, $BE = 5.00$ kN T, $EF = 5.00$ kN T,
　　　　$FG = 4.00$ kN C, $FH = 9.00$ kN T, $GH = 10.1$ kN C

7.34. For the truss shown (Fig. 7-41), pin-supported at A and roller-supported at E, determine the forces in CE and DE when a horizontal load of 3000 lb is applied at D.　　*Ans.*　$CE = 943$ lb T, $DE = 2150$ lb C

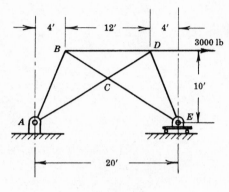

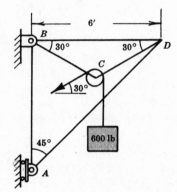

Fig. 7-41　　　　　　　　　　　　　　　　　　　**Fig. 7-42**

7.35. For the truss shown in Fig. 7-42 determine the forces in members BC, AD, and CD. Assume that the pulley is weightless and frictionless. Its diameter is 2 ft. The load of 600 lb is held by a rope inclined 30° with the horizontal.　　*Ans.*　$BC = 600$ lb T, $CD = 1200$ lb T, $AD = 850$ lb C

7.36. A truss that is pin-supported at A and roller-supported at D is inclined 40° with the vertical as shown in Fig. 7-43. The members AC and CD are cables designed for a maximum load of 2000 N (in tension obviously). What is the maximum that M may have?　　*Ans.*　180 kg.

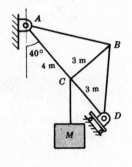

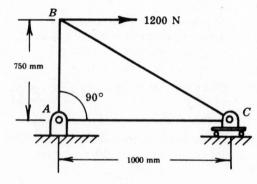

Fig. 7-43　　　　　　　　　　　　　　　　　　　**Fig. 7-44**

7.37. In Fig. 7-44, the truss is pin-supported at A and roller-supported at C. Determine the forces in each member caused by the horizontal 1200-N load.
　Ans.　$AC = 1200$ N T, $AB = 900$ N T, $BC = 1500$ N C

7.38. A truss is pin-supported at A and roller-supported at D as shown in Fig. 7-45. Determine the force in each member caused by the 500-N load.
　Ans.　$AB = BD = 417$ N C, $AC = CD = 333$ N T, $BC = 500$ N T

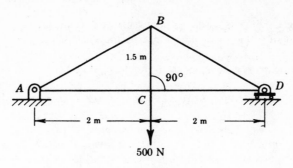

Fig. 7-45

7.39. A cable is suspended between two supports at the same elevation and 200 m apart. The load is 100 N per horizontal meter. The sag is 12 m. Find the length of the cable and the tension at the support.
Ans. $l = 202$ m, $T = 42.9$ kN

7.40. Calculate the sag in a wire 100 ft long held between two supports 99.8 ft apart and at the same level. Use equation (3) in Section 7.3, "Parabolic," neglecting powers of d/a beyond the second.
Ans. $d = 2.73$ ft

7.41. In Problem 7.40, assuming the load on the wire is 0.01 lb per horizontal foot, what tension is necessary to hold it at the support?   *Ans.* $T = 4.59$ lb

7.42. The maximum allowable tension in a cable is 3000 lb. A sag of no more than 10 ft is to be allowed under a load of 100 lb per horizontal foot. How far apart may the supports be placed and how long should the cable be?   *Ans.* $a = 41.6$ ft, $l = 47.4$ ft

7.43. The maximum allowable tension in a cable is 1000 lb. A sag of no more than 4 ft is to be allowed under a load of 1 lb per horizontal foot. How far apart may the supports be placed?   *Ans.* $a = 178$ ft

7.44. The maximum allowable tension in a cable is 90,000 lb. It is to carry a load of 200 lb per horizontal foot when suspended between level supports 300 ft apart. What is the minimum permissible sag?
Ans. $d = 26.6$ ft

7.45. Using the data of Problem 7.40, compute the maximum tension in the wire and its sag assuming the weight is 0.01 lb per lineal foot along the wire.   *Ans.* $T = 4.4$ lb, $d = 2.9$ ft

7.46. A wire having a mass of 0.73 kg/m is stretched between two level supports 48 m apart. If the sag is 12 m, determine the length of the wire and the maximum tension.   *Ans.* $l = 54.9$ m, $T = 268$ N

7.47. A transmission wire 230 m long having a mass of 0.97 kg per lineal meter is suspended between towers of equal elevation 229 m apart. Find the maximum tension and the sag.   *Ans.* $T = 6840$ N, $d = 9$ m

7.48. A transmission line cable is to be strung between two towers at the same elevation and 800 ft apart. The cable weighs 5 lb/ft and its length is 1000 ft. What are the sag and maximum tension?
Ans. $d = 266$ ft, $T_{max} = 3020$ lb

7.49. A rope 50 ft long weighs 0.1 lb per lineal foot. How far apart should supports be placed so that a maximum tension of 10 lb is induced? Use equations (3) and (5) in Section 7.3, "Catenary".

7.50. A wire weighing 0.5 lb/ft is stretched between two level supports 160 ft apart. If the sag is 40 ft, determine the length of the wire and the maximum tension.   *Ans.* $l = 184$ ft, $T = 63$ lb

7.51. A parabolic cable carries a load of 200 lb per horizontal foot. The distance between anchors is 100 ft. The difference in elevation of the two anchors is 20 ft, with a sag, measured from the lower anchor, of 8 ft. Determine the tension at the two anchors.   *Ans.* $T_L = 20$ K, $T_R = 16.7$ K

7.52. A transmission line, weighing 4 lb/ft, is strung between two towers 2000 ft apart that have a 200-ft difference in elevation. The line is strung in such a way that the slope is horizontal at the lower tower. Determine the tensions at the two towers.   *Ans.* $T_{max} = 40.8$ K, $H = 40.0$ K

7.53. During an ice storm, a cylinder of ice forms on a telephone line. The line is strung between poles 80 ft apart. The weight of the clean line is 0.3 lb/ft. How much ice can form if the sag is not to exceed 5 ft and the maximum allowable tension in the line is 6000 lb? Assume the ice to be a solid cylinder with weight density 56 lb/cu ft.   *Ans.*   dia. = 10.8 in

7.54. A rope weighing 1.5 lb/ft is anchored to a wall and passes over a small frictionless drum 40 ft from the wall and at the same elevation. The sag of the rope is 2 ft. What must the length of the hanging rope be to prevent slip on the drum?   *Ans.* $h = 102$ ft

7.55. A cable is to be strung from two towers at the same elevation and 800 ft apart. The weight of the cable is 5 lb/ft and the maximum sag can be 270 ft. What is the length of cable?   *Ans.*   1010 ft

Chapter 8

Forces in Beams

8.1 BEAMS

A beam is a structural member that has a length considerably longer than its cross-sectional dimensions. It carries loads that are usually perpendicular to the longitudinal axis of the beam, and thus the loads are at right angles to the length. The loads may be distributed over a very small distance along the beam, in which case they are called *concentrated,* or they may be distributed over a measurable distance, in which case they are called *distributed.*

Because design criteria usually are concerned with the ability of a beam to withstand shear forces and bending moments, this chapter will not treat external loads that are not perpendicular to the beam. These cause axial forces in the beam.

8.2 TYPES OF BEAMS

(a) *Simple*: Supports are at the ends. See Fig. 8-1(a).

(b) *Cantilever*: On end is mounted in a wall and the other end is free (this is the only type considered here). See Fig. 8-1(b).

(c) *Overhanging*: At least one support is not at the end. See Fig. 8-1(c).

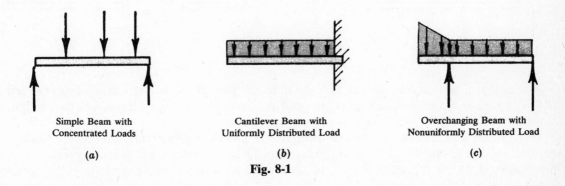

Simple Beam with Concentrated Loads	Cantilever Beam with Uniformly Distributed Load	Overchanging Beam with Nonuniformly Distributed Load
(a)	(b)	(c)

Fig. 8-1

8.3 SHEAR AND MOMENT

Shear and moment at a cross section C–D in a beam are best visualized by dividing the beam into two parts A and B (see Fig. 8-2) to the left and right respectively of C–D. A free-body diagram of part A must show all the external forces acting on A as well as the forces that B exerts on A to hold it in equilibrium.

The equilibrating forces that part B exerts on part A in the free-body diagram are (a) a vertical force V and (b) a set of horizontal distributed forces that, because they have a zero sum, are represented only by their moment M.

Shear at the section C–D is the vertical force V obtained by equaitng the sum of all the vertical forces to zero.

Moment M at the section C–D is obtained by equating to zero the sum of the moments of all forces about a point in the cross section. In the types of problems considered here, any point in the cross section can be used as a moment center.

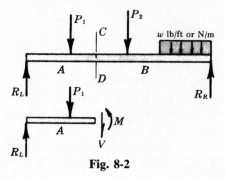

Fig. 8-2

In beam theory it is customary to call the shear V positive if it acts down on the left part A (if the right part B is used as a free-body diagram then positive shear V must act up). Bending moment M at a section is positive if it acts counterclockwise on the left part A (if the right part B is used as a free-body diagram then a positive moment M must act clockwise). Figure 8-3 illustrates these points.

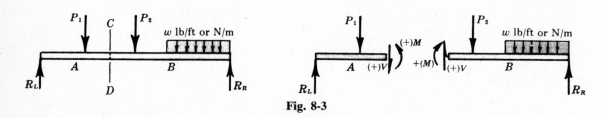

Fig. 8-3

Shear V at a section may now be thought of as the sum of all the vertical forces to the left of the section, where an up force is said to cause positive shear (if the sum of the forces to the right of the section is used a down force is said to cause positive shear).

Moment M at a section may be evaluated as the sum of the moments about that section of all of the forces to the left of the section. An up force to the left of the section causes a positive moment at the section. If the right portion is used an up force contributes positive moment (in contrast to the sign reversal in defining shear).

Naturally, shear and moment will usually vary with the location of the section. Problems 8.1 and 8.2 illustrate these principles.

8.4 SHEAR AND MOMENT DIAGRAMS

Shear and moment diagrams present a graphical picture of the variation of V and M across a beam. Summations of forces and moments of forces to the left of the section (or to the right if desired) can be used to plot V and M directly for simpler problems.

8.5 SLOPE OF THE SHEAR DIAGRAM

The slope of the shear diagram at any section along the beam is the negative of the load per unit length at that point.

Proof: Figure 8-4 shows a portion dx of a beam. The load per unit length is approximately w on this short piece. The shear V and moment M at the left are assumed positive. The shear and moment at the right are assumed to increase to $V + dV$ and $M + dM$, respectively. Then

$$\sum F_v = 0 = +V - w\,dx - (V + dV)$$

from which $dV = -w\,dx$ or $dV/dx = -w$.

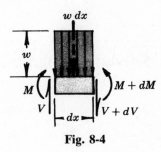

Fig. 8-4

8.6   CHANGE IN SHEAR

The change in shear between two sections of a beam carrying a distributed load equals the negative of the area of the load diagram between the two sections.

Proof: From the preceding proof, use $dV = -w\,dx$ and integrate from x_1 to x_2. Thus,

$$\int_{V_1}^{V_2} dV = \int_{x_1}^{x_2} (-w)\,dx \qquad \text{or} \qquad V_2 - V_1 = -\int_{x_1}^{x_2} (+w)\,dx$$

where $\int_{x_1}^{x_2} (+w)\,dx$ is the area of the load diagram between the two sections.

8.7   SLOPE OF THE MOMENT DIAGRAM

The slope of the moment diagram at any section along the beam is the value of the shear at that section.

Proof: Using the diagram of the preceding proofs, equate the sum of the moments about the right end to zero and obtain

$$-M - V\,dx + w\,dx\,\frac{dx}{2} + M + dM = 0$$

Neglecting differentials of second order, this becomes $dM = V\,dx$ or $dM/dx = V$. This also indicates that the moment is a maximum (or minimum) where the shear diagram crosses the zero line.

8.8   CHANGE IN MOMENT

The change in the moment between two sections of a beam equals the area of the shear diagram between the two sections.

Proof: From the preceding proof, use $dM = V\,dx$ and integrate from x_1 to x_2.

$$\int_{M_1}^{M_2} dM = \int_{x_1}^{x_2} V\,dx \qquad \text{or} \qquad M_2 - M_1 = \int_{x_1}^{x_2} V\,dx$$

where $\int_{x_1}^{x_2} V\,dx$ is the area of the shear diagram between the two sections.

Refer to Problems 8.3–8.5.

Solved Problems

8.1.   In the simple beam shown in Fig. 8-5, determine the shear V and moment M at section C–D 2 m from the left end.

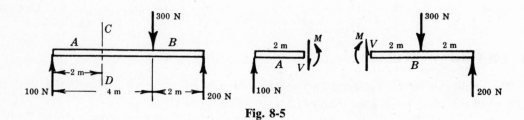

Fig. 8-5

SOLUTION

The reactions are 100 and 200 N as shown. Free-body diagrams of left and right portions of the beam are shown in Fig. 8-5. Using the left part A, the shear is the sum of the forces to the left, or

$$V = +100\,\text{N}$$

If the right portion B is used, the shear is the sum of the forces to the right of the section (but now a down force is positive); hence,

$$V = +100\,\text{N}$$

The moment M at the section can be evaluated as the sum of the moments of all vertical forces to the left. An up force causes positive moment at the section; hence, referring to Fig. 8-5,

$$M = +100(2) = +200\,\text{N} \cdot \text{m}$$

If the right portion is used, an up force contributes positive moment; hence, for the right portion in Fig. 8-5,

$$M = -300(2) + 200(4) = +200\,\text{N} \cdot \text{m}$$

8.2.   Draw shear and moment diagrams for Problem 8.1. It is necessary to use two free-body diagrams as shown in Fig. 8-6.

SOLUTION

The free-body diagram A_1 shows that for $0 < x < 4\,\text{m}$, $V = +100\,\text{N}$ and $M = +100x\,\text{N} \cdot \text{m}$. The free-body diagram A_2 shows that for $4\,\text{m} < x < 6\,\text{m}$, $V = -200\,\text{N}$ and $M = +100x - 300(x - 4) = -200x + 1200\,\text{N} \cdot \text{m}$. If the free-body diagram B_2 is used then $M = +200(6 - x) = -200x + 1200\,\text{N} \cdot \text{m}$ (be sure to use x as the distance from the left end). The value of M at $x = 4\,\text{m}$ is $M = +400\,\text{N} \cdot \text{m}$.

The above information may now be plotted as shear and moment diagrams (see Fig. 8.7).

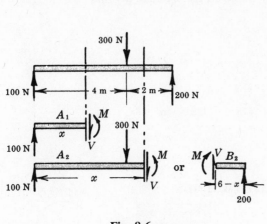

Fig. 8-6

Mathcad

8.3. Determine the shear and moment equations for the beam shown in Fig. 8-8. Draw shear and moment diagrams.

SOLUTION

First determine the right reaction R_R by equating the sum of the moments of all external forces about the left end to zero. The 20-kg/m mass may be replaced (this should only be done to determine the reactions) by a $(20 \times 9.8)(6)$ or 1176-N load at its midpoint.

$$\sum M_{R_L} = 18R_R - 4000 - (3)1176 = 0 \quad R_R = 418 \text{ N}$$

$$\sum M_{R_R} = -18R_L - 4000 + (15)1176 = 0 \quad R_L = 758 \text{ N}$$

In the interval $0 < x < 6$ m the free-body diagram shown in Fig. 8-9 holds.

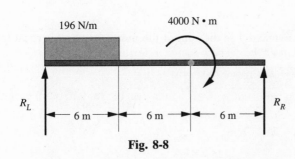

Fig. 8-8

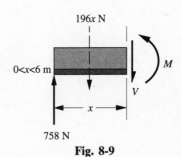

Fig. 8-9

A vertical summation of all forces acting on the free-body diagram yields

$$\sum F_V = 758 - 196x - V = 0 \quad V = 758 - 196x \text{ N}$$

$$\sum M_O = -758x + 196x(x/2) + M = 0 \quad M = 758x - 98x^2 \text{ N} \cdot \text{m}$$

The shear shows that V is the sum of the forces to the left of the section, and M shows that the moment at a section is the sum of the moments of all vertical forces and couples to the left of the section.

The free-body diagrams are shown in Fig. 8-10 for sections (*a*) and (*b*) along the beam. The equations are as follows:

from Fig. 8-10(*a*), $V = -418\,\text{N}$

$$M = -418x + 3528\,\text{N} \cdot \text{m}$$

from Fig. 8-10(b), $V = -418\,\text{N}$

$$M = -418x + 7528\,\text{N} \cdot \text{m}$$

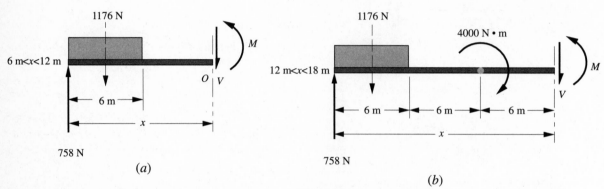

Fig. 8-10

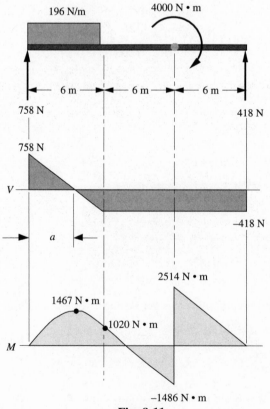

Fig. 8-11

Shear and moment diagrams are shown in Fig. 8-11. To determine the location and value of the maximum moment, remember that the slope of the moment diagram is equal to the shear value at that point (see Sec. 8.7). The moment will be a maximum when the slope (or shear force) is zero. Hence,

$$758 - 196a = 0 \quad a = 3.87 \, \text{m}$$

The maximum moment then becomes the area under the shear curve from 0 to a m, or

$$M_{\text{max}} = 758 \, (3.87)/2 = 1467 \, \text{N} \cdot \text{m}$$

It should be noted that, at the right end of the moment diagram, the value calculated from the shear curve is not exactly zero. This is due to rounding off.

8.4.   The simple beam shown in Fig. 8-12 supports a triangular load and a uniformly distributed load. Derive the shear and moment equations.

SOLUTION

To determine the reactions R_1 and R_2, visualize the resultant of the triangular load $[\frac{1}{2}(120)(9) = 540 \, \text{lb}]$ at the two-thirds point (6 ft from the left end). Then visualize the resultant of the rectangular load $[120(10) = 1200 \, \text{lb}]$ at its midpoint, which is 5 ft from R_2. Sum moments about the left end to obtain

$$-\tfrac{1}{2}(120)(9)(6) - (120)(10)(9 + 5) + 19R_2 = 0$$

From this equation, $R_2 = 1055$ lb.

Now sum moments about the right end to obtain

$$+\tfrac{1}{2}(120)(9)(19 - 6) + (120)(10)(5) - 19R_1 = 0$$

Hence, $R_1 = 685$ lb.

Note that, as a check, $R_1 + R_2 = 1740$ lb, which is the sum of the two loads $[\frac{1}{2}(120)(9) + (120)(10)]$.

Next draw a free-body diagram of a section of the beam and its loading in the interval $0 < x < 6$ ft. This is shown in Fig. 8.-13. Using similar triangles, we get a height for the load at x of $(x/9)(120)$ lb/ft. The total loading is $\frac{1}{2}x(x/9)(120)$, located at $x/3$ to the left of the section. Thus,

$$V = -\frac{1}{2}x\left(\frac{x}{9}\right)(120) + 685 = -\frac{60}{9}x^2 + 685 \, \text{lb}$$

$$M = -\frac{1}{2}x\left(\frac{x}{9}\right)(120)\frac{x}{3} + 685x = -\frac{20}{9}x^3 + 685x \, \text{lb-ft}$$

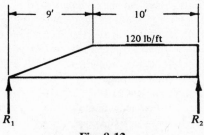

Fig. 8-12

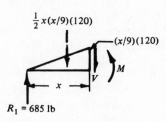

Fig. 8-13

To determine shear and moment for a section that is in the interval $9 \, \text{ft} < x < 19 \, \text{ft}$, draw the free-body diagram shown in Fig. 8-14. The triangular load is equal to $\frac{1}{2}(9)(120) = 540$ lb located 6 ft from the left end or $x - 6$ to the left of the section. The rectangular load is $120(x - 9)$ and is located

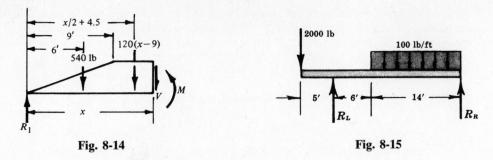

Fig. 8-14   **Fig. 8-15**

$9 + (x - 9)/2 = x/2 + 4.5$ ft from the left end, or $x/2 - 4.5$ ft to the left of the section. Thus,

$$V = 685 - 540 - 120(x - 9) = -120x + 1225 \text{ lb}$$

$$M = 685x - 540(x - 6) - 120(x - 9)\left(\frac{x}{2} - 4.5\right) = -60x^2 + 1225x - 1620 \text{ lb-ft}$$

The reader can check the last equations to see if the moment at the right end is close to zero, as it should be.

8.5.   Draw shear and moment diagrams for the bear shown in Fig. 8-15.

SOLUTION

Summing moments about R_L,

$$+2000(5) - 1400(13) + 20R_R = 0 \quad \text{or} \quad R_R = 410 \text{ lb}$$

Summing moments about R_R,

$$+2000(25) - 20R_L + 1400(7) = 0 \quad \text{or} \quad R_L = 2990 \text{ lb}$$

In drawing the shear diagram (see Fig. 8-16) make use of the fact that the shear at any section is the sum of the forces to the left of the section, an up force contributing positive shear.

For a section a very small distance ε in from the left, $V = -2000$ lb and remains this value until the left reaction is reached. Thus, an ε to the left of the left reaction, $V = -2000$ lb; but an ε to the right of this reaction, $V = +990$ lb. It remains this amount until the distributed load is reached. Then the shear decreases (load is down) at the rate of 100 lb/ft. It reaches zero at 9.9 ft (990/100). Its value an ε to the left of the right end is -410 lb. The right reaction thus closes the shear diagram.

To draw the moment diagram (see Fig. 8-16), first determine the moment at the left end (really an ε to the right of the left end). This value is -2000ε lb-ft or zero at the left end. From $x = 0$ to $x = 5$ ft, the shear is negative (constant value); hence, the slope of the moment diagram is a constant negative value. The moment just to the left of the left reaction equals the moment at the left end (zero) plus the area of the shear diagram from the left end to the left reaction [negative $5(2000) = -10,000$ lb-ft].

The shear is a constant positive value from $x = 5$ ft to $x = 11$ ft; hence, the slope of the moment diagram is positive. The change in moment is the area of the shear diagram from $x = 5$ ft to $x = 11$ ft; or $+6(990) = +5940$. Thus, the moment changes from $-10,000$ to $(-10,000 + 5940) = -4060$ lb-ft.

The shear area is positive over the next 9.9 ft, at which point the moment is a maximum (where $V = 0$). The increase in moment from $x = 11$ ft to $x = 20.9$ ft is the area of the triangular shear diagram, that is, $\frac{1}{2}(990)(9.9) = 4900$ lb-ft.

Thus M at $x = 20.9$ ft is $(-4060 + 4900) = +840$ lb-ft. Since the shear is positive but decreasing in this interval 11 ft $< x <$ 20.9 ft, the moment diagram has a positive but decreasing slope up to the value of 840 lb-ft.

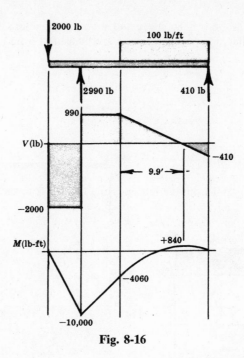

Fig. 8-16

To complete the moment diagram, note that the last shear area is negative but the magnitude of V is increasing. Hence, the moment curve has a negative slope that becomes more negative. Also, the moment curve drops in value an amount equal to the last shear area, which is $\frac{1}{2}(4.1)(410) = 840$ lb-ft. This means the moment at the right end is zero, as it should be.

Supplementary Problems

8.6. A cantilever beam supports a triangular load, and the uniformly distributed load is shown in Fig. 8-17. Derive the shear and moment equations.

Ans. $0 < x < 6$ ft, $V = -\frac{50}{3}x^2$ lb, $M = -\frac{50}{9}x^3$ lb-ft; 6 ft $< x < 12$ ft, $V = +600 - 200x$ lb,
$M = -1200 + 600x - 100x^2$ lb-ft

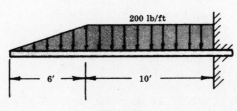

Fig. 8-17

8.7. For Figs. 8-18 to 8-24, derive the shear and moment equations for the beams shown. Also verify the shear and moment diagrams. All x distances are measured from the left end of each beam.

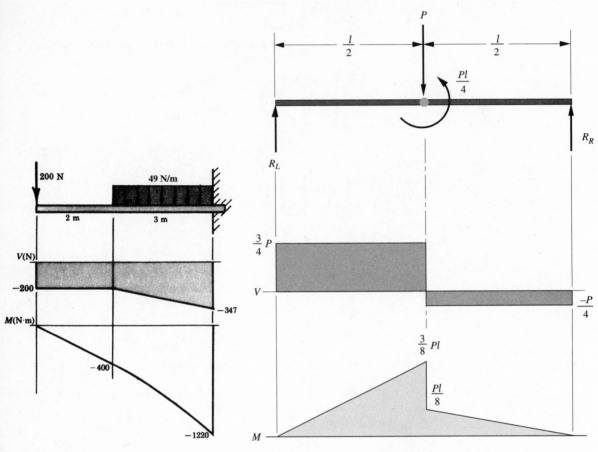

Fig. 8-18   Fig. 8-19

Ans. $0 < x < 2$ m $\begin{cases} V = -200 \text{ N} \\ M = -200x \text{ N} \cdot \text{m} \end{cases}$

$2 \text{ m} < x < 5 \text{ m}$ $\begin{cases} V = -102 - 49x \text{ N} \\ M = -98 - 102x - 24.5x^2 \text{ N} \cdot \text{m} \end{cases}$

Ans. $0 < x < \dfrac{l}{2}$ $\begin{cases} V = \frac{3}{4}P \\ M = \frac{3}{4}Px \end{cases}$

$\dfrac{l}{2} < x < L$ $\begin{cases} V = -P/4 \\ M = -P/4(x - l) \end{cases}$

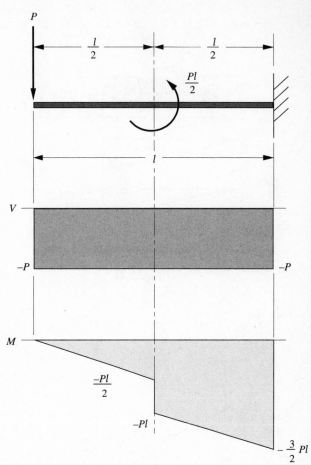

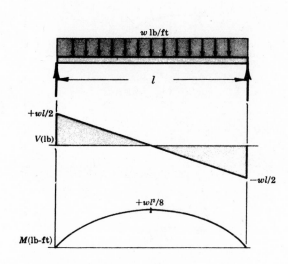

Fig. 8-20

Fig. 8-21

Ans. $\quad 0 < x < L \quad \begin{cases} V = +wl/2 - wx \text{ lb} \\ M = +wlx/2 - wx^2/2 \text{ lb-ft} \end{cases}$

Ans. $\quad 0 < x < \dfrac{l}{2} \quad \begin{cases} V = -P \\ M = -Px \end{cases}$

$\qquad \dfrac{l}{2} < x < l \quad \begin{cases} V = -P \\ M = -P(x + l/2) \end{cases}$

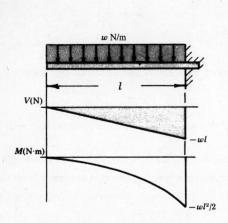

Fig. 8-22

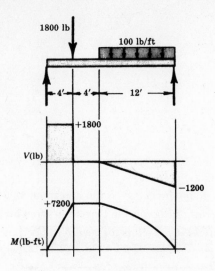

Fig. 8-23

Ans. $0 < x < l$ $\begin{cases} V = -wx \text{ N} \\ M = -wx^2/2 \text{ N} \cdot \text{m} \end{cases}$

Ans. $0 < x < 4 \text{ ft}$ $\begin{cases} V = +1800 \text{ lb} \\ M = +1800x \text{ lb-ft} \end{cases}$

$4 \text{ ft} < x < 8 \text{ ft}$ $\begin{cases} V = 0 \\ M = 7200 \text{ lb-ft} \end{cases}$

$8 \text{ ft} < x < 20 \text{ ft}$ $\begin{cases} V = +800 - 100x \text{ lb} \\ M = +4000 + 800x - 50x^2 \text{ lb-ft} \end{cases}$

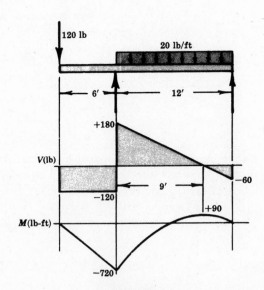

Fig. 8-24

Ans. $0 < x < 6 \text{ ft}$ $\begin{cases} V = -120 \text{ lb} \\ M = -120x \text{ lb-ft} \end{cases}$

$6 \text{ ft} < x < 18 \text{ ft}$ $\begin{cases} V = +300 - 20x \text{ lb} \\ M = -2160 + 300x - 10^2 \text{ lb-ft} \end{cases}$

Chapter 9

Friction

9.1 GENERAL CONCEPTS

1. *Static friction* between two bodies is the tangential force that opposes the sliding of one body relative to the other.

2. *Limiting friction F'* is the maximum value of static friction that occurs when motion is impending.

3. *Kinetic friction* is the tangential force between two bodies after motion begins. It is less than static friction.

4. *Angle of friction* is the angle between the action line of the total reaction of one body on another and the normal to the common tangent between the bodies when motion is impending.

5. *Coefficient of static friction* is the ratio of the limiting friction F' to the normal force N:

$$\mu = \frac{F'}{N}$$

6. *Coefficient of kinetic friction* is the ratio of the kinetic friction to the normal force.

7. *Angle of respose α* is the angle to which an inclined plane may be raised before an object resting on it will move under the action of the force of gravity and the reaction of the plane. This state of impending motion is shown in Fig. 9-1.

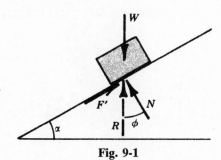

Fig. 9-1

The resultant R of F' and N is shown acting opposite but equal in magnitude to the force of gravity $W = Mg$. Although motion impends, the body is still in equilibrium. By trigonometry, $\alpha = \phi$. Hence, the coefficient of friction μ may be determined by raising the plane to the angle α at which motion impends. At that angle, $\mu = \tan \phi$ and hence $\mu = \tan \alpha$.

9.2 LAWS OF FRICTION

(a) The coefficient of friction is independent of the normal force; however, the limiting friction and kinetic friction are proportional to the normal force.

(b) The coefficient of friction is independent of the area of contact.

(c) The coefficient of kinetic friction is less than that of static friction.

(d) At low speeds, friction is independent of the speed. At higher speeds, a decrease in friction has been noticed.

127

(e) The static frictional force is never greater than that necessary to hold the body in equilibrium. In solving problems involving static friction, the student should assume the frictional force to be an independent unknown unless the problem clearly states that motion is impending. In the latter case, one may use limiting friction $F' = \mu N$, as in (5) above.

9.3 JACKSCREW

The jackscrew is an example of a frictional device. For the square-threaded screw shown in Fig. 9-2, there are essentially two problems: (a) the moment of the force P necessary to raise the load and (b) the moment of the force P necessary to lower the load.

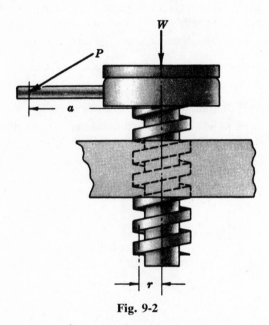

Fig. 9-2

In each case the turning moment is about the longitudinal (vertical in the figure) axis of the screw.

In case (a) the turning moment must overcome friction and raise the load W, whereas in case (b) the load W helps to overcome the friction. To raise the load W, the screw must turn counterclockwise when viewed from the top.

Let β be the lead angle, i.e., the angle whose tangent is equal to the lead divided by the mean circumference (the lead is the distance the screw moves in one turn). Let ϕ be the angle of friction. The formulas for the two cases, where r is the mean radius of the thread, are

$$(a) \quad M = Wr \tan(\phi + \beta)$$

$$(b) \quad M = Wr \tan(\phi - \beta)$$

These formulas hold also if the screw is turning at constant speed. Of course, ϕ is then the angle of kinetic friction. They also apply to a jackscrew in which a cap is added to act as a bearing for a load. To be accurate, another term should be added to the right-hand side of each formula to represent the moment necessary to overcome the friction between the cap and the screw. This extra term is of the form $W\mu r_c$, where W is the load, μ is the coefficient of friction between the cap and the screw, and r_c is the mean radius of the bearing surface between the cap and screw. See Problem 9.14 for an application of this formula. This is an approximation for the more accurate expression for collar frictional moment.

9.4 BELT FRICTION AND BRAKE BANDS

Belt friction and brake bands also illustrate the use of friction. When a belt or band passes over a rough pulley, the tensions in the belt or band on the two sides of the pulley will differ. When slip is about to occur, the following formula applies:

$$T_1 = T_2 e^{\mu\alpha}$$

where T_1 = larger tension
        T_2 = smaller tension
        μ = coefficient of friction
        α = angle of wrap in radians
        e = 2.718 (base of natural logarithms).

9.5 ROLLING RESISTANCE

Rolling resistance occurs because of the deformation of the surface under a rolling load. Figure 9-3 exaggerates this effect. A wheel of weight W and radius r is being pulled out of the depression and over the point A by the horizontal force P. Naturally this is a continuous process as the wheel rolls.

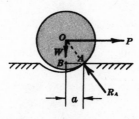

Fig. 9-3

A summation of moments about point A yields the following equation:

$$\sum M_A = 0 = W \times a - P \times (OB)$$

Since the depression is actually very small, the distance (OB) may be replaced by r; hence,

$$P \times r = W \times a$$

The horizontal component of the surface reaction R is equal to P by inspection, and is called rolling resistance. The distance a is called the coefficient of rolling resistance, and is expressed in inches or millimeters. Values of the coefficients for various materials have been tabulated, but the results have not been uniform.

Solved Problems

9.1. The 12-ft-long ladder AB weights 30 lb. It rests against a vertical wall and on the horizontal floor, as is shown in Fig. 9-4(a). What must the coefficient of friction μ be for equilibrium?

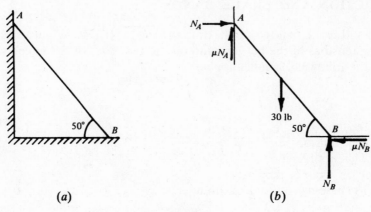

(a)   (b)

Fig. 9-4

SOLUTION

The free-body diagram in Fig. 9-4(b) shows the 30-lb weight, the normal forces at A and B, and the limiting frictional forces μN_A and μN_B. The equations of equilibrium are

$$\sum F_x = 0 = N_A - \mu N_B \tag{1}$$

$$\sum F_y = 0 = N_B + \mu N_A - 30 \tag{2}$$

$$\sum M_A = 0 = -30(6\cos 50°) + N_B(12\cos 50°) - \mu N_B(12\sin 50°) \tag{3}$$

Substitute the value of N_A from equation (1) into equation (2) to find $N_B = 30/(1+\mu^2)$.
Use this value of N_B in equation (3) to get

$$-30(6\cos 50°) + \frac{30}{1+\mu^2}(12\cos 50°) - \frac{30}{1+\mu^2}\mu(12\sin 50°) = 0$$

This yields $\mu = 0.36$.

9.2. Determine the smallest angle θ for equilibrium of a homogeneous ladder of length l leaning against a wall. The coefficient of friction for all surfaces is μ.

SOLUTION

Assuming slipping to impend, draw a free-body diagram of the ladder. There are three unknowns: N_1, N_2 and θ (see Fig. 9-5). The three equations of equilibrium are

$$\sum F_h = 0 = N_1 - \mu N_2 \tag{1}$$

$$\sum F_v = 0 = N_2 - W + \mu N_1 \tag{2}$$

$$\sum M_B = 0 = -W\tfrac{1}{2}l\cos\theta + N_2 l\cos\theta - \mu N_2 l\sin\theta \tag{3}$$

Substitute $N_1 = \mu N_2$ into equation (2) to get $N_2 = W/(1 + \mu^2)$. Substitute this value of N_2 into equation (3) to get $\theta = \tan^{-1}[(1 - \mu^2)/2\mu]$. This is the critical value of θ below which slip will occur.

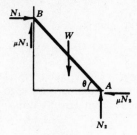

Fig. 9-5

9.3. Determine the value of P that will cause the 70-kg block in Fig. 9-6(a) to move. The coefficient of static friction between the block and the horizontal surface is $\frac{1}{4}$.

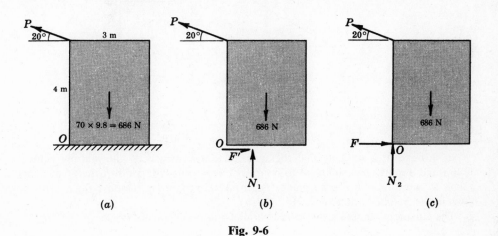

(a) (b) (c)

Fig. 9-6

SOLUTION

Motion may ensue in two ways, assuming that P is applied gradually. The block may slide to the left or it may tip about the forward edge O.

First determine the value of P to cause sliding to the left. In this case limiting friction is used as shown in Fig. 9-6(b). The equations that apply are

$$\sum F_v = 0 = P \sin 20° - 686 + N_1 \qquad (1)$$

$$\sum F_h = 0 = -P \cos 20° + F' \qquad (2)$$

Since $F' = \mu N_1 = \frac{1}{4}N$, substitute into equation (2). The equations are then

$$P \sin 20° + N_1 = 686 \qquad (3)$$

$$-P \cos 20° + \tfrac{1}{4}N_1 = 0 \qquad (4)$$

or

$$P \sin 20° + N_1 = 686 \qquad (5)$$

$$4P \cos 20° - N_1 = 0 \qquad (6)$$

Add equations (5) and (6) to obtain $P = 167$ N.

Refer to Fig. 9-6(c). Next assume that the block will tip about the forward edge at O. Determine the value of P to do this. Note that in this determination no indication of the size of the frictional force is given. It must be labeled F, as shown in Fig. 9-6(c). Since the block is assumed to tip, the normal force is at O. The equations of equilibrium are

$$\sum M_O = 0 = P\cos 20° \times 4 - 686 \times 3/2 \tag{7}$$

$$\sum F_h = 0 = -P\cos 20° + F \tag{8}$$

$$\sum F_v = 0 = +P\sin 20° + N_2 - 686 \tag{9}$$

Equation (7) yields directly $P = 274$ N. The investigation can be stopped here, since to cause sliding P must equal 167 N, whereas to cause tipping P must go to 274 N. It is thus seen that sliding will be the first to occur as P increases steadily from zero to a maximum.

9.4. Will the 200-lb block shown in Fig. 9-7(a) be held in equilibrium by the horizontal force of 300 lb? The coefficient of static friction is 0.3.

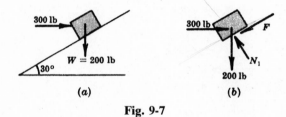

(a)   (b)

Fig. 9-7

SOLUTION

In this case it is unlikely that the force of friction is exactly the limiting value. Assume that the 300-lb force is more than sufficient to hold the block from sliding down the plane. Then it may be large enough to cause motion up the plane. To check this condition, assume that F acts down the plane as shown in the free-body diagram in Fig. 9-7(b).

The equations summing the forces parallel and perpendicular to the plane are

$$\sum F_\parallel = 0 = -F - 200\sin 30° + 300\cos 30° \tag{1}$$

$$\sum F_\perp = 0 = +N_1 - 200\cos 30° - 300\sin 30° \tag{2}$$

Solving these equations, $F = 160$ lb; $N_1 = 323$ lb.

This indicates that the value of F necessary to hold the block from moving up the plane is 160 lb. However, the maximum value obtainable is $F' = 0.3 N_1 = 0.3 \times 323 = 97$ lb. This means that the block will move up the plane. What happens after motion starts is the subject of later chapters.

9.5. What horizontal force P on the wedges B and C is necessary to raise the weight of 20 tons resting on A? See Fig. 9-8(a). Assume μ between the wedges and the ground is $\frac{1}{4}$ and μ between the wedges and A is 0.2. Also assume symmetry of loading.

SOLUTION

By considering the entire setup as a free body, it is evident that the normal force between the ground and each wedge is 10 tons or 20,000 lb.

Draw a free-body diagram of wedge B (see Fig. 9-8(b). Here the normal force of 20,000 lb is shown acting vertically up. The force of friction between the wedge and ground opposes motion, as shown.

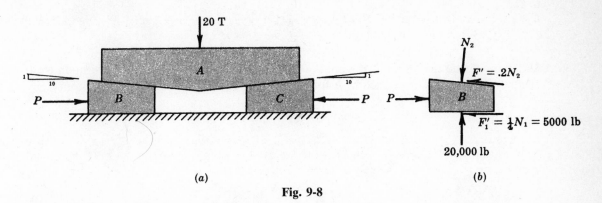

(a)　　　　　　　　　　　　　　　　　　　　　　　(b)

Fig. 9-8

Since motion impends, its value is $\frac{1}{4} \times 20,000\,\text{lb} = 5000\,\text{lb}$. The normal force of A on b is of course perpendicular to their tangent surface, while F' is the limiting value of friction drawn so as to oppose motion. This completes the free-body diagram of B, showing only forces acting on B. The equations of equilibrium are then

$$\sum F_h = 0 = P - 5000 - N_2 \frac{1}{\sqrt{101}} - 0.2N_2 \frac{10}{\sqrt{101}} \tag{1}$$

$$\sum F_v = 0 = 20,000 - N_2 \frac{10}{\sqrt{101}} + 0.2N_2 \frac{1}{\sqrt{101}} \tag{2}$$

From equation (2), $N_2 = 20,500\,\text{lb}$. Substituting into equation (1), $P = 11,100\,\text{lb}$.

9.6. Block B rests on the block A, and is attached by a horizontal rope BC to a wall as shown in Fig. 9-9(a). What force P is necessary to cause motion of A to impend? The coefficient of friction between A and B is $\frac{1}{4}$, and between A and the floor is $\frac{1}{3}$. A has a mass of 14 kg and B has a mass of 9 kg.

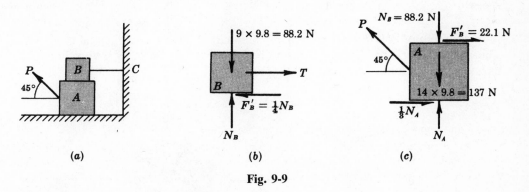

(a)　　　　　　　　　　　　(b)　　　　　　　　　　　　(c)

Fig. 9-9

SOLUTION

Since motion of A impends to the left, it will tend to drag B with it because of friction. A free-body diagram of B shows F'_B acting to the left [see Fig. 9-9(b)].

A summation of forces vertically indicates that $N_B = 88.2\,\text{N}$. Therefore, $F'_B = 22.1\,\text{N}$, and, since B does not move, the tension T is also 22.1 N.

Draw a free-body diagram of body A as shown in Fig. 9-9(c). In this figure F'_B is shown acting to the right, opposing motion. The equations of equilibrium are

$$\sum F_h = 0 = -P \cos 45° + \frac{N_A}{3} + 22.1 \tag{1}$$

$$\sum F_v = 0 = +P \sin 45° + N_A - 137 - 88.2 \tag{2}$$

Since $\sin 45° = \cos 45°$, add the two equations to obtain $N_A = 152$ N. Substitute for N_A in equation (2) to obtain $P = 104$ N.

9.7. What should be the value of the angle θ so that motion of the 40-kg block impends down the plane? The coefficient of friction μ for all surfaces is $\frac{1}{3}$. Refer to Fig. 9-10(a).

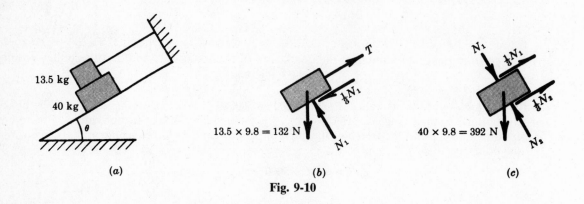

$$(a) \qquad\qquad (b) \qquad\qquad (c)$$

Fig. 9-10

SOLUTION

Draw free-body diagrams of the two weights and sum the forces parallel and perpendicular to the plane. By inspection, $N_1 = 132 \cos \theta$ in the free-body diagram of the 13.5-kg mass. The equations for the 40-kg mass are

$$\sum F_{\parallel} = 0 = -392 \sin \theta + \tfrac{1}{3}N_1 + \tfrac{1}{3}N_2 \qquad (1)$$
$$\sum F_{\perp} = 0 = N_2 - 392 \cos \theta - N_1 \qquad (2)$$

Substitute $N_1 = 132 \cos \theta$ into equations (1) and (2). Then eliminate N_2 to find $\theta = 29.1°$.

9.8. A body weighing 350 lb rests on a plane inclined 30° with the horizontal as shown in Fig. 9-11. The angle of static friction between the body and the plane is 15°. What horizontal force P is necessary to hold the body from sliding down the plane?

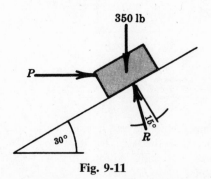

Fig. 9-11

SOLUTION

Since a slight decrease in P would cause motion down the plane, the limiting value of friction F' is used. The reaction R is shown acting at the angle of friction ϕ of 15° with the normal and in such a way as to assist the force P.

It should be noted that the angle of static friction can be used only if motion impends.

The equations of motion are obtained by summing forces parallel and perpendicular to the plane:

$$\sum F_{\parallel} = 0 = P \cos 30° - 350 \sin 30° + R \sin 15°$$

$$\sum F_{\perp} = 0 = R \cos 15° - 350 \cos 30° - P \sin 30°$$

Solving simultaneously by multiplying the first equation by $\cos 15°$ and the second equation by $\sin 15°$, we obtain $P = 93.4$ lb.

9.9. Refer to Fig. 9-12(a). Determine the necessary force P acting parallel to the plane to cause motion to impend. Assume that the coefficient of friction is 0.25 and that the pulley is smooth.

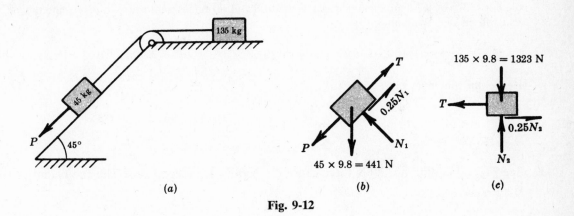

Fig. 9-12

SOLUTION

Draw free-body diagrams of the two masses in Fig. 9-12(b) and (c), indicating the tension in the rope by T. By inspection, $N_2 = 1323$ N; hence, $T = 0.25N_2 = 331$ N. Also by inspection, $N_1 = 441 \cos 45° = 312$ N; thus, $F'_1 = 78$ N. A summation of forces parallel to the inclined plane yields $\sum F = 0 = P + 441 \times 0.707 - 331 - 78$, or $P = 97.2$ N.

9.10. See Fig. 9-13(a). What is the least value of P to cause motion to impend? Assume the coefficient of friction to be 0.20.

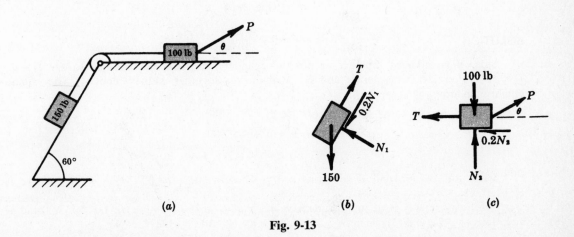

Fig. 9-13

SOLUTION

Draw free-body diagrams of both weights. Note that P is shown at an unknown angle θ with the horizontal. By inspection, $N_1 = 150 \cos 60° = 75$ lb; hence, $T = 0.20 N_1 - 150 \cos 30° = 145$ lb. The equations of equilibrium for the 100-lb weight are

$$\sum F_h = 0 = P \cos \theta - 0.20 N_2 - 145 \tag{1}$$

$$\sum F_v = 0 = P \sin \theta + N_2 - 100 \tag{2}$$

Eliminate N_2 between these two equations to obtain

$$P = \frac{165}{\cos \theta + 0.20 \sin \theta}$$

The value of P will be a minimum when the denominator $(\cos \theta + 0.20 \sin \theta)$ is a maximum. Take the derivative of the denominator with respect to θ and set this equal to zero to determine the value of θ that will make P a minimum:

$$\frac{d}{d\theta} (\cos \theta + 0.20 \sin \theta) = -\sin \theta + 0.20 \cos \theta = 0 \qquad \text{or} \qquad \theta = \tan^{-1} 0.20 = 11°20'$$

Hence, the minimum value of P is

$$P = \frac{165}{\cos 11°20' + 0.20 \sin 11°20'} = 162 \text{ lb}$$

9.11. See Fig. 9-14. Will the 180-N force cause the 100-kg cylinder to slip? The coefficient of friction is 0.25.

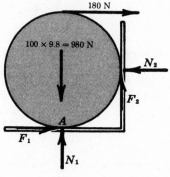

Fig. 9-14

SOLUTION

Since it is unknown whether or not the cylinder slips, it is not possible to say $F_1 = \mu N_1$ and $F_2 = \mu N_2$. Therefore, consider F_1 and F_2 as unknowns together with N_1 and N_2. The equations of equilibrium are

$$\sum F_h = 0 = F_1 - N_2 + 180 \tag{1}$$

$$\sum F_v = 0 = N_1 + F_2 - 980 \tag{2}$$

$$\sum M_A = 0 = -180 \times 2r + F_2 \times r + N_2 \times r \tag{3}$$

Solve for N_1, N_2, and F_1 in terms of F_2. These values are $N_1 = 980 - F_2$, $N_2 = 360 - F_2$, and $F_1 = 180 - F_2$.

Let us assume F_2 is at its maximum value, that is, $0.25 N_2$, and solve for N_2, N_1, and F using equations (1), (2), and (3). Then $N_2 = 288$ N, $N_1 = 908$ N, and $F_1 = 108$ N.

This means that if F_2 assumes its maximum static value then F_1 must be 108 N to hold the system in equilibrium. Since the maximum value of F_1 obtainable is $0.25N_1 = 227$ N, the cylinder will not rotate.

9.12. See Fig. 9-15(a). What must be the coefficient of friction between the gripping surfaces of the tongs and the mass m to prevent slipping?

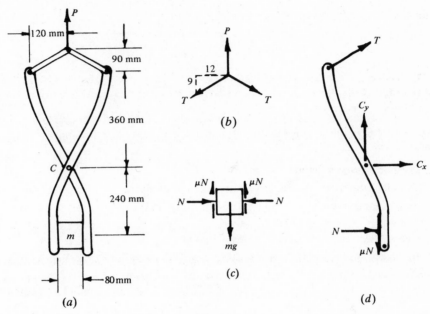

Fig. 9-15

SOLUTION

A free-body diagram of the tongs and mass m would indicate that $P = mg$.
The free-body diagram of the top pin in Fig. 9-15(b) shows that

$$\sum F_y = 0 = P - 2T \times \frac{9}{15} \qquad \text{or} \qquad T = \frac{5mg}{6}$$

The free-body diagram of the mass m in Fig. 9-15(c) shows that

$$\sum F_y = 0 = 2\mu N - mg \qquad \text{or} \qquad \mu N = \frac{mg}{2}$$

Draw a free-body diagram of one arm of the tongs [refer to Fig. 9-15(d)]. Here

$$+\sum M_C = 0 = -T \times \frac{12}{15} \times 360 - T \times \frac{9}{15} \times 120 + N \times 240 - \mu N \times 40$$

Substituting,
$$-\frac{5mg}{6} \times \frac{12}{15} \times 360 - \frac{5mg}{6} \times \frac{9}{15} \times 120 + N \times 240 - \frac{mg}{2} \times 40 = 0$$

Hence,
$$N = \frac{(240 + 60 + 20)mg}{240} = \frac{4mg}{3}$$

Finally,
$$\mu = \frac{\mu N}{N} = \frac{mg/2}{4mg/3} = 0.38$$

9.13. Refer to Fig. 9-16. The coefficient of friction between a copper block A and an aluminum block B is 0.3, and between the block B and the floor is 0.2. The mass of block A is 3 kg and of block B is 2 kg. What force P will cause the motion of block A to impend?

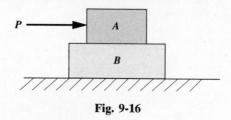

Fig. 9-16

SOLUTION

This represents a class of problem that is characterized as a multiple-mode slip problem, in that there is more than one way for motion to impend. The procedure for analysis can be summarized in four steps.

1. Assume a possible mode of slipping.
2. Calculate the friction necessary to prevent slip at all the other possible slip surfaces.
3. Test whether there is sufficient friction at these other surfaces to prevent slip.
4. If there is enough friction, the problem is solved. If there is not enough friction to prevent slip, choose a second possible mode of slipping and repeat the process.

In this problem let us assume that slip will impend at the surface between A and B, but not between B and the floor. The free-body diagrams are shown in Fig. 9-17. Note that since we assume no impending slip between block B and the floor, $F'_B \neq 0.2 N_B$. The equations of equilibrium are

For A,

$$\sum F_v = N_A - 29.4 = 0$$
$$\sum F_h = P - 0.3 N_A = 0$$

For B,

$$\sum F_v = N_B - N_A - 19.6 = 0$$
$$\sum F_h = 0.3 N_A - F_B = 0$$

Solving yields $F_B = 8.82$ N and $N_B = 49$ N. The largest value of friction before slip impends is $F'_B = 0.2 N_B = 9.8$ N. Since we need only 8.82 N of friction for equilibrium but can generate 9.8 N, block B is not at impending slip and so the initial assumption is correct. If $F_B > F'_B$ then slip would occur and the initial assumption would be incorrect. In that case the process would be repeated with $F'_B = 0.2 N_B$ and $F'_A \neq 0.3 N_A$.

From the equilibrium equation for P, $P = 8.82$ N.

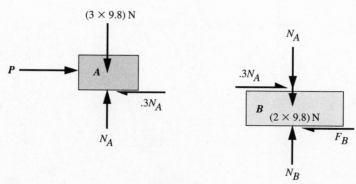

Fig. 9-17

9.14. See Fig. 9-18. Two cylinders, each of weight W and radius r, rest in a box of width $3r$. The sides of the box are smooth. The coefficient of friction between cylinder A and the bottom of the box is 0.12 and between the two cylinders is 0.3. Find the couple M that should be applied to cylinder B for motion of B to impend.

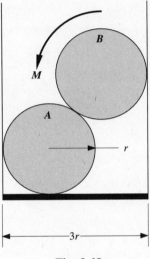

Fig. 9-18

SOLUTION

There are two possible modes of impending slip for cylinder B:

(a) counterclockwise rotation of B, no rotation of A;

(b) counterclockwise rotation of B, clockwise rotation of A.

Choose (a) as the mode of impending motion. For this case, in the free-body diagrams of Fig. 9-19, $F_2 = 0.3N_2$, but $F_3 \neq 0.12N_3$.

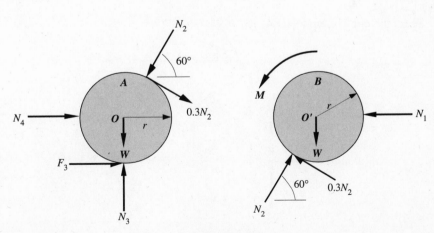

Fig. 9-19

The equilibrium equations become

For A,

$$\sum F_v = N_3 - N_2 \sin 60° - 0.3N_2 \cos 60° - W = 0$$

$$\sum F_h = N_4 + F_3 - N_2 \cos 60° + 0.3N_2 \sin 60° = 0$$

$$\sum M_O = F_3 r - 0.3N_2 r = 0$$

For B,

$$\sum F_v = N_2 \sin 60° + 0.3N_2 \cos 60° - W = 0$$

$$\sum F_h = -N_1 + N_2 \cos 60° - 0.3N_2 \sin 60° = 0$$

$$\sum M_{O'} = M - 0.3N_2 r = 0$$

Solving for F_3 and N_3 yields $F_3 = 0.295W$, $N_3 = 2W$. Now, $F_3' = \mu_3 N_3 = 0.24W$, and since $F_3 > F_3'$, there is insufficient friction between cylinder A and the floor to prevent slip there, and so assumption (a) is false. This means that (b) is the mode of impending motion for cylinder B.

The free-body diagrams are shown in Fig. 9-20. The equilibrium equations become

For A,

$$\sum F_v = N_3 - F_2 \cos 60° - N_2 \sin 60° - W = 0$$

$$\sum F_h = N_4 + 0.12N_3 + F_2 \sin 60° - N_2 \cos 60° = 0$$

$$\sum M_O = 0.12N_3 r - F_2 r = 0$$

For B,

$$\sum F_v = N_2 \sin 60° + F_2 \cos 60° - W = 0$$

$$\sum F_h = -N_1 + N_2 \cos 60° - F_2 \sin 60° = 0$$

$$\sum M_{O'} = M - F_2 r = 0$$

Solving, $F_2 = 0.24W$ and $M = 0.24Wr$.

To check that (b) is indeed the mode of slip, calculate $N_2 = 1.02W$ and $F_2' = 0.306W$. So $F_2' > F_2$ and there is no impending slip between the cylinders.

A possible mode of motion other than the two studied above is that cylinder B will roll up and over cylinder A. For this *not* to happen, N_1 must have a positive value. The equilibrium equation of horizontal forces for cylinder B in Fig. 9-20 gives

$$\sum F_h = -N_1 + N_2 \cos 60° - F_2 \sin 60° = 0 \qquad N_1 = 0.302W$$

This means that cylinder B does not leave the wall.

Finally, the couple M for impending motion is $M = 0.24Wr$.

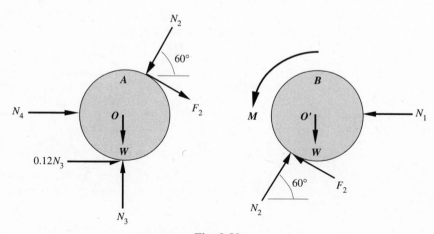

Fig. 9-20

9.15. The mean diameter of the threads of a square-threaded screw is 2 in. The pitch of the thread is $\frac{1}{4}$ in. The coefficient of friction $\mu = 0.15$. What force must be applied at the end of a 2-ft lever that is perpendicular to the longitudinal axis of the screw to raise a load of 4000 lb? What force must be applied to lower the load?

SOLUTION

To raise the load, apply formula (a) under Section 9.3: $M = Wr \tan(\phi + \beta)$.

The turning moment M is equal to the product of the force and the length of the lever. ϕ is the angle whose tangent is 0.15, or $\phi = 8.5°$.

$$\beta = \tan^{-1} \frac{\text{lead}}{\text{mean circumference}} = \tan^{-1} \frac{0.25}{\pi 2} = \tan^{-1} 0.0397 = 2.27°$$

$M = P \times 24 = 4000 \times 1 \tan(8.53° + 2.27°)$. Hence, $P = 31.8$ lb to raise the load.
To determine the force to lower the load, apply the formula $M = Wr \tan(\phi - \beta)$.
$M = P \times 24 = 4000 \times 1 \tan(8.53° - 2.27°)$. Hence, $P = 18.3$ lb to lower the load.

9.16. A jackscrew has 4 threads per inch. The mean radius of the threads is 2.338 in. The mean diameter of the bearing surface under the cap is 3.25 in. The coefficient of friction for all surfaces is 0.06. What turning moment is necessary to raise 1500 lb?

SOLUTION

As indicated in the theoretical discussion, a term must be added to account for the additional moment necessary to overcome the friction between the cap and the screw.

$M = Wr \tan(\phi + \beta) + \mu Wr_c$, where r_c is the mean radius of the bearing surfaces between cap and screw. $\phi = \tan^{-1} 0.06 = 3.43°$. $\beta = \tan^{-1} 0.25/(2\pi \times 2.338) = 1.00°$.
$M = 1500 \times 2.338 \tan(3.43° + 1.00°) + 0.06 \times 1500 \times (3.25/2) = 418$ lb-in.

9.17. Two pulleys, each of 750-mm diameter, are connected by a belt so that each has the belt around half of its circumference. The tension in the tight side of the belt is 200 N. If the coefficient of friction is 0.25, determine the tension in the slack side when the belt is about to slip.

SOLUTION

The angle of wrap for either pulley is 180° or π rad. Using the equation $T_1 = T_2 e^{\alpha\mu}$, we have

$$200 = T_2 \times e^{\pi(0.25)} \qquad \text{or} \qquad T_2 = 91.2 \text{ N}$$

9.18. In Fig. 9-21(a) a drum 635 mm in diameter is encircled by a brake band, which is tightened by a vertical force P of 178 N on the lever AC. Assume that the coefficient of friction between the drum and the band is $\frac{1}{3}$. Neglect friction on all other surfaces. Determine the net braking moment on the drum if rotation is impending clockwise.

SOLUTION

In solving with rotation of the drum impending clockwise, note that friction F on the drum acts to oppose motion, i.e., counterclockwise. This same friction acts on the belt in an opposite or clockwise direction. In considering the free-body diagram of the belt, the band tension T_C must be greater than T_B, since it holds both T_B and F in equilibrium.

Taking moments about C of the forces acting on the lever, we obtain

$$\sum M_C = 0 = +T_B \sin 60° \times 100 - 178 \times 760 \qquad \text{or} \qquad T_B = 1560 \text{ N}$$

The angle of wrap must be found before the formula for tensions can be applied. Figure 9-22 is drawn to show the trigonometry of the problem. First we determine the value of θ.

In Fig. 9-22(a),

$$CD = \frac{DF}{\sin 30°} = \frac{DE - EF}{\sin 30°} = \frac{317.5 - HE \cos 30°}{\sin 30°} = \frac{317.5 - 100 \cos 30°}{\sin 30°} = 462 \text{ mm}$$

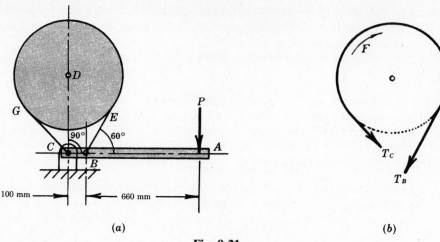

(a)   (b)

Fig. 9-21

From Fig. 9-22(b), $\theta = \sin^{-1} GD/CD = \sin^{-1} 317.5/462 = 43.4°$.
From Fig. 9-22(c), angle of wrap $= 180° + 30° + 43.4° = 253.4° = 4.4227$ rad.
T_C is greater than T_B. Hence, $T_C = T_B e^{\alpha\mu} = 6814$ N.
Hence, the braking moment is $(T_C - T_B) \times 317.5 = (6814 - 1560) \times 317.5 = 1\,670\,000$ N $\cdot$ mm $= 1670$ N $\cdot$ m.

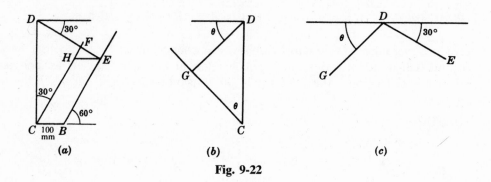

(a)   (b)   (c)

Fig. 9-22

9.19. Four turns of rope around a horizontal post will hold a 1000-lb weight with a pull of 10 lb. Determine the coefficient of friction between the rope and the post.

SOLUTION

Use the equation $T_1 = T_2 e^{\alpha\mu}$, where $T_1 = 1000$ lb and $T_2 = 10$ lb. Hence, $e^{\alpha\mu} = 100$.
But $\alpha =$ the angle of wrap $= 4 \times 2\pi$ rad.
From exponential tables, $e^{4.605} = 100$. Hence, $\alpha\mu = 8\pi\mu = 4.605$, or $\mu = 0.18$.

9.20. What force is necessary to hold a mass of 900 kg suspended on a rope wrapped twice around a post? Assume the coefficient of friction to be $\mu = 0.20$.

SOLUTION

Use the equation $T_1 = T_2 e^{\alpha\mu}$, where T_1 is the larger force $(900 \times 9.8$ N$)$. Hence, the holding force $T_2 = 8820/e^{\alpha\mu}$, where $\alpha = 2 \times 2\pi$ rad and $\mu = 0.20$. The solution is $T_2 = 714$ N.

9.21. A steel wheel 760 mm in diameter rolls on a horizontal steel rail. It carries a load of 500 N. The coefficient of rolling resistance is 0.305 mm. What is the force P necessary to roll the wheel along the rail?

SOLUTION

$$P = \frac{Wa}{r} = \frac{500 \times 0.305}{380} = 0.4 \text{ N}$$

9.22. A circular shaft of diameter D in is resting in a support as shown in Fig. 9-23(a). The shaft supports a weight W lb. Assuming the coefficient of friction is μ, determine the twisting moment M necessary to cause rotary motion to impend.

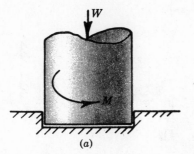 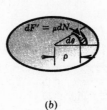

(a)   (b)

Fig. 9-23

SOLUTION

The resisting friction dF acting on the differential area $dA = \rho \, d\rho \, d\theta$ is shown in Fig. 9-23(b). The moment of this friction about the centerline of the shaft is $\rho \, dF'$.

But the normal force on this differential area is the product of the area and the unit load, which is the total weight divided by the total area. Thus, we can write

$$dF' = \mu \, dN = \mu \frac{W}{\frac{1}{4}\pi D^2} \rho \, d\rho \, d\theta$$

or

$$M = \int_0^{D/2} \int_0^{2\pi} \rho \frac{4\mu W}{\pi D^2} \rho \, d\rho \, d\theta$$

$$= \int_0^{D/2} \frac{4\mu W}{\pi D^2} \rho^2 \, d\rho \, (2\pi) = \frac{8\mu W}{D^2} \left[\frac{\rho^3}{3} \right]_0^{D/2} = \frac{\mu WD}{3}$$

Supplementary Problems

9.23. A clamp exerts a normal force of 100 N on three pieces held together as shown in Fig. 9-24. What force P may be exerted before motion impends? The coefficient of friction between the pieces is 0.30
Ans. $P = 60$ N

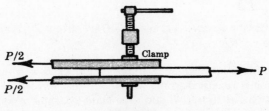

Fig. 9-24

9.24. A homogeneous ladder 18 ft long and weighing 120 lb rests against a smooth wall. The angle between the ladder and the floor is 70°. The coefficient of friction between the floor and the ladder is $\frac{1}{4}$. How far up the ladder can a 180-lb person walk before the ladder slips?   *Ans.*   14.6 ft

9.25. A person can pull horizontally with a force of 100 lb. An 800-lb weight is resting on a horizontal surface for which the coefficient of friction is 0.20. The vertical cable of a crane is attached to the top of the block as shown in Fig. 9-25. What will be the tension in the cable if the person is just able to start the block to the right?   *Ans.* $T = 300$ lb

9.26. Refer to Fig. 9-26. The wedge B is used to raise the load of 200×9.8 N resting on block A. What horizontal force P is required to do this if the coefficient of friction for all surfaces is 0.2?   *Ans.*   1510 N

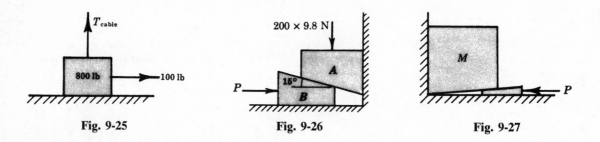

Fig. 9-25   Fig. 9-26   Fig. 9-27

9.27. What force P is needed to drive the 5° wedge to raise the 500-kg mass shown in Fig. 9-27. The coefficient of friction for all surfaces is 0.25.   *Ans.* $P = 3190$ N

9.28. In Fig. 9-28 a 5° wedge is shown splitting a log. The coefficient of friction between the log and the wedge is 0.2. If the blow is equivalent to 200 lb, what is the splitting force normal to the wedge?
Ans.   *Ans.* $N = 410$ lb

9.29. Refer to Fig. 9-29. Block A having a mass of 45 kg rests on block B having a mass of 90 kg, and is tied with a horizontal string to the wall at C. If the coefficient of friction between A and B is $\frac{1}{4}$ and that between B and the surface is $\frac{1}{3}$, what horizontal force P is necessary to move block B?
Ans. $P = 550$ N

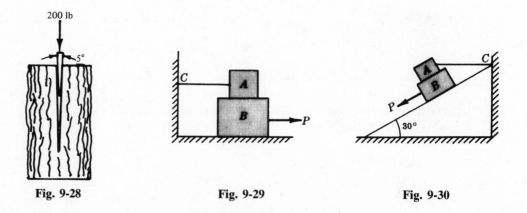

Fig. 9-28   Fig. 9-29   Fig. 9-30

9.30. Refer to Fig. 9-30. Block A weighing 60 lb rests on block B weighing 80 lb. Block A is restrained from moving by a horizontal rope tied to the wall at C. What force P parallel to the plane inclined 30° with the horizontal is necessary to start B down the plane? Assume μ for all surfaces to be $\frac{1}{3}$.
Ans. $P = 40.3$ lb

9.31. Refer to Fig. 9-31. Cube A having a mass of 8 kg is 100 mm on a side. Angle $\theta = 15°$. If the coefficient of friction is $\frac{1}{4}$, will the cube slide or tip as the force P is gradually increased?
Ans.   Slide: $P_{\text{tipping}} = 48.0\,\text{N}$, $P_{\text{sliding}} = 39.2\,\text{N}$

9.32. Refer to Fig. 9-32. The mass of A is 23 kg; the mass of B is 36 kg. The coefficients of friction are 0.60 between A and B, 0.20 between B and the plane, and 0.30 between the rope and the fixed drum. Determine the maximum mass of M before motion impends.   *Ans.* $M = 18.9\,\text{kg}$

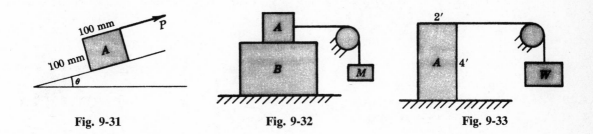

Fig. 9-31   Fig. 9-32   Fig. 9-33

9.33. Refer to Fig. 9-33. The homogeneous body A weighs 120 lb. The coefficients of friction are 0.30 between A and the plane and $2/\pi$ between the rope and the drum. What value of W will cause motion of A to impend?   *Ans.* $W = 81.5\,\text{lb}$

9.34. A rope holding a 50-lb weight E passes over a pulley and is attached to a frame at A as shown in Fig. 9-34. The weight of C is 60 lb. What is the minimum coefficient of friction μ between the outer rope and the pulley for equilibrium?   *Ans.* $\mu = 0.291$

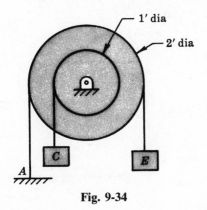

Fig. 9-34

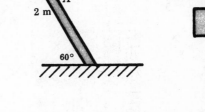

Fig. 9-35

9.35. Refer to Fig. 9-35. The homogeneous bar A having a mass of 18 kg is 2 m long and is inclined 60° with the horizontal plane. The mass M of 7 kg is connected by the rope to the bar. The rope leaving the bar is horizontal. The coefficient of friction between the rope and the drum B is 0.20. What is the minimum coefficient of friction between the bar and the plane for equilibrium?   *Ans.* $\mu = 0.284$

9.36. Refer to Fig. 9-36. The angle of static friction between the block with mass M and the plane is ϕ. For the angles shown in the figure, what is the expression for the force P to move the block up the plane?
Ans. $P = [9.8M \sin(\theta + \phi)]/[\cos(\beta - \phi)]$

9.37. In Problem 9.36, show that the minimum value of P for a given angle θ is $9.8M \sin(\theta + \phi)$.

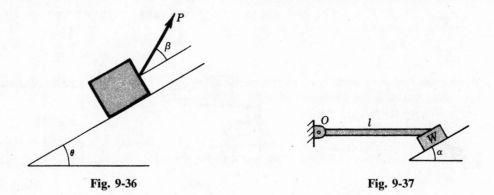

Fig. 9-36   Fig. 9-37

9.38. A homogeneous bar of length l and weight w lb rests horizontally as shown in Fig. 9-37 with its free end on a block of weight W. the block W is at rest on a plane inclined at the angle α with the horizontal. Determine the necessary coefficient of friction μ between the block and the plane for equilibrium. Assume no friction between the bar and the block.   *Ans.* $\mu = (\sin 2\alpha)/(w/W + 2 \cos^2 \alpha)$

9.39. In Fig. 9-38 a cone clutch is drawn showing the mating surfaces. Assume that the mating parts are forced into contact with a normal pressure on the mating surfaces of 10 psi. The area of contact is the product of the 2-in dimension and the mean circumference of the mating parts, that is, $2 \times \pi \times 8$. The total normal force in pounds is 10 times the area just found. If the coefficient of friction is $\mu = 0.35$, determine the force of friction between the mating surfaces.   *Ans.* $F = 176$ lb

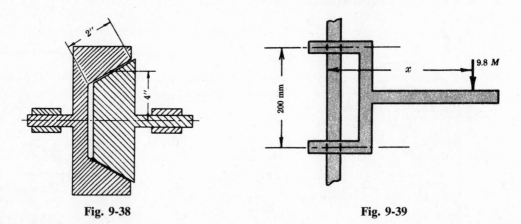

Fig. 9-38   Fig. 9-39

9.40. The lift shown in Fig. 9-39 slides on a vertical shaft 75 mm square. How far out on the platform can a weight $9.8M$ be placed so that the lift will slide down without binding on the shaft? Assume $\mu = 0.25$.
Ans. $x = 438$ mm

9.41. Refer to Fig. 9-40. The homogeneous 360-kg block rests against the vertical wall, for which the coefficient of friction is 0.25. A force P is applied to the midpoint of the right face in the direction shown in the figure. What range of values may P have without disturbing equilibrium?
Ans. $5260 \text{ N} < P < 7890 \text{ N}$

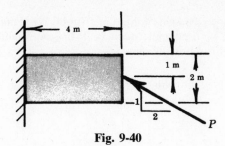

Fig. 9-40

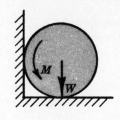

Fig. 9-41

9.42. What couple M is needed to cause the wheel of weight W and radius r shown in Fig. 9-41 to have impending motion? The coefficient of friction for all surfaces is μ.
Ans. $M = \mu W r(1 + \mu)/(1 + \mu^2)$

9.43. A vertical force P exerted on a lever AB holds a 20-kg mass M from falling, as shown in Fig. 9-42. The coefficient of friction between the lever and the 300-mm drum is $\frac{1}{4}$. Neglect the mass of the drum and lever, and determine the force P to hold the mass.   *Ans.* $P = 327$ N

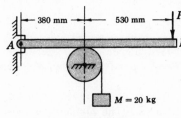

Fig. 9-42

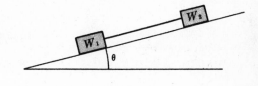

Fig. 9-43

9.44. In Fig. 9-43, W_1 weighs 50 lb and W_2 weighs 30 lb. They are tied together by a rope parallel to the plane. The coefficient of friction between W_1 and the plane is $\frac{1}{4}$, and between W_2 and the plane is $\frac{1}{2}$. Determine the value of the angle θ at which sliding will occur. What is the tension in the rope?
Ans. $\theta = 19.0°$, $T = 4.4$ lb

9.45. Determine the force P to cause motion to impend if the coefficient of friction for both blocks and the plane shown in Fig. 9-44 is 0.25. The force P and the ropes are parallel to the plane. The pulley is frictionless.   *Ans.* $P = 6.6$ N

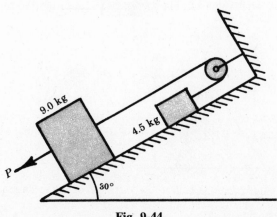

Fig. 9-44

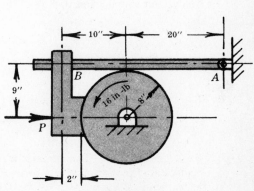

Fig. 9-45

9.46. The drum shown in Fig. 9-45 is subjected to a counterclockwise torque of 16 in-lb. What horizontal force P is necessary to resist motion? The coefficient of friction between both braking members (pinned together at B) and the drum is 0.40. Neglect weight of braking members.   *Ans.* $P = 3.43$ lb

9.47. A body with a mass of 30 kg rests on a plane inclined 45° with the horizontal. The coefficient of friction is $\frac{1}{3}$. What is the range of values of a horizontal force P that will keep the mass from moving up or down the plane?   *Ans.* $147\,\text{N} < P < 588\,\text{N}$

9.48. A plane is inclined at an angle θ with the horizontal. A body can just rest on the plane. Determine the least force P to draw the body up the plane. (*Hint:* $\theta = \alpha$, since θ is the angle of repose; also assume that P acts at angle β with the plane.)   *Ans.* $P = W \sin 2\theta$

9.49. See Fig. 9-46. The uniform bar has a mass of 35 kg. What rightward force P is needed to start the bar moving? The coefficient of friction for all surfaces is 0.30.   *Ans.* $P = 246$ N

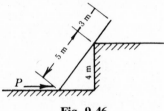

Fig. 9-46

9.50. Two blocks of equal weight W can slide on a horizontal surface. The coefficient of friction between the blocks and the surface is μ. A string of length l is suspended between the blocks and carries a weight M at its midpoint. How far apart will the blocks be for equilibrium?

Ans. $x = \dfrac{l\mu(W + M/2)}{\sqrt{(M/2)^2 + \mu^2(W + M/2)^2}}$

9.51. A prism whose cross section is a regular polygon of n sides rests on a horizontal face. An insect is crawling up the inside. The coefficient of friction between the insect and the inner faces is μ. Show that the highest face that the insect can climb (counting the horizontal bottom face as one) is given by the expression $n/360 \times \tan^{-1}(\mu + 360/n)$.

9.52. A weightless bar A is pinned to a homogeneous 600-lb prism B at point C as shown in Fig. 9-47. A horizontal force P is applied 2 ft above the horizontal plane. If the coefficient of friction between the plane and either the bar or the prism is 0.4, what value of P is necessary to cause motion to impend? Analyze for slipping and tipping of the prism.   *Ans.* $P = 209$ lb

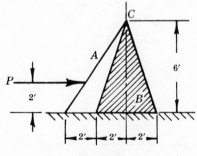

Fig. 9-47

9.53. The uniform, thin slab in Fig. 9-48 weighs 0.25 lb/in. Given the coefficients of friction at the support surfaces as shown, what couple M will cause the motion of the disk to impend counterclockwise. The radius of the disk is 6 in.   *Ans.*   9.72 lb-in

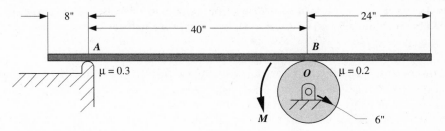

Fig. 9-48

9.54. Repeat Prob. 9.53 with the couple M clockwise and the coefficient of friction at the disk equal to 0.1. *Ans.*   7.56 lb-in

9.55. Applying the results of Prob. 9-53, what horizontal force will cause the motion of the slab to impend to the right?   *Ans.*   3.24 lb

9.56. In Fig. 9-49 a bar of weight 12 lb and length 6 ft is supported 1 ft from its left end and rests on a 6-in radius wheel at its right end. The weight of the wheel is 4 lb. The coefficients of friction are as shown. Find the magnitude of the force P that will cause impending motion of the wheel.   *Ans.*   0.96 lb

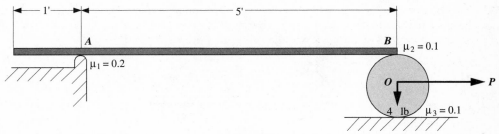

Fig. 9-49

9.57. Solve Prob. 9.56 with $\mu_1 = \mu_2 = 0.15$.   *Ans.*   1.44 lb

9.58. In Fig. 9-50 the block A weighs 15 N and wheel B weighs 20 N. A and B are connected by a weightless link. The coefficient of friction under A is 0.25 and under B is 0.15. The diameter of the wheel is 1 m. What is the value of the couple M to cause impending motion of B?   *Ans.*   1.5 N · m

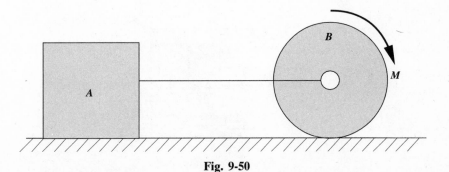

Fig. 9-50

9.59. In Prob. 9.58 what coefficient of friction under wheel B will cause motion to impend at block A? What moment M will then cause motion of the system?   *Ans.* $\mu_B = 0.19$, $M = 1.87\,\text{N} \cdot \text{m}$

9.60. A jackscrew has 3 threads per inch, with the mean thread radius equal to 0.648 in. A lever 18 in long is used to raise or lower a load of 2400 lb. If the coefficient of friction is 0.10, what force perpendicular to the arm is required to (*a*) raise the load and (*b*) lower the load?   *Ans.*   (*a*) 15.9 lb, (*b*) 1.55 lb

9.61. Suppose that the mean radius of the bearing surface between the cap and the screw in Prob. 9.60 is 1.4 in. What value of P is necessary to raise the load if the coefficient of friction between the cap and the screw is 0.07?   *Ans.*   28.9 lb

9.62. A hand press has 5 square threads per inch. The mean diameter is 1.2 in. If the coefficient of friction is 0.08, what force in the press can be exerted by a force of 20 lb applied normal to a lever 20 in long?   *Ans.*   5000 lb

9.63. The screw in a press has 6 threads per inch. The mean dimaeter is 1.38 in and the coefficient of friction is 0.14. If forces of 40 lb are applied as a couple with a moment arm of 20 in, what force does the press exert? See Fig. 9-51.   *Ans.*   6500 lb

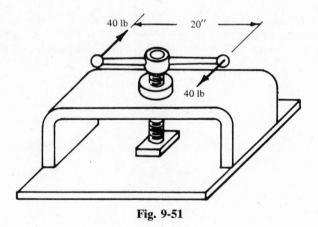

Fig. 9-51

9.64. Refer to Fig. 9-52. What force P applied perpendicular to the handle of the vise is needed to clamp the piece A with a force of 20 lb? The vise screw is square-threaded, with 10 threads per inch. The mean diameter of the screw is 0.438 in. The coefficient of friction is 0.20.   *Ans.* $P = 0.3$ lb

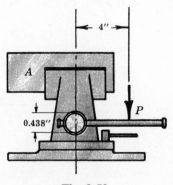

Fig. 9-52

9.65. A square-threaded jackscrew has a pitch of 0.4 in and a mean diameter of 3.0 in. The mean diameter of the bearing surface between the cap and the screw is 3.5 in. The coefficient of friction between all surfaces is 0.10. What force is required at the end of a lever 36 in long to raise 4000 lb? *Ans.* $P = 43.3$ lb

9.66. (a) Determine the load that can be raised with a single-threaded jackscrew having $2\frac{1}{2}$ threads per inch and a mean diameter of 3 in when a force of 200 lb is exerted on a bar 30 in long. Use $\mu = 0.05$. (b) If the mean diameter of the bearing surface between the cap and the screw is $3\frac{1}{2}$ in and the coefficient of friction between these surfaces is 0.12, what load can be raised by the 200-lb force? *Ans.* (a) $W = 43,100$ lb, $W = 17,200$ lb

9.67. A jackscrew has a square thread with a pitch of 0.3 in. The mean thread diameter is 2 in. The cap has an inner diameter of 2 in and an outer diameter of 3 in. If the coefficient of friction for all surfaces is 0.15, what force is necessary to cause motion of a 5000-lb load (a) upward and (b) downward? Use a bar 3 ft long. *Ans.* (a) 53.7 lb to raise, (b) 40.1 lb to lower

9.68. A square-threaded jackscrew has 2 threads per inch and a mean diameter of 2.4 in. Using a coefficient of friction of 0.08, determine the capacity of the jack if a maximum force of 30 lb is recommended at a lever arm of 18 in. *Ans.* $W = 3060$ lb

9.69. Refer to Fig. 9-53. The weight of A is 500 lb and that of B is 100 lb. What force P must be exerted perpendicular to the arm of the jack and 20 in from the jack centerline to raise A? The jackscrew has a lead of 0.32 in and a mean diameter of 2.00 in. The coefficient of friction between the jack and the screw is 0.15. *Ans.* $P = 3.04$ lb

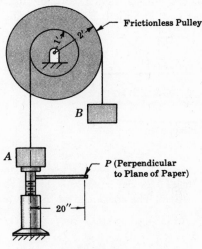

Fig. 9-53

9.70. What force is necessary to hold a mass of 50 kg suspended on a rope wrapped three times around a fixed drum? Assume $\mu = 0.3$. *Ans.* $F = 1.7$ N

9.71. In Prob. 9.70 what force is needed to raise the mass? *Ans.* $F = 140$ kN

9.72. In Fig. 9-54 a mass of 20 kg is attached to a rope wrapped $\frac{1}{4}$ turn around a fixed drum. The coefficient of friction is 0.25. Determine (a) the value of F to hold the mass from falling and (b) the value of F to start raising the mass. *Ans.* (a) 132 N, (b) 290 N

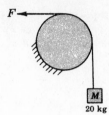

Fig. 9-54

9.73. A weight W is attached to a rope that is wrapped around a fixed drum. If the rope is wrapped twice around the drum, a force of 200 lb is required to support the weight. If the rope is wrapped three times around the drum, the required force to support the weight is 150 lb? What is the weight?   *Ans.* $W = 356$ lb

9.74. A worker lowers a 400-lb boiler section into a pit by means of a rope snubbed $1\frac{1}{4}$ turns around a horizontal pole. If the coefficient of friction is 0.35, what force must be exerted?   *Ans.* $F = 25.6$ lb

9.75. Three turns of a rope around a horizontal post with a pull of 30 N will hold a 100-kg mass. Determine the coefficient of friction between the rope and post.   *Ans.* $\mu = 0.185$

9.76. Refer to Fig. 9-55. The two pulleys are held apart by a spring with compressive force S. The diameter of each pulley is d. The coefficient of friction between the belt and the pulley is μ. Determine the maximum torque that may be transmitted.   *Ans.*   Torque $= \frac{1}{2}Sd(e^{\mu\pi} - 1)/(e^{\mu\pi} + 1)$

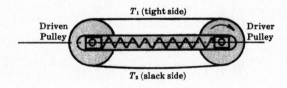

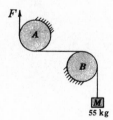

Fig. 9-55   Fig. 9-56

9.77. Refer to Fig. 9-56. A mass of 55 kg is prevented from falling by a rope wrapped $\frac{1}{4}$ turn around drum B and $1\frac{1}{4}$ turns around drum A. Assuming that drum B is smooth and the coefficient of friction between the rope and A is $\frac{1}{4}$, what is the value of the holding force F?   *Ans.* $F = 75.7$ N

9.78. In Prob. 9.77 assume that the coefficient of friction between the rope and the drum B is not zero but $\frac{1}{4}$. What is the value of the holding force F? (*Hint:* Use the free-body diagram of B to determine tension in the rope between A and B.)   *Ans.* $F = 44.8$ N

9.79. A mass of 200 kg is held by passing a rope around a horizontal post and exerting a pull of 220 N. If the coefficient of friction between the post and the rope is $\frac{1}{4}$, how many turns of the rope around the post are necessary?   *Ans.*   1.39 turns

9.80.   Refer to Fig. 9-57. Determine the range of values of tension T for equilibrium. The coefficient of friction between the belt and each fixed drum is $1/\pi$. The tension at the other end of the belt is 15 N.
   Ans. $1.23\,\text{N} < T < 183\,\text{N}$

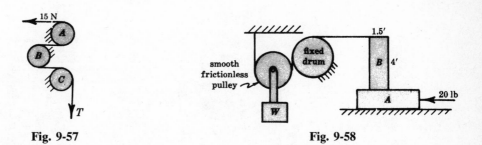

| Fig. 9-57 | Fig. 9-58 |

9.81.   In Fig. 9-58, body A weighs 100 lb and body B weighs 300 lb. The coefficient of friction between A and the plane is 0.20. The coefficient of friction between A and B is 0.20. The coefficient of friction between the rope and the drum is 0.25. Determine the minimum weight W to cause motion of B to impend.
   Ans. $W = 167\,\text{lb}$

9.82.   A platform of negligible weight is held in a horizontal position by a rope fastened to each end and passing over fixed drums, as shown in Fig. 9-59. If the coefficient of friction between the rope and each drum is 0.20, determine how far from the center a 10-lb weight can be placed without upsetting the balance.   *Ans.* $x = 1.82\,\text{ft}$

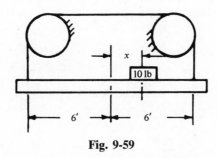

Fig. 9-59

9.83.   What horizontal pull is necessary to move a 1200-ton train along a level track if the coefficient of rolling resistance is 0.009 in and the wheels are 3 ft in diameter?   *Ans.* $P = 1200\,\text{lb}$

9.84.   A brake band encircling drum D is fastened to a horizontal lever at B and C, as shown in Fig. 9-60. The diameter of the drum is 450 mm. The coefficient of friction between the brake band and the drum is $\frac{1}{3}$. Force P is 30 N. What braking moment on the drum is possible (a) if the drum is rotating clockwise, (b) if the drum is rotating counterclockwise?   *Ans.* $M_a = 10.2\,\text{N} \cdot \text{m}$, $M_b = 29.1\,\text{N} \cdot \text{m}$

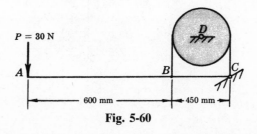

Fig. 5-60

9.85. Refer to Fig. 9-61. What force P is necessary to cause the brake to resist motion of the drum under the action of the moment M? The coefficient of friction between the brake and the drum is μ. The angle of wrap is α.   *Ans.* $P = (M/rc)/[(ae^{\mu\alpha} - b)/(e^{\mu\alpha} - 1)]$

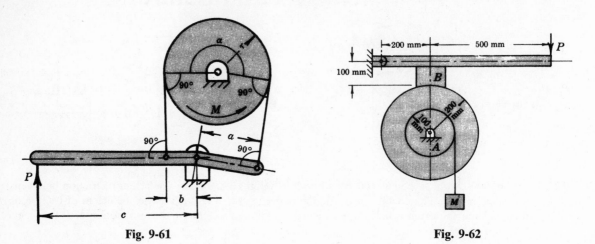

Fig. 9-61   **Fig. 9-62**

9.86. A mass $M = 90$ kg hangs on a compound pulley A free to rotate in frictionless bearings as shown in Fig. 9-62. The coefficient of friction between the facing of brake B and the pulley is 0.25. What minimum force P is needed to prevent rotation?   *Ans.* $P = 567$ N

9.87. A mass M of 1400 kg rests on an oak beam. The beam rests on 200 mm-diameter rollers. Assuming the coefficient of rolling resistance between the beam and the rollers to be 0.89 mm, what horizontal force is necessary to move the load on a level surface?   *Ans.* $P = 122$ N

9.88. A wheel 500 mm in diameter carries a load of 20 000 N. If a horizontal force of 20 N is necessary to move it over a level surface, determine the coefficient of rolling resistance.   *Ans.* $a = 0.25$ mm

9.89. An automobile weighing 3900 lb has 29 in-diameter wheels. Assuming a coefficient of rolling resistance between the tires and the road of 0.02 in, determine the force necessary to overcome rolling friction on a level road.   *Ans.* $P = 5.4$ lb

9.90. A central horizontal force of 1.4 N is necessary to move a drum 900 mm in diameter on a level surface. Assuming that the coefficient of rolling resistance is 0.635 mm, what is the mass of the drum?
Ans. $M = 101$ kg

9.91. The collar bearing shown in Fig. 9-63 supports a load of 680 lb. If the coefficient of friction is 0.20 and if uniform distribution of pressure is assumed, what turning moment M is necessary?
Ans. $M = 212$ lb-in

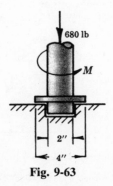

Fig. 9-63

Chapter 10

First Moments and Centroids

10.1 CENTROID OF AN ASSEMBLAGE

The centroid of an assemblage of n similar quantities, Δ_1, Δ_2, Δ_3, ..., Δ_n situated at points $P_1, P_2, P_3, \ldots, P_n$ for which the position vectors relative to a selected point O are $\mathbf{r}_1, \mathbf{r}_2, \mathbf{r}_3, \ldots, \mathbf{r}_n$ has a position vector $\bar{\mathbf{r}}$ defined as

$$\bar{\mathbf{r}} = \frac{\sum\limits_{i=1}^{n} \mathbf{r}_i \Delta_i}{\sum\limits_{i=1}^{n} \Delta_i} \tag{1}$$

where
$\Delta_i = i$th quantity (for example, this could be an element of length, area, volume, or mass)
$\mathbf{r}_i = $ position vector of ith element
$\sum\limits_{i=1}^{n} \Delta_i = $ sum of all n elements
$\sum\limits_{i=1}^{n} \mathbf{r}_i \Delta_i = $ first moment of all elements relative to the selected point O.

In terms of x, y, and z coordinates, the centroid has coordinates

$$\bar{x} = \frac{\sum\limits_{i=1}^{n} x_i \Delta_i}{\sum\limits_{i=1}^{n} \Delta_i} \qquad \bar{y} = \frac{\sum\limits_{i=1}^{n} y_i \Delta_i}{\sum\limits_{i=1}^{n} \Delta_i} \qquad \bar{z} = \frac{\sum\limits_{i=1}^{n} z_i \Delta_i}{\sum\limits_{i=1}^{n} \Delta_i} \tag{2}$$

where
$\Delta_i = $ magnitude of the ith quantity (element)
$\bar{x}, \bar{y}, \bar{z} = $ coordinates of centroid of the assemblage
$x_i, y_i, z_i = $ coordinates of P_i at which Δ_i is concentrated.

10.2 CENTROID OF A CONTINUOUS QUANTITY

The centroid of a continuouus quantity may be located by calculus using infinitesimal elements of the quantity (such as dL of a line, dA of an area, dV of a volume, or dm of a mass). Thus, for a mass m we can write

$$\bar{\mathbf{r}} = \frac{\int \mathbf{r}\, dm}{\int dm} \tag{3}$$

In terms of x, y, and z coordinates the centroid of the continuous quantity has coordinates

$$\bar{x} = \frac{\int x\, dm}{\int dm} = \frac{Q_{yz}}{m} \qquad \bar{y} = \frac{\int y\, dm}{\int dm} = \frac{Q_{xz}}{m} \qquad \bar{z} = \frac{\int z\, dm}{\int dm} = \frac{Q_{xy}}{m} \tag{4}$$

where $Q_{xy}, Q_{yz}, Q_{xz} = $ first moments with respect to xy, yz, xz planes.
The centroid of a homogeneous mass coincides with the centroid of its volume.

The following table indicates the first moments Q of various quantities Δ about the coordinate planes.

Δ	Q_{xy}	Q_{yz}	Q_{xz}	Dimensions
Line	$\int z\,dL$	$\int x\,dL$	$\int y\,dL$	L^2
Area	$\int z\,dA$	$\int x\,dA$	$\int y\,dA$	L^3
Volume	$\int z\,dV$	$\int x\,dV$	$\int y\,dV$	L^4
Mass	$\int z\,dm$	$\int x\,dm$	$\int y\,dm$	mL

where $\quad Q_{xy}, Q_{yz}, Q_{xz}$ = first moments with respect to xy, yz, xz planes
$\qquad\qquad L$ = length
$\qquad\qquad m$ = mass
$\quad dL, dA, dV, dm$ = differential elements of line, area, volume, mass, respectively.

Note that in two-dimensional work, e.g., in the xy plane, Q_{xz} becomes Q_x and Q_{yz} becomes Q_y.

10.3 THEOREMS OF PAPPUS AND GULDINUS

The first theorem states that the area of the surface generated by revolving a plane curve about a nonintersecting axis in its plane is equal to the product of the length of the curve and the distance traveled by the centroid G of the curve during the generation. Suppose, as shown in Fig. 10-1, the curve AB of length L is in the xy plane and is revolved through an angle θ about the x axis to the position $A'B'$. The length dL in moving through the distance $y\theta$ generates a surface $dS = y\theta\,dL$. Then $S = \int dS = \int y\theta\,dL = \theta\int y\,dL = \theta\bar{y}L$. Since $\theta\bar{y}$ is the distance traveled by the centroid G of the curve, the first theorem is proved.

The second theorem states that the volume of the solid generated by revolving a plane area about a nonintersecting axis in its plane is equal to the product of the area and the length of path the centroid travels during the generation. Suppose that $ABCD$ of area A is in the xy plane and is revolved through an angle θ about the x axis to position $AB'C'D$ as shown in Fig. 10-2. The area dA in moving through the distance $y\theta$ generates a volume $dV = y\theta\,dA$. Then $V = \int dV = \int y\theta\,dA = \theta\int y\,dA = \theta\bar{y}A$. Since $\theta\bar{y}$ is the distance traveled by the centroid G of the area, the second theorem is proved.

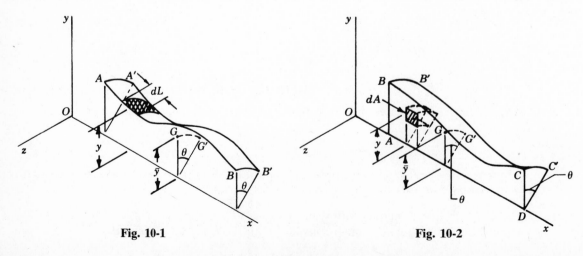

Fig. 10-1   Fig. 10-2

10.4 CENTER OF PRESSURE

When an area is subjected to a pressure, a point in the area exists through which the entire force could be concentrated with the same external effect. This point is called the *center of pressure*. If the pressure is uniformly distributed over an area, the center of pressure coincides with the centroid of the area.

Solved Problems

10.1. Determine Q_x and Q_y for the area bounded by the parabola $y^2 = 4ax$ and the lines $y = 0$, $x = b$.

SOLUTION

To determine Q_x, choose the differential strip parallel to the x axis, as shown in Fig. 10-3. The height of the strip is dy and the width is $b - x$.

$$Q_x = \int y \, dA = \int_0^{2\sqrt{ab}} y(b - x) \, dy$$

Note that the upper limit of y is determined by allowing x to equal b, whence $y = \pm\sqrt{4ab}$. Choose the positive value.

$$Q_x = \int_0^{2\sqrt{ab}} y\left(b - \frac{y^2}{4a}\right) dy = \frac{b(2\sqrt{ab})^2}{2} - \frac{(2\sqrt{ab})^4}{16a} = ab^2$$

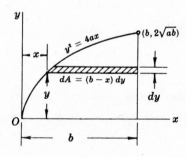

Fig. 10-3

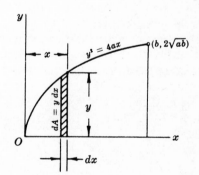

Fig. 10-4

To determine Q_y, choose the differential element parallel to the y axis, as shown in Fig. 10-4. It is at a distance x from the y axis. Hence,

$$Q_y = \int x \, dA$$

But dA is the product of y, the distance from the x axis to the parabola, and dx, the width of the element. Also, x must vary from 0 to b to include the given area.

$$Q_y = \int_0^b xy \, dx = \int_0^b x\sqrt{4ax} \, dx = 2\sqrt{a}\int_0^b x^{3/2} \, dx = \tfrac{2}{5}(2\sqrt{a})x^{5/2}\big]_0^b = \tfrac{4}{5}b^2\sqrt{ab}$$

10.2. Determine Q_x and Q_y in Problem 10.1 using the differential element shown in Fig. 10-5.

SOLUTION

In this case a double integration is involved, as shown in the following:

$$Q_x = \int y \, dA = \int_0^b \int_0^{2\sqrt{ax}} y \, dy \, dx = \int_0^b [\tfrac{1}{2}y^2]_0^{2\sqrt{ax}} \, dx = \int_0^b 2ax \, dx = ab^2$$

$$Q_y = \int x \, dA = \int_0^b \int_0^{2\sqrt{ax}} dy \, x \, dx = \int_0^b [y]_0^{2\sqrt{ax}} x \, dx = \int_0^b 2\sqrt{ax} \, x \, dx = \tfrac{4}{5}b^2\sqrt{ab}$$

The upper limits of the variable y must be expressed in terms of x as shown, because the summation vertically is limited by the curve, which is of varying height.

10.3. Determine the moment of the volume of a right circular cone with respect to its base. Refer to Fig. 10-6.

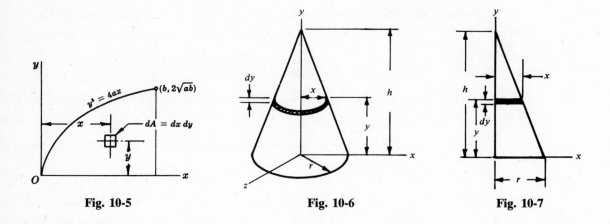

Fig. 10-5   Fig. 10-6   Fig. 10-7

SOLUTION

As shown in Fig. 10-6, choose a differential volume parallel to the base. It is at a distance y above the xz plane, which contains the base.

A cross section in the xy plane yields similar right triangles, as shown in Fig. 10-7. Hence, $x/r = (h - y)/h$ and

$$dV = \pi x^2 \, dy = \frac{\pi r^2}{h^2} (h - y)^2 \, dy$$

To find Q_{xy}, use

$$Q_{xy} = \int y \, dV = \int_0^h y\pi \frac{r^2}{h^2} (h - y)^2 \, dy$$

$$= \frac{\pi r^2}{h^2} \int_0^h (h^2 y - 2hy^2 + y^3) \, dy$$

$$= \frac{\pi r^2}{h^2} \left(h^2 \frac{y^2}{2} - 2h\frac{y^3}{3} + \frac{y^4}{4} \right)_0^h$$

$$= \frac{\pi r^2}{h^2} h^4 (\tfrac{1}{2} - \tfrac{2}{3} + \tfrac{1}{4})$$

$$= \tfrac{1}{12}\pi r^2 h^2$$

10.4. Locate the centroid of the arc of the circle in Fig. 10-8.

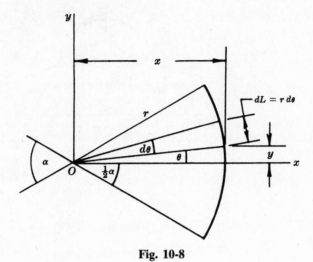

Fig. 10-8

SOLUTION

The x axis was chosen as an axis of symmetry. Polar coordiantes often simplify the integration. From the figure,

$$x = r \cos \theta \qquad y = r \sin \theta$$

$$\bar{x} = \frac{Q_y}{L} = \frac{\int x \, dL}{\int dL} = \frac{\int_{-\alpha/2}^{\alpha/2} xr \, d\theta}{\int_{-\alpha/2}^{\alpha/2} r \, d\theta}$$

$$= \frac{\int_{-\alpha/2}^{\alpha/2} r^2 \cos \theta \, d\theta}{\int_{-\alpha/2}^{\alpha/2} r \, d\theta}$$

$$= \frac{r^2[\sin \tfrac{1}{2}\alpha - \sin (-\tfrac{1}{2}\alpha)]}{r[\tfrac{1}{2}\alpha - (-\tfrac{1}{2}\alpha)]}$$

$$= \frac{r(\sin \tfrac{1}{2}\alpha + \sin \tfrac{1}{2}\alpha)}{\alpha}$$

$$= \frac{2r \sin \tfrac{1}{2}\alpha}{\alpha}$$

Note that α is the subtended angle for the entire arc of the circle.

If $\bar{y}$ were determined by the same method, the integration would yield a cosine term, which when the limits were substituted would disappear. Hence, $\bar{y} = 0$. This can be observed directly, however, because the centroid always lies on the axis of symmetry.

If the arc is a semicircle, α equals 180° or π rad. Then

$$\bar{x} = \frac{2r \sin \tfrac{1}{2}\pi}{\pi} = \frac{2r}{\pi}$$

10.5. Locate the centroid of the bent wire shown in Fig. 10-9.

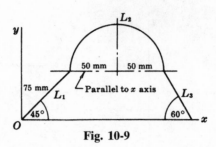

Fig. 10-9

SOLUTION

Let

$$L_1 = \text{75-mm piece at } 45° \text{ with } x \text{ axis}$$

$$L_2 = \text{semicircular piece}$$

$$L_3 = \text{piece at } 60° \text{ with } x \text{ axis}$$

$$\text{Length of } L_3 = \frac{75 \cos 45°}{\sin 60°} = 61.2 \text{ mm}$$

The following table indicates the centroidal distances for each component.

Component	Length	$\bar{x}$	$\bar{y}$
L_1	75	$(75/2) \cos 45° = 26.5$	$(75/2) \sin 45° = 26.5$
L_2	$\pi r = 157$	$75 \cos 45° + 50 = 103$	$75 \sin 45° + 2r/\pi = 84.9$
L_3	61.2	$75 \cos 45° + 100 + (61.2/2) \cos 60° = 168.3$	$(61.2/2) \sin 60° = 26.5$

$$\bar{x} = \frac{L_1\bar{x}_1 + L_2\bar{x}_2 + L_3\bar{x}_3}{L_1 + L_2 + L_3} = \frac{(75 \times 26.5) + (157 \times 103) + (61.2 \times 168.3)}{75 + 157 + 61.2} = 97.1 \text{ mm}$$

$$\bar{y} = \frac{L_1\bar{y}_1 + L_2\bar{y}_2 + L_3\bar{y}_3}{L_1 + L_2 + L_3} = \frac{(75 \times 26.5) + (157 \times 84.9) + (61.2 \times 26.5)}{293.2} = 57.8 \text{ mm}$$

Note that $2r/\pi$, used in determining $\bar{y}_2$, is taken from Problem 10.4.

10.6. Locate the centroid of the bar built up as shown in Fig. 10-10. Assume that the diameter of the bar is negligible compared with the dimensions of the figure.

SOLUTION

The enlarged Fig. 10-11 indicates the trigonometry needed to locate the centroid of the arc. It is on the axis of symmetry, which makes an angle of 75° with the vertical and is at a distance from the center of the arc equal to $(2r/\alpha) \sin \frac{1}{2}\alpha$, where $\alpha = 150\pi/180$ rad. (Refer to Problem 10.4).

Hence, the distance along the radius to the centroid is $[2(1)/2.62] \sin 75° = 0.738$ in. The $\bar{y}$ distance for the arc is therefore $-0.738 \sin 15° = -0.191$ in.

The length of one arc is $r\alpha = 1(2.62) = 2.62$ in.

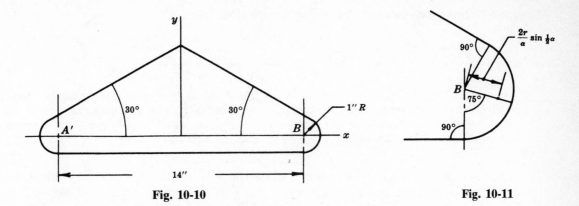

Fig. 10-10　　　　　　　　　　　　　　　　　　　**Fig. 10-11**

Figure 10-12 indicates the method of determining the length of the sloping side.

$$DF = AB = 7 \text{ in}$$
$$BF = BC \cos 30° = 1(0.866) = 0.866 \text{ in}$$
$$FC = BC \sin 30° = 1(0.500) = 0.500 \text{ in}$$
$$DC = DF + FC = 7.5 \text{ in}$$
$$EC = \frac{DC}{\cos 30°} = \frac{7.5}{0.866} = 8.66 \text{ in}$$

The centroid of EC is at G, at a distance above the x axis equal to $\bar{y}_{\text{slope}}$, where

$$\bar{y}_{\text{slope}} = GH + FB = \tfrac{1}{2}(8.66 \sin 30°) + 0.866 = 3.03 \text{ in}$$

The centroid of the horizontal bar is below the x axis. Hence, its $\bar{y}$ is -1 in.
By symmetry, $\bar{x}$ for the composite figure is 0.
To determine $\bar{y}$ for the composite figure, apply the following equation:

$$\bar{y} = \frac{2L_{\text{arc}}\bar{y}_{\text{arc}} + L_{\text{hor}}\bar{y}_{\text{hor}} + 2L_{\text{slope}}\bar{y}_{\text{slope}}}{2L_{\text{arc}} + L_{\text{hor}} + 2L_{\text{slope}}}$$
$$= \frac{2(2.62)(-0.191) + 14(-1) + 2(8.66)(3.03)}{2(2.62) + 14 + 2(8.66)}$$
$$= 1.02 \text{ in}$$

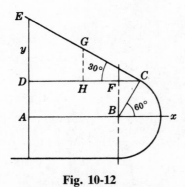

Fig. 10-12

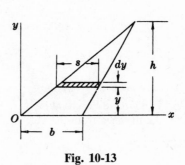

Fig. 10-13

10.7. Locate the centroid of a triangle.

SOLUTION

Choose the differential element of area as shown in Fig. 10-13, where $dA = s \, dy$.

Note that $s/b = (h - y)/h$.

$$\bar{y} = \frac{Q_x}{A} = \frac{\int y \, dA}{A} = \frac{\int_0^h ys \, dy}{\frac{1}{2}bh} = \frac{\int_0^h y\left[\frac{b}{h}(h - y)\right] dy}{\frac{1}{2}bh} = \frac{h}{3}$$

One could now determine $\bar{x}$, but ordinarily it is sufficient to know that the centroid is located at a point whose distance from the base is one-third of the altitude. Draw two lines that are parallel to any two sides and that are at distances from those sides equal to one-third of the altitudes erected on the respective sides. The intersection of these two lines is the centroid.

10.8. Determine the centroid of a sector of a circle where the radius is r and the subtended angle is 2α. The x axis is the axis of symmetry. Solve by (a) using the differential element shown in Fig. 10-14 and (b) using the differential element shown in Fig. 10-15.

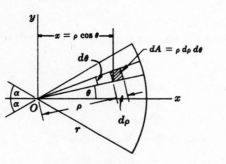

Fig. 10-14

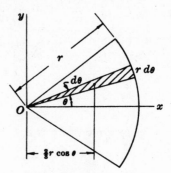

Fig. 10-15

SOLUTION

(a) Using the differential element in Fig. 10-14, we write

$$\bar{x} = \frac{Q_y}{A} = \frac{\int x \, dA}{\int dA} = \frac{\int_{-\alpha}^{\alpha} \int_0^r \rho \cos \theta \rho \, d\rho \, d\theta}{\int_{-\alpha}^{\alpha} \int_0^r \rho \, d\rho \, d\theta} = \frac{\int_{-\alpha}^{\alpha} [\rho^3/3]_0^r \cos \theta \, d\theta}{\int_{-\alpha}^{\alpha} [\rho^2/2]_0^r \, d\theta} = \frac{(r^3/3)[\sin \alpha - \sin(-\alpha)]}{(r^2/2)[\alpha - (-\alpha)]}$$

$$= \frac{2r \sin \alpha}{3\alpha}$$

For a semicircular sector, $2\alpha = \pi$ rad and

$$\bar{x} = \frac{2r \sin \pi/2}{3\pi/2} = \frac{4r}{3\pi}$$

Of course, $\bar{y} = 0$ by the symmetry of the figure.

(b) Using the differential element in Fig. 10-15, we note that the centroid of the triangle is two-thirds of the distance from the vertex (the origin) to the base. Note that $\bar{x}$ for the triangle is $\frac{2}{3}r \cos \theta$.

$$\bar{x} = \frac{Q_y}{A} = \frac{\int_{-\alpha}^{\alpha} (\frac{1}{2}r \, d\theta \, r)(\frac{2}{3}r \cos \theta)}{\int_{-\alpha}^{\alpha} \frac{1}{2}r \, d\theta \, r} = \frac{\frac{1}{3}r^3 \int_{-\alpha}^{\alpha} \cos \theta \, d\theta}{\frac{1}{2}r^2 \int_{-\alpha}^{\alpha} d\theta} = \frac{2r \sin \alpha}{3\alpha}$$

10.9. Determine the centroid for the area bounded by the parabola $y^2 = 4ax$ and the lines $x = 0$, $y = b$.

SOLUTION

Choose the differential strip parallel to the x axis, as shown in Fig. 10-16.

$$\bar{x} = \frac{Q_y}{A} = \frac{\int_0^b (\frac{1}{2}x)x\,dy}{\int_0^b x\,dy} = \frac{\frac{1}{2}\int_0^b x^2\,dy}{\int_0^b x\,dy} = \frac{\frac{1}{2}\int_0^b (y^4/16a^2)\,dy}{\int_0^b (y^2/4a)\,dy} = \frac{3b^2}{40a}$$

Similarly,

$$\bar{y} = \frac{Q_y}{A} = \frac{\int_0^b yx\,dy}{b^3/12a} = \frac{\int_0^b (y^3/4a)\,dy}{b^3/12a} = \frac{b^4/16a}{b^3/12a} = \tfrac{3}{4}b$$

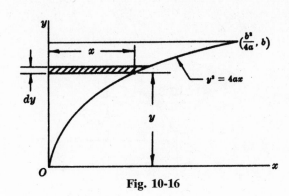

Fig. 10-16

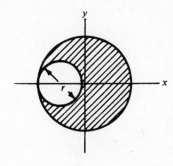

Fig. 10-17

10.10. Determine the centroid of the area remaining after the circle of diameter r is removed from a circle of radius r, as shown in Fig. 10-17.

SOLUTION

By symmetry, $\bar{y} = 0$; i.e., the centroid is on the x axis.

Using the formula for composite areas in which A_L is the area of the large circle and A_S is the area of the small circle, we have

$$\bar{x} = \frac{A_L x_L - A_S x_S}{A_L - A_S} = \frac{\pi r^2(0) - (\pi r^2/4)(-r/2)}{\pi r^2 - \pi r^2/4} = \tfrac{1}{6}r$$

Hence the centroid is one the x axis at a distance $\tfrac{1}{6}r$ to the right of the y axis.

Mathcad

10.11. A semicircular area is removed from the trapezoid as shown in Fig. 10-18. Determine the centroid of the remaining area.

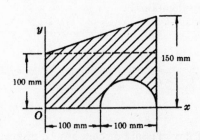

Fig. 10-18

SOLUTION

The shaded area consists of (1) a rectangle plus (2) a triangle minus (3) a semicircular area.

$$\bar{x} = \frac{A_1\bar{x}_1 + A_2\bar{x}_2 - A_3\bar{x}_3}{A_1 + A_2 - A_3} = \frac{2 \times 10^4(100) + 5 \times 10^3(133.3) - \pi(50)^2(150)/2}{2 \times 10^4 + 5 \times 10^3 - \pi(50)^2/2} = 98.6 \text{ mm}$$

$$\bar{y} = \frac{A_1\bar{y}_1 + A_2\bar{y}_2 - A_3\bar{y}_3}{A_1 + A_2 - A_3}$$

$$= \frac{2 \times 10^4(50) + 5 \times 10^3(100 + 50/3) - [\pi(50)^2/2][(4 \times 50)/3\pi]}{21\,070} = 71.2 \text{ mm}$$

10.12. The area in Fig. 10-19(*a*) is revolved about the *y* axis. Determine the centroid of the resulting volume shown in Fig. 10-19(*b*).

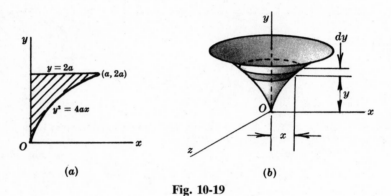

(*a*)   (*b*)

Fig. 10-19

SOLUTION

By symmetry in Fig. 10-19(*b*), $\bar{x} = 0$, $\bar{z} = 0$. Choose a differential volume parallel to the *xz* plane. Its thickness or height is *dy* and it is at a distance *y* above the *xz* plane. Its radius is *x*.

$$\bar{y} = \frac{Q_{xz}}{V} = \frac{\displaystyle\int_0^{2a} y(\pi x^2\, dy)}{\displaystyle\int_0^{2a} \pi x^2\, dy} = \frac{\pi\displaystyle\int_0^{2a} y(y^4/16a^2)\, dy}{\pi\displaystyle\int_0^{2a} (y^4/16a^2)\, dy} = \frac{y^6/6]_0^{2a}}{y^5/5]_0^{2a}} = \tfrac{5}{3}a$$

10.13. Locate $\bar{x}$ for the volume of any pyramid or cone whose base coincides with the *yz* plane. The altitude is *h* and the area of the base is *A*.

SOLUTION

In Fig. 10-20 choose a differential volume at a distance *x* from the *yz* plane. Its area varies with *x*. Call the area A_x and the thickness *dx*.

From geometrical considerations,

$$\frac{A_x}{A} = \frac{(h-x)^2}{h^2}$$

Then, $$\bar{x} = \frac{Q_{yz}}{V} = \frac{\displaystyle\int x\, dV}{\displaystyle\int dV} = \frac{\displaystyle\int_0^h x A_x\, dx}{\displaystyle\int_0^h A_x\, dx} = \frac{\displaystyle\int_0^h x(A/h^2)(h-x)^2\, dx}{\displaystyle\int_0^h (A/h^2)(h-x)^2\, dx} = \tfrac{1}{4}h$$

Thus, the centroid of any pyramid or cone is at a distance from the base equal to one-fourth the altitude.

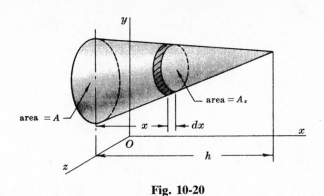

Fig. 10-20

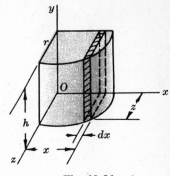

Fig. 10-21

10.14. Locate the centroid of the volume of one-fourth of the right circular cylinder shown in Fig. 10-21.

SOLUTION

It is only necessary to derive the value of $\bar{x}$, since $\bar{z} = \bar{x}$ and of course $\bar{y}$ is one-half the height h. Choose the differential volume dV parallel to the yz plane.

$$dV = zh\,dx$$

However, since the section cut from the solid by any plane parallel to the xz plane yields a quarter circle, the relationship between x and z is $x^2 + z^2 = r^2$.

$$\bar{x} = \frac{Q_{yz}}{V} = \frac{\displaystyle\int x\,dV}{\displaystyle\int dV} = \frac{\displaystyle\int_0^r xzh\,dx}{\displaystyle\int_0^r zh\,dx}$$

$$= \frac{\displaystyle\int_0^r hx(r^2 - x^2)^{1/2}\,dx}{\displaystyle\int_0^r h(r^2 - x^2)^{1/2}\,dx} = \frac{h[-\tfrac{1}{3}(r^2 - x^2)^{3/2}]_0^r}{h[(x/2)(r^2 - x^2)^{1/2} + (r^2/2)\sin^{-1}(x/r)]_0^r} = \frac{4r}{3\pi}$$

Note that the result is exactly the same as for the centroid of the quadrant of a circle. This should be expected, since the only factor that might influence the centroidal position in the solid as compared with the area is the altitude h, which was shown to be a common term in both numerator and denominator.

10.15. A sphere of radius r is cut from a larger sphere of radius R. The distance between their centers is a. Locate the centroid of the remaining volume.

SOLUTION

This is an example to illustrate the technique employed for composite volumes.

$$V_R = \tfrac{4}{3}\pi R^3 \qquad V_r = \tfrac{4}{3}\pi r^3$$

Assume the origin of the x, y, z axes is at the center of the larger sphere and that the positive x axis is the line of centers of the two spheres. Then $\bar{x}_R = 0$ and $\bar{x}_r = a$.

Applying the formula,

$$\bar{x} = \frac{V_R\bar{x}_R - V_r\bar{x}_r}{V_R - V_r} = \frac{\tfrac{4}{3}\pi R^3(0) - \tfrac{4}{3}\pi r^3(a)}{\tfrac{4}{3}\pi R^3 - \tfrac{4}{3}\pi r^3} = \frac{-ar^3}{R^3 - r^3}$$

This means that the centroid is on the line of centers and to the left of the yz plane at a distance of $ar^3/(R^3 - r^3)$.

10.16. Locate the centroid of the composite volume shown in Fig. 10-22. The 40-mm hole is drilled in the center of the top face and is normal to that face.

SOLUTION

From the symmetry of the figure, $\bar{z} = 120$ mm.

Let the parallelepiped be denoted by 1, the triangular piece by 2, and the cylinder by 3. The values needed in the formulas below are given in the following table.

Shape	V	$\bar{x}$	$\bar{y}$
1	1728×10^3	60	30
2	432×10^3	140	20
3	7.54×10^3	60	30

$$\bar{x} = \frac{1728 \times 10^3(60) + 432 \times 10^3(140) - 75.4 \times 10^3(60)}{1728 \times 10^3 + 432 \times 10^3 - 75.4 \times 10^3} = 76.6 \text{ mm}$$

$$\bar{y} = \frac{1728 \times 10^3(30) + 432 \times 10^3(20) - 75.4 \times 10^3(30)}{2084.6 \times 10^3} = 27.9 \text{ mm}$$

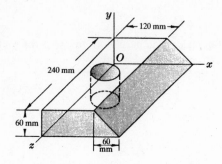

Fig. 10-22

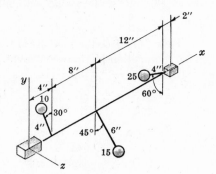

Fig. 10-23

10.17. Three spheres whose volumes are 10, 15, and 25 in³ are located with reference to the shaft as shown in Fig. 10-23. Locate the centroid of the three volumes.

SOLUTION

Assume that the x axis is along the shaft. Use a tabular form to list the data.

V	$\bar{x}$	$\bar{y}$	$\bar{z}$
10	+4	$+4\cos 30°$	$-4\sin 30°$
15	+12	$-6\cos 45°$	$+6\sin 45°$
25	+24	$-4\cos 60°$	$-4\sin 60°$

$$\bar{x} = \frac{(10 \times 4) + (15 \times 12) + (25 \times 24)}{10 + 15 + 25} = 16.4 \text{ in}$$

$$\bar{y} = \frac{(10 \times 4 \times 0.866) - (15 \times 6 \times 0.707) - (25 \times 4 \times 0.500)}{50} = -1.58 \text{ in}$$

$$\bar{z} = \frac{(-10 \times 4 \times 0.500) + (15 \times 6 \times 0.707) - (25 \times 4 \times 0.866)}{50} = -0.86 \text{ in}$$

Incidentally, this same procedure would be followed if the numbers 10, 15, and 25 represented weights or masses concentrated at the centers of the corresponding spheres.

10.18. Determine the centroid of the surface of a hemisphere with respect to its base.

SOLUTION

Refer to Fig. 10-24. The xz plane is chosen in the base of the hemisphere.

The differential strip of area dS is chosen as in the figure. Note that the width dL of this differential surface is *not* vertical but sloping. Let $\theta = \tan^{-1}(y/x)$.

$$\overline{dL^2} = \overline{dx^2} + \overline{dy^2} = \left(\frac{\overline{dx^2}}{dx^2} + \frac{\overline{dy^2}}{dx^2}\right)\overline{dx^2} \quad \text{or} \quad dL = \sqrt{1 + (dy/dx)^2}\, dx$$

$$\bar{y} = \frac{Q_{xz}}{S} = \frac{\displaystyle\int y\, dS}{\displaystyle\int dS} = \frac{\displaystyle\int y(2\pi x\, dL)}{\displaystyle\int 2\pi x\, dL} = \frac{\displaystyle\int_0^r 2\pi y x \sqrt{1 + (dy/dx)^2}\, dx}{\displaystyle\int_0^r 2\pi x \sqrt{1 + (dy/dx)^2}\, dx}$$

In the xy plane the equation of the circle is $x^2 + y^2 = r^2$.

Then $y = \sqrt{r^2 - x^2}$. Differentiating, $dy/dx = -x(r^2 - x^2)^{-1/2}$. Making these substitutions,

$$\bar{y} = \frac{2\pi r \displaystyle\int_0^r x\, dx}{2\pi r \displaystyle\int_0^r x(r^2 - x^2)^{-1/2}\, dx} = \tfrac{1}{2}r$$

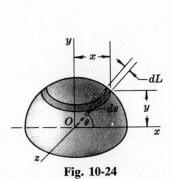

Fig. 10-24

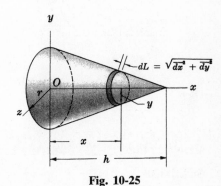

Fig. 10-25

10.19. Determine the centroid of the surface of a right circular cone with respect to its base. The altitude is h. Refer to Fig. 10-25.

SOLUTION

In Fig. 10-25, the x axis is chosen along the altitude of the cone. The differential element of surface dS is $2\pi y\, dL$.

$$\bar{x} = \frac{Q_{yz}}{S} = \frac{\displaystyle\int x\, dS}{\displaystyle\int dS} = \frac{\displaystyle\int x 2\pi y\, dL}{\displaystyle\int 2\pi y\, dL}$$

If r is the radius of the base, then by similar triangles in the xy plane, $y/r = (h - x)/h$.

Hence $dy/dx = -r/h$ and $dL = dx\sqrt{1 + (dy/dx)^2} = dx\sqrt{1 + r^2/h^2}$.

Substituting and simplifying,

$$\bar{x} = \frac{\int_0^h (hx - x^2)\, dx}{\int_0^h (h - x)\, dx} = \frac{h}{3}$$

10.20. Find the center of mass of the right half of the homogeneous ellipsoid $x^2/a^2 + y^2/b^2 + z^2/c^2 = 1$.

SOLUTION

Choose the differential mass parallel to the yz plane as shown in Fig. 10-26. Let the density be δ.
Note that $dV = A\, dx$, where A is the area of an ellipse with axes of length $2y$ and $2z$. The area
$A = \pi yz$.

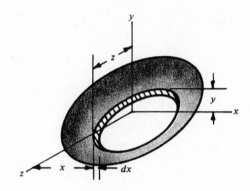

Fig. 10-26

$$dm = \delta\, dV = \delta \pi yz\, dx$$

Then
$$\bar{x} = \frac{Q_{yz}}{m} = \frac{\int x\, dm}{\int dm} = \frac{\delta \pi \int_0^a xyz\, dx}{\delta \pi \int_0^a yz\, dx}$$

To determine y in terms of x, let $z = 0$ in the equation of the ellipsoid. This yields

$$\frac{x^2}{a^2} + \frac{y^2}{b^2} = 1 \qquad \text{or} \qquad y = \frac{b}{a}\sqrt{a^2 - x^2}$$

Similarly, with $y = 0$,

$$\frac{x^2}{a^2} + \frac{z^2}{c^2} = 1 \qquad \text{or} \qquad z = \frac{c}{a}\sqrt{a^2 - x^2}$$

Substituting and simplifying, we have

$$\bar{x} = \frac{\int_0^a \left(x\, \frac{b}{a}\sqrt{a^2 - x^2}\, \frac{c}{a}\sqrt{a^2 - x^2} \right) dx}{\int_0^a \left(\frac{b}{a}\sqrt{a^2 - x^2}\, \frac{c}{a}\sqrt{a^2 - x^2} \right) dx} = \tfrac{3}{8}a$$

Of course, $\bar{y} = \bar{z} = 0$.

10.21. Find the center of mass of a hemisphere whose density varies as the square of the distance
from the base.

SOLUTION

Let the base of the hemisphere be the yz plane as shown in Fig. 10-27. Then the density varies with x^2, or $\delta = Kx^2$.

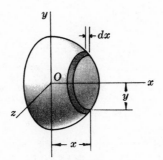

Fig. 10-27

Choose dm parallel to the yz plane and at a distance x from the yz plane.

$$\bar{x} = \frac{Q_{yz}}{m} = \frac{\int x\,dm}{\int dm} = \frac{\int_0^r x\delta\pi y^2\,dx}{\int_0^r \delta\pi y^2\,dx}$$

In the xy plane the coordinates (x, y) lie on a circle of radius r. Hence, $y^2 = r^2 - x^2$. Substituting the values of y^2 and δ, the equation becomes

$$\bar{x} = \frac{\int_0^r xKx^2\pi(r^2 - x^2)\,dx}{\int_0^r Kx^2\pi(r^2 - x^2)\,dx} = \frac{\int_0^r x^3(r^2 - x^2)\,dx}{\int_0^r x^2(r^2 - x^2)\,dx} = \frac{\frac{1}{12}r^6}{\frac{2}{15}r^5} = \frac{5}{8}r$$

10.22. The density at any point of a slender rod varies with the first power of the distance of the point from one end of the rod. Where is the mass center?

SOLUTION

The density is proportional to the distance x along the rod from the one end chosen as origin, that is, $\delta = Kx$ where δ equal s mass per unit length.

To find dm, multiply the differential length dx at the point x by the density at that point. This yields the equation $dm = \delta\,dx = Kx\,dx$. Then

$$\bar{x} = \frac{\int x\,dm}{\int dm} = \frac{\int_0^l xKx\,dx}{\int_0^l Kx\,dx} = \frac{2}{3}l$$

The center of mass is two-thirds of the length from the end chosen as a reference.

10.23. Find the surface area of the annular torus formed by revolving the circle about the x axis in Fig. 10-28. Use the theorems of Pappus and Guldinus.

SOLUTION

The centroid of the circumference is at a distance d from the x axis. Thus in a complete revolution the centroid moves on a circular path of radius d. The distance it travels is $2\pi d$. The length $2\pi r$ of the generating curve is the circumference of the circle. Hence, the surface of the torus is $2\pi d \times 2\pi r = 4\pi^2 rd$.

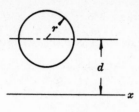

Fig. 10-28

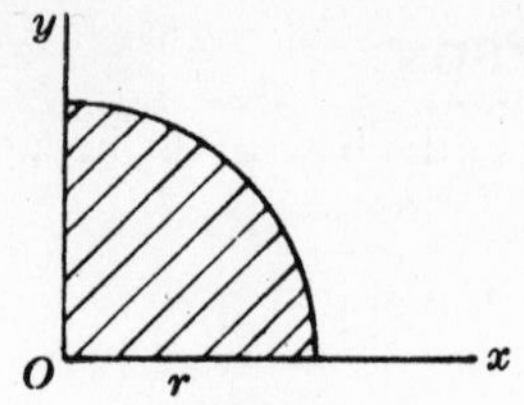

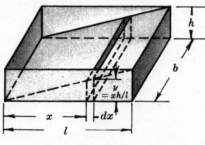

Fig. 10-29

10.24. Determine the centroid of a quadrant of a circle using the theorems of Pappus and Guldinus.

SOLUTION

This is really an application of the theorems in reverse. The area in Fig. 10-29 when revolved generates a hemisphere whose volume is known to be $\frac{2}{3}\pi r^3$.

The length of the path of the centroid is the volume divided by the area of the quadrant:

$$l = \frac{2\pi r^3/3}{\pi r^2/4} = \frac{8}{3}r$$

But the length of the path of the centroid is $2\pi\bar{y}$. Hence, $2\pi\bar{y} = 8r/3$, or $\bar{y} = 4r/3\pi$. This value was derived previously by the methods of the calculus. See Problem 10.8 above.

Of course, $\bar{x} = \bar{y}$ for the quadrant of the circle.

10.25. In Fig. 10-30, a box with dimensions l, b, and h m is shown half full of gravel having a density of δ kg/m^3. Assuming that the height of the gravel varies linearly from zero at the left end to h at the right, determine how far the center of pressure is from the left end.

Fig. 10-30

SOLUTION

At a distance x from the left end select a differential volume as shown. The height $y = xh/l$ and the gravitational force on dV is $dW = g\delta(xh/l)b\,dx$. To locate the center of pressure, use

$$\bar{x} = \frac{\displaystyle\int x\,dW}{\displaystyle\int dW} = \frac{(g\delta hb/l)\displaystyle\int_0^l x^2\,dx}{(g\delta hb/l)\displaystyle\int_0^l x\,dx} = \frac{2}{3}l$$

The center of pressure of the gravel is on a vertical line located $\frac{2}{3}l$ from the left end and $\frac{1}{2}b$ back from the front wall. The entire mass of the gravel if placed along this vertical line would induce in the box supports the same forces as the distributed mass does.

10.26. (*a*) Refer to Fig. 10-31. A beam carries material that weighs w lb/ft³ and has a height y that varies in a known fashion with the distance x from the left end. Determine each right reaction R_R. Assume that b is constant and that the height y for a selected strip does not vary along the b distance.

(*b*) In part (*a*), suppose $w = 150$ lb/ft³ and that the height of the load varies linearly from zero at the left end to 2 ft at the right end as shown in Fig. 10-32. The span is 8 ft and the distance b is 2 ft. Determine each of the two right supports, neglecting the weight of the beam.

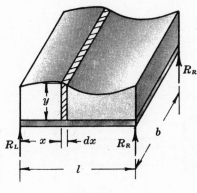

Fig. 10-31

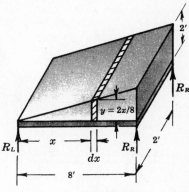

Fig. 10-32

SOLUTION

(*a*) The weight of a differential volume at a distance x from the left end is $dW = wby\,dx$. The moment of this dW with respect to the left end (the vertical plane perpendicular to l) is $x\,dW = wbyx\,dx$.

The sum of the moments of the weights of all such differential volumes about the left end must be balanced by the moments of the two right supports R_R. Thus,

$$\int_0^l wbyx\,dx = 2R_R l$$

The integration may be performed directly if y is a convenient function of x. If not, other means must be employed.

(*b*) The height y of a differential load at a distance x from the left end is $y = 2x/8 = x/4$.

The moment of the entire load relative to the vertical plane containing the two left reactions is

$$M = \int x\,dW = \int x(150\,dV) = \int_0^8 x\left[150\left(\frac{x}{4}\right)(2)\,dx\right] = 12,800 \text{ ft-lb}$$

The moment about the left end of the two right reactions must equal the moment of the load. Hence,

$$2R_R(8) = 12,800 \qquad \text{or} \qquad R_R = 800 \text{ lb}$$

The moment of the load may also be found by using the load in pounds per foot along the beam. If the load at the right end is visualized as 2 ft high, 2 ft back along the beam, and 1 ft in the x direction centered at the right edge, the magnitude of the load will be $2 \times 2 \times 1 \times 150 = 600$ lb per linear foot. Thus, in this problem the load varies from zero at the left end to 600 lb/ft at the right end.

The load on a length dx that is at a distance x from the left end is $p_x\,dx$, where p_x = load per linear foot at x. By similar triangles, $p_s/x = 600/8$ or $p_x = 75x$. Then the moment of the entire load about the left end is

$$M = \int_0^8 xp_x\,dx = \int_0^8 x(75x)\,dx = 12,800 \text{ ft-lb}$$

10.27. The tank shown in Fig. 10-33(a) is full of water having a mass density of 1000 kg/m³. The plate covers a rectangular hole 300 mm high by 600 mm wide. Determine the force induced in each of the two top bolts B by the water acting against the plate. (*Note:* In the solution all dimensions are in meters.)

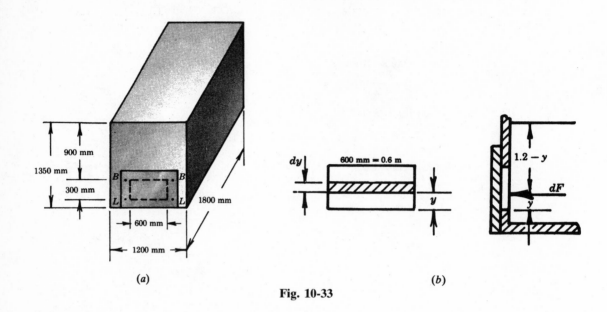

Fig. 10-33

SOLUTION

The diagram in Fig. 10-33(b) shows a differential element of the plate a distance y above the bottom of the hole. since we are interested only in the area of the plate that the water touches, the element of area dA is $0.6\,dy$. [*Note:* If the width of the plate had been a function of y then the differential area would have been $f(y)\,dy$.]

The force dF of the water on the selected differential element is the product of the area dA and the pressure at that distance y above the bottom of the hole.

The pressure p at the point y m above the bottom of the hole is numerically equal to the gravitational force exerted by a column of water $1\,\text{m}^2$ in cross section and $(1.2-y)$ m high: $p = 9.8 \times 1000(1.2-y)\,\text{N/m}^2$. Then the differential force on the small area is $dF = 9.8 \times 1000(1.2-y)(0.6)\,dy$.

The moment of dF about the bottom of the hole is $y\,dF$. The moment of the total water force is $\int y\,dF$. This must be balanced by the moment of the bolt forces B about the bottom of the hole. Hence,

$$2B(0.3) = \int_0^{0.3} y\, 9.8 \times 1000(1.2-y)(0.6)\,dy, \qquad \text{from which} \qquad B = 441\,\text{N}$$

Let the reader show that the force induced in each bottom bolt is 485 N.

10.28. Figure 10-34(a) shows a rectangular gate which separates fluids of two different densities. The gate is hinged at the top and rests against a stop at the bottom. Find d, the greatest difference in depth for which the gate will remain closed.

SOLUTION

Figure 10-34(b) shows the pressure distribution on the gate. The maximum pressure at the bottom at the left is the weight of a 12-ft-high column of fluid (water) on a square foot:

$$p_1 = 62.4 \times 12\,\text{lb/ft}^2$$

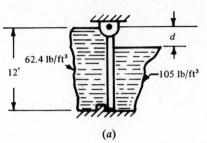

Fig. 10-34

The total force on the left side is

$$F_1 = \tfrac{1}{2}p_1 \times \text{area} = \tfrac{1}{2}(62.4)(12)(12)(a)$$

where a is the distance along the gate (perpendicular to the paper). This force (F_1) acts at the center of pressure, which for a triangular variation is at the centroid of the triangle (8 ft from the top of the gate).

The maximum pressure at the bottom at the right is the weight of a column of fluid on a square foot and $(12 - d)$ ft high:

$$p_2 = 105(12 - d) \text{ lb/ft}^2$$

The total force on the right side is

$$F_2 = \tfrac{1}{2}p_2 \times \text{area} = \tfrac{1}{2}(105)(12 - d)(12 - d)(a)$$

This force F_2 acts at the center of pressure, which is $\tfrac{1}{3} \times \text{height} = \tfrac{1}{3}(12 - d)$ from the bottom.

In order to take moments about the hinge, we use a moment arm of 8 ft for F_1 and a moment arm of $12 - \tfrac{1}{3}(12 - d) = 8 + \tfrac{1}{3}d$. Equate the two moments for equilibrium to obtain

$$\tfrac{1}{2}(62.4)(12)(12)(a)(8) = \tfrac{1}{2}(105)(12 - d)^2(a)(8 + \tfrac{1}{3}d)$$

$$685 = (144 - 24d + d^2)(8 + \tfrac{1}{3}d) \qquad \text{or} \qquad \tfrac{1}{3}d^3 - 144d + 467 = 0$$

A value of $d = 3.33$ ft satisfies this equation.

Supplementary Problems*

10.29. A filament is stretched from a point on the x axis which is 4 in to the right of the origin to a point on the y axis which is 4 in above the origin. Determine the first moment of the filament with respect to the x axis.   *Ans.* $Q = 11.3 \text{ in}^2$

10.30. Find Q_x for the mass m of half the rim of a wheel shown in Fig. 10-35. Assume the thickness of the wheel is small compared with the radius. Use polar coordinates as indicated. Density is δ units of mass per unit length.   *Ans.* $Q_x = 2\delta r^2 = 2rm/\pi$

10.31. Determine the first moment of a half circular area about its diameter.   *Ans.* $Q = \tfrac{2}{3}r^3$

10.32. Find Q_x and Q_y for the area bounded by $y^2 = 4ax$, $x = 0$, $y = b$.   *Ans.* $Q_x = b^4/16a$, $Q_y = b^5/160a^2$

---

* The table of first moments and centroids in Appendix B may be helpful in the solution of numerical problems.

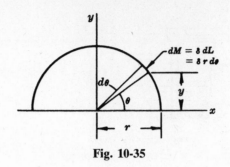

Fig. 10-35

10.33. Determine Q for the volume of a hemisphere with respect to its base.   *Ans.* $Q = \frac{1}{4}\pi r^4$

10.34. Apply the results of Problem 10.33 to determine Q about the base of a hemisphere whose radius is 6 in.   *Ans.* $Q = 1020$ in⁴.

10.35. Determine Q_x and Q_y for one quarter of an ellipse whose major and minor axes are 100 mm and 75 mm, respectively.   *Ans.* $Q_x = 23\,400$ mm³, $Q_y = 31\,300$ mm³

10.36. Apply the results of Problem 10.3 to find the first moment of a right circular cone about its base whose radius is 75 mm. The height is 100 mm.   *Ans.* $Q = 14.7 \times 10^6$ mm⁴

10.37. A uniform bar 18 in long is bent at its midpoint at a 90° angle. Locate its centroid using the sides as axes.   *Ans.* $\bar{x} = \bar{y} = 2.25$ in

10.38. Suppose the bar in Problem 10.37 is bent so that the angle at its midpoint is 70°; how far is its centroid from the line joining its ends?   *Ans.* $d = 3.69$ in

10.39. A wire is bent as shown in Fig. 10-36. Locate its center of gravity.   *Ans.* $\bar{x} = 56.5$ mm, $\bar{y} = 103$ mm

10.40. Find the centroid of the figure composed of lines as shown in Fig. 10-37.
Ans. $\bar{x} = -77.0$ mm, $\bar{y} = 67.6$ mm

10.41. Determine the centroid of a uniform rod bent into triangular shape. See Fig. 10-38.
Ans. $\bar{x} = 2.52$ in, $\bar{y} = 3.11$ in

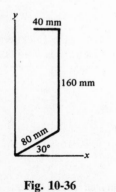

Fig. 10-36

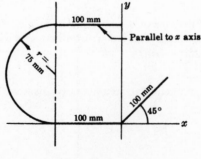
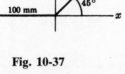

Fig. 10-37

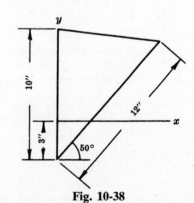

Fig. 10-38

10.42. Show that the following values hold for the indicated areas.

Figure	Area	Q_x	Q_y	Centroid	
				$\bar{x}$	$\bar{y}$
$y = \frac{b}{a^2}x^2$, (a, b)	$\frac{1}{3}ab$	$\frac{1}{10}ab^2$	$\frac{1}{4}a^2b$	$\frac{3}{4}a$	$\frac{3}{10}b$
$y = \frac{b}{a^2}x^2$, (a, b)	$\frac{2}{3}ab$	$\frac{2}{5}ab^2$	$\frac{1}{4}ba^2$	$\frac{3}{8}a$	$\frac{3}{5}b$
$y = \frac{b}{a^n}x^n$, (a, b)	$\frac{ab}{n+1}$	$\frac{ab^2}{2(2n+1)}$	$\frac{a^2b}{n+2}$	$\frac{n+1}{n+2}a$	$\frac{n+1}{2(2n+1)}b$
$y = \frac{b}{a^n}x^n$, (a, b)	$\frac{n}{n+1}ab$	$\frac{n}{2n+1}ab^2$	$\frac{n}{2(n+2)}a^2b$	$\frac{n+1}{2(n+2)}a$	$\frac{n+1}{2n+1}b$
$y = b - \frac{b}{a^2}x^2$	$\frac{2}{3}ab$	$\frac{4}{15}ab^2$	$\frac{1}{4}a^2b$	$\frac{3}{8}a$	$\frac{2}{5}b$
$y = \frac{b}{a}[a^2 - x^2]^{1/2}$	$\frac{\pi ab}{4}$	$\frac{ab^2}{3}$	$\frac{a^2b}{3}$	$\frac{4a}{3\pi}$	$\frac{4b}{3\pi}$

10.43. Determine the centroid for the area bounded by the parabola $y^3 = 4ax$ and the lines $x = 0$, $y = b$.
Ans. $\bar{x} = 3b^2/40a$, $y = \frac{3}{4}b$

10.44. Compute $\bar{x}$ for the area between the semicubical parabola $ay^2 = x^3$ and the line $x = a$.
Ans. $\bar{x} = \frac{5}{7}a$

10.45. Locate the centroid of the area bounded by $y^2 = 2x$, $x = 3$, $y = 0$.   Ans. $\bar{x} = 1.8$, $\bar{y} = \frac{3}{8}\sqrt{6}$

10.46. Locate the centroid of the area between the parabola $y^2 = 4ax$ and the line $y = bx$.
Ans. $\bar{x} = 8a/5b^2$, $\bar{y} = 2a/b$

10.47. Determine the centroid of the area bounded by the curve $x^2 = y$ and the line $x = y$.
Ans. $\bar{x} = 0.5$, $\bar{y} = 0.5$

10.48. Locate the centroid of the area bounded by the parabolas $y^3 = 4x$ and $x^2 = 4y$.   Ans. $\bar{x} = \bar{y} = 1.8$

10.49. Determine the y coordinate of the centroid of the area between the x axis and the curve $y = \sin x$. Use the interval $0 \le x \le \pi$.   Ans. $\bar{y} = \frac{1}{8}\pi$

10.50. Determine $\bar{x}$ for the area bounded by the hyperbola $xy = c^2$ and the lines $x = a$, $x = b$, $y = 0$.
Ans. $\bar{x} = (b - a)/(\log_e b - \log_e a)$

10.51. Locate the centroid for the area between the ellipse $x^2/a^2 + y^2/b^2 = 1$ and the lines $x = a$, $y = b$.
Ans. $\bar{x} = 0.776a$, $\bar{y} = 0.776b$

10.52. Find the centroid of the composite area in Fig. 10-39.   Ans. $\bar{x} = 64.4$ mm, $\bar{y} = 80.6$ mm

10.53. Compute the distance from the top line to the centroid of the T section in Fig. 10-40.
Ans. $d = 2.97$ in

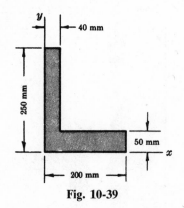

Fig. 10-39

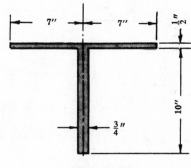

Fig. 10-40

10.54. Locate the centroid of the shaded area formed by removing the triangle from the semicircular area in Fig. 10-41.   Ans. $\bar{x} = 0$, $\bar{y} = 23.4$ mm

10.55. Determine the coordinates of the centroid of the shaded area in Fig. 10-42. The area removed is semicircular.   Ans. $\bar{x} = 3.77$ in, $\bar{y} = 2.69$ in

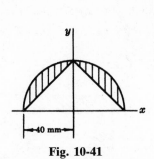

Fig. 10-41

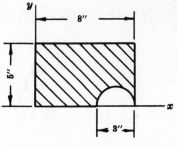

Fig. 10-42

Locate the centroids of the following figures with respect to the axes shown.

10.56.	Fig. 10-43.	*Ans.*	$\bar{x} = 10.0$ mm, $\bar{y} = 110$ *mm*

10.57.	Fig. 10-44.	*Ans.*	$\bar{x} = 0$ in, $\bar{y} = 4.37$ in

10.58.	Fig. 10-45.	*Ans.*	$\bar{x} = 11.9$ *mm*, $\bar{y} = 0$ mm

10.59.	Fig. 10-46.	*Ans.*	$\bar{x} = 1.32$ *in*, $\bar{y} = 3.59$ in

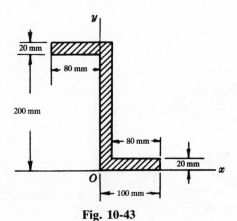

Fig. 10-43

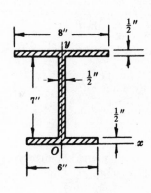

Fig. 10-44

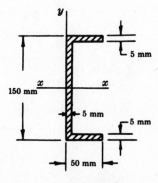

Fig. 10-45

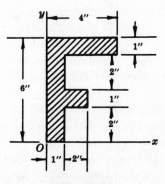

Fig. 10-46

10.60. Fig. 10-47.   *Ans.* $\bar{x} = 99.3$ mm, $\bar{y} = 41.1$ mm

10.61. Fig. 10-48.   *Ans.* $\bar{x} = 3$ in, $\bar{y} = 3.40$ in

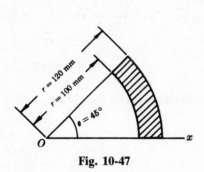

Fig. 10-47

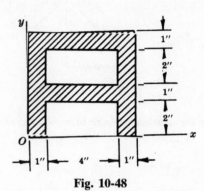

Fig. 10-48

10.62. Show by integration that the centroid of the volume of a right circular cone is at a distance from the base equal to one-fourth the altitude.

10.63. A cone is formed by revolving the area bounded by the lines $by = ax$, $y = 0$ and $x = b$ around the x axis. Determine $\bar{x}$ for the volume of the cone.   *Ans.* $\bar{x} = \frac{3}{4}b$

10.64. Show by integration that the centroid of the volume of a hemisphere is at a distance from the base equal to three-eighths the radius.

10.65. The area formed by the parabola $y^2 = 4ax$ and the lines $x = b$, $y = 0$ is revolved about the x axis. Show that $\bar{x}$ for the volume of the paraboloid so formed equals $\frac{2}{3}b$.

10.66. The first quadrant of the ellipse $x^2/a^2 + y^2/b^2 = 1$ is revolved about the x axis. Show that $\bar{x}$ for the volume generated equals $\frac{3}{8}a$.

10.67. The area bounded by the hyperbola $x^2/a^2 - y^2/b^2 = 1$ and the lines $y = 0$, $x = 2a$ is revolved about the x axis. Compute $\bar{x}$ for the solid so generated.   *Ans.* $\bar{x} = \frac{27}{16}a$

10.68. The area bounded by the curve $y = \sin x$ and lines $y = 0$, $x = 0$, $x = \pi/2$ is revolved about the x axis. Compute $\bar{x}$ for the volume generated.   *Ans.* $\bar{x} = \pi/4 + 1/2\pi$

10.69. The area formed by the curve $y = e^x$ and lines $x = 0$, $x = b$, $y = 0$ is revolved about the x axis. Locate the centroid of the volume formed.   *Ans.* $\bar{x} = (be^{2b} - \frac{1}{2}e^{2b} + \frac{1}{2})/(e^{2b} - 1)$

10.70. The area formed by the curve $x^2/a^2 + y^2/b^2 = 1$ and the lines $x = a$, $y = b$ is revolved about the x axis. Determine $\bar{x}$ for the generated volume.   *Ans.* $\bar{x} = \frac{3}{4}a$

10.71. Determine the centroid of volume for a right circular cone with a base diameter of 100 mm and an altitude of 200 mm.   *Ans.* $d = 50$ mm

10.72. A right circular cone with altitude 8 in and radius of base 6 in is welded to a 12-in-diameter hemisphere so that their bases coincide. Locate the centroid of the total volume.
Ans.   Distance from vertex of cone = 8.55 in

10.73. A right circular cone of altitude 250 mm and a base of diameter 200 mm rests on top of a right circular cylinder of the same base and 300 mm high. Locate the centroid of the composite volume.
Ans.   Distance from vertex of cone = 354 mm

10.74. A hemisphere of radius *a* rests on a right circular cylinder whose base also has radius *a*. If the height of the cylinder is *a*, locate the centroid of the composite volume.
Ans.   Distance from bottom of cylinder = 0.85*a*

10.75. The frustum of a right circular cone has an altitude of 4 ft. The radii of the bases are 2 ft and 4 ft, respectively. Locate the centroid of its volume.   *Ans.*   Distance from large base = 1.57 ft

10.76. From a hemisphere of radius *a* is cut a cone of the same base and altitude. Show that the centroid of the remaining volume is at a distance from the base equal to $\frac{1}{2}a$

10.77. A block is 600 mm × 600 mm × 1200 mm. A hole 150 mm in diameter is cut normal to the top (600-by-600 face) at its center. If the depth of the hole is 460 mm, locate the centroid of the remaining volume.   *Ans.*   Distance from bottom = 593 mm

10.78. Figure 10-49 illustrates a piece turned down in a lathe. Compute its centroid with reference to the left end.   *Ans.*   *d* = 5.48 in

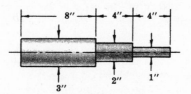

Fig. 10-49

10.79. Two spheres of volumes 5 in³ and 15 in³, respectively, are connected by a small rod whose volume is 10 in³. If the distance between centers is 10 in, how far from the center of the sphere of 5 in³ is the centroid?   *Ans.*   *d* = 6.67 in

10.80. A sphere of radius 400 mm has a spherical cavity of radius 100 mm. If the distance between the centers of the spheres is 200 mm, where is the centroid? Assume that the positive axis passes from the center of the large sphere through the center of the cavity.   *Ans.*   *d* = −3.18 mm.

10.81. Rework Problem 10.19 placing the vertex at the origin of the axes.   *Ans.* $d = -\frac{2}{3}h$

10.82. A tank consists of a cylinder 20 ft in diameter and 20 ft high. It has a hemispherical bottom and a cone-shaped cover 5 ft high. Locate the center of gravity of the empty tank by assuming that it is the same as the center of gravity of the outer surface.   *Ans.* $\bar{y}$ = 17.6 ft from bottom

10.83. A hemisphere whose radius is 75 mm is attached to a cone with the same size base and an altitude of 100 mm. Where is the centroid of the surface, assuming full opening between the two?
Ans.   Distance from vertex = 105 mm

10.84. An open tank is made of a right circular cylinder 36 in high with a diameter of 20 in. The bottom extends below the cylinder for 4 in as a cone with base coinciding with the cylinder. Determine the centroid of the surface.   *Ans.*   Distance from vertex = 19.5 in

10.85. The radius of the base of a right circular cone is 200 mm. Its altitude is 250 mm. Locate the centroid (*a*) of its sloping surface and (*b*) of its volume with respect to its base.
Ans. (*a*) $d = 83.3$ mm, (*b*) $d = 62.5$ mm

10.86. Find the mass center of the first octant of the homogeneous ellipsoid $x^2/a^2 + y^2/b^2 + z^2/c^2 = 1$.
Ans. $\bar{x} = \frac{3}{8}a$, $\bar{y} = \frac{3}{8}b$, $\bar{z} = \frac{3}{8}c$

10.87. Locate the mass center of a hemisphere in which the density at any point varies (*a*) with the first power from the base, and (*b*) with the square of the radial distance from the center.
Ans. (*a*) $d = \frac{8}{15}r$, (*b*) $d = 0.417r$

10.88. Locate the mass center of a right circular cone in which the density at any point varies directly as the distance of the point from the base.   *Ans.* $d = \frac{2}{5}h$

10.89. A mallet has a cylindrical wooden head and a cylindrical wooden handle. The head has a diameter of 4 in and a length of 6 in. The handle has a diameter of $1\frac{1}{4}$ in and is 12 in long. Where is the center of mass with respect to the free end of the handle? The wood weighs 50 lb/ft³.   *Ans.* $d = 12.7$ in

10.90. A cylinder 50 mm in diameter and 50 mm high is cut out of a right circular cone with 200-mm-diameter base and 250-mm height. The base of the cylinder lies in the plane of the base of the cone. If the cone is made of steel of density 7850 kg/m³, how far is the center of mass of the remaining mass above the base?   *Ans.* $d = 64.0$ mm

10.91. A tank 2.5 m in diameter has its center of mass 1.2 m above the bottom when empty. It is filled to a height of 1.8 m with oil (density 880 kg/m³). Locate the center of mass of the tank and oil if the tank has a mass of 3600 kg when empty.   *Ans.* Distance above base = 0.995 m

10.92. A 2-in-diameter hole is drilled 6 in deep in to the center and normal to the top face of a brass cube 8 in on a side. The hole is filled with lead. Brass weighs 525 lb/ft³ and lead weighs 710 lb/ft³. Locate the center of mass of the two metals with respect to the bottom.   *Ans.* 4.04 in

10.93. A steel cylinder has a 4-in-diameter base and a height of 6 in. A hole in the shape of a right circular cone with its base coincident with the base of the cylinder and axis coincident with the axis of the cylinder is filled with lead. The diameter of the base of the cone is 2 in and its height is 1 in. Steel weighs 490 lb/ft³ and lead weighs 710 lb/ft³. Locate the centroid with respect to the base.   *Ans.* $d = 2.97$ in

10.94. In Fig. 10-50, compute the distance from the left side of the composite mass to its center of mass. The large cylinder has a density of 7850 kg/m³, and the small cylinder has a density of 8500 kg/m³.
Ans. $d = 299$ mm

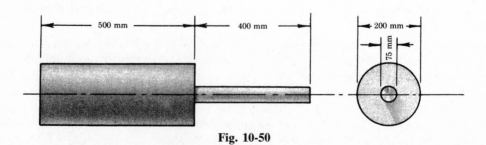

Fig. 10-50

10.95. In Fig. 10-51, determine the surface generated by revolving the arc of the semicircle of radius r about the y axis. Note that the centroid of the arc of the semicircle is at distance $\bar{x} = d + 2r/\pi$.
  Ans. $S = 2\pi^2 rd + 4\pi r^2$

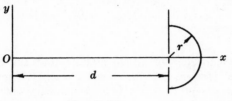

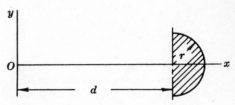

Fig. 10-51   **Fig. 10-52**

10.96. Determine the volume generated by revolving a semicircular area of radius r about the y axis. Refer to Fig. 10-52.   *Ans.* $V = \pi^2 r^2 d + \frac{4}{3}\pi r^3$

10.97. Find the volume generated by revolving the ellipse $x^2/a^2 + y^2/b^2 = 1$ about the line $x = 2a$.
  Ans. $V = 4\pi^2 a^2 b$

10.98. Figure 10-53 shows the flywheel on a small compressor. A cross section of the rim is assumed to be rectangular, with a semicircular groove of $\frac{1}{2}$-in radius cut therein. Compute the distance from the center to the centroid of the cross-sectional area shown at A–A. Using this centroidal distance, determine the volume of the rim only.   *Ans.* $V = 95$ in^3

10.99. A right triangle of base h and altitude r is revolved about its base through 360°. What volume does the triangular area generate?   *Ans.* $V = \frac{1}{3}\pi r^2 h$

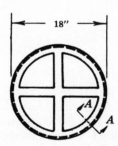

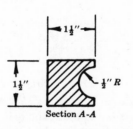

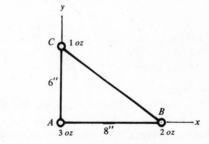

Fig. 10-53   **Fig. 10-54**

10.100. A right triangle ABC is shown in Fig. 10-54.

(a) Locate the centroid of the three lines forming the sides.

(b) Locate the centroid of the enclosed area.

(c) Assume that masses of 3, 2, and 1 oz are concentrated at the points A, B, and C, respectively, and locate the mass center.
  Ans.   (a) $\bar{x} = 3$ in, $\bar{y} = 2$ in; (b) $\bar{x} = 2.67$ in, $\bar{y} = 3$ in; (c) $\bar{x} = 2.67$ in, $\bar{y} = 1$ in

10.101. A wall 6 m high is subjected to water pressure on a vertical face. Where is the center of pressure of the water on the wall?   *Ans.*   4 m below the surface

10.102. In Problem 10.101, what is the turning moment about the base of the water on a section of the wall 12 m long and 1 m thick? Water has a density of 1000 kg/m^3.   *Ans.* 4.23×10^6 N · m or 4.23 MN · m

10.103. A beam 12 ft long carries a uniform unit force of 100 lb per linear ft for a distance of 4 ft from the left end. From that unit force of 100 lb per linear ft the force increases uniformly to a maximum of 300 lb per linear foot at the right end. What are the reactions at the ends?   *Ans.* $R_L = 780$ lb, $R_R = 1220$ lb

10.104. A beam is subjected to a uniform unit force of w lb/ft for one-third the length. The unit force then decreases uniformly to zero at the right end. What is the total force acting, and how is it distributed to the supports?   *Ans.* $P = \frac{2}{3}wl$ lb, $R_L = \frac{23}{54}wl$ lb, $R_R = \frac{13}{54}wl$ lb

10.105. The gate AB in a tank filled with water is inclined as shown in Fig. 10-55 and is 600 mm wide (perpendicular to the view shown). The gate is hinged at B and clamped at A. Determine (*a*) the total normal force acting on the gate due to the water and (*b*) the horizontal clamping force needed at A. *Ans.*   (*a*) 11.4 kN, (*b*) 5.3 kN

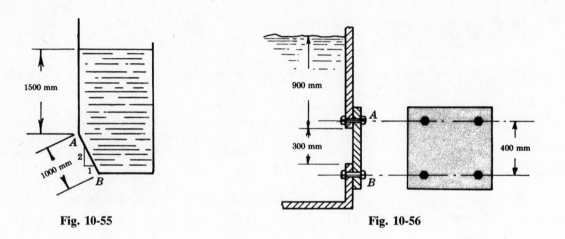

Fig. 10-55   Fig. 10-56

10.106. Refer to Fig. 10-56. A plate covers a 300-mm square hole in the side of the tank, which is filled with oil having a density of 800 kg/m³. Two bolts at A and two at B are used to secure the plate. Find the force in each bolt.   *Ans.*   Force in each bolt at $A = 179$ N, at $B = 192$ N

10.107. Suppose the plate in the preceding problem covers a circular hole 300 mm in diameter. Find the force in each bolt.   *Ans.*   Force in each bolt at $A = 140$ N, at $B = 150$ N

10.108. A concrete dam has the cross section shown in Fig. 10-57. Determine the reaction of the ground on the bottom of a section of the dam 1 ft long. Use the specific weight of water = 62.4 lb/ft³, of concrete = 150 lb/ft³. (*Hint:* Include in the free-body diagram the volume of water above the sloping face of the dam.)   *Ans.* $R_h = 7020$ lb, $R_v = 29,000$ lb at 5.64 ft to the right of A

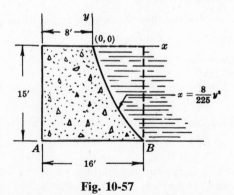

Fig. 10-57

10.109. A semicylindrical trough with a 3-ft radius is filled with fluid having a density of $100 \, \text{lb/ft}^3$. A gate at the end of the trough holds the fluid within the trough. Determine the magnitude of the resultant force on the gate and its location relative to the top.   *Ans.* $F = 1800 \, \text{lb}$, $y = 1.77 \, \text{ft}$ below top

10.110. Solve Prob. 10.108 if the face of the dam is flat instead of parabolic.
Ans. $R_h = 7000 \, \text{lb}$, $R_v = 31,000 \, \text{lb}$ at 5.92 ft to the right of A

10.111. Solve Prob. 10.109 if the cross section of the trough is triangular with a 3-ft depth and 3-ft width at the top.   *Ans.* $F = 450 \, \text{lb}$, $y = 1.5 \, \text{ft}$

10.112. In Fig. 10-58 what is the maximum height of dam and water before overturning is imminent?
Ans. $h = 37.4 \, \text{ft}$

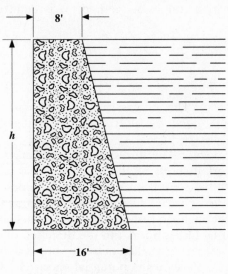

Fig. 10-58

Chapter 11

Virtual Work

11.1 VIRTUAL DISPLACEMENT AND VIRTUAL WORK

A *virtual displacement* δs *of a particle* is any arbitrary infinitesimal change in the position of the particle consistent with the constraints imposed on the particle. This displacement can be imagined; it does not have to take place.

Virtual work δU done by a force is defined as $F_t\,\delta s$, where F_t is the magnitude of the component of the force along the virtual displacement δs.

Virtual work δU done by a couple of moment M is defined as $M\,\delta\theta$, where $\delta\theta$ is the virtual angular displacement.

11.2 EQUILIBRIUM

Equilibrium of a particle: The necessary and sufficient condition for the equilibrium of a particle is zero virtual work done by all the forces acting on the particle during any virtual displacement δs.

Equilibrium of a rigid body: The necessary and sufficient condition for the equilibrium of a rigid body is zero virtual work done by all the external forces acting on the body during any virtual displacement consistent with the constraints imposed on the body.

Equilibrium of a connected system of rigid bodies: Defined as above for rigid bodies. Keep in mind that for a virtual displacement consistent with constraints no work is done by internal forces, by reactions at smooth pins, or by forces normal to the direction of motion. The external forces that do work (including friction if present) are called active or applied forces.

Equilibrium of a system: Exists if the potential energy V^* has a stationary value. Thus, if V is a function of one independent variable such as x then $dV/dx = 0$ will yield the equilibrium value(s) of x.

11.3 STABLE EQUILIBRIUM

Stable equilibrium occurs if the potential energy V is a minimum. If in Fig. 11-1(a) a bead is placed at the bottom of the frictionless wire bent as a circle, intuition indicates this is a position of stable equilibrium with the potential energy of the bead a minimum because any disturbance will be followed by a return to the bottom position. Using the x axis as a standard (datum), the potential energy of the bead any place below the x axis is

$$V = -Wy = -W\sqrt{a^2 - x^2}$$

Set dV/dx equal to zero to find the equilibrium position:

$$\frac{dV}{dx} = +\frac{Wx}{\sqrt{a^2 - x^2}} = 0$$

---

*The potential energy V of a mass m at a distance h above any plane chosen as a reference plane (datum) is mgh. If m is below the datum plane, V is $-mgh$.

The potential energy V of a spring with constant k and that is stretched or compressed a distance x from its unstressed position is $\frac{1}{2}kx^2$.

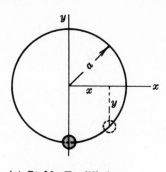

(a) Stable Equilibrium   (b) Unstable Equilibrium

Fig. 11-1

Hence, the solution needed here is $x = 0$ (the bead is then at the bottom). In determining the type of equilibrium, it is necessary to evaluate $d^2V\, dx^2$ at the equilibrium position. Thus,

$$\frac{d^2V}{dx^2} = +\frac{W}{\sqrt{a^2 - x^2}} + \frac{Wx^2}{(a^2 - x^2)^{3/2}}$$

and at $x = 0$, $d^2V/dx^2 = +W/a$ (positive), showing stable equilibrium.

11.4 UNSTABLE EQUILIBRIUM

Unstable equilibrium occurs if the potential energy V is a maximum. If in Fig. 11-1(b) the bead is placed at the top of the wire, intuition indicates that this is a position of unstable equilibrium, with the potential energy of the bead a maximum. Using the x axis as a standard (datum), the potential energy of the bead any place above the x axis is

$$V = +Wy = +W\sqrt{a^2 - x^2}$$

Set dV/dx equal to zero to find the equilibrium position:

$$\frac{dV}{dx} = -\frac{Wx}{\sqrt{a^2 - x^2}} = 0$$

Hence, the solution needed here is $x = 0$ (the bead is then at the top). Note also that

$$\frac{d^2V}{dx^2} = -\frac{W}{\sqrt{a^2 - x^2}} - \frac{Wx^2}{(a^2 - x^2)^{3/2}}$$

and at $x = 0$, $d^2V/dx^2 = -W/a$ (negative), showing unstable equilibrium.

11.5 NEUTRAL EQUILIBRIUM

Neutral equilibrium exists if a system remains in any position in which it is placed. For example, a bead can be located any place on a horizontal wire and will remain there.

11.6 SUMMARY OF EQUILIBRIUM

To determine the value(s) of the variable(s) for which a system is in equilibrium, express the potential energy V of the system as a function of the variable(s). In the foregoing discussion x was the variable. Then set $dV/dx = 0$ to determine the value(s) of x for equilibrium. Evaluate d^2V/dx^2 to ascertain the type of equilibrium:

$$\frac{d^2V}{dx^2} > 0 \qquad \text{equilibrium is stable}$$

$$\frac{d^2V}{dx^2} < 0 \qquad \text{equilibrium is unstable}$$

$$\frac{d^2V}{dx^2} = 0 \qquad \text{equilibrium is neutral}$$

Solved Problems

11.1. A homogeneous ladder having a mass M and length l is held in equilibrium by a horizontal force P as shown in Fig. 11-2. Using virtual work methods, express P in terms of M.

SOLUTION

Assuming that x is positive to the right, the virtual work done by P for an increase δx is $-P\,\delta x$, since P is directed to the left.

Assuming that y is positive up, the virtual work done by the downward force of gravity for an increase δy is $-Mg\,\delta y$.

The total virtual work δU is zero; hence

$$\delta U = -P\,\delta x - Mg\,\delta y = 0 \qquad\qquad (1)$$

From the figure, $x = l \sin\theta$ and $y = \frac{1}{2}l\cos\theta$. Then,

$$\delta x = l(\cos\theta)\,\delta\theta \qquad \text{and} \qquad \delta y = -\tfrac{1}{2}l(\sin\theta)\,\delta\theta$$

Substituting these values into equation (1), gives $P = \frac{1}{2}Mg\tan\theta$.

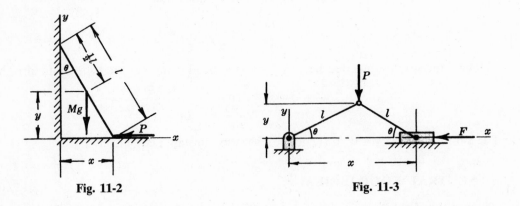

Fig. 11-2   Fig. 11-3

11.2. In the toggle device shown in Fig. 11-3, express the relationship between forces F and P in terms of angle θ.

SOLUTION

Assuming x positive to the right, the virtual work done by F for an increase δx is $-F\,\delta x$, because F is directed to the left. Assuming y positive up, the virtual work done by P for an increase δy is $-P\,\delta y$, because P is directed down. The total virtual work δU is zero; hence,

$$\delta U = -F\,\delta x - P\,\delta y = 0 \tag{1}$$

From the figure, $x = 2l \cos \theta$ and $y = l \sin \theta$. Then

$$\delta x = -2l(\sin \theta)\,\delta\theta \qquad \text{and} \qquad \delta y = l(\cos \theta)\,\delta\theta$$

Substituting these values into equation (1) gives $P = 2F \tan \theta$.

11.3. Using the method of virtual work, determine the relationship between the moment M applied to the crank R and the force F applied to the crosshead in the slider crank mechanism shown in Fig. 11-4.

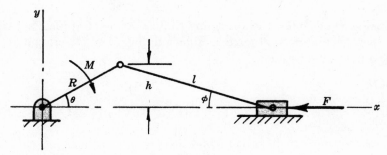

Fig. 11-4

SOLUTION

Assuming θ positive counterclockwise, the virtual work done by M for an increase $\delta\theta$ is $-M\,\delta\theta$. Assuming x positive to the right, the virtual work done by F for an increase δx is $-F\,\delta x$. The total virtual work δU is zero; hence,

$$\delta U = -M\,\delta\theta - F\,\delta x = 0 \tag{1}$$

But from the figure, $x = R \cos \theta + l \cos \phi$. To express ϕ in terms of θ, use $h = R \sin \theta = l \sin \phi$ and obtain $\cos \phi = \sqrt{1 - \sin^2 \phi} = \sqrt{1 - (R/l)^2 \sin^2 \theta}$.

Putting

$$\delta x = -R(\sin \theta)\,\delta\theta - \frac{R^2}{l}\,\frac{\sin \theta \cos \theta}{\sqrt{1 - (R/l)^2 \sin^2 \theta}}\,\delta\theta$$

into equation (1) gives

$$M = FR(\sin \theta)\left[1 + \frac{R \cos \theta}{l\sqrt{1 - (R/l)^2 \sin^2 \theta}}\right]$$

11.4. See Fig. 11-5. Using the method of virtual work, determine the value of F to hold the frame in equilibrium under the action of force P. Each link is of length $2a$, and $\theta = 45°$.

SOLUTION

Let θ be any angle in the analysis. Later the actual value $45°$ will be used. If there is an increase $\delta\theta$, P will rise doing negative work. The *magnitude* of the differential rise is $|\delta(a \cos \theta)| = a \sin \theta\,\delta\theta$. At the same time the F forces will do positive work. The *magnitude* of the differential movement of either F is $\delta(a \sin \theta) = a \cos \theta\,\delta\theta$.

Thus, $\delta U = -P(a \sin \theta\,\delta\theta) + 2F(a \cos \theta\,\delta\theta) = 0$.

The solution is $F = \frac{1}{2}P \tan \theta$; at $\theta = 45°$, $F = 0.5P$.

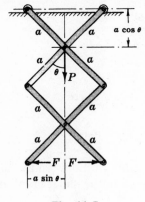

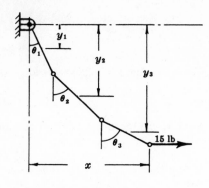

Fig. 11-5   Fig. 11-6

11.5. Three homogeneous links, each weighing 10 lb, and each 6 ft long, are held in the equilibrium position by a 15-lb horizontal force, as shown in Fig. 11-6. Determine the values of θ_1, θ_2, and θ_3 at the equilibrium position.

SOLUTION

For an increase δx, the 15-lb force will do positive work. For an increase δy_1, δy_2, and δy_3, the 10-lb weights will do positive work. Hence,

$$\delta U = +15\,\delta x + 10\,\delta y_1 + 10\,\delta y_2 + 10\,\delta y_3 = 0$$

From the figure,

$$x = 6 \sin \theta_1 + 6 \sin \theta_2 + 6 \sin \theta_3$$

$$y = 3 \cos \theta_1 \qquad y_2 = 6 \cos \theta_1 + 3 \cos \theta_2 \qquad y_3 = 6 \cos \theta_1 + 6 \cos \theta_2 + 3 \cos \theta_3$$

Hence,

$$\delta x = 6 \cos \theta_1 \, \delta\theta_1 + 6 \cos \theta_2 \, \delta\theta_2 + 6 \cos \theta_3 \, d\theta_3$$

$$\delta y_1 = -3 \sin \theta_1 \, \delta\theta_1$$

$$\delta y_2 = -6 \sin \theta_1 \, \delta\theta_1 - 3 \sin \theta_2 \, \delta\theta_2$$

$$\delta y_3 = -6 \sin \theta_1 \, \delta\theta_1 - 6 \sin \theta_2 \, \delta\theta_2 - 3 \sin_3 \, \delta\theta_3$$

Substituting,

$$\delta U = +15(6 \cos \theta_1 \, \delta\theta_1 + 6 \cos \theta_2 \, \delta\theta_2 + 6 \cos \theta_3 \, \delta\theta_3)$$

$$- 10(3 \sin \theta_1 \, \delta\theta_1) - 10(6 \sin \theta_1 \, \delta\theta_1 + 3 \sin \theta_2 \, \delta\theta_2)$$

$$- 10(6 \sin \theta_1 \, \delta\theta_1 + 6 \sin \theta_2 \, \delta\theta_2 + 3 \sin \theta_3 \, \delta\theta_3) = 0$$

To solve for θ_1, allow only $\delta\theta_1$ to exist. This leads to

$$(90 \cos \theta_1 - 30 \sin \theta_1 - 60 \sin \theta_1 - 60 \sin \theta_1)\,\delta\theta_1 = 0$$

from which $\tan \theta_1 = 90/150$ or $\theta_1 = 30°58'$.

To solve for θ_2, allow only $\delta\theta_2$ to exist, and obtain

$$(90 \cos \theta_2 - 30 \sin \theta_2 - 60 \sin \theta_2)\,\delta\theta_2 = 0$$

from which $\theta_2 = 45°$.

Similar reasoning yields $(90 \cos \theta_3 - 30 \sin \theta_3)\,\delta\theta_3 = 0$, from which $\theta_3 = 71°33'$.

Mathcad

11.6. The horizontal spring in Fig. 11-7 is compressed 5 in to the position shown. The coefficient of friction between the 100-lb block and the horizontal plane is 0.3. Using the method of virtual work, determine the moment M that must be applied to the vertical bar pivoted at the top in order to cause motion to the right.

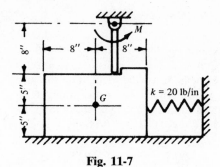

Fig. 11-7

SOLUTION

Assume that the block moves a distance δx to the right. Then the spring compresses from 5 to $5 + \delta x$ in. Also, the bar rotates through an angular distance $\delta x/8$. The normal force that the plane exerts on the block is 100 lb, and hence friction is equal to 30 lb. The total virtual work done is equal to the negative work of the friction, the negative work of the spring, and the positive work of the moment. Hence,

$$\delta U = -30\,\delta x - \tfrac{1}{2}(20)[(5 + \delta x)^2 - 5^2] + M\frac{\delta x}{8} = 0$$

or

$$-30\,\delta x - 10(25 + 10\,\delta x + \delta x^2 - 25) + M\frac{\delta x}{8} = 0$$

Since δx is small, we neglect δx^2, and the expression becomes

$$-30\,\delta x - 100\,\delta x + M\frac{\delta x}{8} = 0$$

Hence, $M = 1040$ lb-in.

11.7. Using the principles of virtual work, determine the components of the pin reactions at A and B in Fig. 11-8(a). Neglect friction at all pins. The force at E is horizontal.

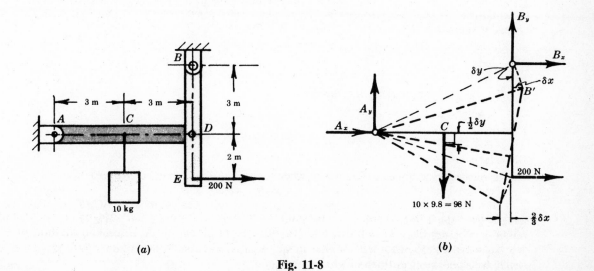

(a)   (b)

Fig. 11-8

SOLUTION

A free-body diagram in Fig. 11-8(b) shows the pin B removed and the frame given a virutal angular displacement $\delta\theta$ about pin A. The virtual displacement of B to B' can be thought of as a displacement δy down and a displacement δx to the right. For the small angular displacement, all points on the member BDE move down an amount δy; hence, point C will undergo one-half the displacement of D. Since the bar BDE rotates through the angle $\delta\theta$, point E undergoes a horizontal displacement which is two-thirds that of point B ($\frac{2}{3}\,\delta x$). The pin reactions at A do no work. The virtual work is

$$\delta U = +98(\tfrac{1}{2}\,\delta y) - B_y\,\delta y + B_x\,\delta x - 200(\tfrac{2}{3}\,\delta x) = 0$$

or
$$\delta U = (49 - B_y)\,\delta y + (B_x - 133)\,\delta x = 0$$

Now since δx and δy are completely independent and arbitrary virtual displacements, δU will be zero if and only if

$$49 - B_y = 0 \qquad \text{and} \qquad B_x - 133 = 0$$

From these, $B_y = 49\,\text{N}$ up and $B_x = 133\,\text{N}$ to right.

The equations of equilibrium can be applied as follows:

$$\sum F_x = A_x + B_x + 200 = 0 \qquad \text{or} \qquad A_x = 333\,\text{N to left}$$

$$\sum F_y = A_y - 98 + B_y = 0 \qquad \text{or} \qquad A_y = 49\,\text{N up}$$

Without the method of virtual work, it would have been necessary to use several free-body diagrams of parts of the frame.

11.8. Two weights W and w are supported as shown in Fig. 11-9 by a bar of negligible weight that is pivoted about an axis at O perpendicular to the plane of the paper. Discuss equilibrium.

DISCUSSION

If $\theta = 0°$ is selected as the standard or reference configuration then, at any angle θ, W has lost potential energy $Wa(1 - \cos\theta)$ and w has gained $wb(1 - \cos\theta)$. The total potential energy V of the system at angle θ is

$$V = -Wa(1 - \cos\theta) + wb(1 - \cos\theta)$$

To study equilibrium, set $dV/d\theta = 0$:

$$\frac{dV}{d\theta} = -Wa \sin\theta + wb \sin\theta = 0$$

This is satisfied when $\sin\theta = 0$ ($\theta = 0°$) or when $wb = Wa$ (θ may then be any angle).

To discuss stability, find $d^2V/d\theta^2 = -Wa \cos\theta + wb \cos\theta$.

Next evaluate $d^2V/d\theta^2$ for the two conditions of equilibrium. At $\theta = 0°$, $d^2V/d\theta^2 = -Wa + wb$. For stable equilibrium, the value must be positive; this obtains when $wb > Wa$. The other state of equilibrium occurs when $wb = Wa$, that is, when $d^2V/d\theta^2 = 0$; this means neutral equilibrium.

To summarize, if $wb > Wa$, the system will be in stable equilibrium when $\theta = 0°$. If $wb = Wa$, the system will stay in any position in which it is placed.

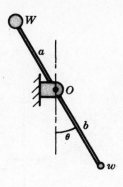

Fig. 11-9

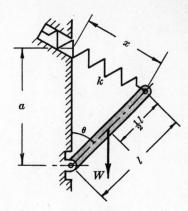

Fig. 11-10

11.9. A uniform bar of length l and weight W is held in equilibrium by a spring with constant k. When $\theta = 0°$, the spring is not stressed. Discuss equilibrium.

DISCUSSION

Refer to Fig. 11-10. Using $\theta = 0°$ as the standard or reference configuration, the bar in position θ has lost potential energy equal to $\frac{1}{2}Wl(1 - \cos \theta)$ and the spring has gained an amount $\frac{1}{2}kx^2$. From the geometry, $x^2 = a^2 + l^2 - 2al \cos \theta$. Hence, the potential energy V for the system at any angle θ is

$$V = -\tfrac{1}{2}Wl(1 - \cos \theta) + \tfrac{1}{2}k(a^2 + l^2 - 2al \cos \theta)$$

and

$$\frac{dV}{d\theta} = -\tfrac{1}{2}Wl \sin \theta + kal \sin \theta$$

which will be zero for $\sin \theta = 0$ ($\theta = 0°$) or for $k = \frac{1}{2}W/a$.

Next find $d^2V/d\theta^2 = -\frac{1}{2}Wl \cos \theta + kal \cos \theta$. At $\theta = 0°$, $d^2V/d\theta^2 = -\frac{1}{2}Wl + kal$; this will be positive if $k > \frac{1}{2}W/a$. At any other angle, $d^2V/d\theta^2 = 0$ if $k = \frac{1}{2}W/a$.

In conclusion, if $k = \frac{1}{2}W/a$, the system will remain in any position in which it is placed.

11.10. In Fig. 11-11, the two identical gears rotate about frictionless pivots. A weightless 2-ft bar rigidly attached to the gear holds the 20-lb weight. The other gear is attached to a vertical spring with constant $k = 12$ lb/in. Determine the angle(s) θ for equilibrium.

SOLUTION

If $\theta = 0°$ is used as the standard or reference configuration and the spring is unstressed, then for any other angle θ the weight loses a potential energy equal to $Wl(1 - \cos \theta)$. The spring gains an amount $\frac{1}{2}k(r\theta)^2$. The total potential energy of the system is

$$V = -Wl(1 - \cos \theta) + \tfrac{1}{2}kr^2\theta^2$$

Then,

$$\frac{dV}{d\theta} = -Wl \sin \theta + kr^2\theta \qquad \frac{d^2V}{d\theta^2} = -Wl \cos \theta + kr^2$$

The first derivative will be zero if $\theta = 0°$ or if $\sin \theta = (kr^2/Wl)\theta$.

At $\theta = 0°$, $d^2V/d\theta^2 = -Wl + kr^2 = -20(2) + (144)(\frac{1}{2})^2 = -4$ (not stable).

The other angle for equilibrium is determined from

$$\sin \theta = \frac{kr^2}{Wl}\theta = \frac{144(\frac{1}{2})^2}{20(2)}\theta = 0.9\theta$$

which, by trial and error, is satisfied by $\theta = 44.1°$. Then $d^2V/d\theta^2 = -20(2) \cos 44.1° + 36 = +7.2$, which indicates stable equilibrium.

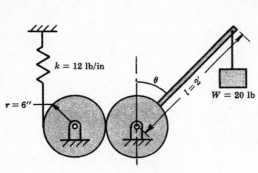

Fig. 11-11   Fig. 11-12

11.11. A uniform rope with mass M is placed on a sphere of radius r as shown in Fig. 11-12. Find the tension T in the rope when it is in a horizontal plane at a vertical distance b below the top. Use the method of virtual work.

SOLUTION

The rope is at a height y above the xz plane. Its length is $l = 2\pi x = 2\pi\sqrt{r^2 - y^2}$.

If the rope is given a virtual displacement δy down, the virtual work done is

$$\delta U = +Mg\,\delta y + T\,\delta l = 0$$

Substituting

$$\delta l = \frac{2\pi(\frac{1}{2})(-2y\,\delta y)}{(r^2 - y^2)^{1/2}} = \frac{-2\pi y\,\delta y}{(2br - b^2)^{1/2}}$$

we obtain

$$T = \frac{Mg(2br - b^2)^{1/2}}{2\pi(r - b)}$$

Supplementary Problems

11.12. Using the method of virtual work, find the force T in the horizontal cross member in terms of the load P and the angle θ. See Fig. 11-13.   *Ans.* $T = \frac{3}{4}P\tan\theta$

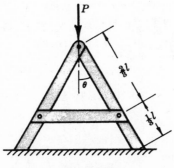

Fig. 11-13

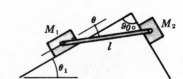

Fig. 11-14

11.13. In Fig. 11-14 the masses M_1 and M_2 are held on the frictionless orthogonal planes by a rigid inextensible bar of length l. Find the equilibrium angle θ.   *Ans.* $\tan\theta = (M_2/M_1)\cos\theta_1$

11.14. Refer to Fig. 11-15. Using the method of virtual work, find the force P to hold the mass M in equilibrium.   *Ans.* $P = \frac{1}{2}Mg$

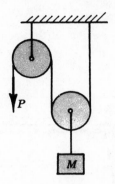

Fig. 11-15

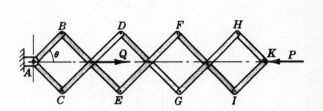

Fig. 11-16

11.15. In Fig. 11-16 the bars AB, AC, HK, and KI are a ft long. The other bars are $2a$ ft long. The bars are connected by frictionless pins. Using the method of virtual work, determine the relationship between P and Q.   *Ans.* $Q = 4P$

11.16. In Fig. 11-17 a spring with its free end at A is stretched 2 in from its unstressed position. At B it is stretched 4 in from its unstressed position. If the spring constant $k = 10$ lb/in, find the work done against the spring force in moving the free end from A to B.   *Ans.* $U = 60$ in-lb

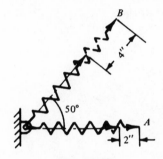

Fig. 11-17

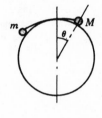

Fig. 11-18

11.17. A chain of weight W is placed on a right circular cone of height h and base radius r. Find the tension T in the chain when it is in a horizontal plane a distance b below the apex.   *Ans.* $T = Wh/2\pi r$

11.18. The two masses m and M are connected by a light inextensible cord, as shown in Fig. 11-18, on the smooth cylindrical surface. If the angle between their respective radii is 90°, determine the angle of equilibrium. Is it stable equilibrium?   *Ans.* $\tan \theta = m/M$; no

11.19. Suppose the masses of the preceding problem are connected as shown in Fig. 11-19. Determine the angle of equilibrium. What type of equilibrium is it?   *Ans.* $\sin \theta = m/M$, unstable

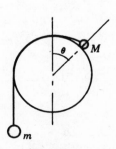

Fig. 11-19

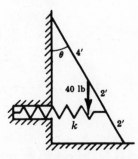

Fig. 11-20

11.20. In Fig. 11-20 the 40-lb homogeneous ladder rests on smooth surfaces. The spring is unstretched when $\theta = 0°$. Study equilibrium conditions if spring constant $k = 50$ lb/ft.
Ans. $\theta = 0°$, stable equilibrium; $\theta = 27.2°$, unstable equilibrium

11.21. Fig. 11-21 shows a Roberval balance in which $r_2 > r_1$ and the weights of the members are considered negligible. What is the relation between R_1 and R_2 for equilibrium?   *Ans.* $R_1 = R_2$

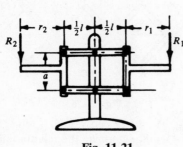

Fig. 11-21

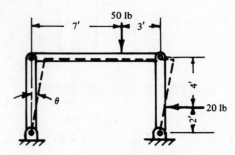

Fig. 11-22

11.22. Determine the angle θ for equilibrium of the three-bar linkage shown in Fig. 11-22. What type of equilibrium exists at that angle?   *Ans.* $\theta = 7.59°$, unstable

11.23. A homogeneous bar has a mass of 50 kg and is 3 m long as shown in Fig. 11-23. What are the values of the spring constant k to insure stable equilibrium? The spring is undeformed in Fig. 11-23.
Ans. $k > 81.7$ N/m

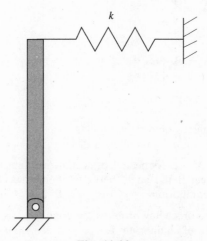

Fig. 11-23

Chapter 12

Kinematics of a Particle

12.1   KINEMATICS

Kinematics is the study of motion without regard to the forces or other factors that influence the motion.

12.2   RECTILINEAR MOTION

Rectilinear motion is motion of a point P along a straight line, which for convenience here will be chosen as the x axis. Vector symbols are omitted in this part.

(a) The *position* of point P at any time t is expressed in terms of its distance x from a fixed origin O on the x axis. This distance x is positive or negative according to the usual sign convention.

(b) The *average velocity* v_{av} of point P during the time interval between t and $t + \Delta t$ during which its position changes from x to $x + \Delta x$ is the quotient $\Delta x / \Delta t$. Mathematically this is written

$$v_{av} = \frac{\Delta x}{\Delta t} \tag{1}$$

(c) The *instantaneous velocity* v of point P at time t is the limit of the average velocity as the increment of time approaches zero as a limit. Mathematically this is written

$$v = \lim_{\Delta t \to 0} \frac{\Delta x}{\Delta t} = \frac{dx}{dt} \tag{2}$$

(d) The *average acceleration* a_{av} of point P during the time interval between t and $t + \Delta t$ during which its velocity changes from v to $v + \Delta v$ is the quotient $\Delta v / \Delta t$. Mathematically this is written

$$a_{av} = \frac{\Delta v}{\Delta t} \tag{3}$$

(e) The *instantaneous acceleration* a of point P at time t is the limit of the average acceleration as the increment of time approaches zero as a limit. Mathematically this is written

$$a = \lim_{\Delta t \to 0} \frac{\Delta v}{\Delta t} = \frac{dv}{dt} = \frac{d^2 x}{dt^2} \tag{4}$$

Also,

$$a = \frac{dv}{dx}\frac{dx}{dt} = v\frac{dv}{dx}$$

(f) For *constant acceleration* $a = k$ the following formulas are valid:

$$v = v_0 + kt \tag{5}$$

$$v^2 = v_0^2 + 2ks \tag{6}$$

$$s = v_0 t + \tfrac{1}{2}kt^2 \tag{7}$$

$$s = \tfrac{1}{2}(v + v_0)t \tag{8}$$

where v_0 = initial velocity
 v = final velocity
 k = constant acceleration
 t = time
 s = displacement

(*g*) *Simple harmonic motion* is rectilinear motion in whch the acceleration is negatively proportional to the displacement. Mathematically this is written

$$a = -K^2 x \qquad (9)$$

As an example, equation (*9*) is satisfied by a point vibrating so that its displacement x is given by the equation

$$x = b \sin \omega t \qquad (10)$$

where b = amplitude in linear measure
       ω = constant *circular* frequency in radians per second
       t = time in seconds

Thus, since $x = b \sin \omega t$, then $v = dx/dt = \omega b \cos \omega t$ and $a = d^2x/dt^2 = -\omega^2 b \sin \omega t = -\omega^2 x$. That is, $a = -K^2 x$, where $K = \omega$, a constant, and the motion is simple harmonic.

12.3 CURVILINEAR MOTION

Curvilinear motion in a plane is motion along a plane curve (path). The velocity and acceleration of a point on such a curve will be expressed in (*a*) rectangular components, (*b*) tangential and normal components, and (*c*) radial and transverse components.

12.4 RECTANGULAR COMPONENTS

The position vector **r** of a point P on such a curve in terms of the unit vectors **i** and **j** along the x and y axes, respectively, is written

$$\mathbf{r} = x\mathbf{i} + y\mathbf{j}$$

As P moves, **r** changes and the velocity **v** can be expressed as

$$\mathbf{v} = \frac{d\mathbf{r}}{dt} = \frac{dx}{dt}\mathbf{i} + \frac{dy}{dt}\mathbf{j}$$

Using $dx/dt = \dot{x}$ and $dy/dt = \dot{y}$ and $d\mathbf{r}/dt = \dot{\mathbf{r}}$ as convenient symbols, we have

$$\mathbf{v} = \dot{\mathbf{r}} = \dot{x}\mathbf{i} + \dot{y}\mathbf{j} \qquad (11)$$

The speed of the point is the magnitude of the velocity **v**; that is,

$$|\mathbf{v}| = \sqrt{\dot{x}^2 + \dot{y}^2}$$

If θ is the angle which the vector **v** makes with the x axis, we can write

$$\theta = \tan^{-1}\frac{\dot{y}}{\dot{x}} = \tan^{-1}\frac{dy/dt}{dx/dt} = \tan^{-1}\frac{dy}{dx}$$

Thus, the velocity vector **v** is tangent to the path at point P (see Fig. 12-1).

The acceleration vector **a** is the time rate of change of **v**; that is,

$$\mathbf{a} = \frac{d\mathbf{v}}{dt} = \frac{d^2\mathbf{r}}{dt^2} = \frac{d^2x}{dt^2}\mathbf{i} + \frac{d^2y}{dt^2}\mathbf{j}$$

Using the symbolic notation $\mathbf{a} = \dot{\mathbf{v}} = \ddot{\mathbf{r}}$, $\ddot{x} = d^2x/dt^2$, and $\ddot{y} = d^2y/dt^2$, we can write

$$\mathbf{a} = \dot{\mathbf{v}} = \ddot{\mathbf{r}} = \ddot{x}\mathbf{i} + \ddot{y}\mathbf{j} \tag{12}$$

The magnitude of the acceleration vector **a** is

$$|\mathbf{a}| = \sqrt{\ddot{x}^2 + \ddot{y}^2}$$

Do not make the mistake of assuming that **a** is tangent to the path at point P.

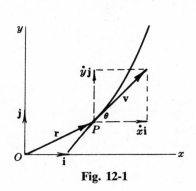

Fig. 12-1

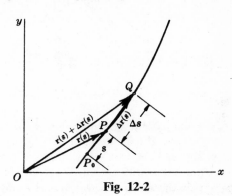

Fig. 12-2

12.5 TANGENTIAL AND NORMAL COMPONENTS

In the preceding discussion the velocity vector **v** and acceleration vector **a** were expressed in terms of the orthogonal unit vectors **i** and **j** along the x and y axes, respectively. The following discussion shows how to express the same vector **v** and the same vector **a** in terms of the unit vector $\mathbf{e}_t$ tangent to the path at point P and the unit vector $\mathbf{e}_n$ at right angles to $\mathbf{e}_t$.

In Fig. 12-2, point P is shown on the curve at a distance s along the curve from a reference point P_0. The position vector **r** of point P is a function of the scalar quantity s. To study this relationship, let Q be a point on the curve near P.

The position vectors $\mathbf{r}(s)$ and $\mathbf{r}(s) + \Delta\mathbf{r}(s)$ for points P and Q, respectively, are shown as well as the change $\Delta\mathbf{r}(s)$, which is the directed straight line PQ. The distance along the curve from P to Q is Δs. The derivative of $\mathbf{r}(s)$ with respect to s is written

$$\frac{d\mathbf{r}(s)}{ds} = \lim_{\Delta s \to 0} \frac{\mathbf{r}(s) + \Delta\mathbf{r}(s) - \mathbf{r}(s)}{\Delta s} = \lim_{\Delta s \to 0} \frac{\Delta\mathbf{r}(s)}{\Delta s}$$

As Q approaches P, the ratio of the magnitude of the straight line $\Delta\mathbf{r}(s)$ to the arc length Δs approaches unity. Also, in direction the straight line $\Delta\mathbf{r}(s)$ approaches the tangent to the path at P. Thus, in the limit, a unit vector $\mathbf{e}_t$ is defined as follows:

$$\frac{d\mathbf{r}(s)}{ds} = \mathbf{e}_t \tag{13}$$

Next consider how $\mathbf{e}_t$ changes with s. As shown in Fig. 12-3(a), the center of curvature C is a distance ρ from P.

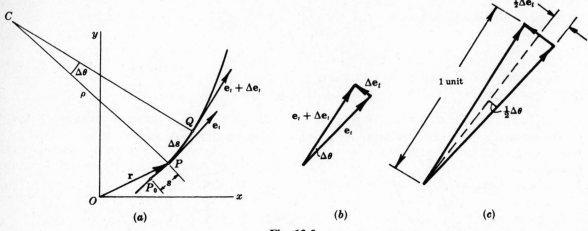

Fig. 12-3

If we assume point Q is close to P, the unit tangent vectors at P and Q are $\mathbf{e}_t$ and $\mathbf{e}_t + \Delta\mathbf{e}_t$, respectively. Since the tangents at P and Q are perpendicular to the radii drawn to C, the angle between $\mathbf{e}_t$ and $\mathbf{e}_t + \Delta\mathbf{e}_t$ as shown in Fig. 12-3(b) is also $\Delta\theta$. Because $\mathbf{e}_t$ and $\mathbf{e}_t + \Delta\mathbf{e}_t$ are unit vectors, $\Delta\mathbf{e}_t$ represents only a change in direction (but not magnitude). Thus the triangle in Fig. 12-3(b) is isosceles and is shown drawn to a larger scale in Fig. 12-3(c). From Fig. 12-3(c) it should be evident that

$$\frac{|\frac{1}{2}\Delta\mathbf{e}_t|}{1} = \sin\frac{1}{2}\Delta\theta \approx \frac{1}{2}\Delta\theta \qquad \text{from which} \qquad |\Delta\mathbf{e}_t| \approx \Delta\theta$$

But from Fig. 12-3(a), $\Delta s = \rho\,\Delta\theta$; hence, we can write $\Delta s \approx \rho\,|\Delta\mathbf{e}_t|$. Thus,

$$\lim_{\Delta s\to 0}\frac{|\Delta\mathbf{e}_t|}{\Delta s} = \frac{1}{\rho}$$

Also, in the limit $\Delta\mathbf{e}_t$ is perpendicular to $\mathbf{e}_t$ and is directed toward the center of curvature C. Let $\mathbf{e}_n$ be the unit vector that is perpendicular to $\mathbf{e}_t$ and directed toward the center of curvature C. Then

$$\frac{d\mathbf{e}_t}{ds} = \lim_{\Delta s\to 0}\frac{|\Delta\mathbf{e}_t|}{\Delta s}\mathbf{e}_n = \frac{1}{\rho}\mathbf{e}_n \qquad (14)$$

The velocity vector $\mathbf{v}$ may now be given in terms of the unit vectors $\mathbf{e}_t$ and $\mathbf{e}_n$. Using equation (13), and noting $ds/dt = \dot{s}$ is the speed of P along the path, we can write

$$\mathbf{v} = \frac{d\mathbf{r}}{dt} = \frac{d\mathbf{r}}{ds}\frac{ds}{dt} = \dot{s}\mathbf{e}_t \qquad (15)$$

The acceleration vector $\mathbf{a}$ is the time derivative of the velocity vector $\mathbf{v}$ defined in equation (15):

$$\mathbf{a} = \frac{d\mathbf{v}}{dt} = \ddot{s}\mathbf{e}_t + \dot{s}\frac{d\mathbf{e}_t}{dt}$$

But,

$$\frac{d\mathbf{e}_t}{dt} = \frac{d\mathbf{e}_t}{ds}\frac{ds}{dt}$$

and from (14) this may be written

$$\frac{d\mathbf{e}_t}{dt} = \frac{\dot{s}}{\rho}\mathbf{e}_n$$

Then

$$\mathbf{a} = \ddot{s}\mathbf{e}_t + \frac{\dot{s}^2}{\rho}\mathbf{e}_n \qquad (16)$$

Note that $\ddot{s}$ along the tangent is the time rate of change of the speed of the point.

12.6 RADIAL AND TRANSVERSE COMPONENTS

The point P on the curve may be located with polar coordinates in terms of any point chosen as a pole. Figure 12-4 shows the origin O as the pole. These polar coordinates are useful in studying the motion of planets and other central force problems. The velocity vector $\mathbf{v}$ and the acceleration vector $\mathbf{a}$ are now derived in terms of unit vectors along and perpendicular to the radius vector. Note that there is an infinite set of unit vectors because any point may be chosen as a pole.

The radius vector $\mathbf{r}$ makes an angle ϕ with the x axis. The unit vector $\mathbf{e}_r$ is chosen outward along $\mathbf{r}$. The unit vector $\mathbf{e}_\phi$ is perpendicular to $\mathbf{r}$ and in the direction of increasing ϕ.

Since the vector $\mathbf{r}$ is r units long in the $\mathbf{e}_r$ direction, we can write

$$\mathbf{r} = r\mathbf{e}_r \qquad (17)$$

The velocity vector $\mathbf{v}$ is the time derivative of the product in equation (17):

$$\mathbf{v} = \dot{\mathbf{r}} = \dot{r}\mathbf{e}_r + r\dot{\mathbf{e}}_r$$

where $\dot{\mathbf{e}}_r = d\mathbf{e}_r/dt$.

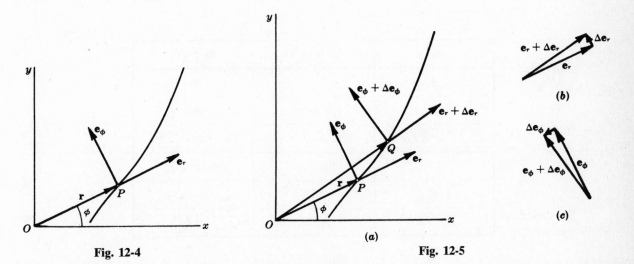

Fig. 12-4   Fig. 12-5

To evaluate $\dot{\mathbf{e}}_r$ and $\dot{\mathbf{e}}_\phi$, allow P to move to a nearby point Q with a corresponding set of unit vectors $\mathbf{e}_r + \Delta\mathbf{e}_r$ and $\mathbf{e}_\phi + \Delta\mathbf{e}_\phi$ as shown in Fig. 12-5(a).

Figures 12-5(b) and (c) illustrate these unit vectors. Since the triangles are isosceles, we can deduce the following conclusions by reasoning similar to that used in the explanation of $\mathbf{e}_t$ and $\mathbf{e}_n$ vectors: $d\mathbf{e}_r$ in the limit has a magnitude $d\phi$ in the $\mathbf{e}_\phi$ direction, and $d\mathbf{e}_\phi$ in the limit has a magnitude $d\phi$ in the negative $\mathbf{e}_r$ direction. Hence,

$$\dot{\mathbf{e}}_r = \frac{d\mathbf{e}_r}{d\phi}\frac{d\phi}{dt} = \dot{\phi}\mathbf{e}_\phi \qquad \text{and} \qquad \dot{\mathbf{e}}_\phi = \frac{d\mathbf{e}_\phi}{d\phi}\frac{d\phi}{dt} = -\dot{\phi}\mathbf{e}_r$$

where $\dot{\phi}$ is the time rate of change of the angle ϕ that the radius vector **r** makes with the x axis.

The velocity vector **v** may now be written

$$\mathbf{v} = \dot{r}\mathbf{e}_r + r\dot{\phi}\mathbf{e}_\phi \tag{18}$$

The acceleration vector **a** is the time derivative of the terms in equation (18):

$$\mathbf{a} = \ddot{r}\mathbf{e}_r + \dot{r}\dot{\mathbf{e}}_r + \dot{r}\dot{\phi}\mathbf{e}_\phi + r\ddot{\phi}\mathbf{e}_\phi + r\dot{\phi}\dot{\mathbf{e}}_\phi$$
$$= \ddot{r}\mathbf{e}_r + \dot{r}\dot{\phi}\mathbf{e}_\phi + \dot{r}\dot{\phi}\mathbf{e}_\phi + r\ddot{\phi}\mathbf{e}_\phi - r\dot{\phi}^2\mathbf{e}_r$$

where $\ddot{\phi}$ is the angular acceleration (time derivative of the angular velocity $\dot{\phi}$). Collecting terms, this becomes

$$\mathbf{a} = (\ddot{r} - r\dot{\phi}^2)\mathbf{e}_r + (2\dot{r}\dot{\phi} + r\ddot{\phi})\mathbf{e}_\phi \tag{19}$$

As a special case of curvilinear motion consider a point moving in a circular path of radius R. Substituting R for r in equations (18) and (19), noting $\dot{R} = \ddot{R} = 0$, we obtain

$$\mathbf{v} = R\dot{\phi}\mathbf{e}_\phi \quad \text{(tangent to the path)} \tag{20}$$
$$\mathbf{a} = -R\dot{\phi}^2\mathbf{e}_r + R\ddot{\phi}\mathbf{e}_\phi \tag{21}$$

Thus, the acceleration has a tangential component of magnitude $R\ddot{\phi}$ and a normal component directed toward the center of magnitude $R\dot{\phi}^2$.

12.7   UNITS

Units have been omitted purposely in the foregoing discussion. The following table lists the units used in the U.S. Customary (so-called engineering) System and in SI.

Symbol	Engineering Units	SI Units
s, ρ, R, x, y	ft	m
$v, \dot{x}, \dot{y}, \dot{s}$	ft/s or fps	m/s
$a, \ddot{x}, \ddot{y}, \ddot{s}$	ft/s^2	m/s^2
θ, ϕ	radians (rad)	radians (rad)
$\omega, \dot{\theta}, \dot{\phi}$	rad/s	rad/s
$\alpha, \ddot{\theta}, \ddot{\phi}$	rad/s^2	rad/s^2

Solved Problems

12.1. A rocket car moves along a straight track according to the equation $x = 3t^3 + t + 2$, where x is in feet and t is in seconds. Determine the displacement, velocity, and acceleration when $t = 4$ s.

SOLUTION

$$x = 3t^3 + t + 2 = 3(4)^3 + 4 + 2 = 198 \text{ ft}$$

$$v = \frac{dx}{dt} = 9t^2 + 1 = 9(4)^2 + 1 = 145 \text{ ft/s}$$

$$a = \frac{dv}{dt} = 18t = 18(4) = 72 \text{ ft/s}^2$$

12.2. In Problem 12.1, what is the average acceleration during the fifth second?

SOLUTION

The velocity at the end of the fifth second is $v = 9(5)^2 + 1 = 226$ ft/sec. Hence, the change in velocity during the fifth second is 226 ft/s − 145 ft/s = 81 ft/s.
The average acceleration is

$$a_{av} = \frac{\Delta v}{\Delta t} = \frac{81 \text{ ft/2}}{1 \text{ s}} = 81 \text{ ft/s}^2$$

12.3. A point moves along a straight line such that its displacement is $s = 8t^2 + 2t$, where s is in meters and t is in seconds. Plot the displacement, velocity, and accelertaion against time. These are called s–t, v–t, a–t diagrams.

SOLUTION

Differentiating $s = 8t^2 + 2t$ yields $v = ds/dt = 16t + 2$ and $a = dv/dt = d^2s/dt^2 = 16$.
This shows that the acceleration is constant, 16 m/s².
To determine values for plotting, use the following tabular form, where t is in seconds, s is in meters, and v is in meters per second.

t	t^2	$8t^2$	$2t$	$s = 8t^2 + 2t$	$16t$	$v = 16t + 2$
0	0	0	0	0	0	2
1	1	8	2	10	16	18
2	4	32	4	36	32	34
3	9	72	6	78	48	50
4	16	128	8	136	64	66
5	25	200	10	210	80	82
10	100	800	20	820	160	162

These data are plotted in the s, v, and a diagrams on the next page. Some valuable relationships may be deduced from these diagrams. The slope of the s–t curve at any time t is the height or ordinate of the v–t curve at time t. This follows since $v = ds/dt$.
Again, the slope of the v–t curve (in this particular case the slope is the same at any point of the straight line, i.e., 16 m/s²) at any time t is the ordinate of the a–t curve at any time t. This follows since $a = dv/dt$.
The two equations just given may also be written as

$$a \, dt = dv \quad \text{and} \quad v \, dt = ds$$

Integration between proper limits yields

$$\int_{t_0}^{t} a \, dt = \int_{v_0}^{v} dv = v - v_0 \quad \text{and} \quad \int_{t_0}^{t} v \, dt = \int_{s_0}^{s} ds = s - s_0 \qquad (1)$$

where $\int_{t_0}^{t} a\, dt$ = area under a–t diagram for time interval from t_0 to t
   $\int_{t_0}^{t} v\, dt$ = area under v–t diagram for time interval from t_0 to t
   $v - v_0$ = change in velocity in same time interval t_0 to t
   $s - s_0$ = change in displacement in same time interval t_0 to t.

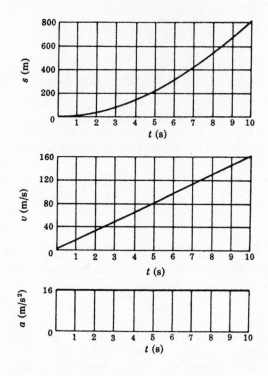

The first equation in (*1*) states that the change in the ordinate of the v–t diagram for any time interval is equal to the area under the a–t diagram within that time interval. A similar statement may be made for the change in the ordinate of the s–t diagram in the second equation in (*1*).

12.4. An automobile accelerates uniformly from rest to 60 m/h in 28 s. Find its constant acceleration and its displacement during this time.

SOLUTION

The following data are given: $v_0 = 0$, $v = 60\,\text{mi/h} = 88\,\text{ft/s}$, $t = 28\,\text{s}$.
To determine the acceleration, which is a constant k, apply the formula $v = v_0 + kt$.

$$k = \frac{v - v_0}{t} = \frac{(88 - 0)\,\text{ft/s}}{28\,\text{s}} = 3.14\,\text{ft/s}^2$$

To determine the displacement using only the original data,

$$s = \frac{v + v_0}{2}\, t = \frac{(88 + 0)\,\text{ft/s}}{2} \times 28\,\text{s} = 1230\,\text{ft}$$

12.5. A particle moves with rectilinear motion. The speed increases from zero to 30 ft/s in 3 s and then decreases to zero in 2 s.

(*a*) Plot the v–t curve

(*b*) What is the acceleration during the first 3 s and during the next 2 s?

(*c*) What is the distance traveled in the 5 s?

(*d*) How long does it take the particle to go 50 ft?

SOLUTION

(*a*) The plot of the v–t curve is shown in Fig. 12-6.

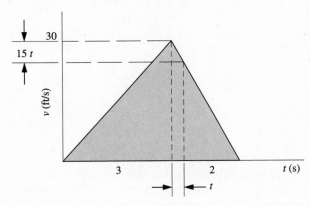

Fig. 12-6

(*b*) The acceleration is the time derivative of the velocity, which is the slope of the v–t curve. Thus,

$$\text{at } t = 3 \text{ s} \quad a = \frac{dv}{dt} = \frac{30}{3} = 10 \text{ ft/s}^2$$

$$\text{at } t = 5 \text{ s} \quad a = \frac{dv}{dt} = -\frac{30}{2} = -15 \text{ ft/s}^2$$

(*c*) The velocity is the time derivative of the displacement. Thus,

$$v = \frac{ds}{dt} \quad ds = v\,dt \quad \Delta s = \int v\,dt$$

The integral of $v\,dt$ is also the area under the v–t curve. Thus,

$$\text{for } t = 5 \quad s = (30)(3)/2 + (30)(2)/2 = 75 \text{ ft}$$

(*d*) The distance traveled in the first 3 s is 45 ft as calculated from the area under the v–t curve. The velocity is given by the equation of the v–t curve for the region $t = 3$ to $t = 5$ s. Thus, for the added 5 ft traveled, the area under the v–t curve is the area under the shaded trapezoid, which is the sum of the rectangle and triangle. Or

$$(30 - 15t)t + (15t)\frac{t}{2} = 5.$$

Solving the resulting quadratic equation, $t = 0.175$ s. Note that the other solution of the quadratic equation is 3.826 s, which is greater than 2 s, the maximum value that t can have. Hence the total time is $T = 3 + 0.175 = 3.175$ s.

12.6. A balloon is rising with a velocity of 2 m/s when a bag of sand is released. If the height at the time of release is 120 m, how long does it take the bag of sand to reach the ground?

SOLUTION

The sand is rising at the same rate as the balloon at the instant of release. Hence

$$v_0 = +2 \text{ m/s} \quad y = 120 \text{ m} \quad g = a = 9.8 \text{ m/s}^2$$

First solve using the ground as the datum ($y = 0$), with up being positive. (*Note:* $y = 0$ as the sand reaches the ground.)

$$y = y_0 + v_0 t + \tfrac{1}{2}at^2$$
$$0 = +120 + 2t + \tfrac{1}{2}(-9.8)t^2$$

This yields $t = 5.16\,\text{s}$

Next solve using the balloon as the datum. Use up as positive. (*Note:* $y = -120$ as the sand reaches the ground.)

$$y = y_0 + v_0 t + \tfrac{1}{2}at^2$$
$$-120 = 0 + 2t + \tfrac{1}{2}(-9.8)t^2$$

This, of course, yields

$$t = 5.16\,\text{s}$$

12.7. A ball is projected vertically upward with a velocity of 80 ft/s. Two seconds later a second ball is projected vertically upward with a velocity of 60 ft/s. At what point above the surface of the earth will they meet?

SOLUTION

Let t be the time after the first ball is projected that the two meet. The second ball will then have been traveling for $t - 2$ s. The displacements for both balls will be the same at time t.

Let s_1 and s_2 be the displacements of the first and second balls, respectively. Then

$$s_1 = (v_0)_1 t - \tfrac{1}{2}gt^2 \quad\text{and}\quad s_2 = (v_0)_2(t - 2) - \tfrac{1}{2}g(t - 2)^2$$

Equating s_1 and s_2 and substituting the given values of $(v_0)_1$ and $(v_0)_2$, we obtain

$$80t - 16.1t^2 = 60(t - 2) - 16.1(t - 2)^2 \quad\text{or}\quad t = 4.15\,\text{s}$$

Substituting this value of t in the equation for s_1 (or s_2), the displacement is

$$s_1 = 80\,\text{ft/s} \times 4.14\,\text{s} - \tfrac{1}{2}(32.2\,\text{ft/s}^2)(4.14\,\text{s})^2 = 54.6\,\text{ft}$$

12.8. A ball is thrown at an angle of 40° to the horizontal. With what initial speed should the ball be thrown in order to land 100 ft away? Neglect air resistance.

SOLUTION

Choose the $x\,y$ axes with the origin at the point where the ball is thrown. By neglecting air resistance the x component of the acceleration is zero. The y component of the acceleration is $-g$.

From equation (7) with $a_x = 0$ and $a_y = -32.2$ f/s,

$$x = v_{0x}t \quad\text{and}\quad y = v_{0y}t - \tfrac{1}{2}(32.2)t^2$$

Given that when $x = 100$, $y = 0$ and $v_{0x} = v_0 \cos 40°$, $v_{0y} = v_0 \sin 40°$, the above equations become

$$100 = v_0 \cos 40°(t)$$
$$0 = v_0 \sin 40°(t) - \tfrac{1}{2}(32.2)t^2$$

Solving the first equation for v_0, substituting in the second equation and solving for t gives $t = 2.28$ s. Substituting this value in the first equation yields

$$v_0 = 57.3\,\text{ft/s}$$

12.9. A particle moves along a horizontal straight line with an acceleration $a = 6\sqrt[3]{s}$. When $t = 2$ s, its displacement $s = +27$ ft and its velocity $v = +27$ ft/s. Calculate the velocity and acceleration of the point when $t = 4$ s.

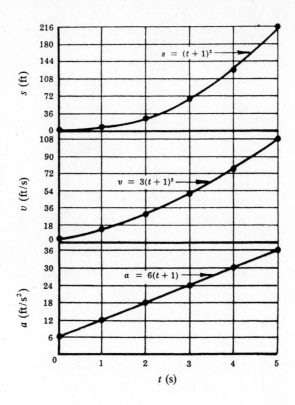

t (s)

SOLUTION

Since the acceleration is given as a function of the displacement, use the differential equation $a\,ds = v\,dv$. Then

$$\int 6s^{1/3}\,ds = \int v\,dv \qquad \text{or} \qquad \tfrac{9}{2}s^{4/3} = \tfrac{1}{2}v^2 + C_1$$

Since $v = +27$ when $s = +27$, $C_1 = 0$ and $v = 3s^{2/3}$.

Next use $v = ds/dt$ to obtain $ds/s^{2/3} = 3\,dt$; from this, $3s^{1/3} = 3t + C_2$. Substitute the condition $s = +27$ when $t = 2$ to obtain $C_2 = 3$ and $s = (t+1)^3$.

The equations are therefore $s = (t+1)^3$, $v = 3(t+1)^2$, $a = 6(t+1)$. When $t = 4$ s, $s = 125$ ft, $v = 75$ ft/s, and $a = 30$ ft/s^2.

A plot of these quantities against time is shown above. Note that the ordinate of the v–t curve at any time t is the slope of the s–t curve at the same time. Also, the ordinate of the a–t curve at any time t is the slope of the v–t curve at that time.

12.10. A particle moves on a vertical line with an acceleration $a = 2\sqrt{v}$. When $t = 2$ s, its displacement $s = 64/3$ ft and its velocity $v = 16$ ft/s. Determine the displacement, velocity, and acceleration of the particle when $t = 3$ s.

SOLUTION

Since $a = dv/dt$, then $2\sqrt{v} = dv/dt$. Separating the variables, $2\,dt = dv/v^{1/2}$. Integrating, $2t + C_1 = 2v^{1/2}$. But $v = 16$ ft/s when $t = 2$ s; hence, $C_1 = 4$.

The equation becomes $t + 2 = v^{1/2}$ or $v = (t+2)^2 = ds/dt$. Then $ds = (t+2)^2\,dt$. Integrating, $s = \tfrac{1}{3}(t+2)^3 + C_2$. But $s = 64/3$ ft when $t = 2$ s; hence, $C_2 = 0$.

The equations are therefore $s = \tfrac{1}{3}(t+2)^3$, $v = (t+2)^2$, and $a = 2(t+2)$.

When $t = 3$ s, $s = 41.7$ ft, $v = 25$ ft/s, and $a = 10$ ft/s^2.

12.11. The acceleration of a point moving on a vertical line is given by the equation $a = 12t - 20$. It is known that its displacement $s = -10$ m at time $t = 0$ and that its displacement $s = +10$ m at time $t = 5$ s. Derive the equation of its motion.

SOLUTION

Integrate $a = dv/dt = 12t - 20$ to obtain $v = 6t^2 - 20t + C_1$. Integrate this once more to obtain $s = 2t^3 - 10t^2 + C_1 t + C_2$.

The constants of integration may now be evaluated. Substitute the known value of s and t:

$$-10 = 2(0)^3 - 10(0)^2 + C_1(0) + C_2 \qquad \text{or} \qquad C_2 = -10$$

$$+10 = 2(5)^3 - 10(5)^2 + C_1(5) - 10 \qquad \text{or} \qquad C_1 = +4$$

The equation of motion is $s = 2t^3 - 10t^2 + 4t - 10$.

12.12. In the system shown in Fig. 12-7(a) determine the velocity and acceleration of block 2 at the instant.

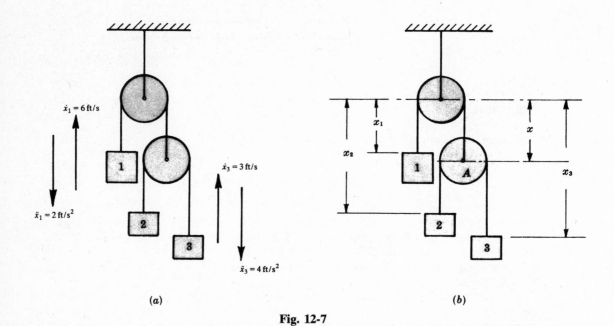

(a)   (b)

Fig. 12-7

SOLUTION

Figure 12-7(b) is drawn to show the position of each weight relative to the fixed support. The length of the cord between the weight 1 and point A is a constant and equals one-half the circumference of the top pulley plus $x_1 + x$. The length of the cord between weights 2 and 3 is a constant and equals one-half the circumference of pulley A plus $x_2 - x + x_3 - x$.

Thus, $x_1 + x = \text{constant}$; $x_2 + x_3 - 2x = \text{constant}$. Time derivatives then show

$$\dot{x}_1 + \dot{x} = 0 \ (1) \qquad \dot{x}_2 + \dot{x}_3 - 2\dot{x} = 0 \ (3)$$

$$\ddot{x}_1 + \ddot{x} = 0 \ (2) \qquad \ddot{x}_2 + \ddot{x}_3 - 2\ddot{x} = 0 \ (4)$$

Calling the upward direction positive and substituting $\dot{x}_1 = 6$ ft/s into equation (1), we find

$\dot{x} = -6$ ft/s. Substituting this value together with $\dot{x}_3 = 3$ ft/s into equation (3), we find $\dot{x}_2 = 2\dot{x} - \dot{x}_3 = 2(-6) - (3) = -15$ ft/s (down).

Similar reasoning for the accelerations shows

$$\ddot{x} = +2 \qquad \text{and} \qquad \ddot{x}_2 = 2\ddot{x} - \ddot{x}_3 = 2(+2) - (-4) = 8 \text{ ft/s}^2$$

12.13. Show that the curvature of a plane curve at point P may be expressed as

$$\frac{1}{\rho} = \frac{\dot{x}\ddot{y} - \ddot{x}\dot{y}}{(\dot{x}^2 + \dot{y}^2)^{3/2}}$$

where ρ is the radius of curvature, $\dot{x}$ and $\dot{y}$ are the x and y components of the speed of P, and $\ddot{x}$ and $\ddot{y}$ are the x and y components of the magnitude of the acceleration of P.

SOLUTION

From the calculus, the curvature of any curve $y = f(x)$ at that point P is

$$\frac{1}{\rho} = \frac{d^2y/dx^2}{[1 + (dy/dx)^2]^{3/2}} \qquad\qquad (1)$$

But $\qquad \dfrac{dy}{dx} = \dfrac{dy}{dt}\dfrac{dt}{dx} = \dfrac{\dot{y}}{\dot{x}} \qquad$ and $\qquad \dfrac{d^2y}{dx^2} = \dfrac{d}{dx}\left(\dfrac{dy}{dx}\right) = \dfrac{d}{dt}\left(\dfrac{dy}{dx}\right)\dfrac{dt}{dx} = \dfrac{d}{dt}\left(\dfrac{\dot{y}}{\dot{x}}\right)\dfrac{1}{\dot{x}} = \dfrac{\dot{x}\ddot{y} - \dot{y}\ddot{x}}{\dot{x}^2\ddot{x}}$

Substituting in (1), we obtain the required equation.

12.14. A particle describes a path $y = 3.6x^2$, where x and y are in meters. The velocity has a constant x component of 2 m/s. Assume that the particle is at the origin at the start of the motion and solve for the components of displacement, velocity, and acceleration in terms of time.

SOLUTION

Since $dx/dt = 2$ m/s, we can integrate to obtain $x = 2t + C_1$. But $x = 0$ at $t = 0$; hence, $C_1 = 0$.

Thus $\qquad\qquad\qquad\qquad\qquad\qquad\qquad x = 2t \text{ m}$

Also, $y = 3.6x^2 = 3.6(2t)^2 = 14.4t^2$ m. Thus,

$$\frac{dy}{dt} = 28.8t \text{ m/s}$$

Finally, $\qquad\qquad\qquad\qquad \dfrac{d^2x}{dt^2} = 0 \qquad$ and $\qquad \dfrac{d^2y}{dt^2} = 28.8 \text{ m/s}^2$

12.15. A particle describes the path $y = 4x^2$ with constant speed v, where x and y are in meters. What is the normal component of the acceleration?

SOLUTION

$$\frac{1}{\rho} = \frac{d^2y/dx^2}{[1 + (dy/dx)^2]^{3/2}} = \frac{8}{[1 + (8x)^2]^{3/2}} \qquad \text{and} \qquad a_n = \frac{v^2}{\rho} = \frac{8v^2}{[1 + 64x^2]^{3/2}} \text{ m/s}^2$$

12.16. A particle moves on a path with a velocity vector of $\mathbf{v} = 3t^2\mathbf{i} - 4t\mathbf{j} + 2\mathbf{k}$ in/s.

(a) Determine the coordinates of its position after 4 s.

   The particle is at the origin when $t = 0$.

(b) Determine the equation of its path.

(c) Determine the projection of the velocity vector in the direction of the vector $\mathbf{n} = 4\mathbf{i} + \mathbf{j} - 3\mathbf{k}$ when $t = 4$ s.

SOLUTION

(a) The position vector is the integral of the velocity vector. This can be seen from the definition of velocity

$$\mathbf{v} = d\mathbf{r}/dt \quad \text{so that} \quad \mathbf{r} = \int \mathbf{v}\, dt = t^3\mathbf{i} - 2t^2\mathbf{j} + 2t\mathbf{k}$$

At $t = 4$ s, $\mathbf{r} = 64\mathbf{i} - 32\mathbf{j} + 8\mathbf{k}$ in. The coordinates of the position at $t = 4$ s are then $x = 64$ in, $y = -32$ in and $y = 8$ in.

(b) At any time t the so-called parametric equations of position are given by the coefficients of $\mathbf{i}$, $\mathbf{j}$, and $\mathbf{k}$ in $\mathbf{r}$. Thus, $x = t^3$, $y = -2t^2$, and $z = 2t$. Eliminating t from these parametric equations yields

$$t = x^{1/3} \qquad y = -2x^{2/3} \qquad \left(\frac{z}{2}\right)^2 = x^{2/3}$$

Combining equations gives

$$y + \left(\frac{z}{2}\right)^2 = -x^{2/3}$$

or

$$x^{2/3} + y + \left(\frac{z}{2}\right)^2 = 0$$

the equation of the path. Let the reader show that at $t = 4$ s this equation is satisfied.

(c) The unit vector in the desired direction is

$$\mathbf{e}_L = \frac{4\mathbf{i} + \mathbf{j} - 3\mathbf{k}}{\sqrt{4^2 + 1^2 + (-3)^2}}$$

Hence, the projection of $\mathbf{v}$ on $\mathbf{n}$ at $t = 4$ s is

$$\mathbf{v} \cdot \mathbf{e}_L = \left[\frac{4\mathbf{i} + \mathbf{j} - 3\mathbf{k}}{\sqrt{26}}\right] \cdot [3(4)^2\mathbf{i} - 4(4)\mathbf{j} + 2\mathbf{k}] = 33.3 \text{ in/s}$$

12.17. A particle moves along the path whose equation is $r = 2\theta$ ft. If the angle $\theta = t^2$ rad, determine the velocity of the particle when θ is $60°$. Use two methods.

SOLUTION

A plot of the path is shown in Fig. 12-8(a) with unit vectors $\mathbf{e}_r$, along r and $\mathbf{e}_\theta$ perpendicular to $\mathbf{r}$ and in the direction of increasing θ.

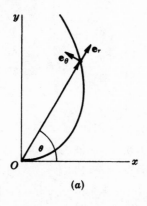

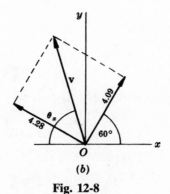

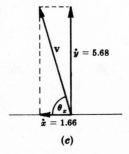

(a)   (b)   (c)

Fig. 12-8

(a) *Polar coordinates.* See Fig. 12-8(b).
Since $\theta = t^2$, $\dot{\theta} = 2t$; since $r = 2\theta = 2t^2$, $\dot{r} = 4t$.
The velocity vector **v** at $\theta = \pi/3$ rad is found as follows:

$$\theta = \frac{\pi}{3} = t^2 \qquad \text{or} \qquad t = 1.023 \text{ s}$$

$$\mathbf{v} = \dot{r}\mathbf{e}_r + r\dot{\theta}\mathbf{e}_\theta = 4(1.023)\mathbf{e}_r + [2(1.047)][2(1.023)]\mathbf{e}_\theta = 4.09\mathbf{e}_r + 4.28\mathbf{e}_\theta$$

and $v = \sqrt{(4.09)^2 + (4.28)^2} = 5.92$ ft/s with $\theta_x = 30° + \tan^{-1}(4.09/4.28) = 73.7°$.

(b) *Cartesian coordinates.* See Fig. 12-8(c).

$$x = r \cos\theta = 2\theta \cos\theta \qquad y = r \sin\theta = 2\theta \sin\theta$$
$$= 2t^2 \cos t^2 \qquad\qquad = 2t^2 \sin t^2$$

Then
$$\dot{x} = 4t \cos t^2 + 2t^2(-\sin t^2)(2t) = -1.66 \text{ ft/s}$$
$$\dot{y} = 4t \sin t^2 + 2t^2(\cos t^2)(2t) = +5.68 \text{ ft/s}$$

at $t = 1.023$ s ($\cos t^2 = \cos \pi/3$, $\sin t^2 = \sin \pi/3$). Hence,

$$v = \sqrt{(-1.66)^2 + (5.68)^2} = 5.92 \text{ ft/s} \qquad \text{with} \qquad \theta_x = \tan^{-1}\frac{5.68}{1.66} = 73.7°$$

12.18. In the preceding problem determine the acceleration of the particle using the same two methods.

SOLUTION

(a) *Polar coordinates.* See Fig. 12-9.
$$\mathbf{a} = (\ddot{r} - r\dot{\theta}^2)\mathbf{e}_r + (2\dot{r}\dot{\theta} + r\ddot{\theta})\mathbf{e}_\theta = -4.77\mathbf{e}_r + 20.94\mathbf{e}_\theta$$

Since $\theta = t^2$, $\dot{\theta} = 2t$, $\ddot{\theta} = 2$; $r = 2\theta = 2t^2$, $\dot{r} = 4t$, $\ddot{r} = 4$; $t = 1.023$ s at $\theta = \pi/3$. Thus,
$$a = \sqrt{(-4.77)^2 + (20.94)^2} = 21.5 \text{ ft/2}^2 \qquad \text{with} \qquad \theta_x = 30° - 12.8° = 17.2°$$

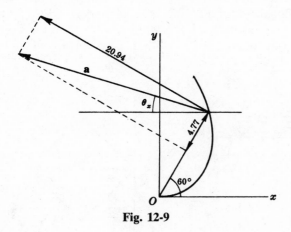

Fig. 12-9

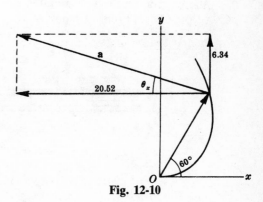

Fig. 12-10

(b) *Cartesian coordinates.* See Fig. 12-10.
Continuing the time derivatives and evaluating at $t = 1.023$ ($t^2 = \pi/3$).

$$\ddot{x} = 4 \cos t^2 + 4t(-\sin t^2)(2t) + 12t^2(-\sin t^2) - 4t^3(\cos t^2)(2t) = -20.52 \text{ ft/s}^2$$

$$\ddot{y} = 4 \sin t^2 + 4t(\cos t^2)(2t) + 12t^2(\cos t^2) + 4t^3(-\sin t^2)(2t) = +6.34 \text{ ft/s}^2$$

Hence, $a = \sqrt{(-20.52)^2 + (6.34)^2} = 21.5$ ft/s^2 with $\theta_x = \tan^{-1}\frac{6.34}{20.52} = 17.2°$

12.19. In the Stotch yoke shown in Fig. 12-11 the crank OA is turning with a constant angular velocity ω rad/s. Derive the expressions for the displacement, velocity, and acceleration of the sliding member.

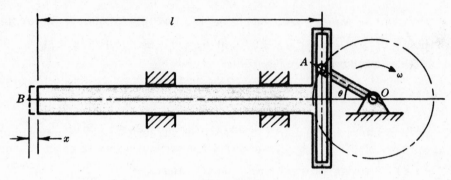

Fig. 12-11

SOLUTION

Let B represent the position of the left end of the slider when $\theta = 0°$. The displacement x is written $x = OB - l - OA \cos \theta$. When the crank is horizontal, $OB = l + OA$, and hence

$$x = l + OA - l - OA \cos \theta = OA(1 - \cos \theta)$$

Let $OA = R$. Also, since the crank is turning with constant angular velocity ω, the expression ωt may be substituted for θ. Differentiating $x = R(1 - \cos \omega t)$ yields

$$v = \frac{dx}{dt} = R\omega \sin \omega t \qquad \text{and} \qquad a = \frac{dv}{dt} = R\omega^2 \cos \omega t$$

12.20. In Fig. 12-12, the oscillating arm OD is rotating clockwise with a constant angular velocity 10 rad/s. Block A slides freely in the slot in the arm OD and is pinned to block B, which slides freely in the horizontal slot in the framework. Determine, for $\theta = 45°$, the total velocity of pin P as a point in block B using (a) Cartesian coordinates and (b) polar coordinates.

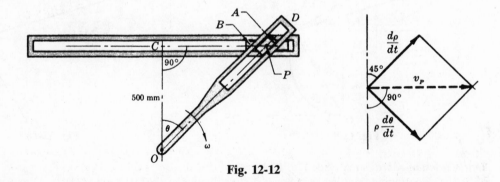

Fig. 12-12

SOLUTION

Note that the total or absolute velocity of P as a point in block B can only be horizontal, since it is a point in B and all points in B move horizontally. But as a point in block A it has this same velocity also.

(a)   Let x = distance of P from C. Then $x = 0.5 \tan \theta$ and $v_P = dx/dt = 0.5(\sec^2 \theta) \, d\theta/dt$. When $\theta = 45°$, $v_P = 0.5(\sec^2 45°)10 = 10$ m/s.

(b)   Let ρ = distance of P from O, which will be used as a pole in studying the motion. Then $\rho = 0.5 \sec \theta$.

The radial component of the velocity along OP is $d\rho/dt = 0.5 \sec \theta \tan \theta \, d\theta/dt$. For $\theta = 45°$, this becomes $0.5 \sec 45° \tan 45° \times 10 = 7.07$ m/s. This component is directed outward along OP.

The transverse component of the velocity is $\rho \, d\theta/dt = 0.5 \sec \theta \, d\theta \, dt$. For $\theta = 45°$, this becomes $0.5 \sec 45° \times 10 = 7.07$ m/s. This component is perpendicular to the arm OP and acts down to the right because ω is clockwise.

The two components are shown to the right of the figure. Hence, $v_P = \sqrt{(d\rho/dt)^2 + (\rho \, d\theta/dt)^2} = 10$ m/s and is horizontal.

12.21. The bar AB shown in Fig. 12-13 moves so that its lowest point A travels horizontally to the right with constant velocity $v_A = 5$ ft/s. What is the velocity of point B when $\theta = 70°$? The length of the bar is 6.24 ft.

SOLUTION

Let x and y be the distances of A and B from point O at any time during the motion. Since $x^2 + y^2 = l^2$, then

$$2x \frac{dx}{dt} + 2y \frac{dy}{dt} = 0$$

and

$$v_B = \frac{dy}{dt} = -\frac{x}{y} \frac{dx}{dt} = -(\cot \theta)(v_A) = -1.82 \text{ ft/s}$$

The minus sign indicates that B is traveling down. Note that v_B is independent of l.

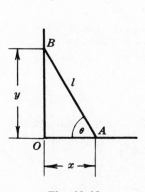

Fig. 12-13

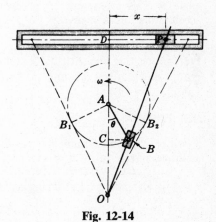

Fig. 12-14

12.22. In the *quick-return mechanism* shown in Fig. 12-14, the crank AB is driven at a constant angular velocity ω rad/s. The slider at B slides along the rod OP, also causing it to oscillate about the hinge O. In turn, OP slides in and out of the slider at P and also moves the second slider pinned at P along the horiozntal slot. A cutting tool attached to this second slider will be subjected to a reciprocating motion. The cutting tool reaches the extremes of its horizontal travel when OP is tangent to the crank circle at B_1 and B_2. Cutting occurs while the crank pin moves from B_2 counterclockwise to B_1. The return stroke occurs in the remaining arc from B_1 to B_2. Since the speed is constant, the times of the working and return strokes are proportional to the angles traversed. The cutting stroke occurs during the larger angle and hence takes the longer time. The return stroke is faster; hence, the name quick-return mechanism is appropriate. Determine the expressions for displacement, velocity, and acceleration of the cutting tool at P.

SOLUTION

From the figure,

$$\frac{x}{BC} = \frac{OD}{OC}$$

Let $OD = l$, $AB = R$, $OA = d$. Then $BC = R \sin \theta$, $OC = OA - AC = d - R \cos \theta$. Substituting in the original expression in x,

$$\frac{x}{R \sin \theta} = \frac{l}{d - R \cos \theta} \qquad \text{or} \qquad x = \frac{Rl \sin \theta}{d - R \cos \theta} \tag{1}$$

and

$$v = \frac{dx}{dt} = \frac{Rl[(d - R \cos \theta) \cos \theta \, d\theta/dt - \sin \theta \, (R \sin \theta \, d\theta/dt)]}{(d - R \cos \theta)^2}$$

Since $d\theta/dt$ is the angular velocity ω, this equation becomes

$$v = \frac{Rl\omega(d \cos \theta - R \cos^2 \theta - R \sin^2 \theta)}{(d - R \cos \theta)^2} \qquad \text{or} \qquad v = Rl\omega \frac{d \cos \theta - R}{(d - R \cos \theta)^2} \tag{2}$$

and

$$a = \frac{dv}{dt} = Rl\omega \frac{(d - R \cos \theta)^2(-d \sin \theta \, d\theta/dt) - (d \cos \theta - R)[2(d - R \cos \theta)R \sin \theta \, d\theta/dt]}{(d - R \cos \theta)^4}$$

$$= \frac{-Rl\omega^2 \sin \theta(d^2 - 2R^2 + Rd \cos \theta)}{(d - R \cos \theta)^3} \tag{3}$$

Equations (1), (2), and (3) give the displacement, velocity, and acceleration for any value of the angle θ. This information is necessary to design the members of the mechanism to withstand the accelerating forces involved.

12.23. Determine the linear displacement, velocity, and acceleration of the crosshead C in the slider crank mechanism for any position of the crank R which is rotating at a constant angular velocity ω rad/s.

SOLUTION

Let C_0 be the extreme left position of the crosshead, which travels horizontally along the centerline as shown in Fig. 12-15. It is evident from the figure that $x = C_0A - CA$ and $CA = CD + DA$.

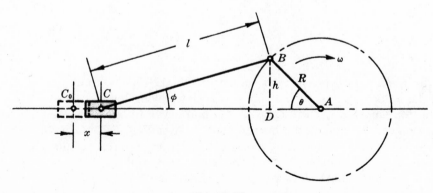

Fig. 12-15

When C is at C_0, B is on the centerline; hence, $C_0A = l + R$. Also, $CA = l \cos \phi + R \cos \theta$. Then

$$x = l + R - l \cos \phi - R \cos \theta$$

The relation between ϕ and θ is derived from the right triangles ADB and DCB:

$$h = l \sin \phi = R \sin \theta$$

Then $\qquad \sin \phi = \dfrac{R}{l} \sin \theta \qquad$ and $\qquad \cos \phi = \sqrt{1 - \sin^2 \phi} = \sqrt{1 - \dfrac{R^2}{l^2} \sin^2 \theta}$

and the displacement is

$$x = l + R - l \sqrt{1 - \frac{R^2}{l^2} \sin^2 \theta} - R \cos \theta$$

Differentiation of this expression with respect to time is somewhat involved because of the radical. However, an approximation of the radical that is sufficiently accurate when $R/l < \frac{1}{4}$ is obtained by using the first two terms of the power series expansion of the square root term:

$$\sqrt{1 - \frac{R^2}{l^2} \sin^2 \theta} \approx 1 - \frac{1}{2} \left(\frac{R^2}{l^2}\right) \sin^2 \theta$$

Making this substitution, the displacement becomes

$$x = l + R - l + \frac{R^2}{2l} \sin^2 \theta - R \cos \theta = R(1 - \cos \theta) + \frac{R^2}{2l} \sin^2 \theta$$

Differentiation yields

$$v = \frac{dx}{dt} = R \sin \theta \frac{d\theta}{dt} + \frac{R^2}{2l} 2 \sin \theta \cos \theta \frac{d\theta}{dt} = R\omega \left(\sin \theta + \frac{R}{2l} \sin 2\theta\right)$$

and $\qquad a = \dfrac{dv}{dt} = R\omega \left(\cos \theta \dfrac{d\theta}{dt} + \dfrac{R}{2l} 2 \cos 2\theta \dfrac{d\theta}{dt}\right) = R\omega^2 \left(\cos \theta + \dfrac{R}{l} \cos 2\theta\right)$

12.24. A point P moves on a circular path with a constant speed (magnitude of its linear velocity) of 12 ft/s. If the radius of the path is 2 ft, study the motion of the projection of the point on a horizontal diameter. Refer to Fig. 12-16.

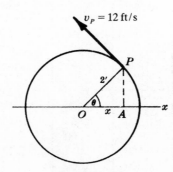

Fig. 12-16

SOLUTION

Point A is the projection on the horizontal diameter of point P. Assume the origin is at the center of the circle. The displacement x of the point A is the projection of the radius vector OP on the x axis (along the horizontal diameter).

Since the line OP sweeps out equal angles in equal times (the angular velocity is constant), the expression for θ may be written $\theta = \omega t$. The x coordinate of P is therefore

$$x = OP \cos \theta = 2 \cos \omega t$$

The angular velocity of the radius is $\omega = v/r = 12/2 = 6$ rad/s. Then $x = 2 \cos 6t$ and

$$v = \frac{dx}{dt} = -12 \sin 6t \qquad \text{and} \qquad a = \frac{dv}{dt} = -72 \cos 6t$$

The equation for a may be rewritten $a = -(36)(2 \cos 6t) = -36x$. But this means that point A moves so that its acceleration a is negatively proportional to its displacement x. Since this is the requirement of simple harmonic motion, it is evident that if point P moves on a circular path with constant speed, the projection of point P on a diameter moves with simple harmonic motion.

12.25. Plot the displacement, velocity, and acceleration of point A in Problem 12.24 against time.

SOLUTION

The amplitude of the displacement x is the maximum value, which occurs when $\cos 6t$ assumes its maximum value of unity (either plus or minus). The amplitude is therefore 2.

The amplitude of velocity v is 12, and the amplitude of acceleration a is 72.

The period is the time T to complete one cycle. It is evident that the motion will repeat itself after the radius vector OP has gone through a complete revolution $\theta = 2\pi$ rad. Equating the function $6t$ to the value of $\theta = 2\pi$ for a cycle yields

$$6 \text{ rad/s} \times T = 2\pi \text{ rad} \qquad \text{or} \qquad T = \frac{2\pi}{6} \text{ s} = 1.05 \text{ s}$$

It is now possible to correlate the angle θ with time t. For example, when $\theta = \pi/2$ or 90°, t is one-fourth of a revolution, that is $\frac{1}{4} \times 1.05$ s $= 0.263$ s. Proceeding in this fashion, draw up a table as follows.

θ	t	$\cos 6t$ or $\cos \theta$	$\sin 6t$ or $\sin \theta$	$x = 2 \cos 6t$	$v = -12 \sin 6t$	$a = -72 \cos 6t$
0°	0	+1.000	0	+2.00	0	−72.0
30°	0.088	+0.866	+0.500	+1.73	−6.00	−62.3
60°	0.175	+0.500	+0.866	+1.00	−10.4	−36.0
90°	0.263	0	+1.000	0	−12.0	0
120°	0.350	−0.500	+0.866	−1.00	−10.4	+36.0
150°	0.438	−0.866	+0.500	−1.73	−6.00	+62.3
180°	0.525	−1.000	0	−2.00	0	+72.0
210°	0.612	−0.866	−0.500	−1.73	+6.00	+62.3
240°	0.700	−0.500	−0.866	−1.00	+10.4	+36.0
270°	0.788	0	−1.000	0	+12.0	0
300°	0.875	+0.500	−0.866	+1.00	+10.4	−36.0
330°	0.962	+0.866	−0.500	+1.73	+6.00	−62.3
360°	1.05	+1.000	0	+2.00	0	−72.0

The plotting of these points provides a visual picture of the motion during one cycle. Naturally, as time increases, these curves duplicate in succeeding cycles. The following graphs indicate the motion during one cycle.

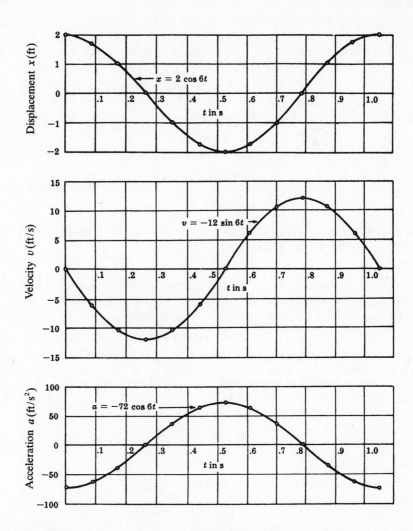

12.26. A flywheel 1.2 m in diameter accelerates uniformly from rest to 2000 rpm in 20 s. What is its angular acceleration?

SOLUTION

In the analysis of this problem, first note that a constant acceleration is involved. This means that the formulas for constant acceleration may be used. These are similar in angular motion to those in rectilinear motion; that is, ω replaces v, α replaces k, and θ replaces s.

The wheel starts from rest; hence, $\omega_0 = 0$. The three known quantities are ω_0, ω, t. The quantity sought is angular acceleration α. The formula involving these four quantities is

$$\omega = \omega_0 + \alpha t$$

A word of warning is in order regarding units: it is recommended in SI to use ω in rad/s and α in rad/s² when t is in seconds

$$\omega_0 = 0 \qquad \omega = 2000 \text{ rpm} = \frac{2000}{60} \frac{\text{rev}}{\text{s}} \times 2\pi \frac{\text{rad}}{\text{rev}} = 209 \frac{\text{rad}}{\text{s}}$$

Hence,
$$\alpha = \frac{\omega - \omega_0}{t} = \frac{209 \text{ rad/s} - 0 \text{ rad/s}}{20 \text{ s}} = 10.5 \frac{\text{rad}}{\text{s}^2}$$

12.27. In Problem 12.26, how many revolutions does the flywheel make in attaining its speed of 2000 rpm?

SOLUTION

To determine the number of revolutions θ, select the equation expressing the relation between θ and the three given quantities ω_0, ω, t. Of course, a formula may be used involving the angular acceleration α just determined, but it is advisable to proceed with data given in the problem to derive the value θ independently of the α, which could by chance have been found incorrectly.

$$\theta = \tfrac{1}{2}(\omega + \omega_0)t = (209 \text{ rad/s} + 0 \text{ rad/s})(20 \text{ s}) = 2090 \text{ rad}$$

To express θ in revolutions,

$$\theta = \frac{2090 \text{ rad}}{2\pi \text{ rad/rev}} = 333 \text{ revolutions}$$

The same result is obtained using ω in rev/s as follows:

$$\theta = \frac{(2000/60) \text{ rev/s} + 0 \text{ rev/s}}{2} \times 20 \text{ s} = 333 \text{ revolutions}$$

12.28. Determine the linear velocity and linear acceleration of a point on the rim of the flywheel in Problem 12.26, 0.6 s after it has started from rest.

SOLUTION

The velocity of a point on the rim is found by multiplying the radius by the angular velocity.
The angular velocity is $\omega = \omega_0 + \alpha t = 0 + (10.5 \text{ rad/s}^2)(0.6 \text{ s}) = 6.30 \text{ rad/s}$.
The magnitude of the linear velocity of a point on the rim when $t = 0.6$ s is

$$v = r\omega = (0.6 \text{ m})(6.30 \text{ rad/s}) = 3.78 \text{ m/s} \qquad \text{(tangent to the rim)}$$

To determine the acceleration completley, use the normal and tangential components. The tangential component a_t is $a_t = r\alpha = (0.6 \text{ m})(10.5 \text{ rad/s}^2) = 6.3 \text{ m/s}^2$. The normal component a_n is $a_n = r\omega^2 = (0.6 \text{ m})(6.3 \text{ rad/s})^2 = 23.8 \text{ m/s}^2$. Figure 12-17 illustrates these components for any point P on the rim.

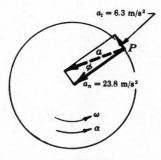

Fig. 12-17

The total acceleration a is the vector sum of the two components a_t and a_n. Let ϕ be the angle between the total acceleration and the radius. The normal acceleration a_n is directed toward the center of the circle.

$$a = \sqrt{a_t^2 + a_n^2} = \sqrt{(6.3)^2 + (23.8)^2} = 24.6 \text{ m/s}^2$$

$$\phi = \tan^{-1}\frac{a_t}{a_n} = \tan^{-1}\frac{6.3}{23.8} = 0.259 \text{ rad} = 14.8°$$

12.29. A uniform slender rod is 4 ft long and rotates on a horizontal plane about a vertical axis through one end. If the rod accelerates uniformly from 40 to 60 rpm in a 5-s interval, what is the linear speed of the center of the rod at the beginning and end of that time interval?

SOLUTION

The speed of the center is $v = r\omega$. Thus the speeds at the beginning and end of the time interval are, respectively,

$$v_B = r\omega_B = (2\text{ ft})\left(\frac{40 \times 2\pi}{60}\text{ rad/s}\right) = 8.38\text{ ft/s}$$

$$v_E = r\omega_E = (2\text{ ft})\left(\frac{60 \times 2\pi}{60}\text{ rad/s}\right) = 12.6\text{ ft/s}$$

12.30. In Problem 12.29, determine the normal and tangential components of the acceleration of the center of the rod 2 s after acceleration begins.

SOLUTION

The uniform acceleration α at any time during the 5-s interval is

$$\alpha = \frac{\omega_E - \omega_B}{t} = \frac{\frac{120}{60}\pi - \frac{80}{60}\pi}{5} = 0.419\text{ rad/s}^2$$

The angular velocity ω after 2 s is

$$\omega = \omega_B + \alpha t = \tfrac{80}{60}\pi + 0.419(2) = 5.03\text{ rad/s}$$

The components of the desired acceleration are

$$a_t = r\alpha = 2(0.419) = 0.838\text{ ft/s}^2$$
$$a_n = r\omega^2 = 2(5.03)^2 = 50.6\text{ ft/s}^2$$

12.31. A wheel 200 mm in diameter coasts to rest from a speed of 800 rpm in 600 s. Determine the angular acceleration.

SOLUTION

Given $\omega_0 = 800$ rpm $= 83.8$ rad/s, $\omega = 0$, $t = 600$ s, then

$$\alpha = \frac{\omega - \omega_0}{t} = \frac{-83.8\text{ rad/s}}{600\text{ s}} = -0.14\text{ rad/s}^2 \qquad \text{(deceleration)}$$

The acceleration is negative. This means that the angular velocity is in one direction, while the angular acceleration is oppositely directed, thereby indicating a slowing down of the wheel.

12.32. A wheel accelerates uniformly from rest to a speed of 200 rpm in $\frac{1}{2}$ s. It then totates at that speed for 2 s before decelerating to rest in $\frac{1}{3}$ s. How many revolutions does it make during the entire time interval?

SOLUTION

From $t = 0$ to $t = \frac{1}{2}$: $\theta_1 = \frac{1}{2}(\omega_0 + \omega)t = \frac{1}{2}(0 + 200/60\text{ rev/s})\quad(\frac{1}{2}\text{ s}) = 0.83$ rev.
From $t = \frac{1}{2}$ to $t = 2\frac{1}{2}$: $\theta_2 = \omega t = (200/60\text{ rev/s})(2\text{ s}) = 6.67$ rev.
From $t = 2\frac{1}{2}$ to rest: $\theta_3 = \frac{1}{2}(\omega_0 + \omega)t = \frac{1}{2}(200/60\text{ rev/s} + 0)(\frac{1}{3}\text{ s}) = 0.56$ rev.
Total number of revolutions $\theta = \theta_1 + \theta_2 + \theta_3 = 8.06$ rev.

12.33. Two friction disks are shown in Fig. 12-18. Derive the expression for the angular velocity ratio in terms of the radii.

SOLUTION

The linear velocities of the mating points A and B on the two wheels are equal. If this were not true, the wheels would slip relative to one another.

The linear velocities of points A and B are, respectively,

$$v_A = R_1 \omega_1 \qquad v_B = R_2 \omega_2$$

But $v_A = v_B$ if the drive is positive, i.e., without slip. Then

$$R_1 \omega_1 = R_2 \omega_2 \qquad \text{or} \qquad \frac{\omega_1}{\omega_2} = \frac{R_2}{R_1}$$

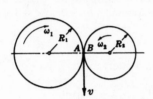

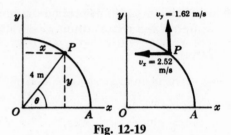

Fig. 12-18   **Fig. 12-19**

12.34. A point P moves on a circular path in a counterclockwise direction so that the length of arc it sweeps out is $s = t^3 + 3$. The radius of the path is 4 m. The units of s and t are m and s, respectively. Determine the axial components of velocity (v_x, v_y) when $t = 1$ s. Refer to Fig. 12-19.

SOLUTION

The distance AP is traversed in 1 s, or $AP = s = 1^3 + 3 = 4$ m. By inspection, $x = 4 \cos \theta$ and $y = 4 \sin \theta$. Differentiating, $v_x = (-4 \sin \theta) \, d\theta/dt$ and $v_P = (4 \cos \theta) \, d\theta/dt$.

These may be evaluated provided that θ is found as a function of time. The relation $s = r\theta$ yields $\theta = s/r = (t^3 + 3)/4$, where θ must be in radians. Differentiate to obtain $d\theta/dt = 0.75t^2$. When $t = 1$ s, $\theta = 1$ rad and $d\theta/dt = 0.75$ rad/s.

Substitution yields $v_x = -2.52$ m/s and $v_y = 1.62$ m/s. The negative sign indicates that the x component of the velocity is directed to the left. The y component of the velocity is directed up.

The total velocity $v = \sqrt{(v_x)^2 + (v_y)^2} = 3.0$ m/s. This could be obtained directly from $v = r \, d\theta/dt = 4(.075t^2)$ when $t = 1$ s, or from $s = t^3 + 3$ and hence $v = ds/dt = 3t^2$.

12.35. In Problem 12-34, determine the axial components of the acceleration a_x and a_y when $t = 1$ s.

SOLUTION

Differentiate the expression for v_x to obtain $a_x = -4 \cos \theta (d\theta/dt)^2 - 4 \sin \theta \, d^2\theta/dt^2$.
Since $d\theta/dt = 0.75t^2$, then $d^2\theta/dt^2 = 1.5t$.
At $t = 1$ s, $a_x = -4(\cos 1)(0.75)^2 - 4(\sin 1)(1.5) = -6.27$ m/s², i.e. to the left.
Similarly, $a_y = -4 \sin \theta (d\theta/dt)^2 + 4 \cos \theta \, d^2\theta/dt^2 = +1.35$ m/s², i.e., up.
The total acceleration $a = \sqrt{(a_x)^2 + (a_y)^2} = 6.41$ m/s².

This could be obtained also by combining the tangential component a_t and the normal component a_n of the acceleration. These are $a_t = r\alpha = r \, d^2\theta/dt^2 = 4(1.5t)$ or 6 m/s², and $a_n = r\omega^2 = r(d\theta/dt)^2 = 4(.075)^2$ or 2.25 m/s². Hence, $a = \sqrt{(a_t)^2 + (a_n)^2} = 6.41$ m/s².

Note that $a_t = d^2s/dt^2 = 6t$ and $a_n = v^2/r = 9t^4/4$ give the same results with $t = 1$.

12.36. The x and y components of the displacement in feet of a point are given by the equations

$$x = 4t^2 - 3t \qquad y = t^3 - 10$$

Determine the velocity and acceleration of the point when $t = 2$ s.

SOLUTION

The velocity components obtained by differentiation are

$$v_x = \frac{dx}{dt} = 8t - 3 \qquad v_y = 3t^2$$

At $t = 2$ s, $v_x = 13$ ft/s and $v_y = 12$ ft/s. Hence,

$$v = \sqrt{(v_x)^2 + (v_y)^2} = 17.7 \text{ ft/s} \qquad \text{and} \qquad \theta_x = \tan^{-1}\frac{v_y}{v_x} = \tan^{-1}\frac{12}{13} = 42.7°$$

where θ_x is the angle between the total velocity and the x axis.

A second differentiation yields the acceleration components: $a_x = dv_x/dt = 8$ and $a_y = dv_y/dt = 6t$. At $t = 2$ s, $a_x = 8$ ft/s^2 and $a_y = 12$ ft/s^2. Hence

$$a = \sqrt{(a_x)^2 + (a_y)^2} = 14.4 \text{ ft/s}^2 \qquad \text{and} \qquad \phi_x = \tan^{-1}\frac{a_y}{a_x} = \tan^{-1}\frac{12}{8} = 56.3°$$

12.37. An automobile is moving south with an absolute velocity of 20 mi/h. An observer at O is stationed 50 ft to the east of the line of travel. When the automobile is directly west of the observer, what is the angular velocity relative to the observer? After moving 50 ft south, what is its angular velocity relative to the observer at O?

SOLUTION

As indicated in Fig. 12-20, the velocity $v_{A/O}$ of the automobile at A relative to O is 20 mi/h or 29.3 ft/s.

However, the linear velocity of A relative to O (in this case it is perpendicular to OA) is the product of the distance OA and the angular velocity of A relative to O. Then

$$v_{A/O} = OA \times \omega_{A/O}$$
$$29.3 \text{ ft/s} = 50 \text{ ft} \times \omega_{A/O} \qquad \text{and} \qquad \omega_{A/O} = 0.588 \text{ rad/s}$$

For the next part of the problem note that the absolute velocity v_B of the vehicle at B is still south 20 mi/h (29.3 ft/s). The component $v_{B/O}$ (velocity of B relative to O) is perpendicular to the arm BO; hence, $v_{B/O} = 29.3 \cos 45° = 20.8$ ft/s. Then

$$v_{B/O} = OB \times \omega_{B/O} \qquad 20.8 \text{ ft/s} = (50\sqrt{2}\text{ ft})(\omega_{B/O}) \qquad \text{and} \qquad \omega_{B/O} = 0.294 \text{ rad/s}$$

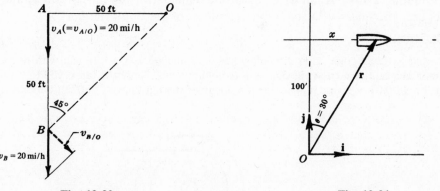

Fig. 12-20   Fig. 12-21

12.38. A boat is traveling 12 mi/h due east. An observer is stationed 100 ft south of the line of travel. Determine the angular velocity of the boat relative to the observer when in the position shown in Fig. 12-21.

SOLUTION

Select unit vectors $\mathbf{i}$ and $\mathbf{j}$ in the east and north directions, respectively. Let $\mathbf{r}$ be the position vector of the boat relative to the observer O. Then

$$\mathbf{r} = x\mathbf{i} + 100\mathbf{j} = 100\tan\theta\,\mathbf{i} + 100\mathbf{j}$$

The velocity $\mathbf{v}$ of the boat is

$$\mathbf{v} = \dot{\mathbf{r}} = 100(\sec^2\theta)(\dot\theta)\mathbf{i} + 0\mathbf{j}$$

Since the speed $v = 12\text{ mi/h} = 17.6\text{ ft/s}$ and $\theta = 30°$.

$$17.6 = 100(\sec^2 30°)\dot\theta \qquad \text{or} \qquad \omega = \dot\theta = 0.132\text{ rad/s} \qquad \text{clockwise}$$

12.39. The motion of a point is described by the following equations:

$$v_x = 20t + 5 \qquad v_y = t^2 - 20$$

In addition it is known that $x = 5$ m and $y = -15$ m when $t = 0$. Determine the displacement, velocity, and acceleration when $t = 2$ s.

SOLUTION

Rewriting the given equations as $v_x = dx/dt = 20t + 5$, $v_y = dy/dt = t^2 - 20$ and integrating, the expressions for x and y are $x = 10t^2 + 5t + C_1$, $y = \frac{1}{3}t^3 - 20t + C_2$.

To evaluate C_1, substitute $x = 5$ and $t = 0$ in the x equation. Then $C_1 = 5$.

To evaluate C_2, substitute $y = -15$ and $t = 0$ in the y equation. Then $C_2 = -15$.

Substituting the values of C_1 and C_2, the equations for displacement become

$$x = 10t^2 + 5t + 5 \qquad y = \frac{1}{2}t^2 - 20t - 15$$

Differentiate v_x and v_y to obtain the equations for acceleration:

$$a_x = \frac{dv_x}{dt} = 20 \qquad a_y = \frac{dv_y}{dt} = 2t$$

Substituting $t = 2$ s in the expressions for displacement, velocity, and acceleration, the following values are obtained: $x = 55$ m, $y = -53$ m; $v_x = 45$ m/s, $v_y = -16$ m/s; $a_x = 20$ m/s^2, $a_y = 4$ m/s^2.

The magnitudes and directions of the total displacement, velocity, and acceleration can be found by combining their components as before.

12.40. The 100-mm-diameter pulley on a generator is being turned by a belt moving 20 m/s and accelerating 6 m/s^2. A fan with an outside diameter of 150 mm is attached to the pulley shaft. What are the linear velocity and acceleration of the tip of the fan?

SOLUTION

In Fig. 12-22 point A on the pulley has the same velocity as the belt with which it coincides at the instant. Hence, the angular velocity ω of the pulley (and also of the fan keyed to the same shaft) is equal to $v/r = 20/0.05 = 400$ rad/s. The linear velocity of the fan tip $v_B = (0.075)(400) = 30$ m/s.

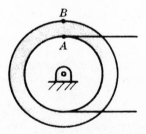

Fig. 12-22

The tangential component of the linear acceleration of point A is equal to the acceleration of the belt; that is, $a_t = r\alpha$ or $6 = 0.05\alpha$. From this, the angular acceleration α of the system is 120 rad/s^2. Then the tangential component of the acceleration of point B is $(0.075)(120) = 9$ m/s^2.

Of course it has a normal component that equals $r\omega^2 = (0.075)(400)^2 = 12\,000$ m/s^2.

Hence, the magnitude of the linear acceleration is $a = \sqrt{(12\,000)^2 + (9)^2} = 12\,000$ m/s^2.

12.41. A ball is thrown at an angle of 40° to the horizontal. What height will the ball reach if it lands 100 ft away? Neglect air resistance.

SOLUTION

Choose the x and y axes with the origin at the point where the ball is thrown. Neglecting air resistance, the x component of the acceleration is zero. The y component of the acceleration is $-g$.

From equation (7) with $a_x = 0$ and $a_y = -32.2$ ft/s,

$$x = v_{0x}t \quad \text{and} \quad y = v_{0y}t - \tfrac{1}{2}(32.2)t^2$$

Given that when $x = 100$, $y = 0$, $v_{0x} = v_0 \cos 40°$ and $v_{0y} = v_0 \sin 40°$, the above equations become

$$100 = v_0 \cos 40° \, (t)$$

$$0 = v_0 \sin 40° \, (t) - \tfrac{1}{2}(32.2)t^2$$

Solving the first equation for v_0, substituting this in the second equation and solving for t gives $t = 2.28$ s. Substituting this value in the first equation yields $v_0 = 57.3$ ft/s. The maximum height occurs at one-half the distance. Therefore, for $t = 1.14$ s, $y_{max} = 57.3 \sin 40° \, (1.14) - \tfrac{1}{2}(32.2)(1.14)^2 = 21.1$ ft.

Supplementary Problems

12.42. If a car moves at the rate of 30 mi/h for 6 min, then 60 mi/h for 10 min, and finally 5 mi/h for 3 min, what is the average velocity in the total interval?		*Ans.*	61.3 ft/s

12.43. A jet-propelled object has straight-line motion according to the equation $x = 2t^3 - t^2 - 2$, where x is in meters and t is in seconds. What is the change in displacement while the speed changes from 4 m/s to 48 m/s?		*Ans.*	$\Delta x = 44$ m

12.44. A body moves along a straight line so that its displacement from a fixed point on the line is given by $s = 3t^2 + 2t$. Find the displacement, velocity, and acceleration at the end of 3 s.
Ans.	33 ft, 20 ft/s, 6 ft/s^2

12.45. The motion of a particle is defined by the relation $s = t^4 - 3t^2 + 2t^2 - 8$, where s is in meters and t is in seconds. Determine the velocity $\dot{s}$ and the acceleration $\ddot{s}$ when $t = 2$ s.
Ans.	$\dot{s} = +4$ m/s, $\ddot{s} = +16$ m/s^2

12.46. A motorbike travels along a straight road between two points at a mean speed of 88 ft/s. It returns at a mean speed of 44 ft/s. What is the mean speed for the round trip?		*Ans.*	58.7 ft/s

12.47. An automobile accelerates uniformly from rest to 72 km/h and then the brakes are applied so that it decelerates uniformly to a stop. If the total time is 15 s, what distance was traveled?
Ans.	$d = 150$ m

12.48. A bullet is fired with a muzzle velocity of 600 m/s. If the length of the barrel is 750 mm, what is the average acceleration?   *Ans.*   240 km/s^2

12.49. An automobile is accelerating from rest at a uniform rate of 8 ft/s^2. How long will it take to reach 30 mi/h and in what distance?   *Ans.*   5.5 s, 121 ft

12.50. A stone is dropped from a balloon that is ascending at a uniform rate of 30 ft/s. If it takes the stone 10 s to reach the ground, how high was the balloon at the instant the stone was dropped?   *Ans.*   1310 ft

12.51. A person in a balloon rising with a constant velocity of 4 m/s propels a ball upward with velocity of 1.2 m/s relative to the balloon. After what time interval will the ball return to the balloon?
Ans. $t = 0.245$ s

12.52. A stone is dropped with zero initial velocity into a well. The sound of the splash is heard 3.63 s later. How far below the ground surface is the surface of the water? Assume that the velocity of sound is 1090 ft/s.   *Ans.* $s = 193$ ft

12.53. Child *A* throws a ball vertically up with a speed of 30 ft/s from the top of a shed 8 ft high. Child *B* on the ground at the same instant throws a ball vertically up with a speed of 40 ft/s. Determine the time at which the two balls will be at the same height above the ground. What is the height?
Ans. $t = 0.8$ s, $h = 21.7$ ft

12.54. A truck, traveling at constant speed, passes a parked police car. The policeman gives chase immediately, accelerating at a constant rate to 80 mi/h in 10 s, after which he maintains a constant speed. If the police car overtakes the truck in one-half mile what is the constant speed of the truck?   *Ans.* $v = 65$ mi/h

12.55. A radar-equipped police car notes a car traveling 70 mi/h. The police car starts pursuit 30 s after the observation and accelerates to 100 mi/h in 20 s. Assuming the speeds are maintained on a straight road, how far from the observation post will the case end?   *Ans.* $s = 13,700$ ft

12.56. An automobile accelerates uniformly from rest on a straight level road. A second automobile starting from the same point 6 s later with initial velocity zero accelerates at 6 m/s^2 to overtake the first automobile 400 m from the starting point. What is the acceleration of the first automobile?
Ans. $a = 2.62$ m/s^2

12.57. Plane *A* leaves an airport and flies north at 120 mi/h. Plane *B* leaves the same airport 20 min later and flies north at 150 mi/h. How long will it take *B* to overtake *A*?   *Ans.* $t = 1.33$ h

12.58. A particle moves on a straight line with the acceleration shown in the graph in Fig. 12-23. Determine the velocity and displacement at time $t = 1, 2, 3,$ and 4 s. Assume that the initial velocity is +3 ft/s and the initial displacement is zero.
Ans. $v_1 = +1$ ft/s, $s_1 = +2$ ft; $v_2 = +3$ ft/s, $s_2 = +4$ ft; $v_3 = -1$ ft/s, $s_3 = +5$ ft; $v_4 = -3$ ft/s, $s_4 = +3$ ft

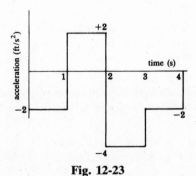

Fig. 12-23

12.59. A particle of dirt falls from an elevator that is moving up with a velocity of 3 m/s. If the particle reaches the bottom in 2 s, how high above the bottom was the elevator when the particle started falling? *Ans.* $s = 13.6$ m

12.60. A bullet is fired vertically upward with a speed of 600 m/s. Theoretically, to what height would the bullet ascend?   *Ans.*   18.4 km

12.61. A ball is thrown vertically upward with a speed of 30 ft/s from the edge of a cliff 50 ft above sea level. What is the highest point above sea level reached? How long does it take the ball to hit the water? With what velocity does it hit the water?   *Ans.* $h = 64.0$ ft, $t = 2.93$ s, $v = 64.3$ ft/s

12.62. A particle moves with an acceleration $a = -6v$, where a is in ft/s^2 and v is in ft/s. When $t = 0$ s, the displacement $s = 0$ and the velocity $v = 9$ ft/s. Find the displacement, velocity, and acceleration when $t = 0.5$ s.   *Ans.* $s = 1.43$ ft, $\dot{s} = v = 0.448$ ft/s, $\ddot{s} = a = 2.69$ ft/s^2

12.63. A body moving with velocity v_0 enters a medium in which the resistance is proportional to the velocity squared. This means $a = -kv^2$. Determine the expression for the velocity in terms of time t. *Ans.* $v = 1/(kt + 1/v_0)$

12.64. The speed of a particle is given by $v = 2t^3 + 5t^2$. What distance does it travel while its speed increases from 7 to 99 ft/s?   *Ans.* $s = 83.3$ ft

12.65. A particle moves to the right from rest with an acceleration of 6 m/s^2 until its velocity is 12 m/s to the right. It is then subjected to an acceleration of 12 m/s^2 to the left until its total distance traveled is 36 m. Determine the total elapsed time.   *Ans.* $t = 4.73$ s

12.66. Water drips from a faucet at the rate of 6 drops per second. The faucet is 8 in above the sink. When one drop strikes the sink, how far is the next drop above the sink?   *Ans.* $h = 7.75$ in

12.67. A particle moving with a velocity of 6 m/s upward is subjected to an acceleration of 3 m/s^2 downward until its displacement is 2 m below its position when the acceleration began. The acceleration then ceases for 3 s. The particle is then subjected to an acceleration of 4 m/s^2 upward for 5 s. Determine the displacement and the distance traveled.   *Ans.* $s = -7.4$ m relative to start, $d = 62.2$ m

12.68. The velocity–time curve for a point moving on a straight line is shown in the graph in Fig. 12-24. How far does the point move in the 2 s?   *Ans.* $x = 5.09$ ft

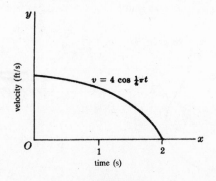

Fig. 12-24

12.69. An object moves on a straight line with constant acceleration 2 m/s^2. How long will it take to change its speed from 5 to 8 m/s? What change in displacement takes place during this time interval?
Ans. $t = 1.5 \text{ s}$, $d = 9.75 \text{ m}$

12.70. The motion of a particle along a straight path is given by the acceleration $a = t^3 - 2t^2 + 7$, where a is in ft/s^2 and t is in seconds. The velocity is 3.58 ft/s when $t = 1 \text{ s}$ and the displacement is $+9.39$ ft when $t = 1 \text{ s}$. Calculate the displacement, velocity, and acceleration when $t = 2 \text{ s}$.
Ans. $s = 15.9 \text{ ft}$, $v = 9.67 \text{ ft/s}$, $a = 7 \text{ ft/s}^2$

12.71. A point moves in rectilinear motion along the x-axis. Given $v = x^{1/2} \text{ m/s}$, determine the position, velocity, and acceleration at $t = 4 \text{ s}$. At $t = 0$, $v = 1 \text{ m/s}$.   *Ans.* $x = 9 \text{ m}$, $v = 3 \text{ m/s}$, $a = \frac{1}{2} \text{ m/s}^2$

12.72. In the system shown in Fig. 12-25, determine the velocity and acceleration of block 3 at the instant considered.   *Ans.* $v_3 = 10.5 \text{ ft/s}$ up, $a_3 = 5.0 \text{ ft/s}^2$ up

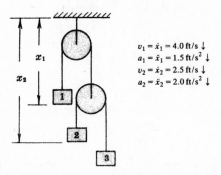

$v_1 = \dot{x}_1 = 4.0 \text{ ft/s} \downarrow$
$a_1 = \ddot{x}_1 = 1.5 \text{ ft/s}^2 \downarrow$
$v_2 = \dot{x}_2 = 2.5 \text{ ft/s} \downarrow$
$a_2 = \ddot{x}_2 = 2.0 \text{ ft/s}^2 \downarrow$

Fig. 12-25

12.73. A point moves along the path $y = \frac{1}{3}x^2$ with a constant speed of 8 ft/s. What are the x and y components of the velocity when $x = 3 \text{ ft}$? What is the acceleration of the point when $x = 3 \text{ ft}$?
Ans. $\dot{x} = 3.58 \text{ ft/s}$, $\dot{y} = 7.16 \text{ ft/s}$, $a = 3.82 \text{ ft/s}^2$

12.74. A particle moves along the curve $y = \frac{1}{3}x^3 + 2$. When $x = 2 \text{ in}$, the x-component of the velocity is 3 in/s. What is the total velocity   *Ans.* $v = 12.4 \text{ in/s}, \angle 76°$

12.75. A point has a normal acceleration of 120 ft/s^2 when traveling on a circular path with a constant peripheral speed of 80 ft/s. What is the radius of the circle?   *Ans.* $r = 53.3 \text{ ft}$

12.76. A point P moves at a constant speed v in a counterclockwise direction along a circle of radius a as shown in Fig. 12-26. Selecting a pole O at the left end of the horizontal diameter, derive expressions for the radial and transverse components of the acceleration. (*Hint:* $r = 2a \cos \theta$.)
Ans. $a_r = -(v^2/a) \cos \theta$, $a_\theta = -(v^2/a) \sin \theta$

Fig. 12-26   **Fig. 12-27**

12.77. In the preceding problem show that total acceleration is v^2/a, which is the normal component for circular motion of a point with constant speed (there is no tangential component).

12.78. In the slider crank mechanism shown in Fig. 12-27, the crank is rotating 200 rpm. Determine the velocity and acceleration of the crosshead when $\theta = 30°$.   *Ans.* $v = 7.6$ m/s, $a = 260$ m/s^2

12.79. In Problem 12.78, determine the velocity of the crosshead when $\theta = 90°$. What is the angular velocity of the member connecting the crosshead and the crank?   *Ans.* $v = 12.6$ m/s, zero

12.80. A particle oscillates with an acceleration $a = -kx$. Determine k if the velocity $v = 2$ ft/s when the displacement $x = 0$, and $v = 0$ when $x = +2$ ft.   *Ans.* $k = +1$

12.81. A particle moves with simple harmonic motion with a frequency of 30 cycles/min and an amplitude of 6 mm. Determine the maximum velocity and acceleration.
Ans. $v_{max} = 18.8$ mm/s, $a_{max} = 59.2$ mm/s^2

12.82. A body having simple harmonic motion has a period of 6 s and an amplitude of 4 ft. Determine the maximum velocity and acceleration of the body.   *Ans.* $\frac{4}{3}\pi$ ft/s, $\frac{4}{9}\pi^2$ ft/s^2

12.83. A particle having a constant speed of 15 ft/s moves around a circle 10 ft in diameter. What is the normal acceleration?   *Ans.*   45 ft/s^2

12.84. The normal acceleration of a point on the rim of a 3 m-diameter flywheel is a constant 15 m/s^2. What is the angular speed of the flywheel?   *Ans.*   3.16 rad/s

12.85. A point moves on a path 3 m in diameter so that the distance traversed is $s = 3t^2$. What is the tangential acceleration at the end of 2 s?   *Ans.*   6 m/s^2

12.86. A particle moves on a 4-in-radius circular path. The distance, measured along the path, is given by $s = 8t^3$ in. What is the magnitude of the total acceleration after the particle has traveled around the circular path once?   *Ans.* $a = 667$ in/s^2

12.87. The flywheel of an automobile acquires a speed of 2000 rpm in 45 s. Find its angular acceleration. Assume uniform motion.   *Ans.*   4.65 rad/s^2

12.88. A rotor with a 0.592-in diameter is spinning 2,000,000 rpm in a high-vacuum chamber. What is the normal component of the acceleration of a point on the rim?   *Ans.* $a_n = 1.08 \times 10^9$ ft/s^2

12.89. a point moves on a circular path with its position from rest defined by $s = t^3 + 5t$, where s and t are measured in meters and seconds, respectively. The magnitude of the acceleration is 8.39 m/s^2 when $t = 0.66$ s. What is the diameter of the path?   *Ans.* $d = 10.8$ m

12.90. A horizontal bar 1.2 m long rotates about a vertical axis through its midpoint. Its angular velocity changes uniformly from 0.5 rad/s to 2 rad/s in 20 s. What is the linear acceleration of a point on the end of the bar 5 s after the speedup occurs?   *Ans.* $a_t = 0.045$ m/s^2, $a_n = 0.46$ m/s^2

12.91. Particle P travels on a circular path with radius 2.5 m as shown in Fig. 12-28. the speed of P is decreasing (the tangential component of the acceleration is thus directed opposite to the velocity vector) at the instant considered. If the total acceleration vector is as shown, determine the velocity of P and the angular acceleration of the line OP at that instant.
Ans. $v = 6.07$ m/s, $\theta_x = 135°$; $\alpha = 3.4$ rad/s^2

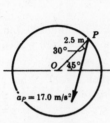

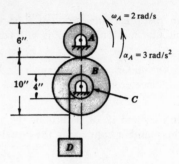

Fig. 12-28   **Fig. 12-29**

12.92. Disk A drives disk B without slip occurring. Determine the velocity and acceleration of the weight D, which is connected by a cord to drum C, which is keyed to disk B as shown in Fig. 12-29.
Ans. $v_D = 2.40$ in/s up, $a_D = 3.60$ in/s^2 up

12.93. The angular acceleration of a rotor is given by $\alpha = Kt^{-1/2}$, where α is in rad/s^2 and t is in seconds. When $t = 1$ s, the angular velocity $v = 10$ rad/s and the angular displacement $\theta = 3.33$ rad. When $t = 0$ s, the angular displacement $\theta = -4$ rad. Determine θ, ω, and α when $t = 4$ s.
Ans. $\theta = 46.7$ rad, $\omega = 18$ rad/s, $\alpha = 2$ rad/s^2

12.94. The displacement in meters of a point is described in terms of x and y components as follows:

$$x = 2t^2 + 5t \qquad y = 4.9t^2$$

Determine the velocity and acceleration at the end of 4 s.
Ans. $v_x = 21$ m/s, $v_y = 39.2$ m/s; $a_x = 4$ m/s^2, $a_y = 9.8$ m/s^2

12.95. A bicycle rider travels 250 m north and then 160 m northwest. What is the rider's displacement? What distance does the rider cover?   *Ans.* $s = 381$ m, 72.7° north of west; $d = 410$ m

12.96. Car A is moving northwest with a speed of 100 km/h. Car B is moving east with a speed of 60 km/h. Determine the velocity of A relative to B. Determine the velocity of B relative to A.
Ans. $v_{A/B} = -131\mathbf{i} + 70.7\mathbf{j}$ km/h, $v_{B/A} = 131\mathbf{i} - 70.7\mathbf{j}$ km/h

12.97. Body A has a velocity of 15 km/h from west to east relative to body B, which in turn has a velocity of 50 km/h from northeast to southwest relative to body C. Determine the velocity of A relative to C.
Ans. $v_{A/C} = 40.8$ km/h, at 60° southwest

12.98. Refer to Fig. 12-30. A rotating spotlight is at a perpendicular distance l from a horizontal floor. The light revolves a constant N revolutions per minute about a horizontal axis perpendicular to the paper. Derive expressions for the velocity and acceleration of the light spot traveling along the floor. Let θ be the angle between the vertical line l and the light beam at time t.
Ans. $\dot{x} = 0.105lN \sec^2 \theta$, $\ddot{x} = 0.022lN^2 \sec^2 \theta \tan \theta$

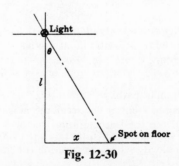

Fig. 12-30

12.99. A particle moves in a straight line such that its v–t curve is as shown in Fig. 12-31. How far has it gone after 10 s? What is the acceleration at 9 s?   *Ans.* $s = 120$ ft, $a = -8$ ft/s^2

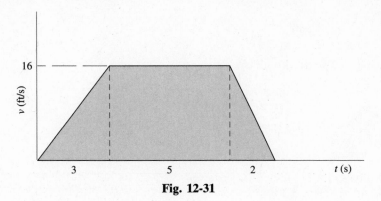

Fig. 12-31

12.100. A particle moves with rectilinear motion. Given the a–s curve shown in Fig. 12-32, determine the velocity after the particle has traveled 30 m if the initial velocity is 10 m/s.   *Ans.* $v = 20$ m/s

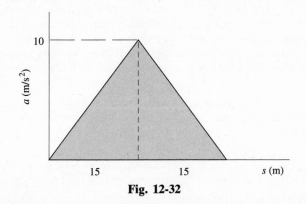

Fig. 12-32

12.101. A particle moves in a straight line with an a–t curve shown in Fig. 12-33. The initial displacement and velocity are zero. At what time and with what displacement will the particle come to rest again?
Ans. $t = 10$ s, $x = 29.3$ m

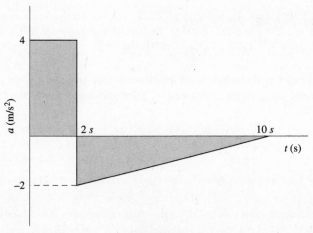

Fig. 12-33

12.102. A particle moves in rectilinear motion such that its a–s curve is as shown in Fig. 12-34. If, initially, $s = 0$ and $v = 4$ m/s, what is the velocity when the position is 8 m? 12 m?   *Ans.* $v = 10.6$ m/s, $v = 12$ m/s

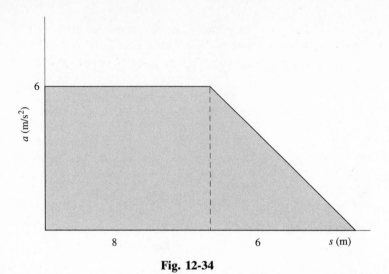

Fig. 12-34

12.103. A particle at rest starts to move on a circle of 2-ft radius. Its tangential component of acceleration, as a function of arc distance, increases linearly from zero to 10 ft/s^2 in one trip around the circle, after which the tangential acceleration remains constant. What is the velocity and what is the normal component of the acceleration after the second trip around the circle?   *Ans.* $v = 19.4$ ft/s, $a_n = 188$ ft/s^2

12.104. A projectile is fired at an elevation angle of 30° with an initial speed of 1500 ft/s. Neglecting air resistance, determine the range, time of flight and maximum altitude of the projectile.
Ans. $R = 60,400$ ft, $t = 46.6$ s, $h = 8760$ ft

12.105. A shell from a mortar is fired onto a 300 ft-high plateau 6000 ft away. How far beyond the edge of the plateau will the shell strike if the initial speed is 800 ft/s and the elevation of the mortar is 60°?
Ans. $x = 11,000$ ft

12.106. In Prob. 12.105 how close to the base of the cliff can the mortar be placed so that the shell will just clear the edge of the cliff?   *Ans.* $x = 176$ ft

12.107. A batter hits a ball, giving it a velocity of 100 ft/s at 40° to the horizontal. One second after the ball is hit, a fielder standing in the direct path of the ball at a distance of 250 ft starts running after the ball at a constant speed. How fast must he run to catch the ball?   *Ans.* $v = 18.7$ ft/s

12.108. A point moves on a path such that its position vector is given by $\mathbf{r} = e^{2t}\mathbf{i} + 40e^{-2t}\mathbf{j}$ m. Determine the velocity and acceleration of the point when $t = 2$ s.   *Ans.* $\mathbf{v} = 109\mathbf{i} - 5.41\mathbf{j}$ m/s, $\mathbf{a} = 218\,\mathbf{i} + 5.41\mathbf{j}$ m/s^2

12.109. A particle moving on a curve has a velocity of $\mathbf{v} = 2\mathbf{i} + (2t + 20)\mathbf{j}$ m/s. At $t = 2$ s the position is $4\mathbf{i} + 75\mathbf{j}$ m. What is the equation of the path?   *Ans.* $y = \frac{1}{4}x^2 + 10x + 31$

12.110. In Prob. 12.109 what is the acceleration at $x = 0$?   *Ans.* $\mathbf{a} = 2\mathbf{j}$ m/s^2

12.111. Given the acceleration vector as $\mathbf{a} = t\mathbf{i} + 2t\mathbf{j} - 3\mathbf{k}$ m/s^2, what is the velocity vector at $t = 2$ s. The velocity vector at $t = 1$ sec is $\mathbf{v} = \mathbf{i} + \mathbf{j} + \mathbf{k}$ m/s.   *Ans.* $\mathbf{v} = \frac{5}{2}\mathbf{i} + 4\mathbf{j} - 2\mathbf{k}$ m/s

12.112. In Prob. 12.111 what is the component of the acceleration vector in the direction of the velocity vector at $t = 2$s? (*Note:* This is the tangential component of the acceleration.)   *Ans.* $a_t = 5.27$ in/s^2

Chapter 13

Dynamics of a Particle

13.1 NEWTON'S LAWS OF MOTION

1. A particle will maintain its state of rest or of uniform motion (at constant speed) along a straight line unless compelled by some force to change that state. In other words, a particle accelerates only if an unbalanced force acts on it.

2. The time rate of change of the product of the mass and velocity of a particle is proportional to the force acting on the particle. The product of the mass m and the velocity $\mathbf{v}$ is the linear momentum $\mathbf{G}$. Thus, the second law states

$$\mathbf{F} = K\frac{d(m\mathbf{v})}{dt} = K\frac{d\mathbf{G}}{dt}$$

If m is constant, the above equation becomes

$$\mathbf{F} = Km\frac{d\mathbf{v}}{dt} = Km\mathbf{a}$$

If suitable units are chosen so that the constant of proportionality $K = 1$, these equations are

$$\mathbf{F} = \frac{d\mathbf{G}}{dt} \qquad \text{or} \qquad \mathbf{F} = m\mathbf{a}$$

3. To every action, or force, there is an equal and opposite reaction, or force. In other words, if a particle exerts a force on a second particle then the second particle exerts a numerically equal and oppositely directed force on the first particle.

13.2 UNITS

(a) Units depend on the system. In most engineering work, the value of K in the above formulas is made equal to unity by proper selection of units. The two fundamental units assigned in the U.S. Customary System are *lb* for force and *ft per sec per sec* or *ft/s²* for acceleration. The unit of mass is then the derived unit in terms of these two. In the case of a freely falling particle near the earth's surface, the only force acting is its weight W. Its acceleration is the acceleration of gravity g (assumed to be 32.2 ft/s² for most localities in the United States). The equation of the second law is then written (vector notation is not used, since this is straight line motion)

$$W = Kma \qquad \text{or} \qquad W = (1)mg$$

Then

$$m = \frac{W \text{ lb}}{g \text{ ft/s}^2} = \frac{W}{g}\frac{\text{lb-s}^2}{\text{ft}}$$

This derived unit of mass is called a *slug*.

(b) In SI units, the value of K is 1 in the above equations because of the coherence of the system. Thus,

$$\mathbf{F} = m\mathbf{a}$$

where m = mass in kilograms
         $\mathbf{a}$ = acceleration in m/s²
         $\mathbf{F}$ = force in newtons

13.3 ACCELERATION

The acceleration of a particle may be determined by the vector equation according to Newton's law

$$\sum \mathbf{F} = m\mathbf{a} = m\ddot{\mathbf{r}}$$

where $\sum \mathbf{F}$ = vector sum of all the forces acting on the particle
   m = mass of the particle
   $\mathbf{a} = \ddot{\mathbf{r}}$ = acceleration

13.4 D'ALEMBERT'S PRINCIPLE

Jean D'Alembert suggested in 1743 that Newton's second law of motion, as given in Section 13.3, could be written

$$\sum \mathbf{F} - m\mathbf{a} = 0$$

Thus, an imaginary force (called an "inertia force"), which is collinear with $\sum \mathbf{F}$ but oppositely sensed and of magnitude *ma*, would if applied to the particle cause it to be in equilibrium. The equations of equilibrium would then apply. Some authors state that the particle is in dynamic equilibrium. Actually the particle is *not* in equilibrium, but the equations of equilibrium can be applied.

13.5 PROBLEMS IN DYNAMICS

The solutions of problems in dynamics vary with the type of force system. Many problems involve forces which are constant; Problems 13.1 through 13.16 are examples of this type. In other problems, the forces vary with position (rectilinear or angular); Problems 13.17 through 13.22 are examples of this type. The subject of vibrations is built on force systems that vary not only with distance but also with velocity. Problems 13.23 and 13.24 deal with forces that vary with the first and second powers of the velocity.

The subject of ballistics is introduced in an elementary manner in Problem 13.25, which deals with the motion of a projectile under the action of the constant force of gravity. To this solution may be added retarding forces that vary with the velocity of the projectile.

An object is said to be moving with central force motion if the force acting on the object is always directed through a central point. Satellites and planets are examples of central force motion. These are discussed in Problems 13.26 through 13.31.

Solved Problems

In solving problems, the vector equation F = *m*a is replaced by scalar equations using components. In the diagrams, vectors are designated by their magnitudes when the directions are apparent.

13.1. A particle weighing 2 lb is pulled up a smooth plane by a force F as shown in Fig. 13-1(*a*). Determine the force of the plane on the particle and the acceleration along the plane.

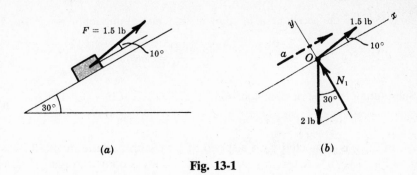

(a)	(b)

Fig. 13-1

SOLUTION

The free-body diagram is shown in Fig. 13-1(b). The acceleration a is shown as a dashed vector acting parallel to the plane and upward. If the value obtained is negative, this indicates that the acceleration acts parallel to the plane but downward.

It is important to keep in mind that the force system shown acting on the particle is not in equilibrium. If it were in equilibrium, the particle would not accelerate.

Applying Newton's laws, there result two equations along the x and y axes chosen, respectively, parallel and perpendicular to the plane.

$$\sum F_x = \frac{W}{g}a_x \quad \text{or} \quad 1.5 \cos 10° - 2 \sin 30° = \frac{2}{32.2}a_x$$

$$\sum F_y = \frac{W}{g}a_y \quad \text{or} \quad 1.5 \sin 10° - 2 \cos 30° + N_1 = 0$$

Assuming that the particle does not leave the plane, its velocity in the y direction is zero. Therefore a_y must also be zero.

The second equation yields the force of the plane on the particle, $N_1 = 1.47$ lb. From the first equation, $a_x = 7.68$ ft/s^2.

13.2. A particle having a mass of 5 kg starts from rest and attains a speed of 4 m/s in a horizontal distance of 12 m. Assuming a coefficient of friction of 0.25 and uniformly accelerated motion, what is the smallest value a constant horizontal force P may have to accomplish this? Refer to Fig. 13-2.

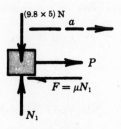

Fig. 13-2

SOLUTION

Equation of motion in the horizontal direction: $\sum F = P - 0.25N_1 = ma$.
By inspection, $N_1 = 9.8 \times 5 = 49$ N.

To determine the acceleration a, apply the kinematic equation $v^2 = v_0^2 + 2as$. Hence,

$$a = \frac{(4\text{ m/s})^2}{2(12\text{ m})} = 0.667\text{ m/s}^2$$

Substituting into the original equation, $P = 5(0.667) + 0.25 \times 49 = 15.6\text{ N}$.

13.3. A mass of 2 kg is projected with a speed of 3 m/s up a plane inclined 20° with the horizontal. Refer to Fig. 13-3(a). After traveling 0.8 m, the mass comes to rest. Determine the coefficient of friction and also the speed as the block returns to its starting position.

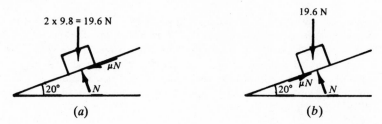

(a) $\qquad\qquad\qquad\qquad\qquad\qquad\qquad\qquad$ (b)

Fig. 13-3

SOLUTION

In the free-body diagram of Fig. 13-3(a), friction is shown acting down the plane. By inspection, the normal force $N = 19.6 \cos 20° = 18.4\text{ N}$. To determine the acceleration a, apply the kinematic equation $v^2 = v_0^2 + 2as$. Hence

$$a = \frac{0 - (+3)^2}{2(0.8)} = -5.63\text{ m/s}^2$$

Summing forces parallel to the plane (up being positive) yields

$$-19.6 \sin 20° - u(18.4) = 2(-5.63)$$

Thus, $u = 0.25$.

To solve for the return speed, refer to Fig. 13-3(b), which shows the frictional force acting up the plane. Using the down direction as positive, the equation of motion becomes

$$+19.6 \sin 20° - 0.25(18.4) = 2(a)$$

Hence, $a = 1.05\text{ m/s}^2$ down the plane.

Finally,

$$v^2 = v_0^2 + 2as \qquad \text{or} \qquad v^2 = 0 + 2(1.05)(0.8)$$

From this,

$$v = 1.3\text{ m/s}$$

13.4. An automobile weighing 1800 lb goes around a 2000-ft curve at a constant speed of 40 mi/h. If the road is not banked, what frictional force must the road exert on the tires so that they will maintain motion along the curve?

SOLUTION

In Fig. 13-4, O is the center of the curve 2000 ft from the automobile. The forces acting on the automobile are its weight W; the normal force N, which is equal to W; and the friction force F. Assuming to the left is positive, the equation of motion becomes

$$\sum F = \frac{W}{g}a_n = \frac{W}{g}\frac{v^2}{r} \quad \text{or} \quad F = \frac{1800 \text{ lb}}{32.2 \text{ ft/s}^2}\frac{(58.7 \text{ ft/s})^2}{2000 \text{ ft}} = 96.3 \text{ lb}$$

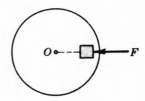

Fig. 13-4   **Fig. 13-5**

13.5. A small block of mass m is on a rotating turntable at a distance r from the center as shown in Fig. 13-5. Assuming a coefficient of friction μ between the mass and the turntable, what is the maximum linear velocity the mass may have without slipping?

SOLUTION

The only force acting horizontally is the friction F, which equals μN.
Sum forces along the radius: $\sum F = ma_n$ or $F = ma_n$.
Since the normal force N between the block and table is equal to mg and $a_n = v^2/r$,

$$\mu mg = \frac{mv^2}{r} \quad \text{or} \quad v = \sqrt{\mu gr}$$

13.6. Figure 13-6 indicates a particle of mass m that can move in a circular path about the y axis. The plane of the circular path is horizontal and perpendicular to the y axis. As the angular velocity ω increases, however, the particle rises, which means that the radius r of its circular path also increases. The mass and cord are called a spherical (sometimes conical) pendulum. Derive the relationship between θ and ω for constant angular velocity, and find the frequency in terms of θ.

SOLUTION

Assume that the constant angular velocity of the particle (or of its cord l) is ω radians per second. The angle θ is the angle between the cord and the y axis. The forces acting on the particle are the force of gravity and the tension T in the cord.
Since the particle is moving with constant angular velocity, its only linear acceleration is the normal component a_n directed toward the center of the path (i.e., toward the intersection of the horizontal plane of travel and the y axis).

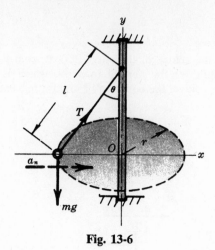

Fig. 13-6

Summing forces along this normal, $\Sigma F_n = T \sin \theta = ma_n$.
Since $a_n = r\omega^2 = (l \sin \theta)\omega^2$, this equation becomes

$$T \sin \theta = m(l \sin \theta)\omega^2 \tag{1}$$

Summing forces in the y direction, $\Sigma F_y = T \cos \theta - mg = ma_y = 0$ or $T = mg/\cos \theta$.
Substituting this value T into equation (1),

$$\frac{mg}{\cos \theta} \sin \theta = m(l \sin \theta)\omega^2 \qquad \text{or} \qquad \omega = \sqrt{\frac{g}{l \cos \theta}}$$

If θ is known, this equation can be solved for the necessary angular velocity ω to maintain θ constant. Or, if ω is known, θ may be found.

Since ω is constant for a given angle θ,

$$\text{frequency } f = \frac{\omega \text{ rad/s}}{2\pi \text{ rad/rev}} = \frac{1}{2\pi} \sqrt{\frac{g}{l \cos \theta}} \text{ Hz}$$

This is the frequency about the y axis.

13.7. A block, assumed to be a particle and weighing 10 lb, rests on a plane which can turn about the y axis. See Fig. 13-7(a). The length of the cord l is 2 ft. What is the tension in the cord when the angular velocity of the plane and block is 10 rev/min?

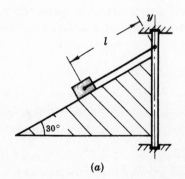

(a)

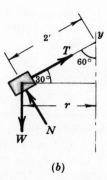

(b)

Fig. 13-7

SOLUTION

From the free-body diagram of the block shown in Fig. 13-7(b), $r = 2 \cos 30° = 1.732$ ft.
The only acceleration present is the normal component a_n directed horizontally toward the y axis:

$$a_n = r\omega^2 = (1.732 \text{ ft})\left(\frac{10 \text{ rev/min} \times 2\pi \text{ rad/rev}}{60 \text{ s/min}}\right)^2 = 1.91 \text{ ft/s}^2$$

Sum forces horizontally along the radius r and along the y axis to obtain the following equations:

$$\sum F_n = T \cos 30° - N \sin 30° = \frac{W}{g} a_n = \frac{10}{32.2} \times 1.91 \tag{1}$$

$$\sum F_y = N \cos 30° + T \sin 30° - 10 = \frac{W}{g} a_y = 0 \tag{2}$$

Solve equation (2) for

$$N = \frac{10}{\cos 30°} - T \frac{\sin 30°}{\cos 30°}$$

Substituting into equation (1),

$$T \cos 30° - \left(\frac{10}{\cos 30°} - T \frac{\sin 30°}{\cos 30°}\right) \sin 30° = \frac{10}{32.2} \times 1.91 \quad \text{or} \quad T = 5.52 \text{ lb}$$

13.8. The 4-lb object at B in Fig.13-8(a) moves in a circular, horizontal path under the action of a
cord AB and a rigid bar BC, which can be considered weightless. At the instant B has a speed
of 6 ft/s, what are the forces in the supporting members AB and BC?

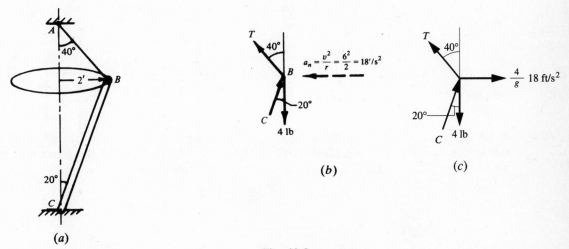

Fig. 13-8

SOLUTION

The free-body diagram in Fig. 13-8(b) shows the object under the action of its weight, tension T,
and compression C. The normal acceleration is $v^2/r = (6)^2/2 = 18 \text{ ft/s}^2$ directed to the left. Summation
of forces yields the following equations:

$$\sum F_v = T \cos 40° + C \cos 20° - 4 = 0$$

$$\sum F_h = T \sin 40° - C \sin 20° = \frac{mv^2}{r} = \frac{4}{g} \times 18 = 2.24$$

The solution of these equations yields

$$T = 4.01\,\text{lb} \qquad \text{and} \qquad C = 0.99\,\text{lb}$$

Another solution, using D'Alembert's Principle and the "inertia force", is shown in Fig. 13-8(c). The figure is the free-body diagram with the inertia force, $ma_n = (4/g)(18)$, acting on the particle outwardly from the center of rotation. Thus,

$$\sum F_v = T \cos 40° + C \cos 20° - 4 = 0$$

$$\sum F_h = -T \sin 40° + C \sin 20° + \frac{4}{g} \times 18 = 0$$

The solution of these equations yields

$$T = 4.01\,\text{lb} \qquad \text{and} \qquad C = 0.99\,\text{lb}$$

In circular motion, the inertia force is called the centrifugal force. The centrifugal force is often erroneously thought of as an actual force. It is not!

13.9. In a device as Atwood's machine, two equal masses M are connected by a very light (negligible mass) tape passing over a frictionless pulley as shown in Fig. 13-9(a). A mass m whose magnitude is much less than M is added to one side, causing that mass to fall and the other of course to rise. The time is recorded by an inked stylus resting on the tape and vibrating. Study the motion.

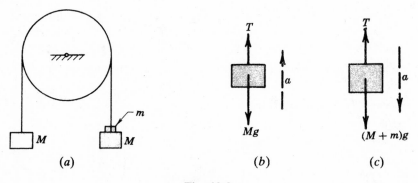

(a)   (b)   (c)

Fig. 13-9

SOLUTION

The free-body diagrams of the two mass systems are shown in Fig. 13-9(b) and (c). The same tension T is acting on each system through the tape because the friction of the pulley is assumed negligible.

The equations of motion, using the same acceleration (otherwise the tape would have broken or become slack) for the two free-body diagrams, are

$$\sum F = T - Mg = Ma \tag{1}$$

$$\sum F = Mg + mg - T = (M + m)a \tag{2}$$

Add equations (*1*) and (*2*) to eliminate the tension T and obtain

$$mg = 2Ma + ma \qquad \text{or} \qquad a = \frac{m}{2M + m} g$$

This expresses the relation between the acceleration of gravity g at the locality where the experiment is performed and the acceleration a of the masses as determined by measurement of distance and time on the tape.

13.10. Figure 13-10 shows a 2-kg mass resting on a smooth plane inclined 20° to the horizontal. A cord which is parallel to the plane passes over a massless, frictionless pulley to a 4-kg mass which will drop vertically when released. What will be the speed of the 4-kg mass 4 s after it is released from rest?

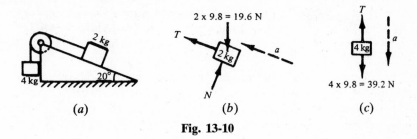

(*a*)   (*b*)   (*c*)

Fig. 13-10

SOLUTION

The free-body diagrams are shown in Fig. 13-10(*b*) and (*c*). Both bodies have the same magnitude of acceleration. The tension T acting on each free body is the same.

The equations of motion are

$$\sum F_b = T - 19.6 \sin 20° = 2a$$
$$\sum F_c = 39.2 - T = 4a$$

Adding, the acceleration is

$$a = 5.42 \text{ m/s}^2$$

The velocity of the 4-kg mass after 4 s is

$$v = v_0 + at = 0 + 5.42(4) = 21.7 \text{ m/s}$$

13.11. Blocks A and B, weighing 20 and 60 lb, respectively, are connected by a weightless rope passing over a frictionless pulley as shown in Fig. 13-11(*a*). Assume a coefficient of friction of 0.30 and determine the velocity of the system 4 s after starting from rest.

SOLUTION

Free-body diagrams are drawn for bodies A and B [see Fig. 13-11(*b*) and (*c*)]. Summing forces perpendicular and parallel to the planes, the equations of motion are

$$\sum F_\perp = N_1 - 20 \cos 30° = \frac{20}{32.2} (0) = 0 \qquad (1)$$

$$\sum F_\parallel = T - 20 \sin 30° - 0.30 N_1 = \frac{20}{32.2} a \qquad (2)$$

$$\sum F_\perp = N_2 - 60 \cos 60° = \frac{20}{32.2} (0) = 0 \qquad (3)$$

$$\sum F_\parallel = 60 \sin 60 - T - 0.30 N2 = \frac{60}{32.2} a \qquad (4)$$

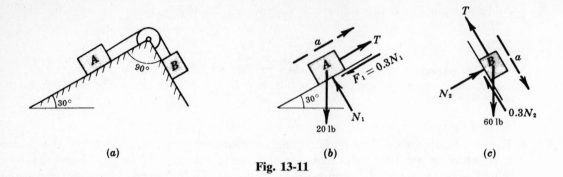

Fig. 13-11

Solve equations (1) and (3) for N_1 and N_2. Substitute these values in equations (2) and (4) and add the two equations to eliminate T. This yields an acceleration $a = 11.1$ ft/s^2.

Applying the kinematics equation $v = v_0 + at$, $v = 0 + 11.1(4) = 44.4$ ft/s.

13.12. Refer to Fig. 13-12(a). Determine the least coefficient of friction between A and B so that slip will not occur. A is a 40-kg mass, B is a 15-kg mass, and F is 500 N, parallel to the plane, which is smooth.

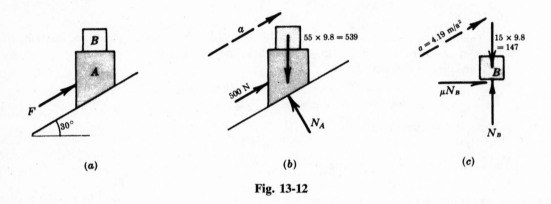

Fig. 13-12

SOLUTION

To determine the acceleration a of the system, draw a free-body diagram of the two masses taken as a unit as shown in Fig. 13-12(b). Summing forces along the plane, $500 - 539 \sin 30° = 55a$ or $a = 4.19$ m/s^2.

Draw a free-body diagram of B as shown in Fig. 13-12(c). Sum forces along the acceleration vector and perpendicular to it to obtain

$$\sum F_\perp = -147 \cos 30° + N_B \cos 30° - \mu N_B \sin 30° = 0 \qquad (1)$$

$$\sum F_\parallel = \mu N_B \cos 30° + N_B \sin 30° - 147 \sin 30° = (15)(4.19) \qquad (2)$$

Multiply the first equation by $\cos 30°$ and the second by $\sin 30°$. Then add to obtain $N_B = 178$ N. Substituting into either equation (1) or (2), $\mu = 0.30$.

13.13. A horizontal force $P = 70$ N is exerted on mass $A = 16$ kg as shown in Fig. 13-13(a). The coefficient of friction between A and the horizontal plane is 0.25. B has a mass of 4 kg and the coefficient of friction between it and the plane is 0.50. The cord between the two masses makes an angle of $10°$ with the horizontal. What is the tension in the cord?

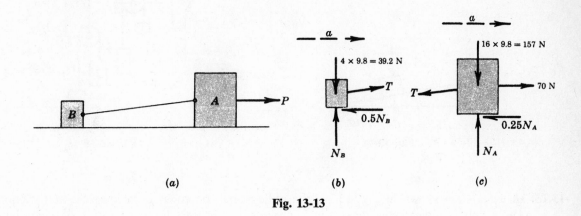

(a)   (b)   (c)

Fig. 13-13

SOLUTION

The equation of motion from the two free-body diagrams are [see Fig. 13-13(b) and (c)]:

$$\sum F_h = 70 - T\cos 10° - 0.25N_A = 16a \qquad (1)$$

$$\sum F_v = N_A - 157 - T\sin 10° = 0 \qquad (2)$$

$$\sum F_h = T\cos 10° - 0.50N_B = 4a \qquad (3)$$

$$\sum F_v = N_B - 39.2 + T\sin 10° = 0 \qquad (4)$$

Substitute N_A in terms of T from equation (2) into equation (1). Substitute N_B in terms of T from equation (4) into equation (3). Eliminate a between these two new equations to obtain $T = 20.5$ N.

13.14. A box is dropped onto a conveyor belt inclined $10°$ with the horizontal and traveling 10 ft/s. If the box is initially at rest and the coefficient of friction between the box and the belt is $\frac{1}{3}$, how long will it take before the box ceases to slip on the belt?

SOLUTION

Refer to Fig. 13-14. The equations of motion for the box are

$$\sum F_\perp = 0 = N - W\cos 10°$$

$$\sum F_\| = \tfrac{1}{3}N - W\sin 10° = \frac{W}{g}a$$

Solving these equations, we get $a = 4.98$ ft/s^2.

To find the time that will elapse before slipping ceases, use

$$v = v_0 + at$$

$$10 = 0 + 4.98t$$

Thus, $t = 2.00$ s

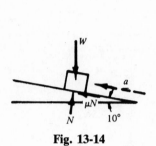

Fig. 13-14

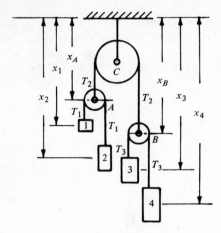

Fig. 13-15

13.15. In the system shown in Fig. 13-15, the pulleys may be considered massless and frictionless. The numbers indicate the masses in kilograms. Determine the acceleration of each mass and the tension in the fixed cord.

SOLUTION

By continuity and the assumptions of no mass for the pulleys and cords, we can state $2T_1 = T_2$ and $2T_3 = T_2$. Also the cords over pulleys A, B, and C are constant in length. Hence,

$$(x_1 - x_A) + (x_2 - x_A) = K_1 \qquad (x_3 - x_B) + (x_4 - x_B) = K_2 \qquad x_A + x_B = K_3$$

where K_1, K_2, and K_3 are constants.

Second derivatives yield

$$a_1 + a_2 = 2a_A \qquad a_3 + a_4 = 2a_B \qquad a_A + a_B = 0$$

The equations of motion, assuming downward as the positive direction, are

$$1 \times 9.8 - T_1 = a_1 \tag{1}$$

$$2 \times 9.8 - T_1 = 2a_2 \tag{2}$$

$$3 \times 9.8 - T_3 = 3a_3 \tag{3}$$

$$4 \times 9.8 - T_3 = 4a_4 \tag{4}$$

Substitute $T_1 = T_2/2$ into equation (1) and then multiply by 2. Substitute $T_1 = T_2/2$ and $a_2 = 2a_A - a_1$ into equation (2). These equations become

$$2 \times 9.8 - \frac{2T_2}{2} = 2a_1 \tag{1'}$$

$$2 \times 9.8 - \frac{T_2}{2} = 4a_A - 2a_1 \tag{2'}$$

Add these two equations to get

$$4 \times 9.8 - 1.5T_2 = 4a_A \tag{5}$$

Substitute $T_3 = T_2/2$ into equation (3) and then multiply by 4. Substitute $T_3 = T_2/2$ and $a_4 = 2a_B - a_3 = -2a_A - a_3$ into equation (4) and multiply by 3. These equations become

$$4 \times 3 \times 9.8 - \frac{4T_2}{2} = 12a_3 \tag{3'}$$

$$3 \times 4 \times 9.8 - \frac{3T_2}{2} = 3[4(-2a_A - a_3)] \tag{4'}$$

Add these two equations to get

$$24 \times 9.8 - 3.5T_2 = -24a_A \qquad\qquad (6)$$

Combine equations (5) and (6) to find $T_2 = 37.6\,\text{N}$.

To find the accelerations of the masses, substitute $T_1 = T_3 = T_2/2 = 18.8$ into equations (1), (2), (3), and (4). These yields

$$a_1 = -9.0\,\text{m/s}^2 \qquad (\text{up})$$

$$a_2 = 0.4\,\text{m/s}^2 \qquad (\text{down})$$

$$a_3 = 3.53\,\text{m/s}^2 \qquad (\text{down})$$

$$a_4 = 5.1\,\text{m/s}^2 \qquad (\text{down})$$

The tension in the fixed cord is $2T_2 = 75.2\,\text{N}$.

13.16. Two masses of 14 kg and 7 kg connected by a flexible inextensible cord rest on a smooth plane inclined 45° with the horizontal as shown in Fig. 13-16(a). When the masses are released, what will be the tension T in the cord? Assume the coefficient of friction between the plane and the 14-kg mass is $\frac{1}{4}$ and between the plane and the 7-kg mass is $\frac{3}{8}$.

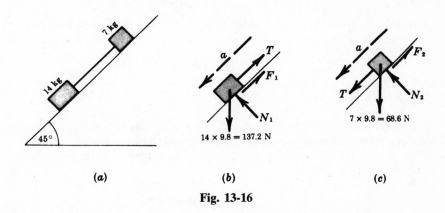

| (a) | (b) | (c) |

Fig. 13-16

SOLUTION

Free-body diagrams of the two masses are shown in Fig. 13-16(b) and (c).

The equations of motion for the 14-kg mass are as follows, where the summations are parallel and perpendicular to the plane:

$$\sum F_{\parallel} = 137.2 \sin 45° - F_1 - T = 14a \qquad\qquad (1)$$

$$\sum F_{\perp} = N_1 - 137.2 \cos 45° = 0 \qquad\qquad (2)$$

The equations of motion for the 7-kg mass are as follows:

$$\sum F_{\parallel} = T + 68.6 \sin 45° - F_2 = 7a \qquad\qquad (3)$$

$$\sum F_{\perp} = N_2 - 68.6 \cos 45° = 0 \qquad\qquad (4)$$

Assume that both masses are moving. Of course, if the friction is great enough, the upper mass may not move.

From equation (2), $N_1 = 137.2 \times 0.707$. Then $F_1 = \frac{1}{4}N_1 = 24.3$.
From equation (4), $N_2 = 68.6 \times 0.707$. Then $F_2 = \frac{3}{8}N_2 = 18.2$.
Substituting these values into equations (1) and (3), the following equations result:

$$137.2 + 0.707 - 24.3 - T = 14a \qquad (5)$$

$$T + 68.6 \times 0.707 - 18.2 = 7a \qquad (6)$$

Multiply equation (6) by 2 and subtract from equation (5) to obtain $T = 4.0$ N.

13.17. A particle of weight W is suspended on a cord of length l as shown in Fig. 13-17(a). Determine the period and frequency of this simple pendulum.

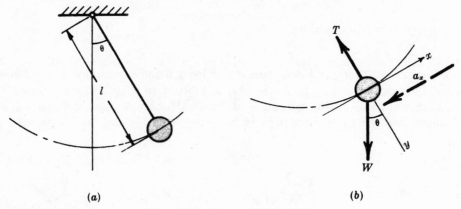

(a)   (b)

Fig. 13-17

SOLUTION

The only forces acting on the particle are its weight vertically down and the tension T in the cord. The position of the particle at any time t may be specified in terms of the angle θ.

Choosing as the x axis the tangent to the path of the particle in the position shown in Fig. 13-17(b), the equation of motion becomes

$$\sum F_x = -W \sin \theta = \frac{W}{g} a_x$$

Thus, the acceleration is zero when θ is zero, i.e., at the lowest position of the particle.

The above differential equation may be solved by noting that a_x is tangent to the path and may therefore be written

$$a_x = l\alpha$$

where α = angular acceleration of the cord and particle. Then

$$-W \sin \theta = \frac{W}{g} l\alpha = \frac{W}{g} l \frac{d^2\theta}{dt^2} \qquad \text{or} \qquad -\frac{g}{l} \sin \theta = \frac{d^2\theta}{dt^2}$$

The solution of this differential equation is simplified by using a series expansion for $\sin \theta$.

$$\sin \theta = \theta - \frac{\theta^3}{3!} + \frac{\theta^5}{5!} - \frac{\theta^7}{7!} + \cdots$$

For small angular displacements, $\sin \theta$ is approximately equal to θ expressed in radians. Thus, for small displacements, the equation of motion becomes

$$-\frac{g}{l}\theta = \frac{d^2\theta}{dt^2}$$

The solution as derived in the theory of differential equations is in the form of sines and cosines:

$$\theta = A \sin \sqrt{\frac{g}{l}}\, t + B \cos \sqrt{\frac{g}{l}}\, t$$

The constants A and B can be evaluated in a given problem by using the boundary conditions.

The frequency ω in radians per second is $\sqrt{g/l}$. The frequency f is $(1/2\pi)\sqrt{g/l}$ Hz (cycles per second).

Incidentally, a cycle is the motion of the particle from a starting point through all possible positions back to the same point. A cycle is complete when the particle moves, let us say, from its top left position to its right position and back to its top left position.

The time to complete one cycle is called the period T. This is then the reciprocal of the frequency f.

$$T = \frac{1}{f(\text{cycles per second})} = \frac{1}{f} \text{ seconds per cycle}$$

13.18. Consider the motion of a particle of mass M resting on a smooth horizontal plane as shown in Fig. 13-18(a). It is attached to a spring that has a constant of K in force units per unit deformation. Displace the mass a distance x_0 from its equilibrium position (spring tension or compression is zero at equilibrium position) and then release it with zero velocity. Study the motion.

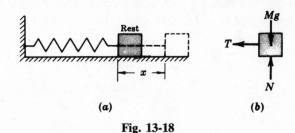

(a)   (b)

Fig. 13-18

SOLUTION

Figure 13-18(b) is a free-body diagram showing the particle in a position a distance x from equilibrium. Acting on the particle in the horizontal direction is the force T in the spring, which is stretched the distance x.

Within the elastic limit of the material, it is assumed that the tension in the spring is proportional to its change in length from the unstressed position. Then

$$T = Kx$$

where T = spring force
  K = constant
  x = change of length

A summation of forces horizontally yields $\sum F_x = -T = Ma_x$.

Note that forces directed to the left, and distances to the left of the rest position, are assumed negative. In this case, the distance x is to the right; hence, a_x is written positive. Tension T is to the left, i.e., negative.

Substituting $T = Kx$ and $a_x = d^2x/dt^2$, the above equation becomes $-Kx = M(d^2x/dt^2)$ (a simple harmonic motion). This differential equation is similar to the one found in Problem 13.17. Its solution is in the form of sines and cosines:

$$x = A \sin \sqrt{\frac{K}{M}}\, t + B \cos \sqrt{\frac{K}{M}}\, t$$

The values of A and B can now be calculated. The value of x is x_0 when t is zero; hence,

$$x_0 = A \sin \sqrt{\frac{K}{M}} 0 + B \cos \sqrt{\frac{K}{M}} 0 = 0 + B \qquad \text{and} \qquad B = x_0$$

and

$$x = A \sin \sqrt{\frac{K}{M}} t + x_0 \cos \sqrt{\frac{K}{M}} t$$

To evaluate A, it is necessary to differentiate x with respect to time, since the other known condition is that the velocity v is zero when the time t is zero:

$$v = \frac{dx}{dt} = A \sqrt{\frac{K}{M}} \cos \sqrt{\frac{K}{M}} t - x_0 \sqrt{\frac{K}{M}} \sin \sqrt{\frac{K}{M}} t$$

When $t = 0$, $v = 0$ and $\sin \sqrt{K/M} \, 0 = 0$. Then $0 = A \sqrt{K/M} \cos 0 - 0$, and $A = 0$. The equation of motion is $x = x_0 \cos \sqrt{K/M} \, t$. Of course, if the initial velocity had some value other than zero, say v_0, then $A = v_0 / \sqrt{K/M}$.

13.19. In Problem 13.18, (a) find the frequency f and period T of the system if the mass M has a weight of 12 oz and the spring constant is 2 oz/in, and (b) find the frequency if the mass M is 0.34 kg and the constant K is 22 N/m.

SOLUTION

(a)
$$f = \frac{1}{2\pi} \sqrt{\frac{K}{M}} = \frac{1}{2\pi} \sqrt{\frac{2/16 \, \text{lb/in} \times 12(32.2) \, \text{in/s}^2}{12/16 \, \text{lb}}} = 1.28 \, \text{Hz}$$

$$T = \frac{1}{f} = \frac{1}{1.28 \, \text{Hz}} = 0.782 \, \text{seconds per cycle}$$

(b)
$$f = \frac{1}{2\pi} \sqrt{\frac{K}{M}} = \frac{1}{2\pi} \sqrt{\frac{22 \, \text{N/m}}{0.34 \, \text{kg}}} = \frac{1}{2\pi} \sqrt{\frac{22(\text{kg} \cdot \text{m/s}^2)/\text{m}}{0.34 \, \text{kg}}} = 1.28 \, \text{Hz}$$

13.20. A particle of mass m rests on the top of a smooth sphere of radius r as shown in Fig. 13-19(a). Assuming that the particle starts to move from rest, at what point will it leave the sphere?

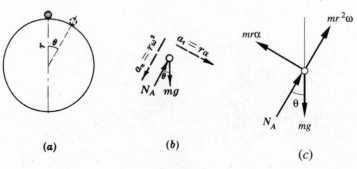

(a)   (b)   (c)

Fig. 13-19

SOLUTION

Let θ be the angular displacement at any time t during its travel. The free-body diagram indicates the only two forces acting on the particle, i.e., the plane reaction along the radius and the weight mg [see Fig. 13-19(b)].

The equations of motion found by summing forces along the radius and the tangent are

$$\sum F_r = -N_A + mg \cos \theta = mr\omega^2 \qquad (1)$$

$$\sum F_t = mg \sin \theta = mr\alpha. \qquad (2)$$

A third equation is necessary for a solution. This is the one expressing a relationship among θ, ω, and α; and for review it is derived here.

Eliminating dt from $\omega = d\theta/dt$ and $\alpha = d\omega/dt$ yields

$$\alpha \, d\theta = \omega \, d\omega \qquad (3)$$

From equation (2), $\alpha = (g/r) \sin \theta$. Substituting into (3), $(g/r) \sin \theta \, d\theta = \omega \, d\omega$. Then

$$\int_0^\theta \frac{g}{r} \sin \theta \, d\theta = \int_0^\omega \omega \, d\omega \qquad \text{or} \qquad \frac{g}{r}(1 - \cos \theta) = \frac{\omega^2}{2} \qquad (4)$$

At the position of departure from the sphere, the normal reaction N_A will be zero. Equation (1) now becomes

$$mg \cos \theta = mr\omega^2 \qquad \text{or} \qquad \frac{g}{r} \cos \theta = \omega^2 \qquad (5)$$

Replacing ω^2 in equation (4) by its value in equation (5), we obtain

$$\frac{g}{r}(1 - \cos \theta) = \frac{g}{2r} \cos \theta \qquad \cos \theta = \tfrac{2}{3} \qquad \theta = 0.841 \text{ rad or } 48.2°$$

For an alternate solution, let Fig. 13.19(c) be the free-body diagram showing the two inertia forces $mr\alpha$ and $mr\omega^2$. The resulting equilibrium equations become

$$\sum F_v = N_A \cos \theta - mg + mr\omega^2 \cos \theta + mr\alpha \sin \theta = 0$$

$$\sum F_h = N_A \sin \theta + mr\omega^2 \sin \theta - mr\alpha \cos \theta = 0$$

When the particle leaves the sphere, $N_A = 0$. Eliminating ω^2 from the simultaneous equations gives $\alpha = (g/r) \sin \theta$. Eliminating α from the equations gives $\omega^2 = (g/r) \cos \theta$. Using Equation (3) and substituting for α gives equation (4). Substituting for ω^2 yields

$$\frac{g}{r}(1 - \cos \theta) = \frac{g}{2r} \cos \theta.$$

This equation gives $\theta = 48.2°$.

13.21. A particle of mass m slides down a frictionless chute and enters a "loop-the-loop" of diameter d. What should be the height h at the start in order that the particle may make a complete circuit in the loop?

SOLUTION

A free-body diagram of the particle at any time in its travel down is shown in Fig. 13-20(b). The equations of motion parallel and perpendicular to the plane are

$$\sum F_\parallel = mg \sin \theta = ma \qquad (1)$$

$$\sum F_\perp = N - mg \cos \theta = 0 \qquad (2)$$

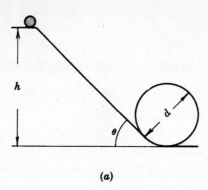

(a)   (b)

Fig. 13-20

From equation (*1*), the acceleration is $a = g \sin \theta$. This value of a is now substituted into the kinematics equation $a\,ds = v\,dv$, where s refers to displacement along the plane. Then $g \sin \theta\,ds = v\,dv$ and

$$\int_0^s g \sin \theta\,ds = \int_0^v v\,dv \qquad \text{or} \qquad g \sin \theta s = \frac{v^2}{2} \tag{3}$$

At the bottom of the plane the speed is found by substituting $h/\sin \theta$ for s in equation (*3*):

$$g \sin \theta \frac{h}{\sin \theta} = \frac{v^2}{2} \qquad \text{or} \qquad v^2 = 2gh \tag{4}$$

This indicates that the speed at the bottom of the smooth incline is independent of the slope of the incline and is the same as if the particle fell vertically downward.

Next draw a free-body diagram of the particle at the top of the loop (see Fig. 13-21).

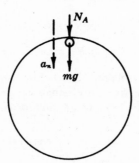

Fig. 13-21

The forces acting are the gravitational force mg and the force of the loop along the vertical radius. Since the minimum speed is the one with which we are concerned, the value of N_A in this case is zero. Expressed somewhat differently, as the speed v increases, the normal acceleration a_n increases. To provide this increasing value of a_n, the reaction N_A must increase also, because $\sum F_n = ma_n$.

In determining the speed at the top of the loop, use the fact just determined that the motion on a smooth path (such as the plane or the side of the loop) is equivalent to vertical motion under the acceleration of gravity only. Hence, the particle loses speed in moving up the side of the loop in an amount equal to the loss in straight-line vertical motion:

$$v_{\text{top}}^2 = v_{\text{bot}}^2 - 2gd$$

But the speed at the bottom v_{bot} has been expressed in equation (4) in terms of height h. Then

$$v_{\text{top}}^2 = 2gh - 2gd = 2g(h - d)$$

A summation of forces along the vertical radius when the particle is at the top results in the following equation, since $N_A = 0$:

$$\sum F_n = mg = \frac{mv_{\text{top}}^2}{d/2} \qquad \text{or} \qquad mg = \frac{4mg}{d}(h - d)$$

Hence, $h = 5d/4$. This means that the particle must leave at this height (or above) in order that a minimum speed be obtained. Otherwise, the particle will not follow the circular path at the top, but rather will "jump across."

13.22. A flexible chain of length l rests on a smooth table with length c overhanging the rounded edge, as shown in Fig. 13-22(a). The system, originally at rest, is released. Describe the motion. The chain weighs w lb/ft. Assume that the chain maintains contact with the table surfaces.

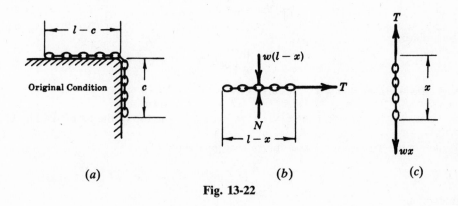

(a)   (b)   (c)

Fig. 13-22

SOLUTION

The free-body diagrams of the two pieces of the chain are drawn with the same tension T shown acting on each piece. See Fig. 13-22(b) and (c). The equations of motion are

$$T = \frac{w(l - x)}{g} a = \frac{w(l - x)}{g} \frac{d^2x}{dt^2} \tag{1}$$

$$wx - T = \frac{wx}{g} a = \frac{wx}{g} \frac{d^2x}{dt^2} \tag{2}$$

Add (1) and (2) to obtain

$$wx = \frac{wl}{g} \frac{d^2x}{dt^2} \qquad \text{or} \qquad \frac{d^2x}{dt^2} = \frac{g}{l} x$$

(This is not simple harmonic motion. Why?)

The solution of this second-order differential equation is $x = ae^{\sqrt{g/l}\, t} + Be^{-\sqrt{g/l}\, t}$.

To evaluate the constants A and B, note the initial conditions, namely $x = c$ and $v = 0$. Substituting $x = c$ when $t = 0$, we get $c = A + B$.

Differentiate x with respect to t: $v = dx/dt = A\sqrt{g/l}\, e^{\sqrt{g/l}\, t} - B\sqrt{g/l}\, e^{-\sqrt{g/l}\, t}$.

Substituting the condition $v = 0$ when $t = 0$, we get $A - B = 0$.

Solving simultaneously $A - B = 0$ and $c = A + B$, we get $A = B = \frac{1}{2}c$.
The solution of the problem is $x = \frac{1}{2}ce^{\sqrt{g/l}\,t} + \frac{1}{2}ce^{-\sqrt{g/l}\,t}$.

The exponential functions should be evaluated for any given time to determine the length x of the overhang. See Appendix C for a computer solution.

13.23. An object of weight W falls in a medium with the resistance R proportional to the velocity. This is approximately true for slowly moving objects. Study the motion.

SOLUTION

The free-body diagram of the object shows the weight W acting down and the retarding force R acting up (see Fig. 13-23). Assuming that x, $\dot{x}$, and $\ddot{x}$ are positive in the downward direction, the differential equation of the motion is

$$+W - k\dot{x} = \frac{W}{g}\ddot{x} = \frac{W}{g}\frac{d\dot{x}}{dt}$$

Rewriting this equation and then integrating,

$$\frac{d\dot{x}}{W/k - \dot{x}} = \frac{kg}{W}dt \quad \text{and} \quad -\ln(W/k - \dot{x}) = \frac{kg}{W}t + C_1$$

Assuming $\dot{x} = \dot{x}_0$ when $t = 0$, $-\ln(W/k - \dot{x}_0) = C_1$ and

$$-\ln(W/k - \dot{x}) = \frac{kg}{W}t - \ln(W/k - \dot{x}_0) \quad \text{or} \quad \frac{W/k - \dot{x}}{W/k - \dot{x}_0} = e^{-(kg/W)t}$$

from which

$$\dot{x} = \frac{W}{k}(1 - e^{-(kg/W)t}) + \dot{x}_0 e^{-(kg/W)t} \tag{1}$$

Since the limiting velocity $\dot{x}_{\max}$ occurs as $t \to \infty$,

$$\dot{x}_{\max} = \frac{W}{k} \tag{2}$$

To determine the distance x as a function of time, write (1) as

$$dx = \frac{W}{k}(1 - e^{-(kg/W)t})\,dt + \dot{x}_0 e^{-(kg/W)t}\,dt$$

Integrating,

$$x = \frac{W}{k}t - \frac{W}{k}(-W/kg)e^{-(kg/W)t} + \dot{x}_0(-W/kg)e^{-(kg/W)t} + C_2$$

Assuming $x = 0$ when $t = 0$, $0 = 0 + W^2/k^2 g - W\dot{x}_0/kg + C_2$ and

$$x = \frac{W}{k}t + \frac{W^2}{k^2 g}e^{-(kg/W)t} - \frac{W\dot{x}_0}{kg}e^{-(kg/W)t} - \frac{W^2}{k^2 g} + \frac{W\dot{x}_0}{kg}$$

Using $W/k = \dot{x}_{\max}$, this expression is regrouped as

$$x = \frac{W}{k}t + \frac{1}{g}(\dot{x}_{\max}\dot{x}_0 - \dot{x}_{\max}^2)(1 - e^{-(kg/W)t}) \tag{3}$$

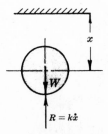

Fig. 13-23

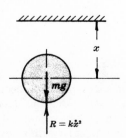

Fig. 13-24

13.24. An object of mass m falls in a medium with the resistance R proportional to the square of the speed. This is approximately true for high-speed objects. Study the motion.

SOLUTION

The free-body diagram of the object (Fig. 13-24) shows the gravitational force mg acting down and the retarding force R acting up. Assuming that x, $\dot{x}$, and $\ddot{x}$ are positive in the downward direction, the differential equation of motion is

$$+mg - k\dot{x}^2 = m\ddot{x} = m\frac{d\dot{x}}{dx}\frac{dx}{dt} = m\frac{d\dot{x}}{dx}\dot{x}$$

Rewriting this equation and integrating,

$$\frac{\dot{x}\,d\dot{x}}{mg/k - \dot{x}^2} = \frac{k}{m}dx \qquad \text{and} \qquad -\tfrac{1}{2}\ln(mg/k - \dot{x}^2) = \frac{k}{m}x + C_1$$

Assuming $\dot{x} = \dot{x}_0$ when $x = 0$, $-\tfrac{1}{2}\ln(mg/k - \dot{x}_0^2) = C_1$ and

$$-\tfrac{1}{2}\ln(mg/k - \dot{x}^2) = \frac{k}{m}x - \tfrac{1}{2}\ln(mg/k - \dot{x}_0^2)$$

Then

$$-\frac{2k}{m}x = \ln\frac{mg/k - \dot{x}^2}{mg/k - \dot{x}_0^2} \qquad \text{or} \qquad \frac{mg}{k} - \dot{x}^2 = \left(\frac{mg}{k} - \dot{x}_0^2\right)e^{-(2k/m)x}$$

and

$$\dot{x}^2 = \frac{mg}{k}(1 - e^{-(2k/m)x}) + \dot{x}_0^2 e^{-(2k/m)x} \qquad (1)$$

The limiting velocity occurs as $x \to \infty$; thus,

$$\dot{x}_{\max} = \sqrt{\frac{mg}{k}} \qquad (2)$$

13.25. A projectile of mass m is given an initial velocity v_0 at an angle θ with the horizontal. Determine the range, the maximum height, and the time of flight, assuming that the projectile hits on the same horizontal plane from which it is fired. Neglect air resistance in this solution.

SOLUTION

The horizontal line is chosen as the x axis. The range is r. The projectile is shown in Fig. 13-25 at some point (x, y) along the path. The only force acting is its weight mg.

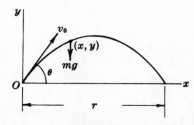

Fig. 13-25

The equations of motion in the x and y directions are

$$\sum F_x = 0 = m\ddot{x} \quad \text{and} \quad \sum F_y = -mg = m\ddot{y}$$

from which $\ddot{x} = 0$ and $\ddot{y} = -g$.

Integrating, $\dot{x} = C_1$ and $\dot{y} = -gt + C_2$. But $\dot{x}$ has constant value $v_0 \cos\theta$; and at $t = 0$, $\dot{y} = v_0 \sin\theta$. Hence,

$$\dot{x} = v_0 \cos\theta \quad \text{and} \quad \dot{y} = -gt + v_0 \sin\theta$$

Another integration yields

$$x = (v_0 \cos\theta)t + C_3 \quad \text{and} \quad y = (v_0 \sin\theta)t - \tfrac{1}{2}gt^2 + C_4$$

Since $x = 0$ and $y = 0$ when $t = 0$, $C_3 = C_4 = 0$, and the equations of motion are

$$x = (v_0 \cos\theta)t \quad \text{and} \quad y = (v_0 \sin\theta)t - \tfrac{1}{2}gt^2$$

To obtain the equation of the path in Cartesian coordinates, eliminate t from the equations. This is most easily done by solving the x equation for t and substituting its value into the y equation.

Substituting $t = x/(v_0 \cos\theta)$ into the y equation yields

$$y = v_0 \sin\theta \frac{x}{v_0 \cos\theta} - \frac{1}{2} g \frac{x^2}{v_0^2 \cos^2\theta} = x\tan\theta - \frac{g}{2v_0^2 \cos^2\theta} x^2$$

which is the equation of a parabola.

To determine the time of flight, equate y to zero. The height is zero at the beginning of the motion and also when the projectile hits the ground.

$$y = 0 = (v_0 \sin\theta)t - \tfrac{1}{2}gt^2 = t(v_0 \sin\theta - \tfrac{1}{2}gt)$$

The values of t that satisfy this equation are $t = 0$ (beginning of flight) and $t = (2v_0 \sin\theta)/g$ (time of flight) obtained from $v_0 \sin\theta = \tfrac{1}{2}gt$.

The range may be calculated by substituting this value of t into the x equation of motion:

$$x = r = v_0 \cos\theta \left(\frac{2v_0 \sin\theta}{g} \right) = \frac{v_0^2}{g} \sin 2\theta \quad \text{(range)}$$

Note that the maximum range for a given initial velocity v_0 is achieved when $\sin 2\theta = 1$, or when $2\theta = 90°$. Thus θ should be 45° for maximum range for a given v_0.

The maximum height can be determined either by assuming that it occurs at half the time of flight (theoretical considerations only) or by solving for the time t when the y component of velocity is equated to zero. Assume that the time is half the time of flight. Substituting this value into the y equation results in

$$h = v_0 \sin\theta \left(\frac{v_0 \sin\theta}{g} \right) - \frac{1}{2} g \left(\frac{v_0 \sin\theta}{g} \right)^2 = \frac{v_0^2 \sin^2\theta}{2g} \quad \text{(maximum height)}$$

13.26. A particle of mass m moves under the attraction of another mass M [see Fig. 13-26(a)], which we shall assume to be at rest. Study the motion of the mass m, noting that the distance between the two particles is not necessarily constant.

SOLUTION

The particle m has radius vector $\mathbf{r}$ with respect to mass M. The unit vectors $\mathbf{e}_r$ and $\mathbf{e}_\phi$ for the polar coordinate system are shown. A free-body diagram of mass m contains only the attractive force $\mathbf{F}$, which is in the negative $\mathbf{e}_r$ direction [see Fig. 13-26(b)]. According to Newton's law of gravitation, this central force is

$$\mathbf{F} = -G\frac{Mm}{r^2}\mathbf{e}_r \qquad (1)$$

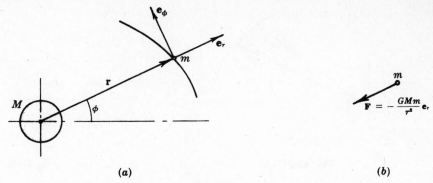

Fig. 13-26

where G is the universal gravitational constant:

$$G = 6.658 \times 10^{-11} \, \text{m}^3/(\text{kg} \cdot \text{s}^2) \qquad \text{in SI}$$
$$G = 3.43 \times 10^{-8} \, \text{lb-ft}^2/(\text{slug})^2 \qquad \text{in the engineering system}$$

In polar coordinates, the acceleration **a** is (see Section 11.6)

$$\mathbf{a} = (\ddot{r} - r\dot{\phi}^2)\mathbf{e}_r + (r\ddot{\phi} + 2\dot{r}\dot{\phi})\mathbf{e}_\phi \tag{2}$$

Substituting from equations (1) and (2) into

$$\mathbf{F} = m\mathbf{a} \tag{3}$$

gives

$$-\frac{GMm}{r^2}\mathbf{e}_r = +m(\ddot{r} - r\dot{\phi}^2)\mathbf{e}_r + m(r\ddot{\phi} + 2\dot{r}\dot{\phi})\mathbf{e}_\phi \tag{4}$$

Equating coefficients of $\mathbf{e}_\phi$ and then of $\mathbf{e}_r$ to zero,

$$r\ddot{\phi} + 2\dot{r}\dot{\phi} = 0 \tag{5}$$

and

$$\frac{-GM}{r^2} = (\ddot{r} - r\dot{\phi}^2) \tag{6}$$

Equation (5) is identical with

$$\frac{1}{r}\frac{d}{dt}(r^2\dot{\phi}) = 0 \tag{7}$$

Integrating (7),

$$r^2\dot{\phi} = C \qquad \text{a constant} \tag{8}$$

Note that in Fig. 13-27 the radius vector **r** sweeps out an area dA as it rotates through the angle $d\phi$. The differential area dA is approximately equal to $\frac{1}{2}r(r\,d\phi) = \frac{1}{2}r^2\,d\phi$. Dividing by dt,

$$\frac{dA}{dt} = \frac{1}{2}r^2\frac{d\phi}{dt} = \frac{1}{2}r^2\dot{\phi} = \tfrac{1}{2}C$$

Thus, the radius vector **r** sweeps out equal areas in equal times.

Next solve equation (6) to determine the path of mass m. For a convenient solution, let $r = 1/u$; then $dr/du = -1/u^2$ and

$$\dot{r} = \frac{dr}{du}\frac{du}{d\phi}\frac{d\phi}{dt} = \frac{-1}{u^2}\frac{du}{d\phi}\dot{\phi}$$

But from equation (8), $\dot{\phi} = C/r^2 = Cu^2$; then

$$\dot{r} = -\frac{1}{u^2}\frac{du}{d\phi}Cu^2 = -C\frac{du}{d\phi}$$

and

$$\ddot{r} = \frac{d\dot{r}}{dt} = \frac{d\dot{r}}{d\phi}\frac{d\phi}{dt} = \left(-C\frac{d^2u}{d\phi^2}\right)Cu^2 = -C^2u^2\frac{d^2u}{d\phi^2}$$

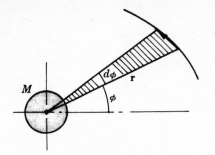

Fig. 13-27   **Fig. 13-28**

Substituting into equation (6), we obtain

$$-GMu^2 = -C^2u^2 \frac{d^2u}{d\phi^2} - C^2u^3$$

or

$$\frac{d^2u}{d\phi^2} + u = \frac{GM}{C^2} \tag{9}$$

As an intelligent guess, the solution of (9) might be of the form

$$u = A \cos \phi + B \tag{10}$$

This form must be substituted into (9) to determine the necessary conditions to make it a solution; thus,

$$\frac{du}{d\phi} = -A \sin \phi \qquad \frac{d^2u}{d\phi^2} = -A \cos \phi$$

and

$$-A \cos \phi + A \cos \phi + B = \frac{GM}{C^2}$$

Hence, $B = GM/C^2$, and the solution of (9) is

$$u = A \cos \phi + \frac{GM}{C^2} \qquad \text{where } u = \frac{1}{r} \tag{11}$$

Equation (11) represents a conic section, which from analytic geometry is the path of a point that moves so that the ratio of its distance from fixed point M (a focus) to its perpendicular distance from a fixed line (the directrix) is a constant e (the eccentricity). Figure 13-28 illustrates the distances involved in the definition

$$e = \frac{r}{d - r \cos \phi} \tag{12}$$

Equation (12) can be solved for $1/r$, yielding

$$\frac{1}{r} = \frac{1}{d} \cos \phi + \frac{1}{ed} \tag{13}$$

The following table lists the types of conics (curves) with their e values.

e	0	<1	1	>1
curve	circle	ellipse	parabola	hyperbola

Comparison of equations (*11*) and (*13*) indicates that

$$\frac{1}{ed} = \frac{GM}{C^2} \quad \text{or} \quad ed = \frac{C^2}{GM} \tag{14}$$

The equation of motion is

$$u = \frac{1}{r} = \frac{1}{d}\cos\phi + \frac{GM}{C^2} \tag{15}$$

13.27. In the preceding problem the path of a planet m around the sun M is elliptical (eccentricity $e < 1$). The sun is at one of the foci of the ellipse. Show that the period T (time for a complete revolution of the planet around the sun) is given by

$$T = \frac{2\pi a^{3/2}}{\sqrt{GM}}$$

where a = semimajor axis of ellipse
           G = gravitational constant
           M = mass of sun

SOLUTION

The path is an ellipse with semimajor axis a and semiminor axis b. Figure 13-29 shows that

$$f + g = 2a \tag{1}$$

Also, from equation (*13*) of Problem 12.26,

$$\frac{1}{r} = \frac{1}{d}\cos\phi + \frac{1}{ed} \tag{2}$$

When $\phi = 0°$, $r = g$; and when $\phi = 180°$, $r = f$. Thus, (*2*) becomes

$$\frac{1}{g} = \frac{1}{d}\cos 0° + \frac{1}{ed} = \frac{1+e}{ed} \quad \text{and} \quad \frac{1}{f} = \frac{1}{d}\cos 180° + \frac{1}{ed} = \frac{1-e}{ed}$$

Solving these two equations for f and g and substituting into (*1*), we find

$$\frac{ed}{1+e} + \frac{ed}{1-e} = 2a \quad \text{or} \quad 1 - e^2 = \frac{ed}{a} \tag{3}$$

But by equation (*14*) of Problem 13.26, $ed = C^2/GM$. Hence, (*3*) becomes

$$1 - e^2 = \frac{C^2}{GMa}$$

The area A of the ellipse is $A = \pi a b = \pi a^2 \sqrt{1 - e^2}$. Thus,

$$A = \pi a^2 \frac{C}{\sqrt{GMa}} = \frac{2\pi a^{3/2}}{\sqrt{GM}} \frac{C}{2} \tag{4}$$

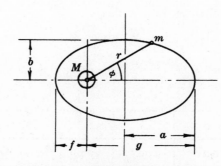

Fig. 13-29

Equation (4) gives the area swept out in one complete revolution, or in time T. This area is also the product of the constant rate dA/dt and time T: $A = (dA/dt)T$. Finally, substituting $dA/dt = C/2$ as determined in Problem 13.26, we have

$$A = \frac{C}{2}T = \frac{2\pi a^{3/2}}{\sqrt{GM}}\frac{C}{2} \quad \text{or} \quad T = \frac{2\pi a^{3/2}}{\sqrt{GM}}$$

13.28. Convert the gravitational constant $G = 6.658 \times 10^{-8}$ cm^3/g·s^2 into the equivalent in engineering units.

SOLUTION

$$G = 6.658 \times 10^{-8}\,\frac{\text{cm}^3}{\text{g}\cdot\text{s}^2} = 6.658 \times 10^{-8}\,\frac{(3.281 \times 10^{-2}\,\text{ft})^3}{(2.205/32.2)10^{-3}\,\text{slug-s}^2} = 3.43 \times 10^{-8}\,\frac{\text{ft}^3}{\text{slug-s}^2}$$

Since 1 slug = 1 lb-s^2/ft,

$$1\,\frac{\text{ft}^3}{\text{slug-s}^2} = 1\,\frac{\text{ft}^3}{(\text{lb-s}^2/\text{ft})\text{s}^2} = 1\,\frac{\text{ft}^4}{\text{lb-s}^4}$$

Thus

$$G = 3.43 \times 10^{-8}\,\frac{\text{ft}^4}{\text{lb-s}^4}$$

13.29. The satellite shown in Fig. 13-30 was fired from point A on the surface of the earth. The burnout point at which all fuel is expended is at P, which is 5000 mi from the earth's center C. Assume the velocity at P is v_0 and that it is perpendicular to the earth's radius extended. Determine the value of v_0 so that the satellite's orbit is (a) circular and (b) parabolic.

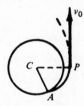

Fig. 13-30

SOLUTION

(a) From equation (8) in Problem 13.26,

$$C = r^2\dot{\phi} = r(r\dot{\phi}) = rv_0 = (5000 \times 5280)v_0 \tag{1}$$

The solution of the differential equation (9) in problem 13.26 is given in equation (11) as

$$u = \frac{1}{r} = A\cos\phi + \frac{GM}{C^2} \tag{2}$$

The solution of equation (13) in the same problem is given as

$$\frac{1}{r} = \frac{1}{d}\cos\phi + \frac{1}{ed} \tag{3}$$

Hence, we see

$$A = \frac{1}{d} \quad \text{and} \quad \frac{1}{ed} = \frac{GM}{C^2} \tag{4}$$

which combine to give

$$e = \frac{AC^2}{GM} \tag{5}$$

where M, the mass of the earth, is 4.09×10^{13} slugs.

Substituting into equation (2) above, we find

$$\frac{1}{5000 \times 5280} = A \cos 0° + \frac{(3.43 \times 10^{-8})(4.09 \times 10^{23})}{(5000 \times 5280 v_0)^2} \tag{6}$$

or

$$A = \frac{1}{5000 \times 5280}\left(1 - \frac{5.31 \times 10^8}{v_0^2}\right) \tag{7}$$

Substitute $e = 0$ (for circular orbit) and the value of A from equation (7) into equation (5) to find

$$0 = \frac{1}{5000 \times 5280}\left(1 - \frac{5.31 \times 10^8}{v_0^2}\right)\frac{C^2}{GM} \tag{8}$$

Equation (8) can only be satisfied if

$$1 - \frac{5.31 \times 10^8}{v_0^2} = 0$$

Thus $v_0 = 23{,}000 \text{ ft/s or } 15{,}700 \text{ mi/h}$

(b) If the path is to be parabolic then $e = 1$. Substituting $e = 1$ into equation (5), we get

$$1 = \frac{1}{5000 \times 5280}\left(1 - \frac{5.31 \times 10^8}{v_0^2}\right)\frac{C^2}{GM}$$

$$= \frac{1}{5000 \times 5280}\left(1 - \frac{5.31 \times 10^8}{v_0^2}\right)\frac{(5000 \times 5280)^2 v_0^2}{(3.43 \times 10^{-8})(4.09 \times 10^{23})}$$

This simplifies to

$$v_0 = 32{,}500 \text{ ft/s or } 22{,}200 \text{ mi/h}$$

Note that a v_0 greater than this last value will yield an eccentricity greater than 1, which means that the satellite will depart on a hyperbolic path and never return to earth.

13.30. A satellite with all fuel expended is at a point 500 mi from the earth's surface and is traveling with a velocity of 36,000 ft/s in a direction perpendicular to an earth radius extended. Determine the eccentricity of the orbit. Assume the earth has a radius of 3940 mi.

SOLUTION

Equation (12) of Problem 13.26 can be written as $ed - er \cos \phi = r$. From this, we can write

$$r = \frac{ed}{1 + e \cos \phi} = \frac{C^2/GM}{1 + e \cos \phi} \tag{1}$$

Our initial conditions are (a) $r = 3940 + 500 = 4440$ mi and (b) $\phi = 0$. Also, $C = r^2\dot{\phi} = r(v_0) = 4440 \times 5280 v_0$, $G = 3.43 \times 10^{-8}$, and $M = 4.09 \times 10^{23}$. Hence, with $v_0 = 36{,}000$ ft/s, we can write equation (1) as

$$4440 \times 5280 = \frac{(4440 \times 5280 \times 36{,}000)^2/(3.43 \times 10^{-8})(4.09 \times 10^{23})}{1 + e}$$

or

$$1 + e = \frac{4440 \times 5280 \times (36{,}000)^2}{(3.43 \times 10^{-8})(4.09 \times 10^{23})}$$

From this, the eccentricity of the orbit is $e = 1 \cdot 17$, and the path is hyperbolic.

13.31. Given that the period T for the passage of the earth around the sun is 1 yr and that the earth at its perihelion is 91,340,000 mi from the sun and at its aphelion is 94,450,000 mi from the sun, determine the mass M of the sun. Assume 1 yr = 365 days, and ignore the effects of the other planets.

SOLUTION

Figure 13-31 shows the earth in both its near and far positions (perihelion and aphelion, respectively). The semimajor axis a of the elliptical orbit is obtained by adding the two distances and dividing by 2; this gives $a = 92.9 \times 10^6$ mi. Using the formula derived in Problem 13.27

$$M = \frac{4\pi^2 a^3}{GT^2} = \frac{4\pi^2 (92.9 \times 10^6 \times 5280)^3}{(3.43 \times 10^{-8})(365 \times 24 \times 3600)^2} = 13.7 \times 10^{28} \text{ slugs}$$

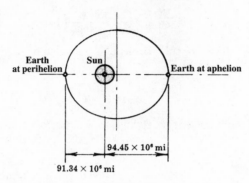

Fig. 13-31

Supplementary Problems

13.32. An automobile weighing 4000 lb is accelerated at the rate of 3 ft/s² along a horizontal roadway. What constant force F (parallel to the ground) is required to produce this acceleration?   *Ans.*   373 lb

13.33. A body is projected up a 25° plane with an initial velocity of 15 m/s. If the coefficient of friction between the body and the plane is 0.25, determine how far the body will move up the plane and the time required to reach the highest point.   *Ans.*   17.7 m, 2.36 s

13.34. A meteorite weighing 1000 lb is found buried 60 ft in the earth. Assuming a striking velocity of 1000 ft/s, what was the average retarding force of the earth on the meteorite?   *Ans.* $F = 265,000$ lb

13.35. Two particles of the same mass are released from rest on a 25° incline when they are 10 m apart. The coefficient of friction between the upper particle and the plane is 0.15, and between the lower one and the plane it is 0.25. Find the time required for the upper one to overtake the lower one. *Ans.* $t = 4.74$ s

13.36. Refer to Fig. 13-32. An automobile weighing 2800 lb and traveling 30 mi/h hits a depression in the road which has a radius of curvature of 50 ft. What is the total force to which the springs are subjected? *Ans.* $N = 6170$ lb

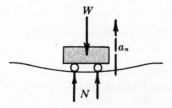

Fig. 13-32

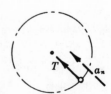

Fig. 13-33

13.37. A small bob of 25-g mass is whirled in an assumed horizontal circular path on a string 400 mm long as shown in Fig. 13-33. What is the tension in the string when the constant angular speed is 30 rad/s? *Ans.* $T = 9.0$ N

13.38. A 90-lb weight has a velocity of 30 ft/s horizontally on a smooth surface. Determine the value of the horizontal force that will bring the weight to rest in 4 s.　　*Ans.*　21.0 lb

13.39. The ball of a conical pendulum weighing 10 lb hangs at the end of an 8-ft string and describes a circular path in a horizontal plane. If the weight is swung so that the string makes an angle of 30° with the vertical, what is the linear speed of the ball?　　*Ans.*　8.63 ft/s

13.40. Two masses of 40 and 35 kg, respectively, are attached by a cord that passes over a frictionless pulley. If the masses start from rest, find the distance covered by either mass in 6 s.　　*Ans.*　11.8 m

13.41. A 40-kg mass is dragged along the surface of a table by means of a cord which passes over a frictionless pulley at the edge of the table and is attached to a 12-kg mass. If the coefficient of friction between the 40-kg mass and the table is 0.15, determine the acceleration of the system and the tension in the cord.　　*Ans.*　1.13 m/s^2, 104 N

13.42. A block *A* has a mass of 8 kg and is at rest on a frictionless horizontal surface. A 4-kg mass *B* is attached to a rope as shown in Fig. 13-34. Determine the acceleration of the mass *B* and the tension in the cord. The pulley is frictionless.　　*Ans.*　$a = 3.27$ m/s^2 down, $T = 26.2$ N

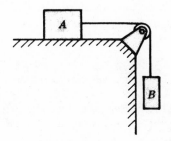

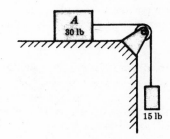

Fig. 13-34　　　　　　　　　　　　　　　　　**Fig. 13-35**

13.43. Two blocks are connected as shown in Fig. 13-35. The coefficient of friction between block *A* and the plane is 0.30. The pulley is frictionless. Determine the tension in the rope.　　*Ans.*　$T = 13.0$ lb

13.44. A block *B* rests on a block *A*, which is being pulled along a smooth horizontal surface by a horizontal force *P*. If the coefficient of friction between the two blocks is μ, determine the maximum acceleration before slipping occurs between *A* and *B*.　　*Ans.*　$a = \mu g$.

13.45. A 20-lb box is dropped onto the body of a truck moving 30 mi/h horizontally. If the coefficient of friction is 0.5, calculate how far the truck will move before the box stops slipping.　　*Ans.*　$s = 120$ ft

13.46. The 90- and 80-lb weights in Fig. 13-36 are attached to ropes passing over pulleys. Neglecting the weight of the pulleys and cords and assuming no friction, compute the tensions in the cords. Note that the length of the cord passing over the pulleys is constant.

$$x_2 + (x_2 - c) + x_1 = \text{constant} - 2 \text{ half-circumferences}$$

Ans.　$T_1 = 52.8$ lb, $T_2 = 106$ lb

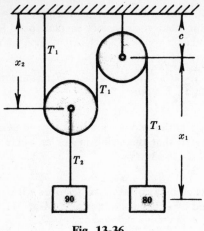

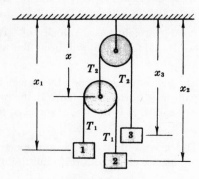

Fig. 13-36   **Fig. 13-37**

13.47. In the system of pulleys and weights shown in Fig. 13-37, let x_1, x_2, x_3 be the positions of the 1-, 2-, and 3-lb weights, respectively, during any phase of the motion after the system is released. Neglect the masses of the pulleys and the cords, and assume no friction. Determine the tensions T_1 and T_2.
Ans. $T_1 = 1.41$ lb, $T_2 = 2.82$ lb

13.48. Refer to Fig. 13-38. The weights A and B are 15 and 55 lb, respectively. Assume that the coefficient of friction between A and the plane is 0.25 and that between B and the plane is 0.10. What is the force between the two as they slide down the incline? In the free-body diagrams, P is the unknown force between A and B.   *Ans.* $P = 1.35$ lb

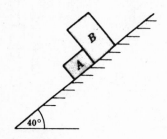

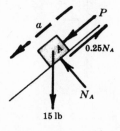

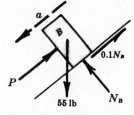

Fig. 13-38

13.49. Two solid blocks resting on a smooth plane are connected by a string as shown in Fig. 13-39. Determine the maximum force P that can be applied to the 8-lb block if the maximum strength of the string is 0.5 lb. Consider the blocks as particles.   *Ans.* $P = 2.35$ lb

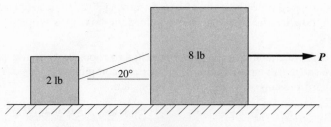

Fig. 13-39

13.50. Do Problem 13-49 using D'Alembert's Method.

13.51. The two blocks A and B in Fig. 13.40 are connected by a flexible weightless cord. The coefficient of friction between the blocks and the turntable is 0.35. The weights of blocks A and B are 10 lb and 20 lb, respectively. If the turntable rotates about a vertical axis with a constant angular speed, determine the speed at which the blocks begin to slide. Also find the tension in the string.
Ans. $\omega = 2.6$ rad/s, $T = 9.8$ lb

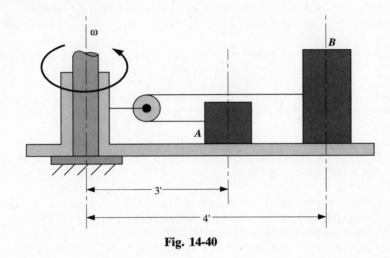

Fig. 14-40

13.52. While the two blocks shown in Fig. 13-41 are at rest, a 5-lb force is applied to the top block. The coefficient of friction between the two blocks is 0.20 and the floor is smooth. (*a*) Determine the acceleration of each block. (*b*) Determine the time that elapses before the right edge of the top block lines up with the right edge of the bottom block.   *Ans.* $a_A = 9.66$ ft/s^2, $a_B = 3.22$ ft/s^2, $t = 0.28$ s

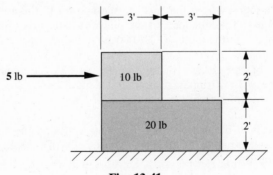

Fig. 13-41

13.53. In Problem 13.52, find the accelerations if the 5-lb force is replaced by a 1.5-lb force.
Ans.   Both blocks have acceleration 1.61 ft/s

13.54. In Problem 13.18, the mass M is displaced a distance x_0 from equilibrium and released with initial velocity v_0. Show that the equation of motion is $x = v_0\sqrt{M/K} \sin \sqrt{M/K}t + x_0 \cos \sqrt{M/K}t$.

13.55. At what height above the earth's surface must a 1.8-m pendulum be in order to have a period of 2.8 s? Assume $g = 9.8$ m/s^2 at the earth's surface and the gravitational force varies inversely as the square of the distance from the earth's center. Take the radius of the earth as 6450 km.   *Ans.*   257 km

13.56. In Problem 13.21, assume the mass is 14 kg and the diameter of the loop is 12 m. Determine the vertical reaction of the track on the mass when the mass is at the top of the loop for (a) $h = 15$ m and (b) $h = 18$ m.   *Ans.*   (a) $N = 0$, (b) $N_T = 137$ N

13.57. A chain 32.2 ft long is stretched with one-half its length on a smooth horizontal table and the other half hanging freely. If the chain starts from rest, find the time for the chain to leave the table. See Fig. 13-22.   *Ans.* $t = 1.32$ s

13.58. A body of mass 1.5 kg falls in a medium where the resistance is $R = k\dot{x}$ and $k = 0.7$ (N · s)/m. What is the terminal velocity?   *Ans.* $\dot{x}_{max} = 21$ m/s

13.59. A particle of mass m is projected vertically upward with a velocity v_0 in a medium whose resistance is kv. Determine the time for the particle to come to rest.   *Ans.* $t = (m/k) \ln (1 + kv_0/mg)$

13.60. A particle of mass m is projected vertically upward with a velocity v_0 in a medium whose resistance is kv^2. Determine the time for the particle to come to rest.   *Ans.* $t = \sqrt{m/kg} \tan^{-1} v_0 \sqrt{k/mg}$

13.61. The planet Mars at its aphelion in its orbit is 154.8×10^6 mi from the sun. At its perihelion it is 128.8×10^6 mi away. Show that the time for one complete revolution by the methods of this chapter is close to the actual value of 687 days.

13.62. The Soviet satellite Sputnik I had a mass of about 83 kg and orbited on a path around the earth that varied between 220 and 900 km above the surface of the earth. Using the radius of the earth as 6340 km, show that the time for a revolution is about 1.5 h.

13.63. The spherical plastic balloon satellite Echo I was orbited on a path that varied from 945 to 1049 mi from the earth's surface. Show that the time for a revolution was initially about 2 h.

13.64. A satellite with all fuel expended is at a point 300 mi from the earth's surface and is traveling with a velocity of 30,000 ft/s in a direction perpendicular to an extended radius of the earth. Determine the type of path which the satellite will follow.   *Ans.* $e = 0.434$; path is elliptical

13.65. At what point on a journey from the earth to the moon will the attractive forces of the two masses on the spaceship be equal? Take the mass of the moon as 0.012 times that of the mass of the earth, and the distance from the earth to the moon as 239,000 mi.   *Ans.* $d = 216,000$ mi from the earth's center

13.66. Given the period of the earth around the sun as 365 days and the perihelion and aphelion as 91.3×10^6 miles and 94.4×10^6 miles, respectively, determine the eccentricity of the earth's orbit.   *Ans.* $e = 0.017$

13.67. A weather satellite is to be placed in a circular orbit around the earth at an altitude of 300 miles. Its initial velocity in orbit is parallel to the earth's surface. What should this initial velocity be?   *Ans.* $v_0 = 17,000$ mi/h

13.68. Referring to the satellite in Problem 13.64, find the maximum altitude and the velocity at that altitude.   *Ans.* $h = 6800$ mi, $v = 8100$ mi/h

13.69. Communication satellites complete a circular orbit around the earth in one day. They are seen to be stationary relative to the earth. Determine the necessary altitude and velocity for a communication satellite.   *Ans.* $h = 22,300$ mi, $v = 6870$ mi/h

Chapter 14

Kinematics of a Rigid Body in Plane Motion

14.1 PLANE MOTION OF A RIGID BODY

Plane motion of a rigid body takes place if every point in the body remains at a constant distance from a fixed plane. In Fig. 14-1, it is assumed that the XY plane is the fixed reference plane. The lamina shown is representative of all laminae that compose the rigid body. The Z distance to any point in the lamina remains constant as the lamina moves.

It is customary to select an arbitrary point B in the body as the origin of a non-rotating reference frame xyz.

The position vector $\mathbf{r}_A$ of any point A (fixed or moving) in the lamina may now be written in terms of the position vector $\mathbf{r}_B$ of B and the vector $\mathbf{BA}$, which is labeled $\boldsymbol{\rho}$. Thus,

$$\mathbf{r}_A = \mathbf{r}_B + \boldsymbol{\rho}$$

If, as a rigid body implies, AB is constant the time rate of change of $\boldsymbol{\rho}$ is that given in equation (20) of Chapter 12. Thus, we can write

$$\mathbf{v}_A = \dot{\mathbf{r}}_A = \dot{\mathbf{r}}_B + \dot{\boldsymbol{\rho}} = \dot{\mathbf{r}}_B + \rho\omega\mathbf{e}_\phi \qquad (1)$$

where $\dot{\mathbf{r}}_B$ = linear velocity of pole B relative to the fixed axes X, Y and Z
   ω = magnitude of angular velocity of $\boldsymbol{\rho}$ about any line parallel to the Z axis
   $\mathbf{e}_\phi$ = unit vector perpendicular to $\boldsymbol{\rho}$ and in the direction of increasing ϕ (as indicated by the right-hand rule, this is counterclockwise about the z axis).

The acceleration $\mathbf{a}_A$ may next be found by applying equation (21) in Chapter 12:

$$\mathbf{a}_A = \dot{\mathbf{v}}_A = \ddot{\mathbf{r}}_A = \ddot{\mathbf{r}}_B - \rho\omega^2\mathbf{e}_r + \rho\alpha\mathbf{e}_\phi \qquad (2)$$

where $\ddot{\mathbf{r}}_B$ = linear acceleration of pole B relative to the fixed axes X, Y and Z
   $\mathbf{e}_r$ = unit vector along $\boldsymbol{\rho}$ directed from B toward A
   $\mathbf{e}_\phi$ = unit vector as specified in equation (1) above
   α = magnitude of angular acceleration of $\boldsymbol{\rho}$ about any line parallel to the Z axis.

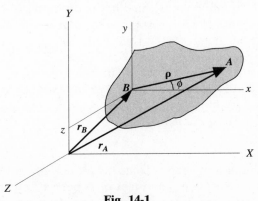

Fig. 14-1

An alternative method of writing equations (*1*) and (*2*) may be developed as follows:

$$\boldsymbol{\omega} = \dot{\phi}\mathbf{k} = \omega\mathbf{k} \qquad \text{and} \qquad \boldsymbol{\alpha} = \ddot{\phi}\mathbf{k} = \dot{\omega}\mathbf{k} = \alpha\mathbf{k}$$

where ω and α will be positive if they are in a counterclockwise direction about the z axis as viewed from the positive end of the axis (right-hand rule).

Then in equation (*1*) the term $\rho\omega\mathbf{e}_\phi$ may be replaced by the cross product $\boldsymbol{\omega} \times \boldsymbol{\rho}$, which is identical with it. (See Problem 14.1.) Likewise, in equation (*2*) the component of the acceleration $\rho\alpha\mathbf{e}_\phi$ is equivalent to $\boldsymbol{\alpha} \times \boldsymbol{\rho}$. The component $-\rho\omega^2\mathbf{e}_r$ is identical with $\boldsymbol{\omega} \times (\boldsymbol{\omega} \times \boldsymbol{\rho})$. These substitutions give

$$\mathbf{v}_A = \mathbf{v}_B + \boldsymbol{\omega} \times \boldsymbol{\rho} \tag{3}$$

$$\mathbf{a}_A = \mathbf{a}_B + \boldsymbol{\omega} \times (\boldsymbol{\omega} \times \boldsymbol{\rho}) + \boldsymbol{\alpha} \times \boldsymbol{\rho} \tag{4}$$

Equations (*3*) and (*4*) may also be written

$$\mathbf{v}_A = \mathbf{v}_B + \mathbf{v}_{A/B} \tag{5}$$

$$\mathbf{a}_A = \mathbf{a}_B + \mathbf{a}_{A/B} \tag{6}$$

where $\mathbf{v}_{A/B}$ and $\mathbf{a}_{A/B}$ are the relative velocity and relative acceleration that A possesses as it rotates around B, which as we have seen is moving with respect to the X, Y, and Z reference frame.

14.2 TRANSLATION

Translation is motion in which the line $\boldsymbol{\rho}$ from B to A does not rotate. Thus, as the lamina moves, every straight line in the lamina is always parallel to its original direction.

14.3 ROTATION

Rotation is motion in which the base point B is fixed. Extending this to the rigid body, there is a line through B parallel to the Z axis that is fixed in the X, Y, and Z system. To describe this motion, the velocity $\mathbf{v}_B$ and acceleration $\mathbf{a}_B$ in equations (*1*) and (*2*) or in equations (*3*) and (*4*) are equated to zero.

14.4 INSTANTANEOUS AXIS OF ROTATION

The instantaneous axis of rotation is that line in a body in plane motion (or the body extended) that is the locus of points of zero velocity. It is perpendicular to the plane of motion (parallel to the Z axis in our system). All other points in the rigid body rotate about that line at that *instant*. It is important to realize that the position of this line of zero velocity in general changes continuously. To locate the instant center I for a lamina that has an angular velocity $\boldsymbol{\omega}$, write the velocity expressions for any two points A and C in the body in terms of I as the base point. (See Fig. 14-2). Thus,

$$\mathbf{v}_A = \mathbf{v}_I + \boldsymbol{\omega} \times \boldsymbol{\rho}_A$$

$$\mathbf{v}_C = \mathbf{v}_I + \boldsymbol{\omega} \times \boldsymbol{\rho}_C$$

But $\mathbf{v}_I$ is zero because I is the instant center. Hence, $\mathbf{v}_A = \boldsymbol{\omega} \times \boldsymbol{\rho}_A$ and $\mathbf{v}_C = \boldsymbol{\omega} \times \boldsymbol{\rho}_C$. These equations mean $\boldsymbol{\rho}_A$ is perpendicular to $\mathbf{v}_A$ (I is on $\boldsymbol{\rho}_A$) and that $\boldsymbol{\rho}_C$ is perpendicular to $\mathbf{v}_C$ (I is on $\boldsymbol{\rho}_C$). Therefore the instant center I is the intersection of the perpendiculars to $\mathbf{v}_A$ and $\mathbf{v}_C$.

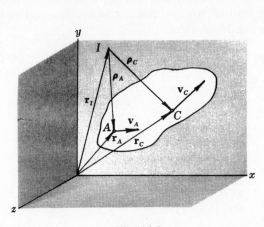

Fig. 14-2

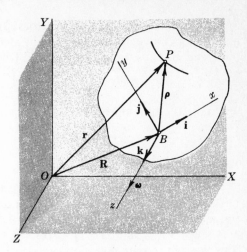

Fig. 14-3

14.5 CORIOLIS' ACCELERATION

In Section 14.1, the velocity and acceleration of a point in the rigid body was derived when the relative reference frame xyz was nonrotating. Figure 14-3 shows the lamina, in plane motion, where the pole B is the origin of a rotating reference frame. The acceleration of P is given by

$$\mathbf{a}_P = \mathbf{a}_{P/\text{path}} + \mathbf{a}_M + 2\boldsymbol{\omega} \times \mathbf{v}_{P/\text{path}} \tag{7}$$

where $\mathbf{a}_{P/\text{path}}$ = acceleration of P relative to the path considered as fixed—use components tangent and normal to the path

$\mathbf{a}_M$ = acceleration of that point M that moves along the path and with which P coincides at the instant

$\mathbf{v}_{P/\text{path}}$ = velocity of P relative to the point M that is on the path and with which P coincides at the instant (this velocity can only be tangent to the path along which the particle is moving in the body)

$\boldsymbol{\omega}$ = angular velocity of the path (or lamina)

$2\boldsymbol{\omega} \times \mathbf{v}_{P/\text{path}}$ = Coriolis' component, the direction of which is obtained by visualizing the rotation in the plane of the lamina of $\mathbf{v}_{P/\text{path}}$ through a right angle in the same sense as $\boldsymbol{\omega}$.

It can be seen in equation (7) that referring the acceleration to rotating axes gives rise to a third term in the acceleration equation. The velocity equation is unchanged. This third term is called the Coriolis component of acceleration, named after the French engineer and mathematician Gustav Coriolis (1792–1843). The Coriolis component of acceleration arises from the fact that $\boldsymbol{\rho}$, measured in the rotating reference frame, is defined with respect to $\mathbf{ijk}$. The unit vectors $\mathbf{ijk}$ change directions and hence have derivatives.

The proof of Coriolis' law is as follows.

In the lamina shown parallel to the fixed XY plane in Fig. 14-3, let

$\mathbf{R}$ = position vector of base point B
$\mathbf{r}$ = position vector of point P, which is moving on some path on the lamina
$\boldsymbol{\rho}$ = radius vector of P with respect to base point B
$\boldsymbol{\omega}$ = angular velocity of the lamina about the Z axis

Let $\mathbf{i}$ and $\mathbf{j}$ be unit vectors along the x and y axes, which are fixed in the lamina and hence rotate with angular velocity $\omega\mathbf{k}$ (note $\boldsymbol{\omega}$ is the same about either z axis or Z axis). Since $\mathbf{i}$ and $\mathbf{j}$ rotate, there is a time change in these unit vectors. In Section 12.6, it was shown that the time rate of change of a unit vector is a vector that is at right angles to the unit vector and has a magnitude equal to the angular speed ω. Thus, $\dot{\mathbf{i}} = \omega\mathbf{j} = \boldsymbol{\omega} \times \mathbf{i}$ and $\dot{\mathbf{j}} = -\omega\mathbf{i} = \boldsymbol{\omega} \times \mathbf{j}$. Now write

$$\mathbf{r} = \mathbf{R} + \boldsymbol{\rho} \qquad (8)$$

Next take the time derivative of equation (8), obtaining

$$\dot{\mathbf{r}} = \dot{\mathbf{R}} + \dot{\boldsymbol{\rho}} \qquad (9)$$

But in terms of the moving axes,

$$\boldsymbol{\rho} = x\mathbf{i} + y\mathbf{j} \qquad (10)$$

and
$$\dot{\boldsymbol{\rho}} = \dot{x}\mathbf{i} + \dot{y}\mathbf{j} + x\dot{\mathbf{i}} + y\dot{\mathbf{j}} = \mathbf{v}_{P/\text{path}} + \boldsymbol{\omega} \times \boldsymbol{\rho}$$

where $\dot{x}\mathbf{i} + \dot{y}\mathbf{j} = \mathbf{v}_{P/\text{path}}$ and $x\dot{\mathbf{i}} + y\dot{\mathbf{j}} = x(\boldsymbol{\omega} \times \mathbf{i}) + y(\boldsymbol{\omega} \times \mathbf{i}) = \boldsymbol{\omega} \times (x\mathbf{i} + y\mathbf{j}) = \boldsymbol{\omega} \times \boldsymbol{\rho}$.

Thus, we can rewrite equation (9) as

$$\dot{\mathbf{r}} = \dot{\mathbf{R}} + \boldsymbol{\omega} \times \boldsymbol{\rho} + \mathbf{v}_{P/\text{path}} \qquad (11)$$

Note that the first two terms on the right-hand side of equation (11) give the absolute velocity of the point M that is fixed on the path but coincides with P at the instant [refer to equation (3)]. Equation (11) may be expressed as

$$\mathbf{v}_P = \mathbf{v}_M + \mathbf{v}_{P/\text{path}} \qquad (12)$$

To derive the expression for the acceleration of P, take the time derivative of equation (9) but express $\dot{\boldsymbol{\rho}}$ as $\dot{x}\mathbf{i} + \dot{y}\mathbf{j} + \boldsymbol{\omega} \times \boldsymbol{\rho}$:

$$\frac{d(\dot{\mathbf{r}})}{dt} = \frac{d(\dot{\mathbf{R}})}{dt} + \frac{d}{dt}(\dot{x}\mathbf{i} + \dot{y}\mathbf{j} + \boldsymbol{\omega} \times \boldsymbol{\rho})$$

Thus,
$$\ddot{\mathbf{r}} = \ddot{\mathbf{R}} + \ddot{x}\mathbf{i} + \ddot{y}\mathbf{j} + \dot{x}\dot{\mathbf{i}} + \dot{y}\dot{\mathbf{j}} + \dot{\boldsymbol{\omega}} \times \boldsymbol{\rho} + \boldsymbol{\omega} \times \dot{\boldsymbol{\rho}}$$

$$= \ddot{\mathbf{R}} + \mathbf{a}_{P/\text{path}} + \boldsymbol{\omega} \times (\dot{x}\mathbf{i} + \dot{y}\mathbf{j}) + \boldsymbol{\alpha} \times \boldsymbol{\rho} + \boldsymbol{\omega} \times (\dot{x}\mathbf{i} + \dot{y}\mathbf{j} + \boldsymbol{\omega} \times \boldsymbol{\rho})$$

$$\ddot{\mathbf{r}} = \ddot{\mathbf{R}} + \mathbf{a}_{P/\text{path}} + \boldsymbol{\omega} \times \mathbf{v}_{P/\text{path}} + \boldsymbol{\alpha} \times \boldsymbol{\rho} + \boldsymbol{\omega} \times \mathbf{v}_{P/\text{path}} + \boldsymbol{\omega} \times (\boldsymbol{\omega} \times \boldsymbol{\rho}) \qquad (13)$$

The first, fourth, and sixth terms on the right-hand side of equation (13) give the absolute acceleration of the point M that is fixed on the path and that coincides with P at the instant [refer to equation (4)]. Equation (13) can be written

$$\ddot{\mathbf{r}} = \mathbf{a}_M + \mathbf{a}_{P/\text{path}} + 2\boldsymbol{\omega} \times \mathbf{v}_{P/\text{path}}$$

which is equivalent to equation (7).

Solved Problems

As in previous chapters, vectors in the diagrams are identified only by their magnitudes when the directions are evident by inspection.

14.1. Determine the linear velocity $\mathbf{v}$ of any point p in the rigid body rotating with angular velocity $\boldsymbol{\omega}$ about an axis as shown in Fig. 14-4.

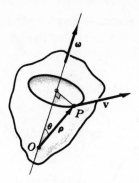

Fig. 14-4

SOLUTION

Select *any* reference point O on the axis of rotation. The radius vector $\boldsymbol{\rho}$ for point P relative to O is shown. The velocity vector $\mathbf{v}$ for point P is tangent to a circle that is in the plane perpendicular to the axis of rotation. The magnitude of the vector $\mathbf{v}$ is the product of the radius of the circle and the angular speed ω. It is apparent from the figure that the radius of the circle is $\rho \sin \theta$. Hence, the magnitude of $\mathbf{v}$ is $\rho\omega \sin \theta$.

By the definition of the cross product, $\boldsymbol{\omega} \times \boldsymbol{\rho}$ has the same magnitude $\rho\omega \sin \theta$ and is perpendicular to the plane containing $\boldsymbol{\rho}$ and $\boldsymbol{\omega}$. Hence, we conclude

$$\mathbf{v} = \boldsymbol{\omega} \times \boldsymbol{\rho}$$

In the plane motion of the lamina shown at the beginning of this chapter, $\boldsymbol{\omega}$ is perpendicular to the plane of the lamina and $\boldsymbol{\rho}$ is in the plane of the laminar; hence, $\mathbf{v} = \boldsymbol{\omega} \times \boldsymbol{\rho}$ is in the plane of the lamina and is perpendicular to $\boldsymbol{\rho}$.

14.2. A flywheel 500 mm in diameter is brought uniformly from rest up to a speed of 300 rpm in 20 s. Find the velocity and acceleration of a point on the rim 2 s after starting from rest.

SOLUTION

This problem illustrates the application of the equations of rotation.

First determine the magnitude of the angular acceleration α that obtains until operating speed is attained.

$$\alpha = \frac{\omega - \omega_0}{t} = \frac{2\pi(300/60)\,\text{rad/s} - 0}{20\,\text{s}} = 1.57\,\text{rad/s}^2$$

Next determine the angular speed of the wheel 2 s after starting.

$$\omega_1 = \omega_0 + \alpha t = 0 + (1.57\,\text{rad/s}^2)(2\,\text{s}) = 3.14\,\text{rad/s}$$

The speed of a point on the rim is

$$v = r\omega = (0.25\,\text{m})(3.14\,\text{rad/s}) = 0.785\,\text{m/s}$$

The magnitude of the normal acceleration of a point on the rim is

$$a_n = r\omega^2 = (0.25\,\text{m})(3.14\,\text{rad/s})^2 = 2.46\,\text{m/s}^2$$

The magnitude of the tangential acceleration of a point on the rim is

$$a_t = r\alpha = (0.25\,\text{m})(1.57\,\text{rad/s}^2) = 0.39\,\text{m/s}^2$$

The magnitude of the total acceleration of a point on the rim is

$$a = \sqrt{a_n^2 + a_t^2} = \sqrt{(2.46)^2 + (0.39)^2} = 2.49\,\text{m/s}^2$$

The angle between the total acceleration vector and the radius to the point is

$$\theta = \cos^{-1}\frac{a_n}{a} = \cos^{-1}\frac{2.46}{2.49} = 0.155 \text{ rad or } 8.9°$$

Figure 14-5 indicates the acceleration result.

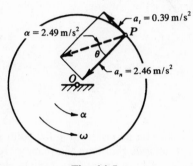

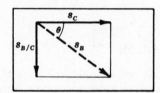

Fig. 14-5   Fig. 14-6

14.3. A ball rolls 2 m across a flat car in a direction perpendicular to the path of the car. In the same time interval during which the ball is rolling, the car moves at a constant speed on a horizontal straight track for a distance of 2.5 m. What is the absolute displacement of the ball?

SOLUTION

The vector equation for the absolute displacement s_B of the ball B in terms of the absolute displacement s_C of the car C is

$$s_B = s_{B/C} + s_C$$

The displacement $s_{B/C}$ of the ball relative to the car is 2 m at right angles to the track. The absolute displacement of the car is 2.5 m along the track. Figure 14-6 indicates these relations.

The absolute displacement s_B of the ball is the sum of the two given vectors. Its magnitude is

$$s_B = \sqrt{(s_{B/C})^2 + (s_C)^2} = \sqrt{(2)^2 + (2.5)^2} = 3.2 \text{ m}$$

The angle that the vector s_B makes with the track is $\theta = \tan^{-1}(s_{B/C}/s_C) = \tan^{-1}(2/2.5) = 0.675$ rad or 38.7°.

14.4. Automobile A is traveling 20 mi/h along a straight road headed northwest. Automobile B is traveling 70 mi/h along a straight road headed 60° south of west. See Fig. 14-7(a). What is the relative velocity of A to B? Of B to A?

SOLUTION

The vector equation relating the velocities is

$$\mathbf{v}_A = \mathbf{v}_{A/B} + \mathbf{v}_B$$

The equation may also be written as follows by subtracting $\mathbf{v}_B$ from both sides:

$$\mathbf{v}_{A/B} = \mathbf{v}_A - \mathbf{v}_B$$

The vector subtraction is performed by adding the negative of $\mathbf{v}_B$ to $\mathbf{v}_A$ as indicated in Fig. 14-7(b).

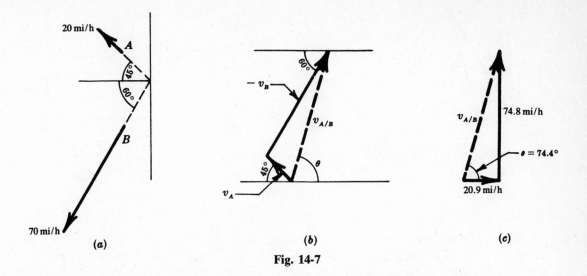

Fig. 14-7

Use the following tabular setup to find the magnitude of the x and y components of $\mathbf{v}_{A/B}$.

Vector	x component	y component
$\mathbf{v}_A$	-20×0.707	$+20 \times 0.707$
$-\mathbf{v}_B$	$+70 \times 0.500$	$+70 \times 0.866$

The magnitude of the x component of $\mathbf{v}_{A/B}$ is $+20.9$ mi/h and the magnitude of the y component is $+74.8$ mi/h. See Fig. 14-7(c). Then

$$v_{A/B} = \sqrt{(20.9)^2 + (74.8)^2} = 77.7 \text{ mi/h}$$

The angle that $\mathbf{v}_{A/B}$ makes with the easterly direction is $\theta = \tan^{-1}(74.8/20.9) = 74.4°$.

To determine the velocity of B relative to A, it is necessary to subtract $\mathbf{v}_A$ from $\mathbf{v}_B$ as indicated in the vector equation

$$\mathbf{v}_{B/A} = \mathbf{v}_B - \mathbf{v}_A$$

Figure 14-8 indicates the subtraction.

Again use a tabular setup.

Vector	x component	y component
$\mathbf{v}_B$	-70×0.500	-70×0.866
$-\mathbf{v}_A$	$+20 \times 0.707$	-20×0.707

The magnitude of the x component of $\mathbf{v}_{B/A}$ is -20.9 mi/h and the magnitude of the y component is -74.8 mi/h. It is apparent that $\mathbf{v}_{B/A}$ is the negative of $\mathbf{v}_{A/B}$.

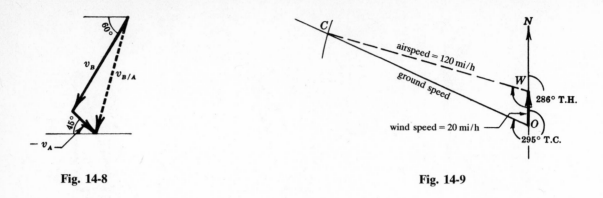

Fig. 14-8   Fig. 14-9

14.5. An airplane pilot determines from a sectional map that the true course over the ground should be 295° to a destination 95 mi away. There is a south wind of 20 mi/h (blowing from the south). Airspeed (speed relative to the air which is moving north 20 mi/h) is 120 mi/h. What should be the pilot's true heading to accomplish the task and how long should it take, assuming no change in the wind?

SOLUTION

Figure 14-9 shows the true course of 295° measured clockwise from the true north line. The wind speed 20 mi/h is drawn to the north from O, the point of departure. A graphical solution is perhaps the easiest. Swing an arc, from the arrow end of the wind vector, with a radius 120 mi/h long. This intersects the true course line in point C.

It is seen from the velocity triangle that $\mathbf{OC} = \mathbf{OW} + \mathbf{WC}$. This is merely the statement that the velocity of the plane along the true course line (its magnitude is called ground speed or absolute speed) is equal to the plane velocity relative to the wind (its magnitude is called airspeed) plus the velocity of the wind to the ground.

From the figure, the angle between the airspeed line WC and the true north is 286°. This is the true heading T.H., or the direction along which the fore and aft line of the plane should be placed.

The ground speed OC scales 127 mi/h. Hence the estimated time of flight is 95 mi divided by 127 mi/h, or 45 min.

14.6. Correcting for magnetic deviation and variation, a pilot calculates the plane's true heading to be 58°. Airspeed is 250 mi/h. Several checkpoints indicate that the plane is making good a true course of 63° at a ground speed of 295 mi/h. Determine the wind direction and magnitude.

SOLUTION

Draw the true heading line at an angle of 58° with the true north line as shown in Fig. 14-10. Along it, draw $\mathbf{OP}$ to scale as 250 mi/h. Draw the true course line at an angle of 63° with the true north line through O. Along it, draw $\mathbf{OW}$ to scale as the ground speed 295 mi/h. But $\mathbf{OW} = \mathbf{OP} + \mathbf{PW}$. This is interpreted as follows: the absolute velocity of the plane equals its velocity relative to the air plus the velocity of the air (wind vector) or $v_{PG} = v_{P/A} + v_{A/G}$.

To scale, $\mathbf{PW}$ is approximately a 50-mi/h wind from the west or from 270°.

14.7. The rod of length l moves so that the velocity of point A is of constant magnitude and directed to the left. Determine the angular velocity $\boldsymbol{\omega}$ and angular acceleration $\boldsymbol{\alpha}$ of the rod when it makes an angle θ with the vertical as shown in Fig. 14-11.

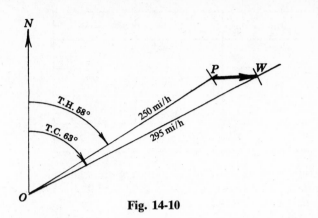

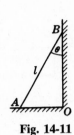

Fig. 14-10   Fig. 14-11

SOLUTION

The absolute velocity $\mathbf{v}_A$ of point A should be expressed in terms of the relative velocity of A to B ($\mathbf{v}_{A/B}$) because this term will introduce the desired quantity $\boldsymbol{\omega}$:

$$\mathbf{v}_A = \mathbf{v}_{A/B} + \mathbf{v}_B$$

The table indicates the known parts of the vectors.

Vector	Direction	Magnitude
$\mathbf{v}_A$	horizontal	v_A
$\mathbf{v}_{A/B}$	$\perp$ rod	$l\omega$ (ω unknown)
$\mathbf{v}_B$	vertical	unknown

The unknowns are the two magnitudes. Draw a vector triangle starting with $\mathbf{v}_A$, which is known completely. Draw a line through one end of $\mathbf{v}_A$ perpendicular to the rod, and draw a vertical line through the other end to close the triangle as shown in Fig. 14-12.

Label the vertical leg v_B and the leg perpendicular to the rod $v_{A/B}$ as indicated. In the right triangle, $v_{A/B} = v_A/\cos\theta$. But $v_{A/B} = l\omega$. Hence, $\omega = v_{A/B}/l = v_A/l\cos\theta$.

Since $\mathbf{v}_{A/B}$ is directed up to the left, A must turn clockwise about B; that is, $\boldsymbol{\omega}$ is clockwise.

Now determine the angular acceleration $\boldsymbol{\alpha}$. The magnitude of the tangential component of A relative to B is $l\alpha$. The equation is $\mathbf{a}_A = (\mathbf{a}_{A/B})_t + (\mathbf{a}_{A/B})_n + \mathbf{a}_B$.

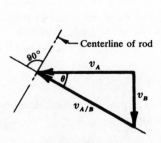

Fig. 14-12

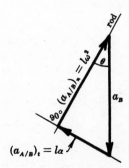

Fig. 14-13

Vector	Direction	Magnitude
$\mathbf{a}_A$	none	0, since v_A = constant
$(\mathbf{a}_{A/B})_t$	$\perp$ rod	$l\alpha$ (α unknown)
$(\mathbf{a}_{A/B})_n$	along the rod	$l\omega^2$ ($\omega = v_A/l\cos\theta$)
$\mathbf{a}_B$	vertical	unknown

The two unknowns are listed. The sum of the three vector quantities on the right side of the vector equation must equal zero. In Fig. 14-13, start with the known $(\mathbf{a}_{A/B})_n$ and through its end points draw two lines, one perpendicular to the rod and the other vertical, i.e., making an angle θ with $(\mathbf{a}_{A/B})_n$. Note that the acceleration $(\mathbf{a}_{A/B})_n$ must be directed from A to B.

In this right triangle, $(a_{A/B})_t = l\alpha = l\omega^2 \tan\theta$. Hence, $\alpha = \omega^2 \tan\theta = (v_A^2 \tan\theta)/(l^2 \cos^2\theta)$. This is clockwise because the tangential component $(\mathbf{a}_{A/B})_t$ indicates that A accelerates around B in a clockwise direction.

14.8. The ladder of length l makes an angle θ with the vertical wall, as shown in Fig. 14-14(a). The foot of the ladder moves to the right with constant speed v_A. Determine $\dot{\theta}$ and $\ddot{\theta}$ in terms of v_A, l, and θ.

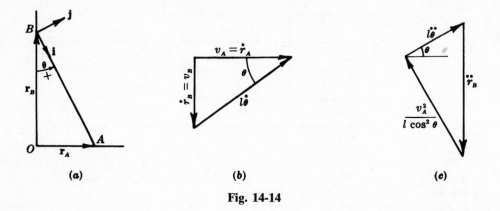

Fig. 14-14

SOLUTION

Let $\mathbf{i}$ and $\mathbf{j}$ be unit vectors along and perpendicular to the ladder; they move with the ladder. In vector notation,

$$\mathbf{r}_A = \mathbf{r}_B + l\mathbf{i}$$

Taking the time derivative, we have

$$\dot{\mathbf{r}}_A = \dot{\mathbf{r}}_B + l\dot{\mathbf{i}}$$

In the above equation, $\dot{\mathbf{r}}_A$ is $\mathbf{v}_A$, which is known completely. Also, $\dot{\mathbf{i}}$ is $\dot{\theta}\mathbf{j}$ (the time derivative of a unit vector was stressed in Chapter 12), and $\dot{\mathbf{r}}_B$ can only be vertical. Figure 14-14(b) shows this relationship. Thus,

$$\cos\theta = \frac{v_A}{l\dot{\theta}} \qquad \text{or} \qquad \dot{\theta} = \frac{v_A}{l\cos\theta} \tag{1}$$

Since $l\dot{\theta}\mathbf{j}$ is in the positive $\mathbf{j}$ direction, $\dot{\theta}$ is positive (ladder is moving counterclockwise).

The derivative of equation (1) with respect to time is

$$\ddot{\theta} = \frac{v_A^2 \tan\theta}{l^2 \cos^2\theta} \tag{2}$$

$\ddot{\theta}$ could also be obtained by differentiating $\dot{\mathbf{r}}_A = \dot{\mathbf{r}}_B + l\dot{\theta}\mathbf{j}$, which yields

$$\ddot{\mathbf{r}}_A = \ddot{\mathbf{r}}_B + l\ddot{\theta}\mathbf{j} - l\dot{\theta}^2\mathbf{i} \tag{3}$$

In (3), $\ddot{\mathbf{r}}_A$ is zero because $\dot{\mathbf{r}}_A$ is constant, $\ddot{\mathbf{r}}_B$ can only be vertical, the $\mathbf{j}$ component (magnitude $l\ddot{\theta}$) is positive, and the $\mathbf{i}$ component [magnitude $l\dot{\theta}^2 = v_A^2/(l\cos^2\theta)$] is negative. Figure 14-14(c) shows these relations. Then

$$\tan\theta = \frac{l\ddot{\theta}}{v_A^2/(l\cos^2\theta)} \qquad \text{or} \qquad \ddot{\theta} = \frac{v_A^2\tan\theta}{l^2\cos^2\theta}$$

which is identical with equation (2). Since $l\ddot{\theta}\mathbf{j}$ is in the positive $\mathbf{j}$ direction, $\ddot{\theta}$ is positive (the ladder is accelerating counterclockwise).

Note that $\dot{\theta}$ in equation (1) agrees (as it should) with the magnitude of $\boldsymbol{\omega}$ in Problem 14.7. Also, the $\ddot{\theta}$ in equation (2) equals the magnitude of $\boldsymbol{\alpha}$ in Problem 14.7.

14.9. A rod 2.5 m long slides down the plane shown in Fig. 14-15 with $v_A = 4$ m/s to the left and $a_A = 5$ m/s^2 to the right. Determine the angular velocity $\boldsymbol{\omega}$ and the angular acceleration $\boldsymbol{\alpha}$ of the rod when $\theta = 30°$.

SOLUTION

As in Problem 14.7,

$$\mathbf{v}_A = \mathbf{v}_{A/B} + \mathbf{v}_B$$

The table indicates what is known.

Vector	Direction	Magnitude
$\mathbf{v}_A$	horizontal	4 m/s to left
$\mathbf{v}_{A/B}$	$\perp$ rod	$l\omega$ (ω unknown)
$\mathbf{v}_B$	along the 45° line	unknown

Draw the vector triangle to fit the vector equation given above (see Fig. 14-16).

Measurement yields $v_{A/B} = 2.93$ m/s. Hence $\omega = v_{A/B}/l = 1.17$ rad/s. ω is clockwise.

To determine $\boldsymbol{\alpha}$, use the vector equation $\mathbf{a}_A = (\mathbf{a}_{A/B})_t + (\mathbf{a}_{A/B})_n + \mathbf{a}_B$ with the following table.

Vector	Direction	Magnitude
$\mathbf{a}_A$	horizontal	5 m/s^2 to right
$(\mathbf{a}_{A/B})_t$	$\perp$ rod	$l\alpha$ (α unknown)
$(\mathbf{a}_{A/B})_n$	along the rod from A to B	$l\omega^2$ $2.5(1.17)^2 = 3.42$ m/s^2
$\mathbf{a}_B$	along the 45° line	unknown

Draw the vector polygon to fit the above table. First draw $\mathbf{a}_A$. (See Fig. 14-17.) Then through the tail of this vector draw $(\mathbf{a}_{A/B})_n$. Through the head of the vector $\mathbf{a}_A$ draw a line along a 45° line, and through the head of the vector $(\mathbf{a}_{A/B})_n$ draw a perpendicular to the rod in the 30° plane.

The value of $(a_{A/B})_t$ is 2.75 m/s^2. Hence, $\alpha = 2.75/2.5 = 1.1$ rad/s^2. $\boldsymbol{\alpha}$ is counterclockwise.

Note: A computer solution to Problem 14.9 is available in Appendix C.

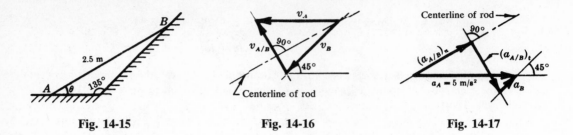

| Fig. 14-15 | Fig. 14-16 | Fig. 14-17 |

Mathcad

14.10. In the slider crank mechanism shown in Fig. 14-18, the crank is rotating at a constant speed of 120 rpm. The connecting rod is 24 in long and the crank is 4 in long. For an angle of 30°, determine the absolute velocity of the crosshead P.

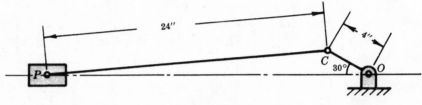

Fig. 14-18

SOLUTION

The angular speed of the crankpin C is 120 rpm. Then $\omega = 2\pi(120/60) = 4\pi$ rad/s.

The linear speed of the crankpin C is therefore $v = r\omega = (4/12)(4\pi) = 4.19$ ft/s.

Figure 14-19 indicates this velocity in a direction perpendicular to the crank.

The component of this velocity along the connecting rod PC will next be determined.

Angle θ must first be found. The figure indicates that $\theta = 90° - 30° - \beta$. But β may be found by using the sine law in the triangle PCO:

$$\frac{4}{\sin \beta} = \frac{24}{\sin 30°}$$

from which $\beta = 4.8°$ and $\theta = 60° - 4.8° = 55.2°$. Hence, the component of the velocity of C along the connecting rod is $4.19 \cos 55.2° = 2.39$ ft/s.

But all points on the connecting rod must have the same velocity along the rod otherwise the rod would be either crushed or pulled apart. Hence, point P, which is a point on the rod, has a velocity component of 2.39 ft/s along the rod. However, its total velocity is along the line of travel of the crosshead. Then

$$v_P = \frac{2.39 \text{ ft/s}}{\cos \beta} = \frac{2.39 \text{ ft/s}}{0.9965} = 2.40 \text{ ft/s}$$

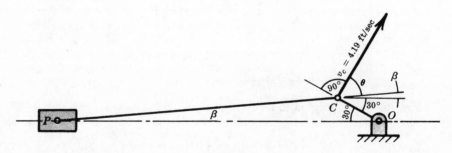

Fig. 14-19

14.11. In Problem 14.10, determine the velocity of the crosshead by graphical means.

SOLUTION

Use the vector equation $\mathbf{v}_P = \mathbf{v}_{P/C} + \mathbf{v}_C$.

In Fig. 14-19, $\mathbf{v}_C$ is shown perpendicular to the crank at C. Its length is 4.19 ft/s.

The absolute velocity of point P is along the line of travel of the crosshead (horizontal in this figure). The velocity $\mathbf{v}_{P/C}$ of P to C is perpendicular to the line joining P and C (the connecting rod). The vector equation for this problem contains three vectors, each of course with magnitude and direction. If four of the six quantities (counting direction and magnitude as two quantities per vector) are known, the other two may be found.

Vector	Direction	Magnitude
$\mathbf{v}_P$	horizontal	?
$\mathbf{v}_{P/C}$	$\perp$ rod	?
$\mathbf{v}_C$	$\perp$ crank	4.19 ft/s

The table indicates that the magnitude of $\mathbf{v}_P$ may be found, since four of the six quantities are known.

First draw the one vector $\mathbf{v}_C$ that is known both in magnitude and direction. Through one end of $\mathbf{v}_C$ draw a horizontal line and through the other end draw a line perpendicular to the connecting rod. The choice of ends is immaterial. Both are shown in Fig. 14-20.

Measurement to scale yields v_P equal to 2.40 ft/s.

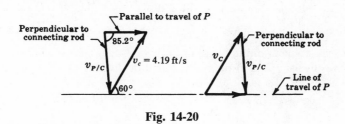

Fig. 14-20

14.12. In Problem 14.10, determine the velocity of the crosshead by use of instant centers.

SOLUTION

The instant center of the connecting rod relative to the frame is the point about which all points in the rod appear to rotate at that instant. Point C is a point on the crank and on the rod. Its absolute velocity (relative to the frame) is the same whether it is a point on the crank or the rod. However, as a point on the crank, its velocity is perpendicular to the crank. Therefore, since its velocity is the same when it is considered as a point on the rod, the instant center for the rod is somewhere along the crank extended. (The velocity of a point in rotation is perpendicular to the radius drawn to it from the center of rotation.)

Similarly, point P is a point on the crosshead and on the connecting rod. As a point on the crosshead, its velocity is horizontal. Hence, as a point on the rod, its velocity is the same (i.e., horizontal). The instant (instantaneous) center for the rod is therefore on a perpendicular to the line of travel of the crosshead, i.e., on a vertical line through P.

The instant center I is at the intersection of the vertical line through P and the crank extended, as shown in Fig. 14-21.

Since I is the center of rotation for all points on the rod, it follows that the linear velocity of a particular point is perpendicular to the line joining I and the point, and the magnitude of the velocity is proportional to the distance of the point from I, i.e., center of rotation.

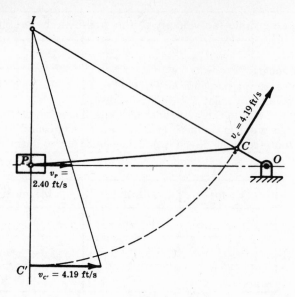

Fig. 14-21

These facts are utilized in a graphical solution as follows. Swing an arc with radius IC until it intersects the line connecting I with the point whose velocity is desired, i.e., P. IC' equals IC. Draw the vector 4.19 ft/s perpendicular to IC' at C'. Draw a gauge line from I to the tip of this vector. The velocity of point P is drawn from P perpendicular to IC' out to the gauge line. Its value is 2.40 ft/s.

14.13. Determine the angular velocity of the connecting rod, referring to Problem 14.11.

SOLUTION

Direct measurement of the figure in Problem 14.11 yields $v_{P/C} = 3.68$ ft/s.
Then the angular speed of the rod is found by dividing $v_{P/C}$ by the length of the rod.

$$\omega = \frac{v_{P/C}}{l} = \frac{3.68 \text{ ft/s}}{2 \text{ ft}} = 1.84 \text{ rad/s}$$

ω is counterclockwise.

14.14. Referring to Problem 14.10, find the acceleration of the crosshead in the slider crank mechanism.

SOLUTION

Since the angular velocity of the crank is constant, the linear acceleration of point C consists of only the normal component directed toward the center O. Its magnitude is $r\omega^2$.

$$(a_C)_n = r\omega^2 = \tfrac{4}{12}(4\pi)^2 = 52.6 \text{ ft/s}^2$$

The acceleration of P, which is horizontal, is determined by the following vector equation:

$$(\mathbf{a}_P) = (\mathbf{a}_{P/C}) + (\mathbf{a}_C)$$

The acceleration $\mathbf{a}_{P/C}$ of point P relative to point C is one of rotation. It is well to write it in terms of its tangential and normal components, which are respectively perpendicular and parallel to the connecting rod. The equation now becomes

$$(\mathbf{a}_P) = (\mathbf{a}_{P/C})_n + (\mathbf{a}_{P/C})_t + (\mathbf{a}_C)$$

where n and t denote normal and tangential components. Eight elements are involved, and the equation may be solved if no more than two of these elements are unknown. Tabulate the elements as follows.

Vector	Direction	Magnitude
$\mathbf{a}_C$	along crank	52.6 ft/s²
$(\mathbf{a}_{P/C})_t$	⊥ connecting rod	?
$(\mathbf{a}_{P/C})_n$	along connecting rod	(length of rod) × $(\omega_{\text{rod}})^2$
$\mathbf{a}_P$	horizontal	?

Note that the magnitude of the normal component $(a_{P/C})_n$ is actually

$$(a_{P/C})_n = (2 \text{ ft})(1.84 \text{ rad/s})^2 = 6.77 \text{ ft/s}^2$$

The vector diagram is now drawn starting with the known vectors (both magnitude and direction) $\mathbf{a}_C$ and $(\mathbf{a}_{P/C})_n$ (see Fig. 14-22).

Through the tail end of $\mathbf{a}_C$ draw a horizontal line, and through the arrow end of $(\mathbf{a}_{P/C})_n$ draw a line perpendicular to the connecting rod. These lines meet in a point M, which determines the length of $\mathbf{a}_P$ and $(\mathbf{a}_{P/C})_t$. By measurement, $a_P = 50.0 \text{ ft/s}^2$.

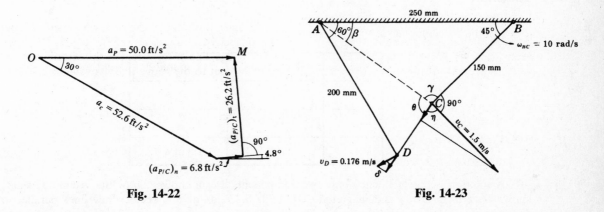

Fig. 14-22   Fig. 14-23

14.15. If the angular velocity of BC is as shown in the quadric crank mechanism in Fig. 14-23, determine the angular velocity of AD and the velocity of point D for the phase indicated.

SOLUTION

The velocity of point C as a point on BC is perpendicular to BC and its magnitude is found to be

$$v_C = BC \times \omega_{BC} = (0.15 \text{ m})(10 \text{ rad/s}) = 1.5 \text{ m/s}$$

To determine the velocity of D by resolution of velocities, first analyze Fig. 14-23 to find angles. By the cosine law,

$$AC = \sqrt{(AB)^2 + (BC)^2 - 2AB \times BC \cos 45°} = \sqrt{(250)^2 + (150)^2 - 2 \times 250 \times 150 \times \cos 45°} = 179 \text{ mm}$$

By the sine law,

$$\frac{BC}{\sin \beta} = \frac{AC}{\sin 45°} = \frac{AB}{\sin \gamma}$$

Hence,

$$\sin \beta = \frac{BC \sin 45°}{AC} = \frac{150 \sin 45°}{179} \qquad \beta = 36.3°$$

$$\sin \gamma = \frac{AB \sin 45°}{AC} = \frac{250 \sin 45°}{179} \qquad \gamma = 99.0°$$

In triangle ADC, angle $DAC = 60° - 36.3° = 23.7°$. Applying the cosine law in triangle ADC,

$$CD = \sqrt{(AD)^2 + (AC)^2 - 2AD \times AC \cos 23.7°}$$

$$= \sqrt{(200)^2 + (179)^2 - 2(200)(179) \cos 23.7°} = 80.5 \text{ mm}$$

By the sine law, $CD/\sin 23.7° = AD/\sin \theta = AC/\sin D$ or $80.5/\sin 23.7° = 200/\sin \theta = 179/\sin D$. Solving, $\theta = 87.0°$ and $D = 63.4°$.

It is now evident that the angle η between CD and the velocity $\mathbf{v}_c$ is

$$\eta = 360° - (90° + \theta + \gamma) = 360° - (90° + 87.0° + 99.0°) = 84.0°$$

The component of this velocity along the bar CD is $1.5 \cos 84.0° = 0.157 \text{ m/s}$. Note that this component is directed from C toward D. This is also the component of the velocity of D along CD.

The angle δ between velocity vector of D and bar CD is $\delta = 180° - (90° + 63.4°) = 26.6°$.

The magnitude of the velocity of D is $v_D = 0.157/\cos 26.6° = 0.176 \text{ m/s}$. Note that this velocity is directed such that the arm AD turns clockwise whereas the arm BC turns counterclockwise.

The angular speed of AD is $\omega_{AD} = v_D/AD = (0.176 \text{ m/s})/0.2 \text{ m} = 0.88 \text{ rad/s}$ clockwise.

14.16. Solve Problem 14.15 graphically. See Fig. 14-24.

SOLUTION

The vector equation involved is $\mathbf{v}_D = \mathbf{v}_{D/C} + \mathbf{v}_C$.
List the six components.

Vector	Direction	Magnitude
$\mathbf{v}_D$	$\perp AD$	?
$\mathbf{v}_{DC}$	$\perp DC$	?
$\mathbf{v}_C$	$\perp BC$	1.5 m/s

A solution is possible, since only two components are unknown. Draw the vector $\mathbf{v}_C$, since it is known both in direction and magnitude. Through the ends of this vector draw lines parallel to the direction of $\mathbf{v}_D$ and $\mathbf{v}_{D/C}$ until they meet. This determines the velocity $\mathbf{v}_D$ when measured to the scale of the drawing. Hence, $v_D = 0.176 \text{ m/s}$.

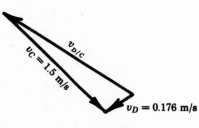

Fig. 14-24

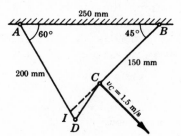

Fig. 14-25

14.17. Solve Problem 14.15 by use of instant centers.

SOLUTION

Points C and D are points on the arm CD. The instant center of CD relative to the frame is the intersection of the lines drawn perpendicular to the absolute velocities of C and D. The latter velocities, however, are perpendicular to BC and AD. Hence, the instant center I is at the intersection of BC and AD as shown in Fig. 14-25.

The velocity of C is shown perpendicular to BC and of magnitude 1.5 m/s. In Fig. 14-26 draw an arc with IC as radius and I as center cutting the line AD in C'.

Draw a vector at C' perpendicular to AD and of magnitude 1.5 m/s. Draw the gauge line IE. The velocity of D is found by erecting a perpendicular to AD at D. Its length to the gauge line is 0.176 m/s.

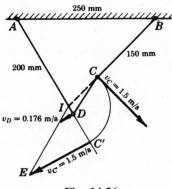

Fig. 14-26

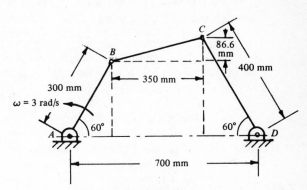

Fig. 14-27

14.18. Figure 14-27 shows a quadric crank mechanism with various lengths given (or calculated). If the crank AB is rotating 3 rad/s counterclockwise, determine the linear velocities of points B and C and the angular velocities of BC and DC.

SOLUTION

Since point B is on the rotating crank AB, its velocity is

$$\mathbf{v}_B = \boldsymbol{\omega}_{AB} \times \boldsymbol{\rho}_{AB} \qquad \text{or} \qquad 3\mathbf{k} \times (300 \times 0.5\mathbf{i} + 300 \times 0.866\mathbf{j}) = 450\mathbf{j} - 779\mathbf{i} \text{ mm/s}$$

Note that the magnitude of $\mathbf{v}_B$ can be found directly from $v_B = r\omega = 300(3) = 900$ mm/s. Since this vector is perpendicular to the crank, it is directed to the left and up. It can be written

$$900(-\cos 30° \,\mathbf{i} + \sin 30° \,\mathbf{j}) = 450\mathbf{j} - 779\mathbf{i} \text{ mm/s}$$

This is, of course, the same expression as that found by using the vector cross product.

To determine the motion of BC, we shall use

$$\mathbf{v}_C = \mathbf{v}_B + \mathbf{v}_{C/B} = \mathbf{v}_B + \boldsymbol{\omega}_{BC} \times \boldsymbol{\rho}_{BC} \tag{1}$$

To use equation (1), assume that DC is rotating counterclockwise. This means that point C moves to the left and down. Since a 30° angle is involved, we write

$$\mathbf{v}_C = -v_C \cos 30° \,\mathbf{i} - v_C \sin 30° \,\mathbf{j} = -0.866 v_C \mathbf{i} - 0.5 v_C \mathbf{j}$$

Next, assume that BC is rotating counterclockwise; hence, we can write

$$\boldsymbol{\omega}_{BC} = \omega_{BC} \mathbf{k}$$

Also note that

$$\boldsymbol{\rho}_{BC} = 350\mathbf{i} + 86.6\mathbf{j}$$

Make these substitutions in equation (1) to get

$$-0.866 v_C \mathbf{i} - 0.5 v_C \mathbf{j} = -779\mathbf{i} + 450\mathbf{j} + \omega_{BC} \mathbf{k} \times (350\mathbf{i} + 86.6\mathbf{j})$$

or

$$-0.866 v_C \mathbf{i} - 0.5 v_C \mathbf{j} = -779\mathbf{i} + 450\mathbf{j} + 350\omega_{BC}\mathbf{j} - 86.6\omega_{BC}\mathbf{i} \tag{2}$$

Equating the $\mathbf{i}$ terms and then the $\mathbf{j}$ terms will yield

$$-0.866 v_C = -779 - 86.6\omega_{BC} \tag{3}$$

$$-0.5 v_C = 450 + 350\omega_{BC} \tag{4}$$

Multiply equation (*3*) by 350/86.6 and add to equation (*4*). This yields $v_C = 675$ mm/s, and ω_{BC} is then found to be -2.25 rad/s. This means that BC is rotating clockwise instead of counterclockwise as assumed at the beginning.

To find ω_{DC}, we note that

$$\mathbf{v}_C = \boldsymbol{\omega}_{DC} \times \boldsymbol{\rho}_{DC} = \omega_{DC}\mathbf{k} \times (-200\mathbf{i} + 346.4\mathbf{j}) \quad \text{or} \quad \mathbf{v}_C = -675 \times 0.866\mathbf{i} - 675 \times 0.5\mathbf{j}$$
$$= -200\omega_{DC}\mathbf{j} - 346.4\omega_{DC}\mathbf{i} \tag{5}$$

Equate either the $\mathbf{i}$ terms or the $\mathbf{j}$ terms to find $\omega_{DC} = +1.69$ rad/s (counterclockwise as assumed).

Note: The following is a different method that can be used to solve this problem. It involves the use of the law of sines to solve the relative velocity part of equation (*1*) or

$$\mathbf{v}_C = \mathbf{v}_{C/B} + \mathbf{v}_B$$

We know that $\mathbf{v}_B$ has a magnitude of 900 mm/s directed to the left and up at an angle of 30° with the horizontal. We also know that $\mathbf{v}_C$ has a magnitude of 400 mm/s, with ω_{DC} unknown in this second solution of the problem. Thus, $\mathbf{v}_C$ is along a line making an angle of 30° with the horizontal, although its sense is unknown. Also known is the fact that $\mathbf{v}_{C/B}$ is perpendicular to BC, with a magnitude equal to the product of ω_{BC} and the length BC. BC is equal to $\sqrt{(350)^2 + (86.6)^2} = 361$ mm. The velocity of C to B (since it is perpendicular to BC) makes an angle with respect to the vertical given by $\theta = \tan^{-1}(86.6/350) = 13.9°$.

Figure 14-28(*a*) shows $\mathbf{v}_B$, which is completely known, drawn from an arbitrary point O to point R, with a length of 900 mm/s and at an angle of 30°.

From R draw a line, making an angle of 13.9° with the vertical. Then draw a line from O, making an angle of 30° with the horizontal. These lines meet at point S.

The figure is redrawn in Fig. 14-28(*b*) to make the law of sines easier to apply. From Fig. 14-28(*b*), we can see that

$$\frac{v_{C/B}}{\sin 60°} = \frac{v_C}{\sin 46.1°} = \frac{900}{\sin 73.9°}$$

This yields $v_C = 675$ mm/s, and leads to $\omega_{DC} = 1.69$ rad/s clockwise. Also, $v_{C/B} = 812$ mm/s, and leads to $\omega_{BC} = 2.25$ rad/s clockwise.

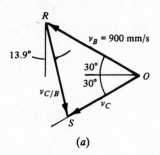
(a)

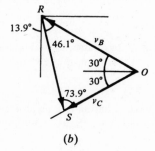
(b)

Fig. 14-28

14.19. In Problem 14.18, determine the linear accelerations of points B and C and the angular accelerations of BC and CD.

SOLUTION

The mechanism is shown in Fig. 14-29, with components of accelerations some of which are known completely and some of which are known only in direction.

The acceleration of C as a point on BC is expressed as

$$\mathbf{a}_C = (\mathbf{a}_{C/B})_t + (\mathbf{a}_{C/B})_n + \mathbf{a}_B$$

Note that $\mathbf{a}_B$ has only a normal component. If there were an α_{AB}, there would be a tangential component to add to the equation.

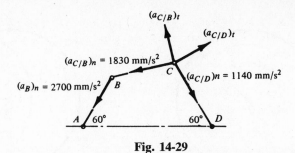

Fig. 14-29

Of course, C is a point on DC, and its acceleration is then written $\mathbf{a}_C = (\mathbf{a}_{C/D})_t + (\mathbf{a}_{C/D})_n$. Hence, we write

$$(\mathbf{a}_{C/D})_t + (\mathbf{a}_{C/D})_n = (\mathbf{a}_{C/B})_t + (\mathbf{a}_{C/B})_n + \mathbf{a}_B \qquad (1)$$

Three of the terms in equation (1) are known completely. They are the normal components of the accelerations, because all angular velocities were found in Problem 14.18.

To determine the normal acceleration of B, use equation (4) in Section 14.1 or $\boldsymbol{\omega}_{AB} \times \boldsymbol{\omega}_{AB} \times \boldsymbol{\rho}_{AB}$ or use $AB \times (\omega_{AB})^2$. B has an acceleration of 2700 mm/s² directed from B toward A as shown.

Similarly, $(\mathbf{a}_{C/B})_n$ is directed from C toward B and has a magnitude $BC \times (\omega_{BC})^2 = 361(2.25)^2 = 1830$ mm/s². Also, $(\mathbf{a}_{C/D})_n$ is directed from C toward D and has a magnitude $DC \times (\omega_{DC})^2 = 400(1.69)^2 = 1140$ mm/s².

The $(\mathbf{a}_{C/B})_t$ is perpendicular to BC and is assumed to act to the left and up. A positive value would mean that α_{BC} is counterclockwise.

The $(\mathbf{a}_{C/D})_t$ is perpendicular to DC and is assumed to act to the right and up. A positive value would mean α_{DC} is clockwise.

The five acceleration vectors in equation (1) are

$$\mathbf{a}_B = 2700(-\cos 60° \, \mathbf{i} - \sin 60° \, \mathbf{j}) = -1350\mathbf{i} - 2340\mathbf{j}$$

$$(\mathbf{a}_{C/D})_n = 1140(\cos 60° \, \mathbf{i} - \sin 60° \, \mathbf{j}) = 570\mathbf{i} - 987\mathbf{j}$$

$$(\mathbf{a}_{C/B})_n = 1830(-\cos 13.9° \, \mathbf{i} - \sin 13.9° \, \mathbf{j}) = -1780\mathbf{i} - 440\mathbf{j}$$

$$(\mathbf{a}_{C/D})_t = 400\alpha_{DC}(\cos 30° \, \mathbf{i} + \sin 30° \, \mathbf{j}) = 346\alpha_{DC}\mathbf{i} + 200\alpha_{DC}\mathbf{j}$$

$$(\mathbf{a}_{C/B})_t = 361\alpha_{BC}(-\sin 13.9° \, \mathbf{i} + \cos 13.9° \, \mathbf{j}) = -86.7\alpha_{BC}\mathbf{i} + 350\alpha_{BC}\mathbf{j}$$

Equation (1) now becomes

$$346\alpha_{DC}\mathbf{i} + 200\alpha_{DC}\mathbf{j} + 570\mathbf{i} - 987\mathbf{j} = -86.7\alpha_{BC}\mathbf{i} + 350\alpha_{BC}\mathbf{j} - 1780\mathbf{i} - 440\mathbf{j} - 1350\mathbf{i} - 2340\mathbf{j} \qquad (2)$$

Equating coefficients of the $\mathbf{i}$ and $\mathbf{j}$ terms in equation (2), we have

$$346\alpha_{DC} + 570 = -86.7\alpha_{BC} - 1780 - 1350 \qquad (3)$$

$$200\alpha_{DC} - 987 = 350\alpha_{BC} - 440 - 2340 \qquad (4)$$

These simplify to

$$346\alpha_{DC} + 86.7\alpha_{BC} = -3700 \qquad (3')$$

$$200\alpha_{DC} - 350\alpha_{BC} = -1790 \qquad (4')$$

Hence, α_{DC} has a magnitude of 10.5 rad/s². Since its sign is negative, it is counterclockwise (opposite to the direction originally assumed).

Also, α_{BC} has a magnitude of 0.87 rad/s². Since its sign is negative, it is clockwise (opposite to the direction originally assumed).

14.20. A wheel 3 m in diameter rolls to the right on a horizontal plane with an angular velocity of 8 rad s (clockwise of course) and an angualr acceleration counterclockwise of 4 rad/s² as shown in Fig. 14-30. The latter merely indicates that the angular velocity of the wheel is decreasing. Determine the linear velocity and acceleration of the top point B on the wheel.

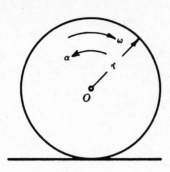

Fig. 14-30

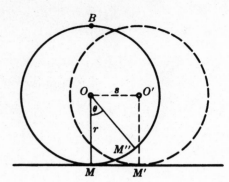

Fig. 14-31

SOLUTION

Draw Fig. 14-31 illustrating the problem. For convenience, the center O is chosen as the base point in this relative motion.

It is necessary to show first that the velocity and acceleration of the center O of a rolling wheel may be expressed in terms of the given ω and α, respectively, and the distance from O to the surface on which the wheel rolls (1.5 m in this case).

In Fig. 14-31, O is displaced to O' through a distance s. Since rolling takes place, MM'' on the wheel contacts MM' on the horizontal surface. Hence, the arc MM'', which is $r\theta$, equals MM', which is equal to OO' or s.

Therefore, $s = r\theta$, where s is the magnitude of the linear displacement of center O, r is the radius of the wheel, and θ is the magnitude of the angular displacement of the wheel.

Differentiation yields

$$v_O = r\frac{d\theta}{dt} = r\omega \qquad \text{and} \qquad a_O = r\frac{d\omega}{dt} = r\alpha$$

These facts may now be applied to the present problem:

$$v_O = (1.5\ \text{m})(8\ \text{rad/s}) = 12\ \text{m/s}$$
$$a_O = (1.5\ \text{m})(4\ \text{rad/s}^2) = 6\ \text{m/s}^2$$

$\mathbf{v}_O$ is directed to the right and $\mathbf{a}_O$ is directed to the left.

The velocity vector equation that will be used is

$$\mathbf{v}_B = \mathbf{v}_{B/O} + \mathbf{v}_O$$

The velocity of B relative to O is perpendicular to the radius OB and to the right (since OB is moving clockwise). Its magnitude is

$$v_{B/O} = OB \times \omega = (1.5\ \text{m})(8\ \text{rad/s}) = 12\ \text{m/s}$$

The absolute velocity $\mathbf{v}_B$ is therefore made up of two components ($\mathbf{v}_{B/O}$ and $\mathbf{v}_O$) each horizontal to the right and each 12 m/s. Hence, $\mathbf{v}_B$ has magnitude 24 m/s and is directed horizontally to the right.

To determine the absolute acceleration $\mathbf{a}_B$, apply the vector equation

$$\mathbf{a}_B = (\mathbf{a}_{B/O})_t + (\mathbf{a}_{B/O})_n + \mathbf{a}_O$$

The relative acceleration $(\mathbf{a}_{B/O})_t$ is directed horizontally to the left (since the angular acceleration of OB is counterclockwise). Its magnitude is equal to OB times the magnitude of the angular acceleration α, or

$$(a_{B/O})_t = (1.5\ \text{m})(4\ \text{rad/s}^2) = 6\ \text{m/s}^2 \qquad \text{(directed left)}$$

The $(a_{B/O})_n$ is directed toward O at the instant considered and is equal in magnitude to OB times the square of the angular speed ω:

$$(a_{B/O})_n = (1.5\ \text{m})(8\ \text{rad/s})^2 = 96\ \text{m/s}^2 \qquad \text{(directed down)}$$

Figure 14-32 presents these facts more clearly.

The acceleration of B may be found graphically or analytically. For an analytical solution, note that the horizontal component of $6+6=12\,\text{m/s}^2$ to the left. The vertical component is $96\,\text{m/s}^2$ down. Therefore,

$$a_B = \sqrt{(a_h)^2 + (a_v)^2} = \sqrt{(12)^2 + (96)^2} = 96.7\,\text{m/s}^2$$

and $\tan\phi = 12/96$ or $\phi = 0.124\,\text{rad}\ (7.12°)$.

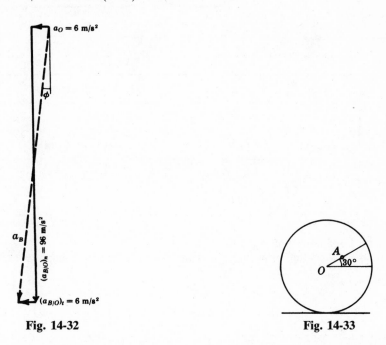

Fig. 14-32   Fig. 14-33

14.21. In Problem 14.20, what are the velocity and acceleration of point A, 0.6 m from the center and on a line making an angle of 30° above the horizontal radius? See Fig. 14-33.

SOLUTION

The usual vector equations apply.

$$\mathbf{v}_A = \mathbf{v}_{A/O} + \mathbf{v}_O \tag{1}$$
$$\mathbf{a}_A = (\mathbf{a}_{A/O})_t + (\mathbf{a}_{A/O})_n + \mathbf{a}_O \tag{2}$$

Tabulate the six elements in the velocity equation (1).

Vector	Direction	Magnitude
$\mathbf{v}_A$	?	?
$\mathbf{v}_{A/O}$	$\perp OA$	$OA \times \omega = (0.6\,\text{m})(8\,\text{rad/s}) = 4.8\,\text{m/s}$
$\mathbf{v}_O$	horizontal	12 m/s to the right

The velocity equation (1) is represented by Fig. 14-34.

The magnitude of the horizontal component of $\mathbf{v}_A = 4.8\sin 30° + 12 = 14.4\,\text{m/s}$.

The magnitude of the vertical component of $\mathbf{v}_A = 4.8\cos 30° = 4.16\,\text{m/s}$.

Hence, $v_A = \sqrt{(14.4)^2 + (4.16)^2} = 15\,\text{m/s}$ and $\phi = \tan^{-1}(4.16/14.4) = 0.281\,\text{rad}$ or 16.1°.

Next tabulate the eight elements in the acceleration equation (2).

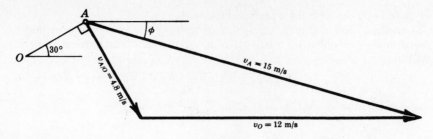

Fig. 14-34

Vector	Direction	Magnitude
$\mathbf{a}_A$	?	?
$(\mathbf{a}_{A/O})_t$	$\perp OA$	$OA \times \alpha = 0.6\,\text{m} \times 4\,\text{rad/s}^2 = 2.4\,\text{m/s}^2$
$(\mathbf{a}_{A/O})_n$	along OA	$OA \times \omega^2 = 0.6\,\text{m} \times (8\,\text{rad/s})^2 = 38.4\,\text{m/s}^2$
$\mathbf{a}_O$	horizontal	$6\,\text{m/s}^2$ to the left

Figure 14-35 represents equation (2).

To obtain the value of $\mathbf{a}_A$, note that the magnitudes of its components horizontally and vertically are

$$(a_A)_h = -6 - 38.4 \cos 30° - 2.4 \sin 30° = -40.5\,\text{m/s}^2$$

$$(a_A)_v = -38.4 \sin 30° + 2.4 \cos 30° = -17.1\,\text{m/s}^2$$

Hence, $a_A = \sqrt{(-40.5)^2 + (-17.1)^2} = 44.0\,\text{m/s}^2$ with $\phi = \tan^{-1}(-17.1/-40.5) = 0.4$ rad or $22.9°$.

Note that ϕ is below the negative x axis.

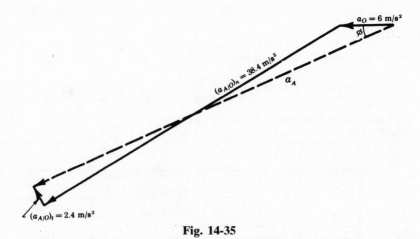

Fig. 14-35

14.22. A cylinder and axle roll under the influence of a weight W as shown in Fig. 14-36. What is the displacement s_O of the center of the cylinder when the weight is displaced 10 ft down? The pulley is assumed to work in frictionless bearings.

SOLUTION

Point I is the instant center between the cylinder and the surface on which it is rolling. Hence, the magnitude of the absolute displacement of A, which is equal to that of the displacement of W, may be expressed as 5θ, where the radius $AI = 5$ ft and θ is the magnitude of the angular displacement. Since $5\theta = 10$, $\theta = 2$ rad and $s_O = 3\theta = 6$ ft (directed to the right).

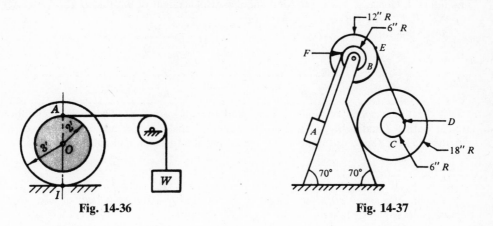

Fig. 14-36   **Fig. 14-37**

14.23. Figure 14-37 shows a weight A on a frictionless plane. The weight is attached to a cord that is wrapped around the small step of pulley B, which turns on frictionless bearings. Another cord, which is wrapped around the large step of pulley B, is in turn wrapped around the smaller of two cylinders C, which are integrally connected. The center of cylinder C is moving with a speed of 12 in/s and an acceleration of 18 m/s², both down the plane. Determine the velocity and acceleration of the weight A. Assume that the cords are parallel to the planes as shown.

SOLUTION

Point E on cord DE has the same speed as point D, which is on the cylinder. As a point on the cylinder, D has a speed of $[(18 + 6)/18] \times 12 = 16$ in/s. The mating point on the pulley B has the same speed as E (16 in/s). Then the speed of point F is $6/12 \times 16 = 8$ in/s. This then is the speed of weight A; i.e., 8 in/s up along the plane.

The component of the acceleration of point D that is parallel to the plane is $24/18 \times 18 = 24$ in/s². This, then, is also the magnitude of the tangential component of the acceleration of point E.

Finally, the tangential component of the acceleration of point F is $6/12 \times 24 = 12$ in/s². Thus, the acceleration of weight A is 12 in/s² up along the plane.

14.24. Solve Problem 14.23 if the cord ED is unwrapping from the bottom of the smaller cylinder instead of from the top. Move the pulley so that the cord ED is parallel to the plane.

SOLUTION

The speed of point D in its new location is $[(18 - 6)/18] \times 12 = 8$ in/s. The speed of point F is $6/12 \times 8 = 4$ in/s. Thus, the velocity of the weight A is 4 in/s up along the plane.

The component of the acceleration of point D in its new location is $12/18 \times 18 = 12$ in/s². Then the acceleration of point F is $6/12 \times 12 = 6$ in/s². The acceleration of weight A is thus 6 in/s² up along the plane.

14.25. In Fig. 14-38, the cylinder of radius r rolls on the surface of radius R. Study the motion.

SOLUTION

Let OGP be the original position and $OG'B$ the position after some time has elapsed. Point P has then moved to point P' and, since pure rolling is assumed, the arc BCP' of the cylinder must equal the arc PB on the surface.

The angle θ is the angular displacement of the $G'P'$ with respect to its original position GP (or $G'C$, which is parallel to GP). The angle ϕ is the angular displacement of the line OGP in the same time interval. Hence, $\theta + \phi$ is the total angular displacement of the line $G'P'$.

Arc BCP' = arc PB, or $r(\theta + \phi) = R\phi$.

Solve this for

$$\theta = \frac{R - r}{r}\phi$$

Hence,
$$\frac{d\theta}{dt} = \frac{R-r}{r}\frac{d\phi}{dt} \quad \text{and} \quad \frac{d^2\theta}{dt^2} = \frac{R-r}{r}\frac{d^2\phi}{dt^2}$$

If the linear speed of point G is v_G and the magnitude of its linear acceleration tangent to its circular path (of radius $R - r$) is $(a_G)_t$, the angular speed $d\phi/dt$ and angular acceleration magnitude $d^2\phi/dt^2$ of G are found as follows (note again that G moves on a circular path of radius $R - r$):

$$\frac{d\phi}{dt} = \frac{v_G}{R - r} \quad \text{and} \quad \frac{d^2\phi}{dt^2} = \frac{(a_G)_t}{R - r}$$

The angular speed $d\theta/dt$ and angular acceleration magnitude $d^2\theta/dt^2$ of any point on the cylinder relative to its center are then

$$\frac{d\theta}{dt} = \frac{R-r}{r}\frac{d\phi}{dt} = \frac{v_G}{r} \quad \text{and} \quad \frac{d^2\theta}{dt^2} = \frac{(a_G)_t}{r}$$

The absolute velocity and the absolute acceleration of any point on the cylinder may now be found by referring the motion of the point first to the center and then adding the motion of the center.

For example, the velocity $\mathbf{v}_B$ of the contact point B equals the sum of the velocity of B relative to the center (magnitude is $r\, d\theta/dt$) and the velocity of the center $\mathbf{v}_G$. If the wheel is rolling up the plane, $d\theta/dt$ is clockwise, and $r\, d\theta/dt$ is thus tangent to the cylinder and directed down to the left. Its magnitude is $r\, d\theta/dt$ or $r(v_G/r) = v_G$. Add this to the velocity of G, which is of the same magnitude, parallel to it but directed up to the right. The absolute velocity of B is found to be zero; i.e., B is the instant center.

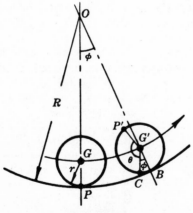

Fig. 14-38

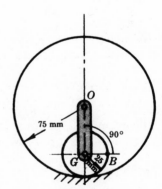

Fig. 14-39

14.26. In an epicyclic gear train, the arm is moving 6 rad/s clockwise and has an acceleration of 10 rad/s² counterclockwise. Determine the linear velocity and acceleration of point B in the phase shown in Fig. 14-39.

SOLUTION

In the previous problem, it was shown that the magnitudes of the angular velocity $\boldsymbol{\omega}$ and angular acceleration $\boldsymbol{\alpha}$ of the small wheel about its own center are given by

$$\omega = \frac{d\theta}{dt} = \frac{R-r}{r}\frac{d\phi}{dt} \quad \text{and} \quad \frac{d^2\theta}{dt^2} = \frac{R-r}{r}\frac{d^2\phi}{dt^2}$$

where ϕ is the angular change of the arm OG. Hence, $\omega = [(75-25)/25] \times 6 = 12 \text{ rad/s}$ (counterclockwise) and $\alpha = [(75-25)/25] \times 10 = 20 \text{ rad/s}^2$ (clockwise). By inspection, clockwise motion of the arm means counterclockwise motion of the small gear.

The velocity $\mathbf{v}_B$ of point B is the vector sum of its relative velocity to G and the velocity of G. $\mathbf{v}_B = \mathbf{v}_{B/G} + \mathbf{v}_G$. The table follows.

Vector	Direction	Magnitude
$\mathbf{v}_B$	?	?
$\mathbf{v}_{B/G}$	vertically up	$r\omega = 300 \text{ mm/s}$
$\mathbf{v}_G$	horizontally to left	$(R-r)6 = 300 \text{ mm/s}$

The sum shown graphically in Fig. 14-40 is 424 mm/s.

The acceleration $\mathbf{a}_B$ of point B is the vector sum of its relative acceleration to G (two components: normal and tangent) and the acceleration of G (two components also).

$$\mathbf{a}_B = (\mathbf{a}_{B/G})_t + (\mathbf{a}_{B/G})_n + (\mathbf{a}_G)_t + (\mathbf{a}_G)_n$$

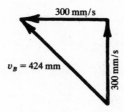

Fig. 14-40

The table follows.

Vector	Direction	Magnitude
$(\mathbf{a}_{B/G})_t$	vertically down	$r\alpha = 500 \text{ mm/s}^2$
$(\mathbf{a}_{B/G})_n$	horizontally to left	$r\omega^2 = 3600 \text{ mm/s}^2$
$(\mathbf{a}_G)_t$	horizontally to right	$(R-r)10 = 500 \text{ mm/s}^2$
$(\mathbf{a}_B)_n$	vertically up	$(R-r)6^2 = 1800 \text{ mm/s}^2$

A vector diagram need not be drawn. The vertical summation is a net amount of $1800 - 500 = 1300 \text{ mm/s}^2$ up. The horizontal summation is a net amount of 3100 mm/s^2 to the left. The resultant absolute acceleration of B is thus to the left and up of magnitude 3360 mm/s^2.

Note: The absolute velocity $\mathbf{v}_B$ may be determined also by using the instant center, the common point of tangency of the two circles (see preceding problem). Although the absolute velocity of the instant center is zero, its absolute acceleration is not.

14.27. In Fig. 14-41, the washer is sliding outward on the rod with a velocity of 4 m/s when its distance from point O is 2 m. Its velocity along the rod is increasing at the rate of 3 m/s². The angular velocity of the rod is 5 rad/s counterclockwise and its angular acceleration is 10 rad/s² clockwise. Determine the absolute acceleration of point P on the washer.

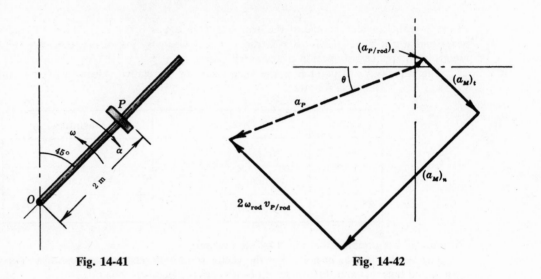

Fig. 14-41   **Fig. 14-42**

SOLUTION

According to Coriolis' law, the absolute acceleration of P is expressed as

$$\mathbf{a}_P = (\mathbf{a}_{P/\text{rod}})_t + (\mathbf{a}_{P/\text{rod}})_n + (\mathbf{a}_M)_t + (\mathbf{a}_M)_n + 2\boldsymbol{\omega} \times \mathbf{v}_{P/\text{rod}}$$

where $(\mathbf{a}_{P/\text{rod}})_t$ = acceleration of P along its path relative to the rod, i.e., 3 m/s² outward along the rod

$(\mathbf{a}_{P/\text{rod}})_n$ = acceleration of P normal to its path along the rod, i.e., zero here since it is moving on a straight-line path

$(\mathbf{a}_M)_t$ = tangential component of the acceleration of the point M on the rod that coincides with P at the instant involved; magnitude $r\alpha = 2(10) = 20$ m/s², down to the right

$(\mathbf{a}_M)_n$ = normal component of the acceleration of the point M on the rod that coincides with P at the instant involved; magnitude $r\omega^2 = 2(5^2) = 50$ m/s², along the rod toward O

$2\boldsymbol{\omega} \times \mathbf{v}_{P/\text{rod}}$ = supplementary or Coriolis' acceleration with the magnitude indicated and in a direction obtained by rotating the vector $\mathbf{v}_{P/\text{rod}}$ through a right angle in the same sense as $\boldsymbol{\omega}$ (counterclockwise in this problem); magnitude $2(5)(4) = 40$ m/s², up to the left

The vector diagram indicates each of these accelerations and their vector sum $\mathbf{a}_P$ (see Fig. 14-42). Thus, $a_P = 51$ m/s², with $\theta = 22°$.

14.28. One vane of an impeller wheel has its center of curvature C located as shown in Fig. 14-43 at a given instant. A particle P, 8 in from the center O, has a velocity 10 m/s and an acceleration 20 in/s² directed outward and tangent to the vane. Determine the acceleration of P if the angular velocity of the wheel is 2 rad/s counterclockwise and the angular acceleration is 3 rad/s² clockwise.

SOLUTION

According to Coriolis' law, the absolute acceleration of P is

$$\mathbf{a}_P = (\mathbf{a}_{P/\text{vane}})_t + (\mathbf{a}_{P/\text{vane}})_n + (\mathbf{a}_M)_t + (\mathbf{a}_M)_n + 2\boldsymbol{\omega}_{\text{vane}} \times \mathbf{v}_{P/\text{vane}}$$

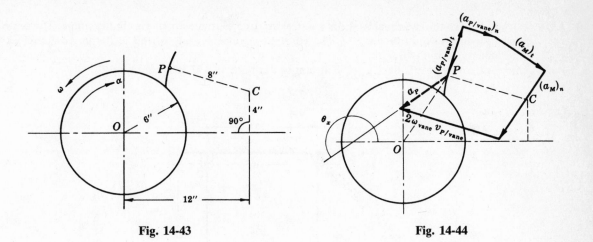

Fig. 14-43   **Fig. 14-44**

where $(\mathbf{a}_{P/\text{vane}})_t = 20$ in/s^2 outward and $\perp$ to PC

$(\mathbf{a}_{P/\text{vane}})_n = (v_{P/\text{vane}})^2/PC = (10)^2/8 = 12.5$ m/s^2 directed from P to C

$(\mathbf{a}_M)_t$ = tangential component of the acceleration of the point M on the vane that coincides with P at the instant; magnitude $= OP \times \alpha = 8 \times 3 = 24$ in/s^2, directed down to the right and $\perp$ to OP

$(\mathbf{a}_M)_n$ = normal component of the acceleration of the point M on the vane that coincides with p at the instant; Magnitude $= OP \times \omega^2 = 8(2)^2 = 32$ in/s^2, directed from P to O

$(\mathbf{v}_{P/\text{vane}}) = 10$ in/s, outward $\perp$ to PC

$(\boldsymbol{\omega}_{\text{vane}}) = 2$ rad/s, counterclockwise

$2\boldsymbol{\omega}_{\text{vane}} \times \mathbf{v}_{P/\text{vane}} = 2(2)(10) = 40$ in/s^2 directed from C toward P; the direction is obtained by rotating $\mathbf{v}_{P/\text{vane}}$, which is $\perp$ to CP and outward, through 90° in the sense of $\boldsymbol{\omega}_{\text{vane}}$, i.e., counterclockwise in the plane of the paper.

The vector sum of these components yields $a_P = 21$ in/s^2 and $\theta_x = 215°$. See Fig. 14-44.

Supplementary Problems

14.29. A rigid body is rotating 12 rad/s about an axis through the origin and has direction cosines 0.421, 0.365, and 0.831 with respect to the x, y, and z axes, respectively. What is the velocity of a point in the body defined by the position vector (with respect to the origin) $\mathbf{r} = -2\mathbf{i} + 3\mathbf{j} - 4\mathbf{k}$?
Ans. $\mathbf{v} = -47.4\mathbf{i} + 0.26\mathbf{j} + 23.9\mathbf{k}$ m/s

14.30. A rigid body is rotating 200 rpm about the line $\mathbf{i} - 3\mathbf{j} + 4\mathbf{k}$. The origin is on the line. What is the linear velocity of the point $P(3, 3, -1)$?   *Ans.* $\mathbf{v} = -37.0\mathbf{i} + 53.6\mathbf{j} + 49.3\mathbf{k}$ m/s

14.31. Determine the angular velocities, in rad/s, of the second and minute hands of an old-fashioned watch.   *Ans.*   0.105 rad/s; 0.0018 rad/s

14.32. A rigid body is rotating at a rate of 60 rpm about a line from the origin to the point $(3, 0, 5)$, where the coordinates are in feet. Determine the linear velocity, in ft/s, of the point $(1, -2.2)$ in the body.   *Ans.* $10.8\mathbf{i} - 1.08\mathbf{j} - 6.47\mathbf{k}$ ft/s

14.33. Refer to Fig. 14-45. The equal bars AB and CD are free to rotate about pins in the frame. The bar BD is equal in length to the distance AC. The constant angular velocity of AB is 10 rpm counterclockwise. Determine the motion of BD.

Ans. All points on BD have $v = 753$ in/min to the right; $a = 47,300$ in/min² up

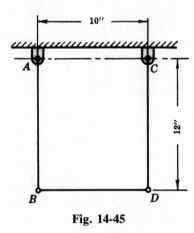

Fig. 14-45

14.34. At a given time, a shaft is rotating 50 rpm about a fixed axis; 20 s later, it is rotating 1050 rpm. What is the average angular acceleration α in rad/s²?   *Ans.* $\alpha = 5.23$ rad/s²

14.35. A flywheel with diameter 500 mm starts from rest with constant angular acceleration 2 rad/s². Determine the tangential and normal components of a point on the rim 3 s after the motion began.
Ans. $a_t = 500$ mm/s², $a_n = 9000$ mm/s²

14.36. A particle is at rest on a phonograph turntable at a radius r from the center. Assuming that the turntable starts from rest and accelerates with a uniform angular acceleration C, determine the magnitudes of the tangential and normal components of the acceleration of the particle at time t.
Ans. $a_n = C^2/rt^2$, $a_t = Cr$

14.37. A rotor decreases uniformly from 1800 rpm to rest in 320 s. Determine the angular deceleration and the number of radians before coming to rest.   *Ans.* $\alpha = -0.589$ rad/s², $\theta = 30,100$ rad

14.38. A bar pivoted at one end moving 5 rad/s clockwise is subjected to a constant angular deceleration. After a certain time interval the bar has an angular displacement of 8 rad counterclockwise and has moved through a total angle of 20.5 rad. What is the angular velocity at the end of the time interval?
Ans. $\omega = 7.58$ rad/s

14.39. the drum shown in Fig. 14-46 is used to hoist the weight W a distance 6 ft. The drum accelerates uniformly from rest to 15 rpm in 1.5 s and then moves at a constant speed of 15 rpm. What is the total elapsed time?   *Ans.* $t = 6.48$ s

14.40. Refer to Fig. 14-47. The hoisting mechanism consists of a drum 1200 mm in diameter around which is wrapped the cable. Integral with the drum is a gear with pitch diameter 900 mm. This gear is driven by a pinion with a 300-mm pitch diameter. The acceleration of the mass M is 6 m/s² up. What is the angular acceleration of the pinion?   *Ans.* $\alpha = 30$ rad/s² clockwise

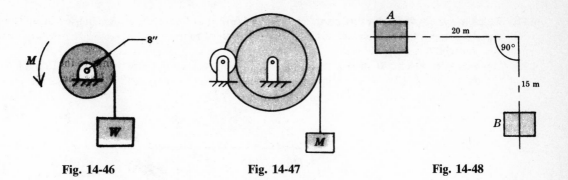

Fig. 14-46   Fig. 14-47   Fig. 14-48

14.41. Rain is falling vertically at 5 ft/s. A person is walking on level ground with a speed of 4 ft/s. What is the velocity of the rain relative to the person and at what angle forward of the vertical should an umbrella be held?   *Ans.* 6.4 ft/s, 38.6°

14.42. A person is walking due east at 3 km/h. The wind appears to be coming from the north. When the pace is dropped to 1 km/h, the wind appears to be coming from the northwest. What is the speed of the wind?   *Ans.* $v = 3.61$ km/h

14.43. Mass A moves to the right at 15 m/s from the position shown in Fig. 14-48. Mass B starts from B at the same instant moving vertically upward at 20 m/s. Determine the relative velocity of A to B $\frac{1}{2}$ s after motion begins.   *Ans.* $v_{A/B} = 25$ m/s, $\theta_x = 307°$

14.44. Refer to Fig. 14-49. Points O and P are on a thin lamina that has motion in the xy plane. Point O is known to have the velocity shown. Determine the absolute velocity of point P.   *Ans.* $v_P = 0$

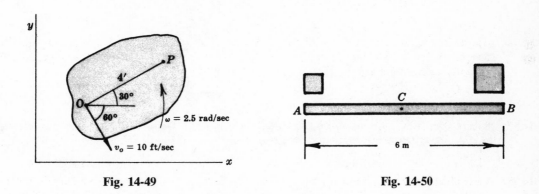

Fig. 14-49   Fig. 14-50

14.45. The iron bar in Fig. 14-50 is attracted by two large but unequal magnets in such a way that end A has a vertical acceleration of 4 m/s² and end B has a vertical acceleration of 6 m/s². Determine the acceleration of the center and the angular acceleration of the bar.
Ans. $a_C = 5$ m/s² up, $\alpha = 0.333$ rad/s² counterclockwise

14.46. A ladder of length l leans against a vertical wall. The bottom moves away from the wall along a horizontal floor with a constant velocity v_0. Determine the velocity and acceleration of the top of the ladder. (*Hint:* $x = v_0 t$.)

Ans. $\dot{y} = \dfrac{-v_0 t}{\sqrt{l^2 - v_0^2 t^2}}$, $\ddot{y} = \dfrac{-v_0^2 l^2}{(l^2 - v_0^2 t^2)^{3/2}}$

14.47. A ladder of length l makes an angle θ with the horizontal floor and leans against a vertical wall. Show that the center of the ladder moves on a circular path of radius $\frac{1}{2}l$.

14.48. The bar AB in Fig. 14-51 slides so that its bottom point A has a velocity of 400 mm/s to the left along the horizontal plane. What is the angular velocity of the bar in the phase shown? Use the instantaneous center method.   *Ans.* $\omega = 0.293$ rad/s clockwise

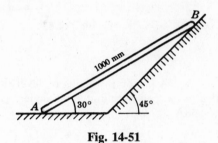

Fig. 14-51

14.49. The top of a ladder 10 ft long slides down a smooth wall while its bottom slides along a smooth plane perpendicular to the wall. Show that the velocity of its midpoint is directed along the ladder when the ladder makes an angle of 45° with the horizontal plane.

Mathcad

14.50. In Fig. 14-52, the velocity of point A is 5 m/s to the right and its acceleration is 8 m/s² to the right. What are the velocity and acceleration of point B?   *Ans.* $v_B = 8.66$ m/s down, $a_B = 33.9$ m/s² down

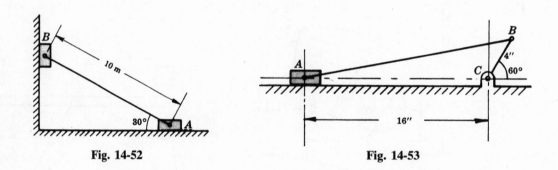

Fig. 14-52   **Fig. 14-53**

14.51. The crank CB of the slider crank mechanism is rotating at a constant 30 rpm clockwise. Determine the velocity of the crosshead A in the phase shown in Fig. 14-53. Use two methods of solution.
Ans. $v_A = 9.68$ in/s to the right

14.52. In the slider crank mechanism shown in Fig. 14-54, the crank is turning clockwise at 120 rpm. What is the velocity of the crosshead when the crank is in the 60° phase° Use two methods of solution.
Ans.   10.3 ft/s toward the right

14.53. The linear velocity of the crosshead (slider) in Fig. 14-55 is 2.4 m/s to the left along the horizontal plane. Using the instant center method, determine the angular velocity of the crank AB in the phase shown.   *Ans.* $\omega = 2.44$ rad/s counterclockwise

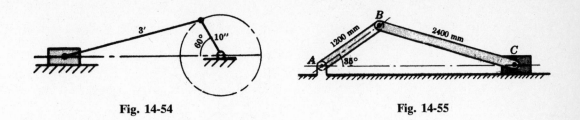

| Fig. 14-54 | Fig. 14-55 |

14.54. A bar slides on a vertical post and is attached to a block A that is moving to the right with a constant velocity C. See Fig. 14-56. Determine the angular velocity $\dot{\theta}$ of the bar.   *Ans.* $\dot{\theta} = -(C/a)\sin^2\theta$

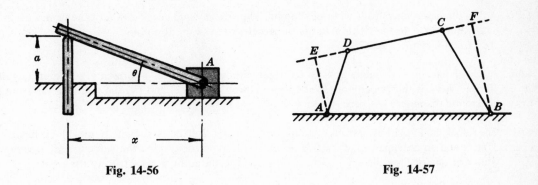

| Fig. 14-56 | Fig. 14-57 |

14.55. In the linkage shown in Fig. 14-57, the pins A and B are fixed. Link AD rotates with angular speed ω_A. Prove that the angular speed ω_B of link BC is given by the expression $\omega_B = \omega_A(AE/BF)$, where AE and BF are perpendicular to DC.

14.56. In the four-bar linkage (quadric crank mechanism) shown in Fig. 14-58, the angular velocity of AB is 8 rad/s clockwise. Determine the angular velocity of CD and the angular velocity of BC.
Ans. $\omega_{CD} = 12.0$ rad/s clockwise, $\omega_{BC} = 0$

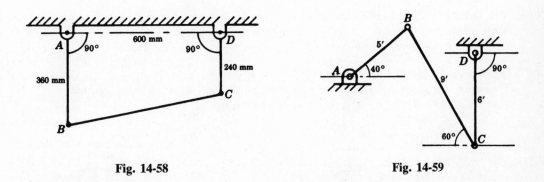

| Fig. 14-58 | Fig. 14-59 |

14.57. Crank AB is moving 2 rpm clockwise in the phase shown in Fig. 14-59. What is the angular velocity of the arm CD? Use two methods of solution.   *Ans.* $\omega_{CD} = 3.28$ rpm counterclockwise

14.58. In the linkage shown in Fig. 14-60, bar AB is constrained to move horizontally and bar CD rotates about point D. If the left end of the horizontal bar has a velocity of 24 in/s to the left and an acceleration of 40 in/s^2 to the right, what are the angular velocity and acceleration of CD?
Ans. $\omega = 5.33$ rad/s clockwise, $\alpha = 73.3$ rad/s^2 clockwise

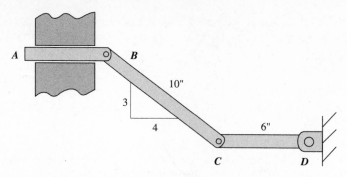

Fig. 14-60

14.59. A 3-m diameter wheel rolls without slipping. The velocity and acceleration of the center of the wheel are 8 m/s and 5 m/s² to the right. What is the acceleration of the top point.
Ans. $a = 43.8 \text{ m/s}^2$, $\theta_x = 283°$

14.60. A wheel 300 mm in diameter rolls to the right without slipping on a horizontal plane. Its angular speed is 30 rpm. What is the velocity of (*a*) the top point on the wheel and (*b*) the point on the front end of the horizontal diameter?   *Ans.*   (*a*) $v = 0.942 \text{ m/s}$, $\theta_x = 0°$; (*b*) $v = 0.666 \text{ m/s}$, $\theta_x = 315°$

14.61. The block in Fig. 14-61 moves on two 4-in-diameter rollers as shown. If the block has a velocity of 3.0 ft/s and an acceleration of 2.0 ft/s², both to the right, determine the velocity and acceleration of the center of one of the rollers.   *Ans.* $v = 1.5 \text{ ft/s}$ to the right, $a = 1.0 \text{ ft/s}^2$ to the right

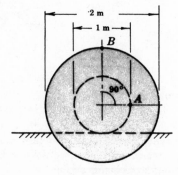

Fig. 14-61

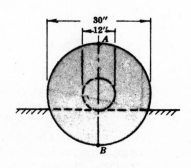

Fig. 14-62

14.62. The disk in Fig. 14-62 rolls without slipping on the horizontal plane. Its center has an acceleration of 4 ft/s² directed horizontally to the left. At the instant that its center has a velocity of 3 ft/s to the right, determine the acceleration of point *P*.   *Ans.* $a = 7.32 \text{ ft/s}^2$, $\theta_x = 183°$

14.63. A composite wheel rolls without slipping with angular velocity 30 rpm clockwise. See Fig. 14-63. Determine the absolute velocities of points *A* and *B*. Use two methods of solution.
Ans. $v_A = 2.22 \text{ m/s}$, $\theta_x = -45°$; $v_B = 4.71 \text{ m/s}$ to the right

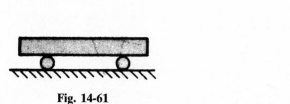

Fig. 14-63   **Fig. 14-64**

14.64. The composite wheel shown in Fig. 14-64 rolls without slipping on the horizontal plane. If the angular velocity of the wheel is 20 rad/s clockwise, determine the absolute velocities of the top and bottom points of the wheel.   *Ans.* $v_A = 35$ ft/s to the right, $v_B = 15$ ft/s to the left

14.65. The wheel in Fig. 14-65 moves such that its center has a velocity of 4 m/s horizontally to the right. The angular velocity of the wheel is 4 rad/s clockwise. Determine the absolute velocity of the points P and Q.   *Ans.* $v_P = 5.32$ m/s, $\theta_x = 25.0°$; $v_Q = 0$

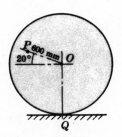

Fig. 14-65

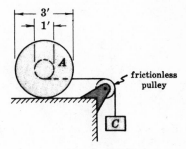

Fig. 14-66

14.66. The composite wheel A in Fig. 14-66 rolls without slipping on the horizontal plane. The cord wrapped around the axle is attached to the weight C as shown. The velocity of C changes uniformly from 2 ft/s downward to 6 ft/s downward in 2 s. Determine the angular displacement of A during the interval.
Ans. $\theta = 8$ rad clockwise

14.67. The cylinder C shown in Fig. 14-67 is 500 mm in diameter and rolls without slipping on the horizontal plane. The pulley B is frictionless. If the displacement of A is 100 mm down, what is the angular displacement of C?   *Ans.* $\theta = 0.4$ rad clockwise

14.68. In the preceding problem, the velocity and acceleration of A are 100 mm/s and 50 mm/s^2, respectively. What are the angular velocity and angular acceleration of the cylinder C?
Ans. $\omega = 0.4$ rad/s clockwise, $\alpha = 0.2$ rad/s^2 clockwise

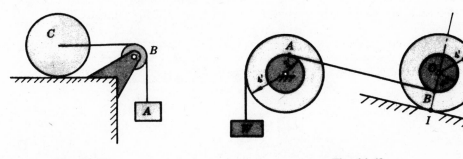

Fig. 14-67   **Fig. 14-68**

14.69. Weight W in Fig. 14-68 is suspended from the pulley, which turns in frictionless bearings. As the pulley turns, the cord AB from the axle of the cylinder is wrapped upon the pulley. The weight descends from rest with a constant acceleration of 16 ft/s^2. Determine the displacement, velocity, and acceleration of the center O of the cylinder after 3 s. Cord AB is parallel to the inclined plane.
Ans. $s_0 = 72$ ft, $v_0 = 48$ ft/s, $a_0 = 16$ ft/s^2

14.70. Solve Problem 14.69 if the cord AB is unwrapping from the top of the axle instead of the bottom. Move the pulley so that AB is still parallel to the plane.   *Ans.* $s_0 = 12$ ft, $v_0 = 16$ ft/s, $a_0 = 5.33$ ft/s^2

14.71. In Fig. 14-69, the valve A is actuated by the eccentric rotating 30 rpm counterclockwise. Express the velocity and acceleration of the valve in terms of the angle ϕ.
Ans. $\dot{x} = 7.85 \sin \phi$ in/s to the left, $\ddot{x} = 24.6 \cos \phi$ in/s^2 to the left

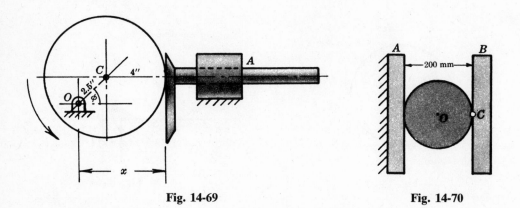

Fig. 14-69   **Fig. 14-70**

14.72. Refer to Fig. 14-70. The gear rack A is stationary whereas the gear rack B has a velocity of 600 mm/s down and an acceleration of 450 mm/s^2 down. Determine the velocity of the gear center, the acceleration of the gear center, and the acceleration of the contact point C on the gear.
Ans. $v_0 = 300$ mm/s down, $a_0 = 225$ mm/s^2 down, $a_C = 1000$ mm/s^2 at $\theta_x = 207°$

14.73. In Fig. 14-71, the upper plate is moving to the right with a speed of 10 ft/s and a linear acceleration of 4 ft/s^2. The lower plate is also moving to the right, with a speed of 5 ft/s and a linear acceleration of 3 ft/s^2. What are the angular velocity and acceleration of the 4-ft-diameter disk. There is no slip between the disk and the plates.   *Ans.* $\omega = \frac{5}{4}$ rad/s clockwise, $\alpha = \frac{1}{4}$ rad/s^2 clockwise

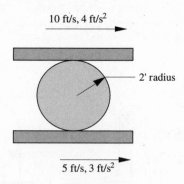

10 ft/s, 4 ft/s^2

2' radius

5 ft/s, 3 ft/s^2

Fig. 14-71

14.74. Solve Problem 14.73 if the lower plate moves to the left with a speed of 5 ft/s and an acceleration of 3 ft/s^2 *Ans.* $\omega = \frac{15}{4}$ rad/s clockwise, $\alpha = \frac{7}{4}$ rad/s^2 clockwise

14.75. In Fig. 14-72, the disk rolls without slipping on the horizontal plane with an angular velocity of 10 rpm clockwise and an angular acceleration of 6 rad/s^2 counterclockwise. The bar AB is attached as shown. The line OA is horizontal. Point B moves along the horizontal plane shown. Determine the velocity and acceleration of point B for the phase shown.
Ans. $v_B = 1.1$ m/s to the right, $a_B = 7.47$ m/s^2 to the left

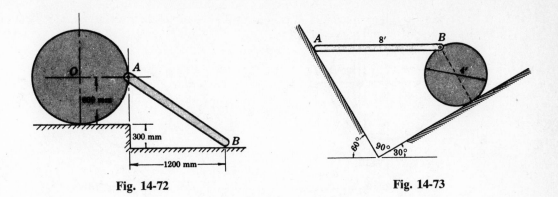

Fig. 14-72   Fig. 14-73

14.76. The 8-ft bar AB in Fig. 14-73 is connected by a frictionless pin at B to the 4-ft-diameter cylinder, which is rolling with constant center velocity 12 ft/s down the plane inclined 30° with the horizontal. The end A slides on the frictionless plane, which makes an angle of 60° with the horizontal. For the phase shown with the bar horizontal, determine the angular velocity and acceleration of the bar.
Ans. $\omega = 6.0$ rad/s clockwise, $\alpha = 62.4$ rad/s² counterclockwise

14.77. The crank OB in Fig. 14-74 is rotating with constant angular velocity 6 rad/s clockwise. Disk C rolls without slipping inside the fixed circle. Determine (a) the angular velocity of disk C and (b) the absolute acceleration of point P in the phase shown, where crank OB is horizontal and P is the top point on disk C.   *Ans.*   (a) $\omega_C = 18$ rad/s counterclockwise, (b) $a_P = 341$ ft/s² at $\theta_x = 252°$

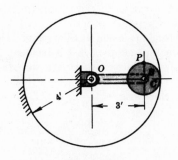

Fig. 14-74

14.78. A bead moves along a rod such that $r = 0.5t^2$. At the same time, the rod moves such that $\theta = t^2 + t$. Determine the radial and transverse components of the velocity and acceleration of the bead at time $t = 2$ s. Assume that r is in meters.   *Ans.* $v_r = 2$ m/s, $v_\theta = 10$ m/s, $a_r = -49$ m/s², $a_\theta = 24$ m/s²

14.79. A rod is rotating in a horizontal plane at a constant rate of 2 rad/s clockwise about a vertical axis through one end. A washer is sliding along the rod with a constant speed of 4 ft/s. Determine the radial and transverse components of the acceleration at the instant when the washer is 2 ft from the vertical axis.   *Ans.* $a_r = -8$ ft/s², $a_\theta = 16$ ft/s²

14.80. In Fig. 14-75, the ball P moves with a constant velocity 2 m/s down along the smooth slot cut as shown in the disk rotating with angular velocity 3 rad/s clockwise and angular acceleration 8 rad/s² counterclockwise. Determine the acceleration of P in the phase shown.   *Ans.* $a = 14.9$ m/s², with $\theta_x = 171°$

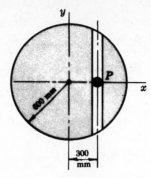

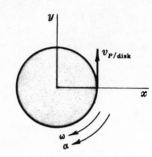

Fig. 14-75   **Fig. 14-76**

14.81. A particle moves with speed 600 mm/s around the circumference of a 1200-mm-diameter disk that is rotating in the opposite direction with angular velocity of 2 rad/s (clockwise). See Fig. 14-76. The disk has angular acceleration 4 rad/s² clockwise. Determine the acceleration of the particle in the phase shown.   *Ans.* $a = 2.47$ m/s² with $\theta_x = 256°$

14.82. A bug moves with constant speed v ft/s along the circumference of a disk of radius r ft. The disk rotates in the opposite direction with constant angular velocity ω rad/s. What is the absolute acceleration of the bug?
Ans. $(v - r\omega)^2/r$

14.83. A bug moves with constant speed v along a radius r of a disk rotating with constant angular velocity ω. What is the absolute acceleration of the bug as it reaches the rim of the disk?
Ans. $a = \omega\sqrt{r^2\omega^2 + 4v^2}$

14.84. In Fig. 14-77, the member AD slides inside a collar pinned to PC at P, which is at a fixed distance from C. If M is the fixed point on the member AD coincident with P at the instant considered, then the velocity and acceleration of P relative to M are horizontal. Detemrine the angular velocity and acceleration of member CP.
Ans. $\omega_{CP} = 10$ rad/s counterclockwise, $\alpha_{CP} = 75$ rad/s² counterclockwise

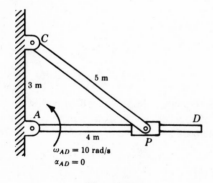

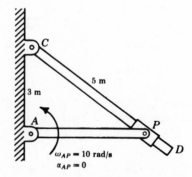

Fig. 14-77   **Fig. 14-78**

14.85. In Fig. 14-78, the member CD slides inside a collar pinned to AP at P, which is a fixed distance from A. In this problem, M is the point in CD that is coincident with P; thus, the velocity and acceleration of P relative to M are along the line CD. Determine the angular velocity and acceleration of the member CD.
Ans. $\omega_{CD} = 6.4$ rad/s counterclockwise, $\alpha_{CD} = 13.4$ rad/s² counterclockwise

14.86. The rod AB in Fig. 14.79 is pinned at B and rests on the 8-in-radius wheel. The wheel rolls without slipping with an angular velocity of 12 rad/s clockwise and an angular acceleration of 3 rad/s^2 counterclockwise. What are the angular velocity and acceleration of the rod AB?
Ans. $\omega = 1.6$ rad/s clockwise, $\alpha = 19.7$ rad/s^2 counterclockwise

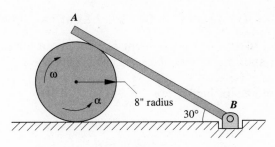

Fig. 14-79

14.87. The two rods AB and CD in Fig. 14-80 rotate about A and D, respectively. The 30-in rod is free to slide in a collar at B and is pinned at C. If the angular speed of AB is constant at 10 rad/s clockwise and the angular speed of CD is constant at 8 rad/s counterclockwise, what are the angular speed and acceleration of the 30-in rod?   *Ans.* $\omega = 0.353$ rad/s clockwise, $\alpha = 0.78$ rad/s^2 counterclockwise

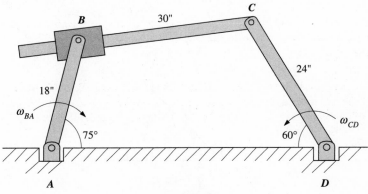

Fig. 14-80

Chapter 15

Moments of Inertia

15.1 AXIAL MOMENT OF INERTIA OF AN ELEMENT OF AREA

The axial moment of inertia I of an element of area about an axis **in** its plane is the product of the area of the element and the square of its distance from the axis. The moment of inertia is also called the second moment of area.

In Fig. 15-1, the moments of inertia are

$$dI_x = y^2 \, dA$$

$$dI_y = x^2 \, dA$$

15.2 POLAR MOMENT OF INERTIA OF AN ELEMENT OF AREA

The polar moment of inertia J of an element about an axis **perpendicular** to its plane is the product of the area of the element and the square of its distance from the axis. This can also be thought of as the moment of inertia about the z axis.

In Fig. 15-1, the polar moment of inertia is

$$dJ = \rho^2 \, dA = (x^2 + y^2) \, dA = dI_y + dI_x$$

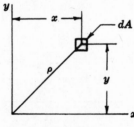

Fig. 15-1

15.3 PRODUCT OF INERTIA OF AN ELEMENT OF AREA

The product of inertia of an element of area in the figure is defined as follows:

$$dI_{xy} = xy \, dA$$

15.4 AXIAL MOMENT OF INERTIA OF AN AREA

The axial moment of inertia of an area is the sum of the axial moments of inertia of its elements:

$$I_x = \int y^2 \, dA$$

$$I_y = \int x^2 \, dA$$

15.5   RADIUS OF GYRATION OF AN AREA

The radius of gyration of an area with respect to an axis is given by the equation $k = \sqrt{I/A}$.

15.6   POLAR MOMENT OF INERTIA OF AN AREA

The polar moment of inertia of an area is the sum of the polar moments of inertia of its elements:

$$J = \int \rho^2 \, dA$$

15.7   PRODUCT OF INERTIA OF AN AREA

The product of inertia of an area is the sum of the products of inertia of its elements:

$$I_{xy} = \int xy \, dA$$

15.8   PARALLEL AXIS THEOREM

The parallel axis theorem states that the axial or polar moment of inertia of an area about any axis equals the axial or polar moment of inertia of the area about a parallel axis through the centroid of the area plus the product of the area and the square of the distance between the two parallel axes.

In Fig. 15-2, x and y are any axes through O, while x' and y' are coplanar parallel axes through the centroid G.

$$I_x = \bar{I}_{x'} + Am^2$$
$$I_y = \bar{I}_{y'} + An^2$$
$$I_O = \bar{J} + Ar^2$$

The product of inertia of an area with respect to any two axes equals the product of inertia about two parallel centroidal axes plus the product of the area and the distances from respective axes:

$$I_{xy} = \bar{I}_{x'y'} + Amn$$

where m and n are the coordinates of G relative to the (x, y) axes through O or the coordinates of O relative to the (x', y') axes through G. In the first case, m and n are positive, whereas in the second case, they are negative. In either case, their product is positive. See Fig. 15-2.

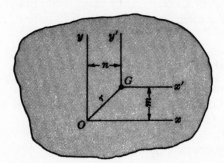

Fig. 15-2

15.9   COMPOSITE AREA

The axial or polar moment of inertia, or product of inertia, of a composite area is the sum of the axial or polar moments of inertia, or products of inertia, of the component areas making up the whole.

Units of any of the foregoing are the fourth power of length. In the U.S. Customary system, the units are usually in^4. In SI, the units are m^4 or mm^4. In using mm^4, it is convenient to use 10^6 mm^4 (particularly in tables listing properties in SI units).

15.10 ROTATED SET OF AXES

The moment of inertia of any area with respect to a rotated set of axes (x', y') may be expressed in terms of the moments and product of inertia with respect to the (x, y) axes as follows:

$$I_{x'} = \tfrac{1}{2}(I_x + I_y) + \tfrac{1}{2}(I_x - I_y)\cos 2\theta - I_{xy}\sin 2\theta$$
$$I_{y'} = \tfrac{1}{2}(I_x + I_y) - \tfrac{1}{2}(I_x - I_y)\cos 2\theta + I_{xy}\sin 2\theta$$
$$I_{x'y'} = \tfrac{1}{2}(I_x - I_y)\sin 2\theta + I_{xy}\cos 2\theta$$

where I_x, I_y = moments of inertia with respect to the (x, y) axes

$I_{x'}, I_{y'}$ = moments of inertia with respect to the (x', y') axes, which have the same origin as the (x, y) axes but are rotated through an angle θ

I_{xy} = product of inertia with respect to the (x, y) axes

$I_{x'y'}$ = product of inertia with respect to the (x', y') axes

For a proof of this, refer to Problem 15.21.

Maximum moments of inertia of any area occur with respect to principal axes: a special set of (x', y') axes for which $2\theta' = \tan^{-1}\left[-2I_{xy}/(I_x - I_y)\right]$.

$$I_{x'} = \tfrac{1}{2}(I_x + I_y) \pm \sqrt{\tfrac{1}{4}(I_x - I_y)^2 + I_{xy}^2}$$
$$I_{y'} = \tfrac{1}{2}(I_x + I_y) \mp \sqrt{\tfrac{1}{4}(I_x - I_y)^2 + I_{xy}^2}$$

For details, refer to Problem 15.22.

15.11 MOHR'S CIRCLE

Mohr's circle is a device that makes it unnecessary to memorize the formulas connected with rotation of axes. Refer to Problems 15.23 and 15.24.

15.12 AXIAL MOMENT OF INERTIA OF AN ELEMENT OF MASS

The axial moment of inertia of an element of mass is the product of the mass of the element and the square of the distance of the element from the axis.

15.13 AXIAL MOMENT OF INERTIA OF A MASS

The axial moment of inertia of a mass is the sum of the axial moments of all its elements. Thus, for a mass of which dm is one element with coordinates (x, y, z), the following definitions hold (see Fig. 15-3):

$$I_x = \int (y^2 + z^2)\, dm$$

$$I_y = \int (x^2 + z^2)\, dm$$

$$I_z = \int (x^2 + y^2)\, dm$$

where I_x, I_y, I_z = axial moments of inertia (with respect to the x, y, and z axes, respectively).

For a thin plate essentially in the xy plane, the following relations hold (see Fig. 15-4):

$$I_x = \int y^2 \, dm$$

$$I_y = \int x^2 \, dm$$

$$J_O = \int \rho^2 \, dm = \int (x^2 + y^2) \, dm = I_x + I_y$$

Table 15-1   Moments of Inertia of Areas

Figure	Area and moment	Figure	Area and moment
RECTANGLE 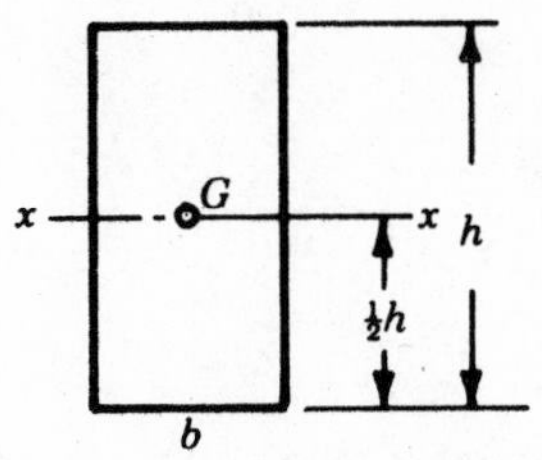	$A = bh$ $I_x = \frac{1}{12}bh^3$ $I_b = \frac{1}{3}bh^3$	TRIANGLE 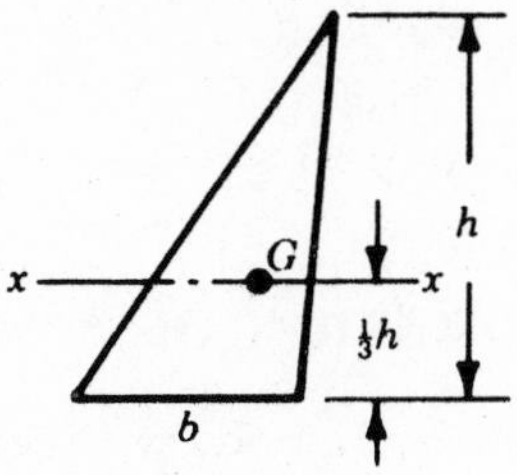	$A = \frac{1}{2}bh$ $I_x = \frac{1}{36}bh^3$ $I_b = \frac{1}{12}bh^3$
HOLLOW RECTANGLE 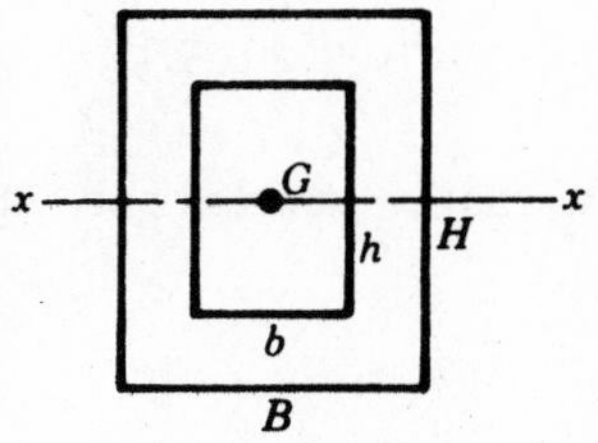	$A = BH - bh$ $I_x = \frac{1}{12}(BH^3 - bh^3)$	ELLIPSE 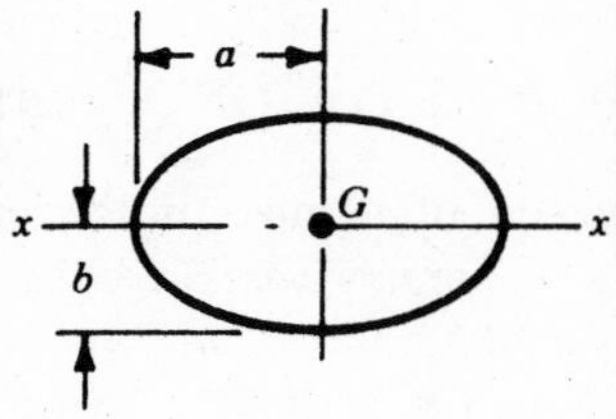	$A = \pi ab$ $I_x = \frac{1}{4}\pi ab^3$
CIRCLE 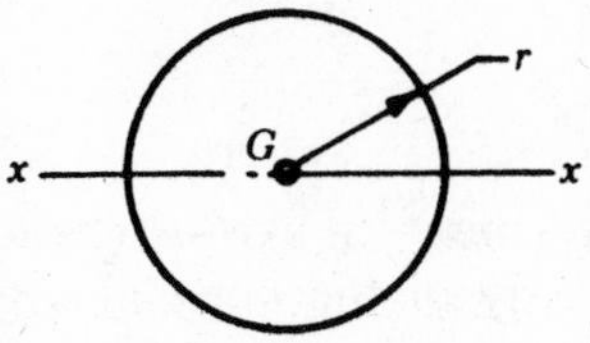	$A = \pi r^2$ $I_x = \frac{1}{4}\pi r^4$	SEMIELLIPSE 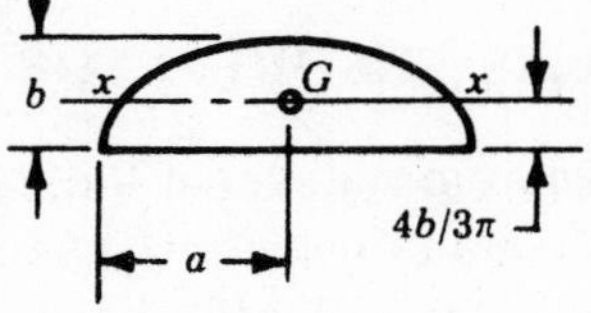	$A = \frac{1}{2}\pi ab$ $I_x = 0.11ab^3$
SEMICIRCLE 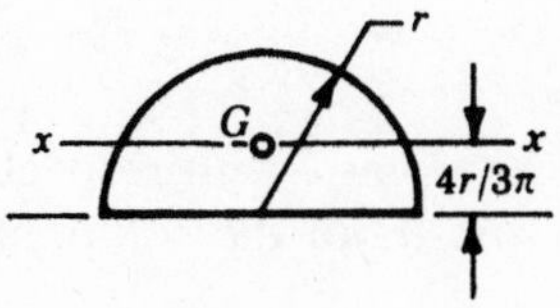	$A = \frac{1}{2}\pi r^2$ $I_x = 0.11r^4$	QUARTER CIRCLE 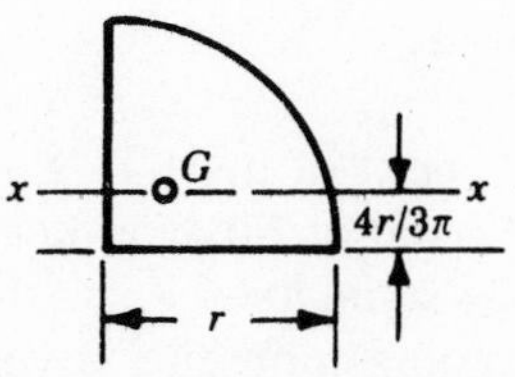	$A = \frac{1}{4}\pi r^2$ $I_x = 0.055r^4$

where I_x, i_y = axial moments of inertia about the x and y axes, respectively, and J_O = axial moment of inertia about the z axis.

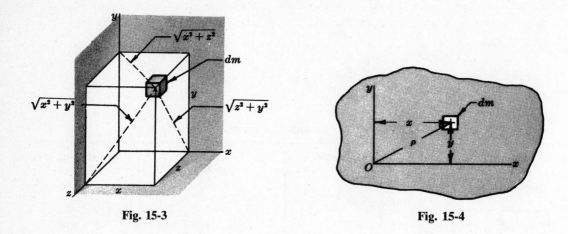

Fig. 15-3   Fig. 15-4

15.14 RADIUS OF GYRATION OF A MASS

The radius of gyration k of a body with respect to an axis is $k = \sqrt{I/m}$, that is, the square root of the quotient of its moment of inertia divided by the mass.

15.15 PRODUCT OF INERTIA OF A MASS

The product of inertia of a mass is the sum of the products of inertia of its elements (see Fig. 15-4):

$$I_{xy} = \int xy\, dm$$

The product of inertia is zero if one, or both, of the reference axes is an axis of symmetry.

15.16 PARALLEL AXIS THEOREM FOR A MASS

The parallel axis theorem states that the moment of inertia of a body about an axis is equal to the moment of inertia $\bar{I}$ about a parallel axis through the center of gravity of the body plus the product of the mass of the body and the square of the distance between the two parallel axes.

15.17 COMPOSITE MASS

The axial or polar moment of inertia, or product of inertia of a composite mass, with respect to an axis, is the sum of the axial or polar moments of inertia, or products of inertia of the component masses, with respect to the same axes.

Units of all the foregoing moments involve those of mass and square of length. In the U.S. Customary system, the units are slug-ft^2 or lb-s^2-ft. In SI, the units are kg $\cdot$ m^2.

Solved Problems

15.1. Determine the axial moment of inertia of the rectangle with base b and altitude h about a centroidal axis parallel to the base. Refer to Fig. 15.5.

SOLUTION

Choose an element of area dA parallel to the base and at a distance y from the centroidal x axis as in Fig. 15-6:

$$I_x = \bar{I} = \int y^2 \, dA = \int_{-h/2}^{h/2} y^2 b \, dy = \tfrac{1}{12} bh^3$$

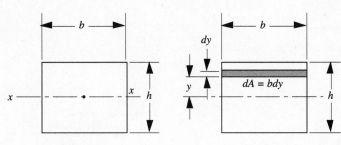

Fig. 15-5   Fig. 15-6   Fig. 15-7

15.2. Determine the axial moment of inertia of the rectangle in Problem 15.1 with respect to the base. Refer to Fig. 15-7.

SOLUTION

$$I_x = \int y^2 \, dA = \int_0^h y^2 b \, dy = \tfrac{1}{3} bh^3$$

15.3. Determine the axial moment of inertia of a rectangle with respect to its base by means of the parallel axis theorem. See Fig. 15-8. Assume that the results of Problem 15.1 are known.

SOLUTION
$$I_x = \bar{I} + A(\tfrac{1}{2}h)^2 = \tfrac{1}{12}bh^3 + bh(\tfrac{1}{4}h^2) = \tfrac{1}{3}bh^3$$

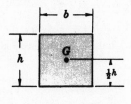

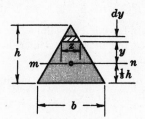

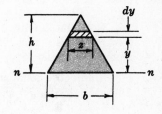

Fig. 15-8   Fig. 15-9   Fig. 15-10

15.4. Determine the axial moment of inertia for a triangle of base b and altitude h about a centroidal axis parallel to the base. Refer to Fig. 15-9.

SOLUTION

$$I_{mn} = \int y^2 \, dA = \int_{-h/3}^{2h/3} y^2 z \, dy = \int_{-h/3}^{2h/3} y^2 \frac{b}{h} (\tfrac{2}{3}h - y) \, dy = \tfrac{1}{36} bh^3$$

since by similar triangles, $b/h = z/(\tfrac{2}{3}h - y)$ or $z = (b/h)(\tfrac{2}{3}h - y)$.

15.5. Determine the axial moment of inertia for a triangle of base b and altitude h about the base. Refer to Fig. 15-10,

SOLUTION

$$I_b = \int y^2 \, dA = \int_0^h y^2 z \, dy = \int_0^h y^2 \frac{b}{h}(h - y) \, dy = \tfrac{1}{12}bh^3$$

since, by similar triangles, $z/(h - y) = b/h$ or $z = (b/h)(h - y)$.

15.6. Knowing the results of Problem 15.5 (a comparatively simple integration problem), find the axial moment of inertia of the triangle about a centroidal axis parallel to the base. See Fig. 15-11.

SOLUTION

This is really an application of the parallel axis theorem in reverse.

$$\bar{I} = I_b - A(\tfrac{1}{3}h)^2 = \tfrac{1}{12}bh^3 - (\tfrac{1}{2}bh)(\tfrac{1}{3}h)^2 = \tfrac{1}{36}bh^3$$

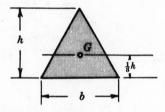

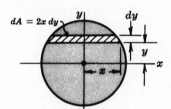

Fig. 15-11   Fig. 15-12

15.7. Determine the axial moment of inertia for a circle of radius r about a diameter. Refer to Fig. 15-12.

SOLUTION

$$I_x = \int y^2 \, dA = 2\int_0^r y^2(2x \, dy) = 4\int_0^r y^2\sqrt{r^2 - y^2} \, dy$$
$$= 4[-\tfrac{1}{4}y\sqrt{(r^2 - y^2)^3} + \tfrac{1}{8}r^2\{y\sqrt{r^2 - y^2} + r^2 \sin^{-1}(y/r)\}]_0^r$$
$$= 4[0 + \tfrac{1}{8}r^2(0 + r^2 \sin^{-1} 1) + 0 - \tfrac{1}{8}r^2(0 + 0)] = 4(\tfrac{1}{8}r^4)(\tfrac{1}{2}\pi) = \tfrac{1}{4}\pi r^4$$

Note that this integral could have been evaluated from $-r$ to r instead of as twice the integral from 0 to r. This is permissible since the moment of inertia of the two halves of the area is equal to the moment of inertia of the whole.

15.8. Determine the axial moment of inertia of a circle of radius r about a diameter, using the differential area dA shown in Fig. 15-13.

SOLUTION

$$I_x = \bar{I} = \int y^2 \, dA$$

where $y = \rho \sin \theta$ and $dA = \rho \, d\rho \, d\theta$. Then,

$$I_x = \bar{I} = \int_0^{2\pi} \int_0^r \rho^3 \, d\rho \sin^2 \theta \, d\theta = \int_0^{2\pi} \sin^2 \theta \, d\theta \left[\frac{\rho^4}{4}\right]_0^r$$
$$= \tfrac{1}{4}r^4[\tfrac{1}{2}\theta - \tfrac{1}{4}\sin 2\theta]_0^{2\pi} = \tfrac{1}{4}r^4(\pi - \tfrac{1}{4}\sin 4\pi - 0 + \tfrac{1}{4}\sin 0) = \tfrac{1}{4}\pi r^4$$

This problem illustrates the ease of finding moments of inertia if the proper choice of element of area is made.

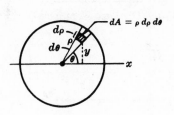

Fig. 15-13

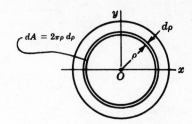

Fig. 15-14

15.9. Determine the polar moment of inertia for a circle of radius r about an axis through its center and perpendicular to the plane of the circle. Refer to Fig. 15-14.

SOLUTION

Choose as the differential area an annular ring of radius ρ and thickness $d\rho$. Hence, dA equals the circumference, $2\pi\rho$, times the thickness, $d\rho$.

$$\bar{J} = \int \rho^2 \, dA = \int_0^r \rho^2 2\pi\rho \, d\rho = 2\pi[\tfrac{1}{4}\rho^4]_0^r = \tfrac{1}{2}\pi r^4$$

At this point, it is possible to derive the value of the axial moment of inertia about a diameter. Since $\bar{J} = I_x + I_y$ and $I_x = I_y$, $I_x = \tfrac{1}{2}\bar{J} = \tfrac{1}{4}\pi r^4$.

15.10. Determine the axial and polar moments of inertia for the ellipse shown in Fig. 15-15.

SOLUTION

$$I_x = \int y^2 \, dA = \int_{-b}^b y^2(2x \, dy)$$

where $x^2/a^2 + y^2/b^2 = 1$ or $x = (a/b)\sqrt{b^2 - y^2}$.
Then

$$I_x = \int_{-b}^b y^2(2a/b)\sqrt{b^2 - y^2} \, dy$$
$$= (2a/b)[-\tfrac{1}{4}y\sqrt{(b^2 - y^2)^3} + \tfrac{1}{8}b^2\{y\sqrt{b^2 - y^2} + b^2 \sin^{-1}(y/b)\}]_{-b}^b = \tfrac{1}{4}\pi ab^3$$

A similar integration would yield $I_y = \tfrac{1}{4}\pi a^3 b$. Of course, $\bar{J} = I_x + I_y = \tfrac{1}{4}\pi ab(a^2 + b^2)$.

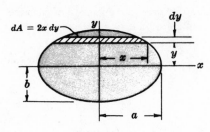

Fig. 15-15

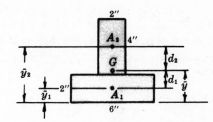

Fig. 15-16

15.11. Determine the axial moment of inertia of the T section shown in Fig. 15-16 about the centroidal axis parallel to the base.

SOLUTION

The first step is to locate the centroid G using the two subdivisions of area shown.

Subdivision	Area	Distance from base to centroid
A_1	12	1
A_2	8	4

Hence, $(A_1 + A_2)\bar{y} = A_1\bar{y}_1 + A_2\bar{y}_2$, $(12 + 8)\bar{y} = 12 \times 1 + 8 \times 4$, and $\bar{y} = 2.2$ in.

Next determine I for each subdivision about its own centroidal axis parallel to the base, using $I = \frac{1}{12}bh^3$:

$$I_1 = \tfrac{1}{12}(6)(2^3) = 4 \text{ in}^4 \qquad I_2 = \tfrac{1}{12}(2)(4^3) = 10.7 \text{ in}^4$$

The final step is to transfer from each subdivision's centroidal axis to the axis through G to determine $\bar{I}$ for the entire area. By the parallel axis theorem ($d_1 = 2.2 - 1 = 1.2$; $d_2 = 4 - 2.2 = 1.8$),

$$\bar{I} = (I_1 + A_1 d_1^2) + (I_2 + A_2 d_2^2) = (4 + 12 \times 1.44) + (10.7 + 8 \times 3.24) = 57.9 \text{ in}^4$$

15.12. Determine the axial moment of inertia about a centroidal axis parallel to the base of the composite area shown in Fig. 15-17.

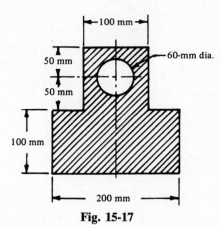

Fig. 15-17

SOLUTION

The first step is the location of the centroid of the composite area. Let T represent the top rectangular area, B the bottom rectangular area, and C the circular area. Using the base as the reference line, we have

$$\bar{y} = \frac{A_T \bar{y}_T + A_B \bar{y}_B - A_C \bar{y}_C}{A_T + A_B - A_C}$$

$$= \frac{(100 \times 100)(150) + (200 \times 100)(50) - [\pi(60)^2/4](150)}{(100 \times 100) + 200 \times 100 - \pi(60)^2/4} = 76.4 \text{ mm}$$

The distance d_T from the centroid of the top area to the common centroid is

$$150 - 76.4 = 73.6 \text{ mm}$$

Similarly,

$$d_B = 76.4 - 50 = 26.4 \text{ mm}$$

and

$$d_C = 150 - 76.4 = 73.6 \text{ mm}$$

The values of I for each component area about that area's centroidal axis parallel to the base of the composite area are as follows:

$$I_T = \tfrac{1}{12}b_T h_T^3 = \tfrac{1}{12}(100)(100)^3 = 8.33 \times 10^6 \text{ mm}^4$$

$$I_B = \tfrac{1}{12}b_B h_B^3 = \tfrac{1}{12}(200)(100)^3 = 16.67 \times 10^6 \text{ mm}^4$$

$$I_C = \tfrac{1}{4}\pi r^4 = \tfrac{1}{4}\pi(30)^4 = 0.64 \times 10^6 \text{ mm}^4$$

Finally,

$$I = (I_T + A_T d_T^2) + (I_B + A_B d_B^2) - (I_C + A_C d_C^2)$$

$$= [8.33 \times 10^6 + (100)(100)(73.6)^2] + [16.67 \times 10^6 + (200 \times 100)(26.4)^2]$$

$$- \left[0.64 \times 10^6 + \frac{\pi(60)^2}{4}(73.6)^2\right] = 77.1 \times 10^6 \text{ mm}^4$$

15.13. Determine the axial moment of inertia for the channel shown in Fig. 15-18 about a centroidal axis parallel to the base b.

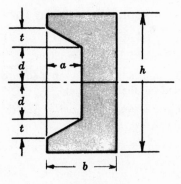

Fig. 15-18

SOLUTION

In this case the centroidal axis is, by symmetry, at half the height.

Consider the channel as made up of a rectangle of base b and altitude h from which have been deleted two triangles of altitude t and base a together with a rectangle of base a and height $2d$. Refer to Fig. 15-19.

To determine I_x for the channel, subtract I_x of the triangles and the smaller rectangle from I_x for the large rectangle. That is,

$$I_x = I_1 - (I_2 + I_3 + I_4)$$

Then

$$I_1 = \tfrac{1}{12}bh^3$$

$$I_2 = I_3 = \tfrac{1}{36}at^3 + \tfrac{1}{2}at(d + \tfrac{1}{3}t)^2 = \tfrac{1}{36}at^3 + \tfrac{1}{2}atd^2 + \tfrac{1}{3}at^2d + \tfrac{1}{18}at^3$$

$$I_4 = \tfrac{1}{12}a(2d)^3 = \tfrac{2}{3}ad^3$$

and

$$I_x = \tfrac{1}{12}bh^3 - atd^2 - \tfrac{2}{3}at^2d - \tfrac{1}{6}at^3 - \tfrac{2}{3}ad^3$$

Of course, this result will be less formidable when numerical values are assigned.

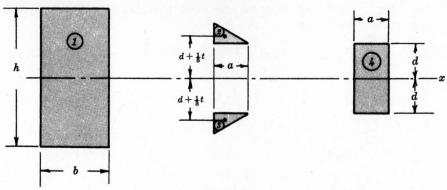

Fig. 15-19

15.14. A column is built up of 2-in planks (actual size) as shown in Fig. 15-20. Determine the axial moment of inertia about a centroidal axis parallel to a side.

SOLUTION

The centroid is located by inspection at the midpoint. To determine $\bar{I}$, it is only necessary to double the summation of the axial moments of areas *1* and *2* about a line through the midpoint.

$$I_1 = \tfrac{1}{12}b_1h_1^3 = \tfrac{1}{12}(2)(8^3) = 85.3 \text{ in}^4$$

$$I_2 = \tfrac{1}{12}b_2h_2^3 + A_2d_2^2 = \tfrac{1}{12}(4)(2^3) + (8)(3^2) = 74.7 \text{ in}^4$$

$$\bar{I} = 2(85.3 + 74.7) = 320 \text{ in}^4$$

Another technique is to subtract I for the inner square from I for the outer square, both about an axis through the midpoint and parallel to a side:

$$\bar{I}_1 = I_O - I_i = \tfrac{1}{12}b_oh_O^3 - \tfrac{1}{12}b_ih_i^3 = \tfrac{1}{12}(8)(8^3) - \tfrac{1}{12}(4)(4^3) = 320 \text{ in}^4$$

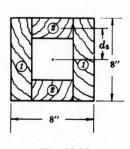

Fig. 15-20

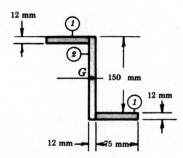

Fig. 15-21

15.15. Determine the axial moment of inertia about the horizontal centroidal axis of the Z section shown in Fig. 15-21. What is the radius of gyration?

SOLUTION

The centroidal axis is an axis of symmetry in this case. The axial moment of inertia $\bar{I}$ is then equal to the sum of the axial moments of area *2* and two areas *1*.

$$A = (162)(12) + 2(75 \times 12) = 3.74 \times 10^3 \text{ mm}^2$$

$$I_1 = \tfrac{1}{12}b_1h_1^3 + A_1d_1^2 = \tfrac{1}{12}(75)(12)^3 + (75 \times 12)(75)^2 = 5.07 \times 10^6 \text{ mm}^4$$

$$I_2 = \tfrac{1}{12}b_2h_2^3 = \tfrac{1}{12}(12)(162)^3 = 4.25 \times 10^6 \text{ mm}^4$$

$$\bar{I}_2 = 2I_1 + I_2 = 14.4 \times 10^6 \text{ mm}^4$$

$$k = \sqrt{\frac{\bar{I}}{A}} = \sqrt{\frac{14.4 \times 10^6}{3.74 \times 10^3}} = 62.1 \text{ mm}$$

15.16. What is the product of inertia about two adjacent sides of a rectangle of base b and altitude h?

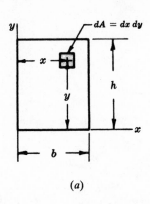
(a)

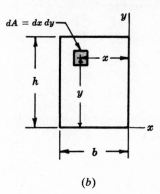
(b)

Fig. 15-22

SOLUTION

As shown in Fig. 15-22(a), the x and y axes are along two adjacent sides.
Denote the product of inertia as I_{xy}. Then

$$I_{xy} = \int xy \, dA = \int_0^h \int_0^b xy \, dx \, dy = [\tfrac{1}{2}x^2]_0^b [\tfrac{1}{2}y^2]_0^h = \tfrac{1}{4}b^2h^2$$

Next choose the y axis as the right side of the rectangle [see Fig. 15-22(b)]. Then

$$I_{xy} = \int_0^h \int_{-b}^0 xy \, dx \, dy = [\tfrac{1}{2}x^2]_{-b}^0 [\tfrac{1}{2}y^2]_0^h = (0 - \tfrac{1}{2}b^2)(\tfrac{1}{2}h^2 - 0) = -\tfrac{1}{4}b^2h^2$$

This indicates that the product of inertia may be positive or negative, depending on the location of the area relative to the axes.

15.17. Determine the product of inertia about two centroidal axes parallel to the sides of a rectangle of base b and altitude h.

SOLUTION

In Fig. 15-23, it is seen that the x' limits of integration are from $-b/2$ to $b/2$. The y' limits of integration are from $-h/2$ to $h/2$. Hence,

$$I_{x'y'} = \int x'y' \, dA = \int_{-h/2}^{h/2} \int_{-b/2}^{b/2} x'y' \, dx' \, dy' = 0$$

This could also be deduced from the result of Problem 15.16 using the parallel axis theorem for product of inertia:

$$I_{xy} = I_{x'y'} + A(\tfrac{1}{2}b)(\tfrac{1}{2}h)$$

where $\tfrac{1}{2}b$ and $\tfrac{1}{2}h$ are the perpendicular distances between the x, y and x', y' axes.

$$I_{x'y'} = \tfrac{1}{4}b^2h^2 - bh(\tfrac{1}{2}b)(\tfrac{1}{2}h) = 0$$

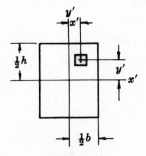

Fig. 15-23

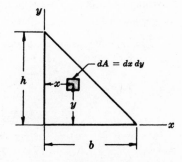

Fig. 15-24

15.18. Determine the product of inertia with respect to the base and altitude of a right triangle. Refer to Fig. 15-24.

SOLUTION

By definition,

$$I_{xy} = \int_0^h \int_0^x xy \, dx \, dy$$

The upper limit of the x integration depends on y. Therefore it must be evaluated from the equation of the sloping line, which is

$$y = -\frac{h}{b}x + h \qquad \text{or} \qquad x = -\frac{b}{h}(y - h)$$

$$I_{xy} = \int_0^h \int_0^{-(b/h)(y-h)} xy \, dx \, dy = \int_0^h [\tfrac{1}{2}x^2]^{-(b/h)(y-h)} y \, dy = \int_0^h \frac{b^2}{2h^2}(y^2 - 2yh + h^2)y \, dy = \tfrac{1}{24}b^2h^2$$

15.19. Determine the product of inertia with respect to the bounding radii of a quadrant of a circle of radius r. Use (*a*) the element of the area shown in Fig. 15-25, and (*b*) the element of the area shown in Fig. 15-26.

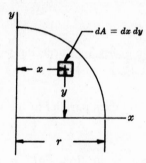

Fig. 15-25

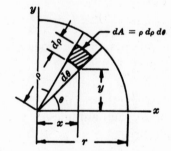

Fig. 15-26

SOLUTION

(*a*) By definition,

$$I_{xy} = \int_0^r \int_0^x xy \, dx \, dy$$

Here the double integration means a summation first with respect to the variable x, which depends on y according to the equation $x^2 + y^2 = r^2$. Substituting,

$$I_{xy} = \int_0^r \int_0^{\sqrt{r^2-y^2}} xy \, dx \, dy = \int_0^r [\tfrac{1}{2}x^2]_0^{\sqrt{r^2-y^2}} y \, dy = \int_0^r \tfrac{1}{2}(r^2 - y^2)y \, dy = \tfrac{1}{8}r^4$$

(*b*) As before,

$$I_{xy} = \int \int xy \, dA = \int_0^r \int_0^{\pi/2} \rho \cos \theta \rho \sin \theta \rho \, d\rho \, d\theta$$

$$I_{xy} = \int_0^{\pi/2} [\tfrac{1}{4}\rho^4]_0^r \cos \theta \sin \theta \, d\theta = \tfrac{1}{4}r^4[\tfrac{1}{2}\sin^2 \theta]_0^{\pi/2} = \tfrac{1}{8}r^4$$

15.20. In Problem 15.19, determine the product of inertia for a quadrant of a circle of radius r equal to 50 mm.

SOLUTION

$$I_{xy} = \tfrac{1}{8}r^4 = \tfrac{1}{8}(50)^4 = 7.81 \times 10^5 \text{ mm}^4$$

15.21. Show that the moments of inertia of an area with respect to the rotated set of axes (x', y') may be expressed as follows:

$$I_{x'} = \tfrac{1}{2}(I_x + I_y) + \tfrac{1}{2}(I_x - I_y)\cos 2\theta - I_{xy}\sin 2\theta$$
$$I_{y'} = \tfrac{1}{2}(I_x + I_y) - \tfrac{1}{2}(I_x - I_y)\cos 2\theta + I_{xy}\sin 2\theta$$
$$I_{x'y'} = \tfrac{1}{2}(I_x - I_y)\sin 2\theta + I_{xy}\cos 2\theta$$

where　　I_x, I_y = moments of inertia with respect to (x, y) axes
　　　　$I_{x'}, I_{y'}$ = moments of inertia with respect to (x', y') axes
　　　　I_{xy} = product of inertia with respect to (x, y) axes
　　　　$I_{x'y'}$ = product of inertia with respect to (x', y') axes

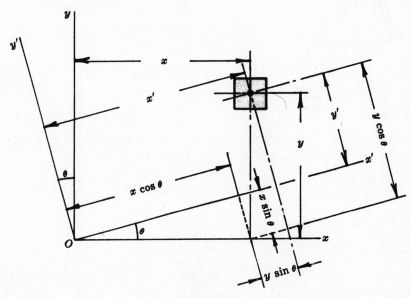

Fig. 15-27

SOLUTION

Figure 15-27 indicates an element dA of the area. By definition,

$$I_{x'} = \int y'^2 \, dA \qquad I_{y'} = \int x'^2 \, dA \qquad I_{x'y'} = \int x'y' \, dA \qquad (1)$$

But from the figure, $x' = x\cos\theta + y\sin\theta$ and $y' = -x\sin\theta + y\cos\theta$. Squaring,

$$x'^2 = x^2\cos^2\theta + 2xy\cos\theta\sin\theta + y^2\sin^2\theta$$

$$y'^2 = x^2\sin^2\theta - 2xy\sin\theta\cos\theta + y^2\cos^2\theta$$

Also,　　　　　　　$x'y' = -x^2\cos\theta\sin\theta - xy\sin^2\theta + xy\cos^2\theta + y^2\cos\theta\sin\theta$

Substituting into equation (1),

$$I_{x'} = \int x^2 \sin^2 \theta \, dA - \int xy \sin 2\theta \, dA + \int y^2 \cos^2 \theta \, dA$$

$$I_{y'} = \int x^2 \cos^2 \theta \, dA + \int xy \sin 2\theta \, dA + \int y^2 \sin^2 \theta \, dA$$

$$I_{x'y'} = \int -x^2 \cos \theta \sin \theta \, dA + \int y^2 \sin \theta \cos \theta \, dA + \int xy(\cos^2 \theta - \sin^2 \theta) \, dA$$

Since the integration over the area is independent of θ, the above equations may be written (using $I_x = \int y^2 \, dA$, $I_y = \int x^2 \, dA$, and $I_{xy} = \int xy \, dA$) as

$$I_{x'} = I_y \sin^2 \theta - I_{xy} \sin 2\theta + I_x \cos^2 \theta$$

$$I_{y'} = I_y \cos^2 \theta + I_{xy} \sin 2\theta + I_x \sin^2 \theta$$

$$I_{x'y'} = \tfrac{1}{2}(I_x - I_y) \sin 2\theta + I_{xy} \cos 2\theta$$

Using $\cos^2 \theta = \tfrac{1}{2}(1 + \cos 2\theta)$ and $\sin^2 \theta = \tfrac{1}{2}(1 - \cos 2\theta)$, we finally obtain

$$I_{x'} = \tfrac{1}{2}(I_x + I_y) + \tfrac{1}{2}(I_x - I_y) \cos 2\theta - I_{xy} \sin 2\theta$$

$$I_{y'} = \tfrac{1}{2}(I_x + I_y) - \tfrac{1}{2}(I_x - I_y) \cos 2\theta + I_{xy} \sin 2\theta$$

$$I_{x'y'} = \tfrac{1}{2}(I_x - I_y) \sin 2\theta + I_{xy} \cos 2\theta$$

15.22. Determine the values of $I_{x'}$ and $I_{y'}$ in Problem 15.21 with respect to the principal axes (these axes yield maximum or minimum values of I).

SOLUTION

To determine the value of θ that will make $I_{x'}$ a maximum, take the derivative of $I_{x'}$ with respect to θ and equate the resulting expression to zero. Thus,

$$\frac{dI_{x'}}{d\theta} = \tfrac{1}{2}(I_x - I_y)(-2 \sin 2\theta) - I_{xy}(2 \cos 2\theta) = 0 \qquad \text{or} \qquad \tan 2\theta' = -\frac{2I_{xy}}{I_x - I_y}$$

This the value θ' of θ that will make $I_{x'}$ a maximum (or minimum). It is necessary to evaluate $\sin 2\theta'$ and $\cos 2\theta'$; this is most easily done from Fig. 15-28. Thus,

$$\cos 2\theta' = \pm \frac{I_x - I_y}{\sqrt{(I_x - I_y)^2 + 4I_{xy}^2}} \qquad \sin 2\theta' = \mp \frac{2I_{xy}}{\sqrt{(I_x - I_y)^2 + 4I_{xy}^2}}$$

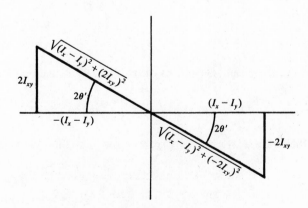

Fig. 15-28

Substituting these values and simplifying, we obtain

$$I_{x'} = \tfrac{1}{2}(I_x + I_y) \pm \sqrt{[\tfrac{1}{2}(I_x - I_y)]^2 + I_{xy}^2}$$

By taking the derivative of $I_{y'}$ with respect to θ, it can be seen that the same value θ' makes $I_{y'}$ a maximum (or minimum). Substituting the values of $\sin 2\theta'$ and $\cos 2\theta'$ yields

$$I_{y'} = \tfrac{1}{2}(I_x + I_y) \mp \sqrt{[\tfrac{1}{2}(I_x - I_y)]^2 + I_{xy}^2}$$

Since $I_{y'}$ has a negative sign before the radical for the value of θ' that gives $I_{x'}$ a positive sign before the radical, we can consolidate the results by stating that, with respect to the principal axes (x', y'), one value, say $I_{x'}$, is maximum while the other value, $I_{y'}$, is minimum. Note further that for this particular θ' (principal axes),

$$I_{x'y'} = \tfrac{1}{2}(I_x - I_y)\sin 2\theta + I_{xy}\cos 2\theta = 0$$

15.23. Given I_x, I_y, and I_{xy} for an area with respect to (x, y) axes, determine graphically, using Mohr's circle, the values $I_{x'}$, $I_{y'}$, and $I_{x'y'}$ for a set of (x', y') axes located counterclockwise at an angle θ with respect to the (x, y) set of axes.

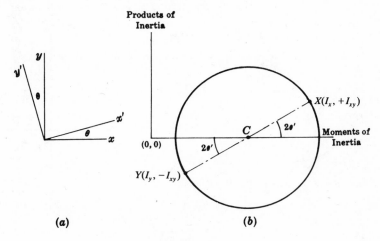

(a)   (b)

Fig. 15-29

SOLUTION

Figure 15-29(a) shows the orientation of the axes. In Fig. 15-29(b) an orthogonal set of lines is drawn. Any moment of inertia I will be located to the right of the vertical line. Any product of inertia will be located above or below the horizontal line.

Assume that $I_x > I_y$ and I_{xy} is positive. Locate the point X that has coordinates $(I_x, +I_{xy})$ and the point Y that has coordinates $(I_y, -I_{xy})$. Draw the line XY that crosses the horizontal line at C as shown. Draw a circle with C as center and containing X and Y (see Fig. 15-30).

Next draw a line at an angle 2θ from the XY line in a counterclockwise direction. The coordinates of X' and Y' are the values of $I_{x'}$, $I_{y'}$, and $I_{x'y'}$.

These values can be seen as follows. Draw vertical lines XA and YB. The distance $BC = CA = \tfrac{1}{2}(I_x - I_y)$. The distance XA is I_{xy}. The radius of the circle CX is the hypotenuse of the right triangle ACX. As such, any radius equals $\sqrt{[\tfrac{1}{2}(I_x - I_y)]^2 + I_{xy}^2}$.

Now the I coordinate of X' equals the distance from O to the center $\tfrac{1}{2}(I_x + I_y)$ plus the projection CX' on the horizontal. CX' (a radius) makes an angle $(2\theta + 2\theta')$ with the horizontal I axis. The $2\theta'$ is the angle referred to in Problem 15.22 that makes $I_{x'}$ a maximum.

$$\text{Projection of } CX' = \sqrt{[\tfrac{1}{2}(I_x - I_y)]^2 + I_{xy}^2}[\cos(2\theta + 2\theta')]$$

$$= \sqrt{[\tfrac{1}{2}(I_x - I_y)]^2 + I_{xy}^2}[\cos 2\theta \cos 2\theta' - \sin 2\theta \sin 2\theta'] \qquad (1)$$

Substituting

$$\cos 2\theta' = \frac{CA}{CX} = \frac{\frac{1}{2}(I_x - I_y)}{\sqrt{[\frac{1}{2}(I_x - I_y)]^2 + I_{xy}^2}} \quad \text{and} \quad \sin 2\theta' = \frac{AX}{CX} = \frac{I_{xy}}{\sqrt{[\frac{1}{2}(I_x - I_y)]^2 + I_{xy}^2}}$$

into equation (1) and simplifying, we find

$$\text{projection of } CX' = \frac{1}{2}(I_x - I_y)\cos 2\theta - I_{xy}\sin 2\theta$$

Thus, $\qquad I_{x'} = \frac{1}{2}(I_x + I_y) + \frac{1}{2}(I_x - I_y)\cos 2\theta - I_{xy}\sin 2\theta$

Similarly, $\qquad I_{y'} = \frac{1}{2}(I_x + I_y) - \frac{1}{2}(I_x - I_y)\cos 2\theta + I_{xy}\sin 2\theta$

Note that the maximum I occurs at the right end of the diameter of the circle with a value equal to the distance to the center plus a radius. Hence,

$$I_{max} = \frac{1}{2}(I_x + I_y) + \sqrt{[\frac{1}{2}(I_x - I_y)]^2 + I_{xy}^2}$$

The minimum value is at the left end of the diameter and is

$$I_{min} = \frac{1}{2}(I_x + I_y) - \sqrt{[\frac{1}{2}(I_x - I_y)]^2 + I_{xy}^2}$$

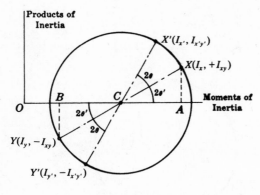

Fig. 15-30

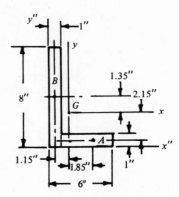

Fig. 15-31

15.24. Determine the principal centroidal moments of inertia for the unequal angle shown in Fig. 15-31.

SOLUTION

First locate the centroid of the angle using the (x'', y'') axes as shown. The angle is divided into the parts A and B.

$$\bar{x}'' = \frac{5(1)(3.5) + 8(1)(0.5)}{5(1) + 8(1)} = 1.65 \text{ in}$$

$$\bar{y}'' = \frac{5(1)(0.5) + 8(1)(4)}{13} = 2.65 \text{ in}$$

Next draw (x, y) axes through this centroid G whose coordinates have been found. They determine I_x and I_y by transferring from the parallel centroidal axes of A and B:

$$I_x = \tfrac{1}{12}(5)(1)^3 + 5(2.15)^2 + \tfrac{1}{12}(1)(8)^3 + 8(1.35)^2 = 80.8 \text{ in}^4$$

$$I_y = \tfrac{1}{12}(1)(5)^3 + 5(1.85)^2 + \tfrac{1}{12}8(1)^3 + 8(1.15)^2 = 38.8 \text{ in}^4$$

In order to determine the principal moments of inertia, the value of the product of inertia I_{xy} must be found. This also is transferred from the parallel centroidal axes of A and B, keeping in mind that the products of inertia about the centroidal axes of A and B are both zero because these centroidal axes are axes of symmetry. Hence, in terms of the transfer distances, the value of I_{xy} becomes

$$I_{xy} = 0 + (8 \times 1)(-1.15)(+1.35) + 0 + (5 \times 1)(+1.85)(-2.15) = -32.3 \text{ in}^4$$

Note: The signs of the transfer distances are important in determining the product of inertia. In the above equation, the values 1.85 and −2.15 are the coordinates of the centroid of A relative to the (x, y) axes. Likewise, the values −1.15 and +1.35 are the coordinates of the centroid of B relative to the (x, y) axes.

The Mohr's circle analysis will now be used to determine the values of the principal moments and their axes. The points x and y are located as shown in Fig. 15-32. Then, using xy as the diameter, draw a circle.

The distance to the center of the cirlce is $\frac{1}{2}(I_x + I_y) = 59.8$, and the radius of the circle is $\sqrt{(21)^2 + (32.3)^2} = 38.5$. Thus, the maximum I is $59.8 + 38.5 = 98.3 \text{ in}^4$ and occurs clockwise from the y axis at an angle θ' defined by $\theta' = \frac{1}{2}\tan^{-1}(32.3/21) = 28.5°$. The minimum value of I is $59.8 - 38.5 = 21.3 \text{ in}^4$ and is located clockwise from the x axis as shown in Fig. 15-33.

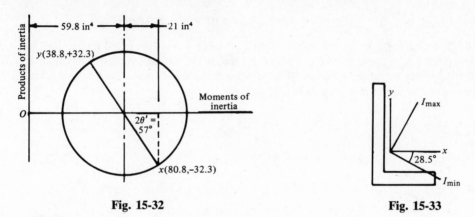

Fig. 15-32   **Fig. 15-33**

15.25. Calculate the principal moments of inertia for the L-section about its centroid. Refer to Fig. 15-34.

SOLUTION

First locate the centroid of the L-section, which is split into two areas, A and B:

$$\bar{x}'' = \frac{-(25)(125)(125/2) - (25)(100)(25/2)}{(25)(125) + (25)(100)} = -40.3 \text{ mm}$$

$$\bar{y}'' = \frac{+(25)(125)(25/2) + (25)(100)(75)}{5625} = +40.3 \text{ mm}$$

The centroid is 40.3 mm above the base and 40.3 mm to the left of its right side.

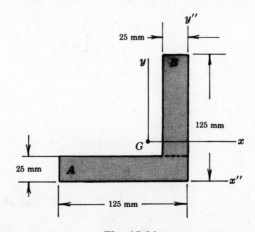

Fig. 15-34

Next determine the values I_x, I_y, and I_{xy} for the (x, y) axes through the centroid of the entire area. This will be done by transferring each value from the individual centroids of parts A and B:

$$I_x = \tfrac{1}{12}(125)(25)^3 + 125 \times 25(40.3 - 12.5)^2 + \tfrac{1}{12}(25)(100)^3 + 25 \times 100(75 - 40.3)^2 = 7.67 \times 10^6 \text{ mm}^4$$

$$I_y = \tfrac{1}{12}(25)(125)^3 + 25 \times 125(62.5 - 40.3)^2 + \tfrac{1}{12}(100)(25)^3 + 100 \times 25(40.3 - 12.5)^2 = 7.67 \times 10^6 \text{ mm}^4$$

The second computation checks the first; they should be equal for an L-section with equal legs:

$$I_{xy} = 0 + (125)(25)(+22.2)(+27.8) + 0 + (100)(25)(-27.8)(-34.7) = 4.34 \times 10^6 \text{ mm}^4$$

Using a Mohr's circle construction with point X as I_x and I_{xy}, that is, 7.67×10^6 and $+4.34 \times 10^6$, and point Y as I_y and $-I_{xy}$, that is, 7.67×10^6 and -4.34×10^6, it is evident that the radius is 4.34×10^6. See Fig. 15-35. Hence,

$$I_{\max} = 7.67 \times 10^6 + 4.34 \times 10^6 = 12.0 \times 10^6 \text{ mm}^4$$

$$I_{\min} = 7.67 \times 10^6 - 4.34 \times 10^6 = 3.33 \times 10^6 \text{ mm}^4$$

The principal axes are at $2\theta = 90°$ in Mohr's circle or $45°$ in the actual figure. The maximum value in Mohr's circle is clockwise from the X point. Hence, in the actual figure, it is $45°$ clockwise from the x axis. Similarly, the minimum value is $90°$ clockwise from the Y point, and in the actual figure is $45°$ clockwise from the y axis as shown in Fig. 15-36.

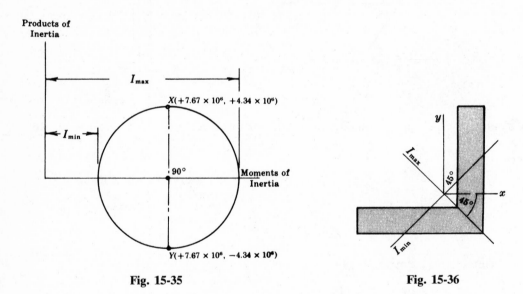

Fig. 15-35   Fig. 15-36

15.26. Derive the expression for the moment of inertia about a centroidal axis perpendicular to a bar of length l, mass m, and small cross section as shown in Fig. 15-37. Find its radius of gyration.

SOLUTION

By definition, $I_y = \int x^2 \, dm$.
But dm is the mass of a portion of the bar of length dx. Its mass dm is dx/l of the entire mass m, that is, $(dx/l)m = dm$.
Hence,

$$I_y = \int_{-l/2}^{l/2} x^2 \frac{m}{l} \, dx = \tfrac{1}{12}ml^2$$

The radius of gyration $k = \sqrt{I_y/m} = l/\sqrt{12}$.

Centroidal moments of inertia of masses for some common shapes are listed below with reference to problems in which these results are derived.

Table 15-2   Centroidal Moments of Inertia of Masses

Problem	Figure	Name	I_x	I_y	I_z
15.26		Slender bar	—	$\frac{1}{2}ml^2$	$\frac{1}{12}ml^2$
15.28		Rectangular parallelepiped	$\frac{1}{12}m(a^2 + b^2)$	$\frac{1}{12}m(a^2 + c^2)$	$\frac{1}{12}m(b^2 + c^2)$
15.29		Thin circular disk	$\frac{1}{4}mR^2$	$\frac{1}{4}mR^2$	$\frac{1}{2}mR^2$
15.32 15.33		Right circular cylinder	$\frac{1}{12}m(3R^2 + h^2)$	$\frac{1}{2}m(3R^2 + h^2)$	$\frac{1}{2}mR^2$
15.34		Sphere	$\frac{2}{5}mR^2$	$\frac{2}{5}mR^2$	$\frac{2}{5}mR^2$
15.36		Right circular cone	$\frac{3}{10}mR^2$	$\frac{3}{5}m(\frac{1}{4}R^2 + h^2)$	$\frac{3}{5}m(\frac{1}{4}R^2 + h^2)$

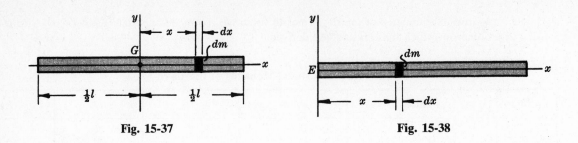

Fig. 15-37   Fig. 15-38

15.27. Derive the expression for the moment of inertia of a bar about an axis through one end and perpendicular to the bar, whose length is l. Assume that the mass is m and the cross section is small in comparison with the length. Refer to Fig. 15-38.

SOLUTION

As in Problem 15.26, write

$$I_y = \int x^2 \, dm = \int_0^l x^2 \frac{m}{l} \, dx = \tfrac{1}{3}ml^2$$

The same result could be obtained by use of the parallel axis theorem.

$$I_E = \bar{I} + m(\tfrac{1}{2}l)^2 = \tfrac{1}{12}ml^2 + \tfrac{1}{4}ml^2 = \tfrac{1}{3}ml^2$$

15.28. What is the moment of inertia about a centroidal axis perpendicular to a face of a rectangular parallelepiped (block)?

SOLUTION

As can be seen in Fig. 15-39, the moment of inertia of the block about the z axis is equal to the sum of a series of thin plates each of thickness dz, cross section b by c, and mass dm.

First determine I_z for a thin plate with cross section b by c, thickness dz, and mass dm (see Fig. 15.-40).

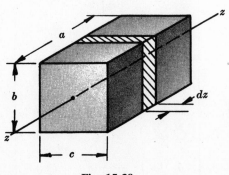   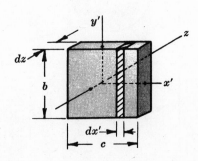

Fig. 15-39   Fig. 15-40

Since $I_z = I'_x + I'_y$, find I'_x and I'_y to obtain the result. However, I'_x is really the sum of the centroidal moments of a series of bars of mass dm' and of negligible cross section (dx' by dz) and height b. According to Problem 15.26, this may then be written

$$I'_x = \int \tfrac{1}{12} dm' \, b^2 = \tfrac{1}{12} b^2 \, dm$$

Similar reasoning yields

$$I'_y = \int \tfrac{1}{12} dm' \, c^2 = \tfrac{1}{12} c^2 \, dm$$

It follows from this that I_z for a thin plate of mass dm is equal to $I'_x + I'_y$, or $I_z = \tfrac{1}{12} dm(b^2 + c^2)$. For the entire block, it is seen that

$$I_z = \int \tfrac{1}{12} dm \, (b^2 + c^2) = \tfrac{1}{12} m(b^2 + c^2)$$

or, more rigorously, since $dm = (dz/a)m$,

$$I_z = \int_0^a \frac{1}{12} \frac{m}{a} (b^2 + c^2) \, dz = \tfrac{1}{12} m(b^2 + c^2)$$

15.29. Determine the moment of inertia about a diameter for a homogeneous thin circular disk of radius r and density δ.

(a) Consider the disk to be made of thin bars dx by t in cross section and of varying heights $2y$, as shown in Fig. 15-41.

(b) Consider the differential element as that shown in Fig. 15-42.

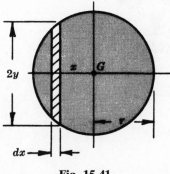

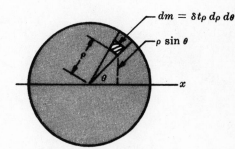

Fig. 15-41   Fig. 15-42

SOLUTION

(a) The mass of the chosen strip is $dm = 2\delta t y \, dx$. Using Problem 15.26, its moment of inertia is $\tfrac{1}{12} dm \, (2y)^2$. For the entire disk,

$$I_x = \int_{-r}^{r} \tfrac{1}{12}(2)\delta t y \, dx \, (2y)^2 = \int_{-r}^{r} \tfrac{2}{3}\delta t y^3 \, dx$$

But $y = \sqrt{r^2 - x^2}$, and substituting, we have

$$I_x = \int_{-r}^{r} \tfrac{2}{3}\delta t (\sqrt{r^2 - x^2})^3 \, dx = \tfrac{2}{3}\delta t \left[\tfrac{1}{4}x(\sqrt{r^2 - x^2})^3 - \tfrac{3}{8}r^2 x \sqrt{r^2 - x^2} + \tfrac{3}{8}r^4 \sin^{-1} \frac{x}{r} \right]_{-r}^{r}$$

This yields

$$I_x = \tfrac{2}{3}\delta t \tfrac{3}{8} r^4 [\tfrac{1}{2}\pi - (-\tfrac{1}{2}\pi)] = \tfrac{1}{4}(\pi r^2 \delta t)r^2 = \tfrac{1}{4}mr^2$$

(b)

$$I_x = \int \rho^2 \sin^2 \theta \, dm = \int_0^{2\pi} \int_0^r \rho^2 \sin^2 \theta \, \delta t \rho \, d\rho \, d\theta = \delta t \int_0^{2\pi} \int_0^r \rho^3 \sin^2 \theta \, d\rho \, d\theta$$

$$= \delta t \int_0^{2\pi} [\tfrac{1}{4}\rho^4]_0^r \sin^2 \theta \, d\theta = \delta t (\tfrac{1}{4}r^4)[\tfrac{1}{2}\theta - \tfrac{1}{4}\sin \theta]_0^{2\pi}$$

$$= \delta t (\tfrac{1}{4}r^4)(\tfrac{1}{2} \times 2\pi) = \tfrac{1}{4}\delta t \pi r^4 = \tfrac{1}{4}(\delta t \pi r^2)r^2 = \tfrac{1}{4}mr^2$$

15.30. Assume that the disk in Problem 15.29 is made of steel with a density of 490 lb/ft^3. Its thickness is 0.02 in and its diameter is 4 in. Determine the moment of inertia with respect to a diameter.

SOLUTION

The weight of the disk is

$$\pi r^2 t \delta = \frac{\pi (2)^2 (0.02) \times 490}{1728} = 0.071 \text{ lb}$$

Its mass is $0.071/32.2 = 0.0022$ slugs, and its moment of inertia is

$$\tfrac{1}{4}mr^2 = \tfrac{1}{4}(0.0022)(\tfrac{2}{12})^2 = 15 \times 10^{-6} \text{ slug-ft}^2$$

15.31. Show that the polar moment of inertia for the disk in Problem 15.29 is $\tfrac{1}{2}mr^2$. What is its radius of gyration? Refer to Fig. 15-43.

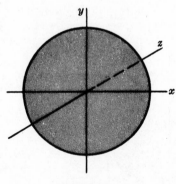

Fig. 15-43

SOLUTION

Since $I_y = I_x = \tfrac{1}{4}mr^2$, the polar moment of inertia I_z is

$$I_z = I_x + I_y = \tfrac{1}{2}mr^2$$

The radius of gyration is

$$k = \sqrt{\frac{I_z}{m}} = \sqrt{\frac{\tfrac{1}{2}mr^2}{m}} = \frac{r}{\sqrt{2}}$$

15.32. Determine the moment of inertia about a geometrical axis of a right circular cylinder of radius R and mass m. See Fig. 15-44.

SOLUTION

Consider the cylinder to be made up of a series of thin disks of height dz as shown.
For a thin disk (Problem 15.31), $I_z = \frac{1}{2}(\text{mass})R^2 = \frac{1}{2}(m\,dz/h)R^2$.
For the entire cylinder,

$$I_z = \int_0^h \frac{1}{2}\frac{m\,dz}{h}R^2 = \frac{1}{2}mR^2$$

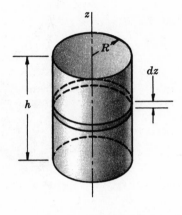

Fig. 15-44

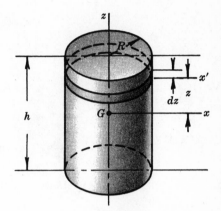

Fig. 15-45

15.33. Determine the moment of inertia for a right circular cylinder of radius R and mass m about the centroidal x axis shown in Fig. 15-45. What is the radius of gyration?

SOLUTION

As in Problem 15.32, consider the cylinder as made up of a series of thin disks of height dz and mass $m\,dz/h$. The moment of inertia of the thin disk about its x' axis parallel to the x axis is given by Problem 15.29 as $I_{x'} = \frac{1}{4}(m\,dz/h)R^2$. Transferring to the x axis by the paralelel axis theorem,

$$I_x = I_{x'} + \frac{m\,dz}{h}z^2 = \frac{1}{4}\frac{m\,dz}{h}R^2 + \frac{m\,dz}{h}z^2$$

To determine I_x for the entire cylinder, sum the I_x for all disks:

$$I_x = \frac{mR^2}{4h}\int_{-h/2}^{h/2} dz + \frac{m}{h}\int_{-h/2}^{h/2} z^2\,dz = \frac{1}{4}mR^2 + \frac{1}{12}mh^2 = \frac{1}{12}m(3R^2 + h^2)$$

$$k = \sqrt{\frac{I_x}{m}} = \sqrt{\tfrac{1}{12}(3R^2 + h^2)}$$

15.34. Determine the moment of inertia about a diameter of a sphere of mass m and radius R. What is its radius of gyration?

SOLUTION

Choose a thin disk parallel to the xz plane as shown in Fig. 15-46. Assume density δ.
The moment of inertia of a thin disk of radius x about the y axis is $\frac{1}{2}(\text{mass})x^2$. To find I_y for the entire sphere, add the individual moments just indicated, where $dm = \delta\,dV = \delta(\pi x^2\,dy)$:

$$I_y = \int_{-R}^{R} \frac{1}{2}(\delta\pi x^2\,dy)x^2 = \frac{1}{2}\pi\delta\int_{-R}^{R} x^4\,dy$$

But, from the equation of the cross section of the sphere in the xy plane (a circle), $x^2 + y^2 = R^2$.. Hence,

$$I_y = \tfrac{1}{2}\pi\delta \int_{-R}^{R} (R^2 - y^2)^2 \, dy = \tfrac{8}{15}\delta\pi R^5$$

Since the mass is $m = \tfrac{4}{3}R^3\delta$, we have

$$I_y = (\tfrac{4}{3}\pi R^3\delta)(\tfrac{2}{5}R^2) = \tfrac{2}{5}mR^2$$

The radius of gyration $k = \sqrt{I_y/m} = \sqrt{\tfrac{2}{5}}R$.

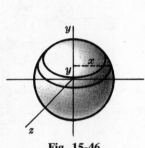

Fig. 15-46

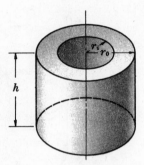

Fig. 15-47

15.35. Determine the moment of inertia of a homogeneous hollow right circular cylinder with respect to its geometric axis. Refer to Fig. 15.47

SOLUTION

For the outer cylinder, $(I_z)_o = \tfrac{1}{2}m_o r_o^2 = \tfrac{1}{2}(\pi r_o^2 h\delta)r_o^2$

For the inner cylinder, $(I_z)_i = \tfrac{1}{2}m_i r_i^2 = \tfrac{1}{2}(\pi r_i^2 h\delta)r_i^2$

For the tube, $I_z = (I_z)_o - (I_z)_i = \tfrac{1}{2}\pi\delta h(r_o^2 + r_i^2)(r_o^2 - r_i^2)$

Expanding,

$$I_z = (\tfrac{1}{2}h\pi\delta r_o^2 - \tfrac{1}{2}h\pi\delta r_i^2)(r_o^2 + r_i^2) = (\tfrac{1}{2}m_o - \tfrac{1}{2}m_i)(r_o^2 + r_i^2) = \tfrac{1}{2}m(r_o^2 + r_i^2)$$

where m refers to the mass of the tube.

15.36. Determine the moments of inertia about the x and y axes of the right circular cone of mass m and dimensions shown in Fig. 15.48. If the cone has a mass of 500 kg, a radius $R = 250$ mm, and a height $h = 500$ mm, show that $I_x = 9.38$ kg $\cdot$ m^2 and $I_y = 79.7$ kg $\cdot$ m^2.

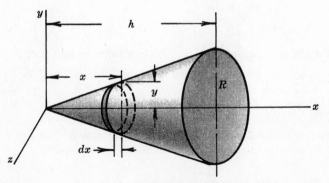

Fig. 15-48

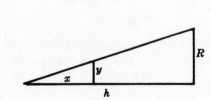

Fig. 15-49

SOLUTION

To find I_x, choose a thin lamina perpendicular to the x axis as shown in Fig. 15-48. Assume density δ.

The moment of inertia about the x axis of this lamina of radius y is $\frac{1}{2}(\text{mass})y^2$. To find I_x for the entire cone, add the individual moments just indicated, noting that the mass of the chosen lamina is $dm = \delta\,dV = \delta(\pi y^2\,dx)$. Since $y = Rx/h$ (Fig. 15-49.)

$$I_x = \int_0^h \tfrac{1}{2}\delta(\pi y^2\,dx)y^2 = \int_0^h \tfrac{1}{2}\delta\pi\left(\frac{Rx}{h}\right)^4 dx = \tfrac{1}{10}\delta\pi R^4 h$$

But the mass of the entire cone is $\frac{1}{3}\pi\delta R^2 h$. Thus, we can write

$$I_x = (\tfrac{1}{3}\pi R^2 h\delta)(\tfrac{3}{10}R^2) = \tfrac{3}{10}mR^2$$

To find I_y, which equals I_z, it is necessary to apply the transfer theorem to obtain the moment of inertia for the lamina relative to the y axis. Since $y = Rx/h$,

$$I_y = \int (\tfrac{1}{4}\,dm\,y^2 + dm\,x^2) = \int_0^h (\delta\pi y^2\,dx)(\tfrac{1}{4}y^2 + x^2)$$

$$= \int_0^h \tfrac{1}{4}\pi\delta\left(\frac{R^4}{h^4}\right)x^4\,dx + \int_0^h \pi\delta\left(\frac{R^2}{h^2}\right)x^4\,dx = \tfrac{1}{20}\pi\delta R^4 h + \tfrac{1}{5}\pi\delta R^2 h^3$$

Using $m = \frac{1}{3}\pi\delta R^2 h$, this expression becomes $I_y = \frac{3}{20}mR^2 + \frac{3}{5}mh^2 = \frac{3}{5}m(\frac{1}{4}R^2 + h^2)$.

For the numerical part,

$$I_x = 0.3(500)(0.25)^2 = 9.38\,\text{kg}\cdot\text{m}^2 \qquad \text{and} \qquad I_y = \tfrac{3}{5}(500)[\tfrac{1}{4}(0.25)^2 + (0.5)^2] = 79.7\,\text{kg}\cdot\text{m}^2$$

15.37. Compute the moment of inertia of the cast-iron flywheel shown in Fig. 15-50. Cast iron weighs 450 lb/ft³.

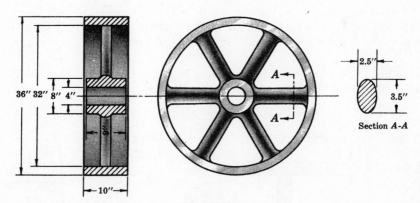

Fig. 15-50

SOLUTION

In analyzing the problem, consider the hub and the rim as hollow cylinders and the spokes as slender rods. First determine the weights of the components:

$$W_{\text{hub}} = (\pi r_o^2 - \pi r_i^2)h\delta = \pi[(4/12)^2 - (2/12)^2](9/12)(450) = 88.2\,\text{lb}$$

$$W_{\text{rim}} = (\pi r_o^2 - \pi r_i^2)h\delta = \pi[(18/12)^2 - (16/12)^2](10/12)(450) = 556\,\text{lb}$$

For one spoke, elliptical in cross section,

$$W_{\text{spoke}} = \pi ab l\delta = \pi(2.5/24)(3.5/24)(12/12)(450) = 21.5\,\text{lb}$$

Next determine I for each component about the axis of rotation. For the spokes, this will entail a transfer from the parallel centroidal axis.

$$I_{\text{hub}} = \tfrac{1}{2}m(r_o^2 + r_i^2) = \tfrac{1}{2}(88.2/32.2)[(4/12)^2 + (2/12)^2] = 0.19 \text{ lb-s}^2\text{-ft}$$

$$I_{\text{rim}} = \tfrac{1}{2}m(r_o^2 + r_i^2) = \tfrac{1}{2}(556/32.2[(18/12)^2 + (16/12)^2] = 34.8 \text{ lb-s}^2\text{-ft}$$

$$I_{\text{spokes}} = 6(\tfrac{1}{12}ml^2 + md^2) = 6[\tfrac{1}{12}(21.5/32.2)(12/12)^2 + (21.5/32.2)(10/12)^2] = 3.11 \text{ lb-s}^2\text{-ft}$$

$$I_{\text{wheel}} = (0.19 + 34.8 + 3.11) \text{ lb-s}^2\text{-ft} = 38.1 \text{ lb-s}^2\text{-ft}$$

Supplementary Problems*

15.38. A rectangle has a base of 2 in and a height of 6 in. Calculate its moment of inertia about an axis through the center of gravity and parallel to the base.   *Ans.* 36 in^4

15.39. Determine the moment of inertia of an isosceles triangle with base of 150 mm and sides of 125 mm about its base.   *Ans.* $12.5 \times 10^6 \text{ mm}^4$

15.40. Determine the moment of inertia of a circle of radius 2 ft about a diameter.   *Ans.* 12.6 ft^4

15.41. Find the moment of inertia with respect to the y axis of the plane area between the parabola $y = 9 - x^2$ and the x axis.   *Ans.* $\frac{324}{5}$

15.42. Determine the moment of inertia with respect to each coordinate axis of the area between the curve $y = \cos x$ from $x = 0$ to $x = \tfrac{1}{2}\pi$ and the x axis.   *Ans.* $I_x = \tfrac{2}{9}$, $I_y = \tfrac{1}{4}\pi^2 - 2$.

15.43. Find the moment of inertia with respect to each coordinate axis of the area between the curve $y = \sin x$ from $x = 0$ to $x = \pi$ and the x axis.   *Ans.* $I_x = \tfrac{4}{9}$, $I_y = \pi^2 - 4$

15.44. Refer to Fig. 15-51. Determine the moment of inertia of the composite figure about an axis through its center of gravity and parallel to the base. What is the radius at gyration?   *Ans.* 46.3 in^4, 2.24 in

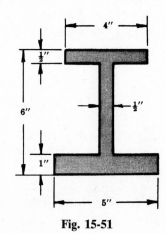

Fig. 15-51

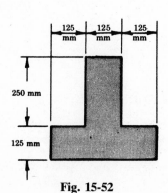

Fig. 15-52

---

* Table 15.1 may be helpful in the solution of numerical problems.

15.45. Referring to Fig. 15-52, determine the moment of inertia of the composite figure about a horizontal centroidal axis.　　*Ans.*　883×10^6 mm⁴

15.46. Refer to Fig. 15-53. Compute the moment of inertia of the composite figure about a centroidal axis parallel to the 250-mm side. What is the radius at gyration?　　*Ans.*　1.35×10^6 mm⁴, 21.1 mm

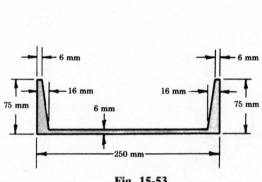

Fig. 15-53

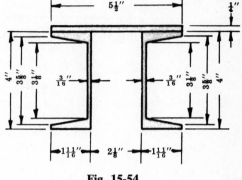

Fig. 15-54

15.47. Refer to Fig. 15-54. Compute the moment of inertia of the composite figure about a horizontal centroidal axis.　　*Ans.*　12.8 in⁴

15.48. What is the polar moment of inertia of a circle 80 mm in diameter about an axis through its center of gravity and perpendicular to its plane?　　*Ans.*　4.02×10^6 mm⁴

15.49. Calculate the product of inertia of a rectangle with a base of 100 mm and a height of 80 mm about two adjacent sides.　　*Ans.*　16×10^6 mm⁴

15.50. Calculate the product of inertia of a rectangle having a base of 150 mm and height of 100 mm about two adjacent sides.　　*Ans.*　56.3×10^6 mm⁴

Mathcad

15.51. Determine the values of I_x, I_y, and I_{xy} for the area bounded by the x axis, the line $x = a$, and the curve $y = (b/a^n)x^n$.　　*Ans.*　$I_x = ab^3/3(3n + 1)$, $I_y = a^3b/(n + 3)$, $I_{xy} = a^2b^2/2(2n + 2)$

15.52. In the preceding problem the area becomes triangular when $n = 1$. Check the values by direct integration.　　*Ans.*　$I_x = \frac{1}{12}ab^3$, $I_y = \frac{1}{4}a^3b$, $I_{xy} = \frac{1}{8}a^2b^2$

15.53. Determine the values of I_x, I_y, and I_{xy} for the area bounded by the y axis, the line $y = b$, and the curve $y = (b/a^n)x^n$.　　*Ans.*　$I_x = nab^3/(3n + 1)$, $I_y = na^3b/3(n + 3)$, $I_{xy} = na^2b^2/4(n + 1)$

15.54. The figures in problems 15.51 and 15.53 form a rectangle when added together. Add the values of I_x and check with Problem 15.1 to see if you obtain the moment of inertia of a rectangle about its base. Repeat the process for I_{xy} and check your result with Problem 15.16.

15.55. Determine I_y and I_{xy} for Fig. 15-52, where the x, y axes pass through the centroid.
Ans.　$I_y = 590 \times 10^6$ mm⁴, $I_{xy} = 0$

15.56. Determine I_x, I_y, and I_{xy} for the x, y axes that pass through the centroid of the unequal angle shown in Fig. 15-55.　　*Ans.*　$I_x = 63.5 \times 10^6$ mm⁴, $I_y = 114 \times 10^6$ mm⁴, $I_{xy} = -46.9 \times 10^6$ mm⁴

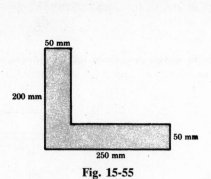

Fig. 15-55

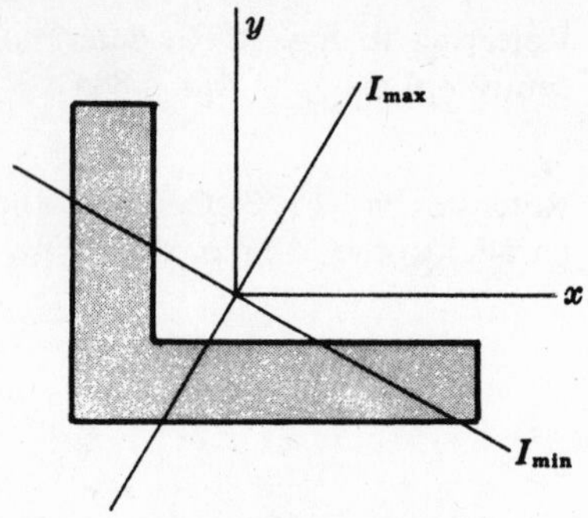

Mathcad

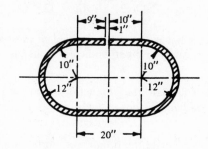

Fig. 15-56

15.57. In the preceding problem use Mohr's circle to locate the principal axes and determine the principal moments of inertia (see Fig. 15-56).
Ans. $I_{max} = 142 \times 10^6$ mm^4 at 30.9° clockwise from y axis, $I_{min} = 35.5 \times 10^6$ mm^4 at 30.9° clockwise from x axis

15.58. Locate the principal axes through the centroid of the area shown in Fig. 15-57. Next determine the principal moments of inertia for those axes.
Ans. $I_{max} = 373$ in^4 at 28.8° clockwise from y axis, $I_{min} = 45.0$ in^4 at 28.8° clockwise from x axis

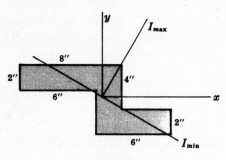

Fig. 15-57

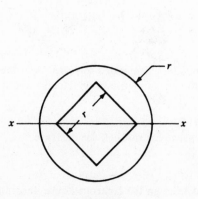

Fig. 15-58

15.59. Find $\bar{I}_x$ for the shaded area in Fig. 15-58.   *Ans.*   17,900 in^4

15.60. Determine $\bar{I}_x$ for the area formed by subtracting the square of side r from the circle of radius r as shown in Fig. 15-59.   *Ans.* $\bar{I}_x = 0.702r^4$

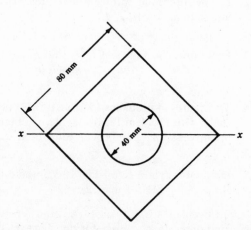

Fig. 15-59

Fig. 15-60

15.61. Determine $\bar{I}_x$ for the area formed by subtracting the circle of radius 20 mm from the square of side 80 mm as shown in Fig. 15-60.   *Ans.* $\bar{I}_x = 3.29 \times 10^6 \text{ mm}^4$

15.62. Determine the moments of inertia of the thin elliptical disk of mass m shown in Fig. 15-61. Refer to Problem 15.29.   *Ans.* $I_x = \frac{1}{4}mb^2$, $I_y = \frac{1}{4}ma^2$, $I_z = \frac{1}{4}m(a^2 + b^2)$

15.63. Determine the moments of inertia of the ellipsoid of revolution with mass m shown in Fig. 15-62.
Ans. $I_x = \frac{2}{5}mb^2$, $I_y = I_z = \frac{1}{5}m(a^2 + b^2)$

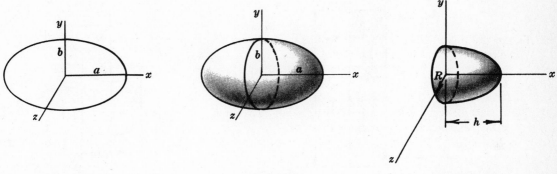

Fig. 15-61   Fig. 15-62   Fig. 15-63

15.64. Determine the moments of inertia of the paraboloid of revolution with mass m shown in Fig. 15-63. The equation in the xy plane is $y^2 = -(R^2/h^2)x^2 + R^2$.   *Ans.* $I_x = \frac{2}{3}mR^2$, $I_y = I_z = \frac{1}{3}m(R^2 + h^2)$

15.65. Show that the moment of inertia about a diameter for a thin hollow sphere of mass m is $\frac{2}{3}mR^2$.

15.66. Determine the moment of inertia about a diameter for a hollow sphere of mass m with inner and outer radii R_i and R_o respectively.   *Ans.* $I = \frac{2}{5}m(R_o^5 - R_i^5)/(R_o^3 - R_i^3)$

15.67. Show that the moment of inertia about a centroidal axis parallel to a side for a cube of mass m is $I = \frac{1}{6}ma^2$, where a is the length of a side.

15.68. Find the moment of inertia of a 4-ft-long, $\frac{1}{2}$-in-diameter steel rod about an axis through one end and perpendicular to the rod. The steel weighs 490 lb/ft³.   *Ans.*   0.44 slug-ft² or lb-s²-ft

15.69. Determine the moment of inertia of a 10-ft-long steel pipe with a 3.50-in outside diameter and a 2.89-in inside diameter with respect to its longitudinal axis. The steel weighs 490 lb/ft³.
Ans.   0.058 slug-ft²

15.70. Find the moment of inertia of a brass cylindrical shaft 75 mm in diameter and 3 m long with respect to its geometric axis of rotation. Use a density of 8500 kg/m³.   *Ans.*   0.079 kg · m²

15.71. In Problem 15.70, what is the moment of inertia of the mass about an axis that is (*a*) centroidal and perpendicular to the geometric axis and (*b*) through the end and perpendicular to the geometric axis?
Ans.   (*a*) 84.5 kg · m², (*b*) 338 kg · m²

15.72. Calculate the moment of inertia of a rectangular prism that is 6 in high, 4 in wide, and 10 in long, with respect to its longitudinal centroidal axis. Use a weight of 40 lb/ft³. *Ans.* 0.005 slug-ft²

15.73. Find the moment of inertia of an aluminum sphere 200 mm in diameter with respect to a centroidal axis. Aluminum has a density of 2560 kg/m³. *Ans.* 0.043 kg · m²

15.74. Determine the moment of inertia of the flywheel shown in Fig. 15-64 (solid web) with respect to its axis of rotation. Cast iron weighs 4500 lb/ft³. *Ans.* 2.12 slug-ft²

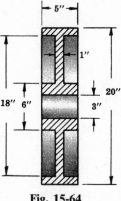

Fig. 15-64

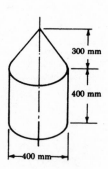

Fig. 15-65

15.75. As shown in Fig. 15-65, a brass cone is mounted on the top of an aluminum cylinder. The density of brass is 8500 kg/m³ and that of aluminum is 2560 kg/m³. Determine the moment of inertia for the system about the vertical geometric axis. *Ans.* $\bar{I}_y = 3.86$ kg · m²

15.76. A steel shaft and a steel disk are joined as shown in Fig. 15-66. The density of steel is 7850 kg/m³. Determine the moment of inertia of the system about the y axis through the end.
Ans. $I_y = 5.38 \times 10^{-4}$ kg · m²

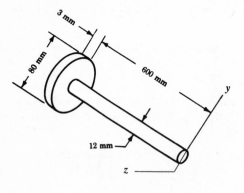

Fig. 15-66

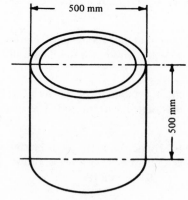

Fig. 15-67

15.77. Fig. 15-67 shows in schematic form an open steel bucket made with a wall and base that are 6 mm thick. The bucket is half full of concrete mix. The density of steel is 7850 kg/m³ and that of the concrete mix is 2400 kg/m³. Determine the total moment of inertia about a vertical centroidal axis.
Ans. 5.79 kg · m²

15.78. A 10-kg homogeneous sphere is 1 m in diameter. Two slender bars are attached diametrically opposite to one another in a horizontal line. Each bar has a mass of 2 kg and is 1.5 m long. What is the moment of inertia of the three masses about a vertical centroidal axis?   *Ans.* $I = 8 \text{ kg} \cdot \text{m}^2$

15.79. A dumbell-like weight consists of two 4-in-diameter solid spheres attached to each end of a 36-in slender rod of 1-in diameter. The spheres and rod are of copper with a density of 560 lb/ft³. What is the mass moment of inertia about an axis perpendicular to the rod at its midpoint?
Ans. $I = 2.17 \text{ slug-ft}^2$

15.80. Figure 15-68 shows a simulated communications satellite. The solid central hub has four identical arms each weighing 2 lb, attached 90° apart, as shown. The hub has a weight of 342 lb. What is the mass moment of inertia about the y-centroidal axis?   *Ans.* $I_y = 2.84 \text{ slug-ft}^2$

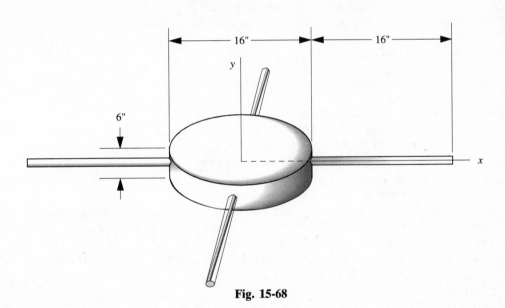

Fig. 15-68

15.81. In Problem 15.80, what is the mass moment of inertia with respect to the x-centroidal axis?
Ans. $I_x = 1.64 \text{ slug-ft}^2$

Chapter 16

Dynamics of a Rigid Body in Plane Motion

16.1 VECTOR EQUATIONS OF PLANE MOTION

In Section 14.1, plane motion of a rigid body was defined as that motion in which every point in the body remains at a constant distance from a fixed plane. In the case of kinetics of a rigid body, there is a further condition, namely the body shall have a plane of symmetry. This is more restrictive than necessary, but it does simplify the moment equation.

The vector equations of plane motion can be written

$$\sum \mathbf{F} = m\bar{\mathbf{a}}$$

$$\sum \mathbf{M}_O = I_O \alpha \mathbf{k} + m\mathbf{r}_{G/O} \times \mathbf{a}_O = (I_O \alpha + m\bar{x}a_{Oy} - m\bar{y}a_{Ox})\mathbf{k}$$

where
$\sum \mathbf{F}$ = resultant of the external forces acting on the body
$\sum \mathbf{M}_O$ = resultant of the external moments acting on the body
m = mass of the body
$\bar{\mathbf{a}}$ = acceleration of the mass center of the body
$\mathbf{a}_O$ = acceleration of reference point O
α = angular acceleration of the body
I_O = moment of inertia of the body relative to the reference point O
$\bar{x}, \bar{y}$ = coordinates of the mass center relative to the reference point O
$\mathbf{r}_{G/O}$ = position vector of the mass center relative to the reference point O
a_{Ox}, a_{Oy} = magnitude of the components of the acceleration of the reference point O along the x and y axes

A right-hand set of coordinate axes is assumed in the above vector equations. This means that if the x and y axes are chosen positive to the right and up then counterclockwise rotation must be chosen positive to be consistent with the right-hand system. Problem 16.1 illustrates this point.

16.2 SCALAR EQUATIONS OF PLANE MOTION

The moment equation in the preceding section may be simplified by the proper choice of the reference point O. One such choice is to use the mass center as O. Then $\bar{x}$ and $\bar{y}$ are zero.

With this choice, the scalar equations of plane motion are

$$\sum F_x = m\bar{a}_x \qquad \sum F_y = m\bar{a}_y \qquad \sum \bar{M} = \bar{I}\alpha$$

where
$\sum F_x, \sum F_y$ = algebraic sums of the magnitudes of the components of the external forces along the x and y axes, respectively
m = mass of the body
$\bar{a}_x, \bar{a}_y$ = components of the linear acceleration of the mass center in x and y directions, respectively
$\sum \bar{M}$ = algebraic sum of the moments of the external forces about the mass center
$\bar{I}$ = moment of inertia of the body about the mass center
α = magnitude of the angular acceleration of the body

Note that the moment equation can also be written as $\sum M_O = I_O \alpha$, provided that the point O is a point whose acceleration either is zero or is directed through the mass center of the body.

Note also that the component of a force acting in the same direction as that chosen for the mass center acceleration is assigned a positive sign (if opposite in sense, it should be assigned a negative sign).

Similarly, moments of forces are considered positive if they have the same sense as that assigned to the angular acceleration α.

It will be seen later that translation and rotation are special cases of Plane Motion.

16.3 PICTORIAL REPRESENTATION OF THE EQUATIONS

A pictorial representation of the equations can be used to emphasize that plane motion is a combination of translation and rotation. The drawing in Fig. 16.1 shows an object with all external forces acting on it equated to the object with the effective forces $m\bar{a}$ and with the moment $\bar{I}\alpha$ of the effective forces. It is apparent that $\sum F_x = m\bar{a}_x$, $\sum F_y = m\bar{a}_y$, and $\sum \bar{M} = \bar{I}\alpha$. Keep in mind that moments are taken relative to the mass center G.

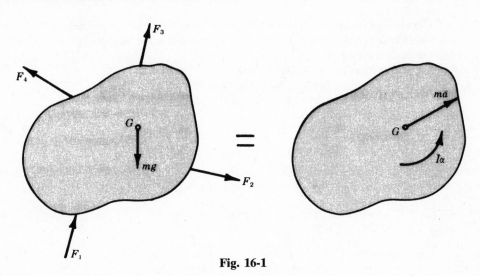

Fig. 16-1

16.4 TRANSLATION OF A RIGID BODY

Translation of a rigid body is defined as the motion in which all particles of the body have the same acceleration. Then, the scalar equations of Section 16.2 become

$$\sum F_x = ma_x \qquad \sum F_y = ma_y \qquad \sum \bar{M} = 0$$

where $\sum F_x, \sum F_y$ = algebraic sums of the components of the external forces in the x and y directions, respectively

   m = mass of the body

   a_x, a_y = components of the acceleration of the body in the x and y directions, respectively

   $\sum \bar{M}$ = sum of the moments of the external forces about the mass center of the body

16.5 ROTATION OF A RIGID BODY

Rotation of a rigid body about a fixed axis is defined as the motion in which all particles along the fixed axis are at rest and all other particles of the body move on circular paths with centers along the axis of rotation.

(a) If a body has a plane of symmetry and rotates about a fixed axis perpendicular to this plane, then, from Section 16.2, the scalar equations of motion of the body under the action of an unbalanced force system are

$$\sum F_n = m\bar{r}\omega^2$$
$$\sum F_t = m\bar{r}\alpha$$
$$\sum M_O = I_O\alpha$$

where $\sum F_n$ = algebraic sum of the components of all external forces (which are the applied forces F_1, F_2, F_3, etc., the gravitational force on the body, and the reaction R of the axis on the body) along the n axis, which is the line drawn between the center of rotation O and the mass center G; note that the positive sense is from G toward O because $\bar{a}_n = \bar{r}\omega^2$ has that sense

$\sum F_t$ = algebraic sum of the components of the external forces along the t axis, which is perpendicular to the n axis at O; note that the positive sense along this axis agrees with that of $\bar{a}_t = \bar{r}\alpha$

$\sum M_O$ = algebraic sum of the moments of the external forces about the axis of rotation through O; note that positive sense agrees with the assumed sense of the angular acceleration α

m = mass of the body

G = center of mass of the body

$\bar{r}$ = distance from the center of rotation O to the mass center G

I_O = moment of inertia of the body about the axis of rotation

ω = angular speed of the body

α = magnitude of the angular acceleration of the body

This type of rotation is called *non-centroidal rotation*.

(b) If the rotation is about a fixed axis through G (i.e., if G and O coincide) then $F = 0$ and moments are taken about the mass center. The equations of motion, again from Section 16.2, become

$$\sum F_x = 0 \qquad \sum F_y = 0 \qquad \sum \bar{M} = \bar{I}\alpha$$

where $\sum F_x$ = algebraic sum of the components of the external forces along any axis chosen as the x axis

$\sum F_y$ = algebraic sum of the components of the external forces along the y axis

$\sum \bar{M}$ = algebraic sum of the moments of the external forces about the axis of rotation through the mass center G (axis of symmetry)

$\bar{I}$ = moment of inertia of the body about the axis of rotation through the mass center G

α = magnitude of the angular acceleration of the body

This type of rotation is called *centroidal rotation*.

16.6 CENTER OF PERCUSSION

The center of percussion is that point P on the n axis in Fig. 16-2 through which the resultant of the effective forces act. It is at a distance q from the center of rotation O. The distance q is given by

$$q = k_O^2/\bar{r}$$

where k_O^2 = square of the radius of gyration of the body with respect to the axis of rotation through O; note that $k_O^2 = I_O/m$, I_O being the mass moment of inertia of the body about O and m its total mass

$\bar{r}$ = distance from the center of rotation O to the mass center G

16.7 THE INERTIA-FORCE METHOD FOR RIGID BODIES

In Chapter 13, D'Alembert's Principle was applied to *particles* in motion. Similarly, D'Alembert's Principle can be applied to *rigid bodies* in motion. In the case of rigid body motion, not only must a force equal and opposite to $m\bar{a}$ be applied to the body at the center of mass but also

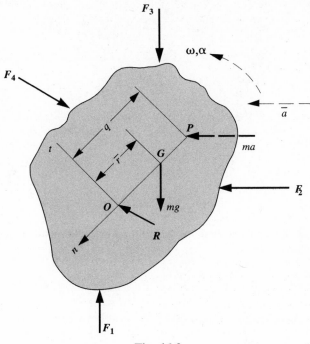

Fig. 16-2

a couple equal and opposite to $\bar{I}\alpha$ must be applied to the free body. In Fig. 16-3, then, the reversed effective force and the reversed effective couple will balance out the external forces and couples. Hence,

$$\sum \mathbf{F} - m\bar{\mathbf{a}} = 0$$

$$\sum \mathbf{M} - I\alpha\mathbf{k} = 0$$

The advantage of the inertia-force method, based on D'Alembert's Principle, is that it converts a dynamics problem into an equivalent problem in equilibrium. This allows moments to be conveniently taken about any axis and not only centroidal axes.

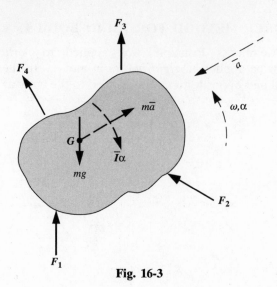

Fig. 16-3

Solved Problems

Vectors are designated in the diagrams by their magnitudes when the directions are evident by inspection.

General Plane Motion

16.1. A ring of negligible mass and radius r has attached to it three small masses as shown in Fig. 16-4(a). What is the angular acceleration immediately after the ring is placed on the horizontal plane? Assume no angular velocity and that sufficient friction exists so that there is no slipping.

SOLUTION

The free-body diagram in Fig. 16-4(b) shows the plane reactions (a normal force N and a frictional component F). A nonrotating set of axes through the geometric center O of the ring is shown. The total gravitational force, which acts through the mass center G of the three masses in Fig. 16-4(a), is $4mg$. Note that the coordinates of G are

$$\bar{x} = \frac{mg(-r) + mg(0) + 2mg(0.866r)}{4mg} = 0.183r$$

$$\bar{y} = \frac{m(0) + mg(-r) + 2mg(0.5r)}{4mg} = 0$$

No slipping is assumed, and thus the geometric center O has acceleration components.

$$a_{Ox} = r\alpha \qquad \text{and} \qquad a_{Oy} = 0$$

To be consistent with the sense of a_O, shown acting to the right in Fig. 16-4(b), α must be shown

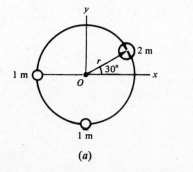

(a)

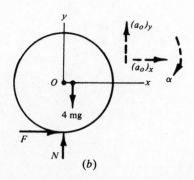

(b)

Fig. 16-4

acting clockwise. The moment equation is given in vector form in Section 16.1. In scalar form, this becomes

$$\sum M_O = I_O \alpha + m\bar{x}(a_O) - m\bar{y}(a_O)$$

This equation is based on a right-hand set of axes with α positive in the counterclockwise direction. In using the equation, we must substitute $-\alpha$ for α because we assumed its sense to be clockwise.

Using the free-body diagram, the equations of motion are

$$\sum F_x = m\bar{a}_{Ox} \quad \text{or} \quad F = 4mr\alpha$$
$$\sum F_y = m\bar{a}_{Oy} \quad \text{or} \quad N - 4mg = 0$$

The moment equation becomes

$$+Fr - 4mg(+0.183r) = 4mr^2(-\alpha) + 4m(+0.183r)(0) - 4m(0)r\alpha$$

Using $F = 4mr\alpha$, the moment equation becomes

$$(4mr\alpha)r - 0.732mgr = -4mr^2\alpha$$

Thus $\alpha = (0.0915g/r) \text{ rad/s}^2$. If $g = 32.2 \text{ ft/s}^2$ then $\alpha = (2.95/r) \text{ rad/s}^2$.

16.2. A 600-lb wheel with a 30-in diameter rolls without slipping down a plane inclined at an angle of 25° with the horizontal. Determine the friction force F and the acceleration of the mass center.

SOLUTION

Figure 16-5 shows the force system acting on the wheel.

Considerable difficulty is usually experienced in indicating the direction of the friction force F. (F may have any value between $-\mu N$ and μN). In this case, the friction must act up the plane; otherwise, the wheel would slip down the plane. Also, friction is the only force that has a moment about the mass center, and therefore is the force causing the angular acceleration ($\sum \bar{M} = \bar{I}\alpha$).

Choose the x axis parallel to the plane, with the positive direction down. The y axis is positive up. The mass $m = 600/32.2 = 18.6 \text{ lb-s}^2/\text{ft}$ or slugs. The moment of inertia is

$$\bar{I} = \tfrac{1}{2}mr^2 = \tfrac{1}{2}(18.6 \text{ lb-s}^2/\text{ft})(15/12 \text{ ft})^2 = 14.5 \text{ lb-s}^2\text{-ft or slug-ft}^2$$

The above units have been commonly used in engineering textbooks.

The equations of motion are (1) $\sum F_x = m\bar{a}_x$, (2) $\sum F_y = m\bar{a}_y$, (3) $\sum \bar{M} = \bar{I}\alpha$. Substituting values, these become

$$600 \sin 25° - F = 18.6\bar{a}_x \tag{1'}$$
$$N - 600 \cos 25° = 18.6\bar{a}_y = 0 \tag{2'}$$
$$F(15/12) = 14.5\alpha \tag{3'}$$

These three equations involve four unknown quantities; hence, another equation in the unknowns is needed. Since this is an example of rolling, we shall use $\bar{a}_x = r\alpha = (15/12)\alpha$.

Substitute $\alpha = \tfrac{4}{5}\bar{a}_x$ into equation (3') to obtain $F = 9.27\bar{a}_x$.

Substitute $F = 9.27\bar{a}_x$ into equations (1') to obtain $\bar{a}_x = 9.19 \text{ ft/s}^2$. Then $F = 84.3 \text{ lb}$.

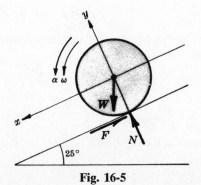

Fig. 16-5

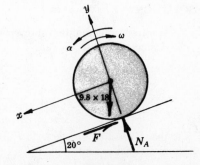

Fig. 16-6

16.3. The center of a wheel having a mass of 18 kg and 600 mm in diameter is moving at a certain instant with a speed of 3 m/s up a plane inclined 20° with the horizontal (see Fig. 16-6). How long will it take to reach the highest point of its travel?

SOLUTION

The free-body diagram shows the friction F acting up the plane. Here, as in Problem 16-2, friction is the only force with a moment about the mass center. It therefore causes the angular acceleration, which must be counterclockwise. Note that the angular velocity ω is clockwise until the wheel stops at its highest point. Of course, on the way down the α and ω will be in the same direction—counterclockwise.

The equations of motion are (1) $\sum F_x = m\bar{a}_x$, (2) $\sum F_y = m\bar{a}_y$, and (3) $\sum \bar{M} = \bar{I}\alpha$.

Substitute values to obtain the following equations (assume the wheel is a disk, so that $\bar{I} = \frac{1}{2}mr^2$):

$$9.8 \times 18 \sin 20° - F = 18\bar{a}_x \qquad (1')$$

$$N_A - 9.8 \times 18 \cos 20° = 18\bar{a}_y = 0 \qquad (2')$$

$$F \times 0.3 = \tfrac{1}{2}18(0.3)^2\alpha \qquad (3')$$

As in Problem 16.1, $\bar{a}_x = r\alpha = 0.3\alpha$.

From $(3')$, $F = 9\bar{a}_x$. Substitute into $(1')$ to obtain $\bar{a}_x = 2.23$ m/s². The value of $\bar{a}_x$ is positive, i.e., down the plane.

To determine the time to come to rest, i.e., to reach its highest point after the initial speed of 3 m/s was observed, apply the kinematics scalar equation $v = v_0 + at$.

Keep in mind that the downward direction is positive. The final speed v is zero, the initial speed v_0 is up the plane and is therefore -3 m/s. The acceleration $\bar{a}_x$ is down the plane and is therefore $+2.23$ m/s².

Then $v = v_0 + at$, $0 = -3 + 2.23t$, and $t = 1.35$ s.

16.4. Study the motion of a homogeneous cylinder of radius R and mass m that is acted upon by a horizontal force P applied at various positions along a vertical centerline as shown in Fig. 16-7. Assume movement upon a horizontal plane.

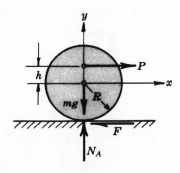

Fig. 16-7

SOLUTION

The free-body diagram shows the force P applied at a distance h above the center.

Assume F acts to the left. The equations of motion are

$$\sum F_x = P - F = m\bar{a}_x \qquad (1)$$

$$\sum F_y = N_A - mg = 0 \qquad (2)$$

$$\sum \bar{M} = P \times h + F \times R = \tfrac{1}{2}mR^2\alpha \qquad (3)$$

Note that P must be greater than F for motion to ensue to the right. Is it possible for motion to occur with friction F equal to zero?

Substitute $\bar{a}_x = R\alpha$ into (3) to obtain

$$P \times h + F \times R = \tfrac{1}{2}mR\bar{a}_x \tag{3'}$$

Divide equation (3') by $\tfrac{1}{2}R$ to obtain

$$\frac{2Ph}{R} + 2F = m\bar{a}_x \tag{3''}$$

Equating the left-hand sides of (3'') and (1), $2Ph/R + 2F = P - F$ or $3F = P(1 - 2h/R)$.

It is evident that F will be zero if the term $(1 - 2h/R)$ is zero, i.e., if $h = \tfrac{1}{2}R$. Thus, if the force P is applied at a point one-half the radius above the center, the frictional force F is zero.

Also note that if $h = R$, the equation becomes $3F = P(1 - 2R/R) = -P$. The friction has now reversed itself and acts to the right. With $F = -\tfrac{1}{3}P$, equation (3) becomes, for the case where $h = R$,

$$PR - \tfrac{1}{3}PR = \tfrac{1}{2}mR^2\alpha \qquad \text{or} \qquad \tfrac{2}{3}P = \tfrac{1}{2}mR\alpha$$

This indicates that α is positive or the cylinder rolls to the right.

Next assume that h becomes smaller until finally P is applied through the mass center, where h is zero. Under these conditions, $3F = P[1 - (2 \times 0)/R] = P$.

Naturally, at any time in the previous discussion, if P becomes too large, F will tend to increase. As soon as the maximum possible value of the friction is exceeded, the cylinder will slip. A new assumption must then be made, namely, that the friction F is now equal to the product of the coefficient of friction and the normal force N_A. The equations of motion are now

$$\sum F_x = P - \mu N_A = m\bar{a}_x \tag{4}$$

$$\sum F_y = N_A - mg = 0 \tag{5}$$

$$\sum \bar{M} = Ph + \mu N_A R = \tfrac{1}{2}mR^2\alpha \tag{6}$$

These equations indicate the existence of both sliding (linear acceleration $\bar{a}_x$) and rolling (angular acceleration α). Refer to Problems 16.6 and 16.22.

16.5. A homogeneous sphere with a mass of 20 kg has a peripheral slot cut in it as shown in Fig. 16-8. A force of 40 N is exerted on a string wrapped in the slot. If the sphere rolls without slipping, determine the acceleration of its mass center and the frictional force F. Neglect the effect of the slot.

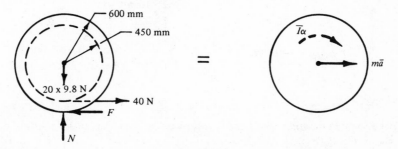

Fig. 16-8

SOLUTION

Assume rolling to the right. The angular acceleration will be clockwise, and the mass center acceleration will be to the right.

Assuming that friction acts to the left, the equations of motion are

$$\sum F = m\bar{a} \qquad \text{or} \qquad 40 - F = 20\bar{a} \qquad\qquad (1)$$

$$\sum M = I\alpha \qquad \text{or} \qquad F \times 0.6 - 40 \times 0.45 = \tfrac{2}{5}(20)(0.6)^2\alpha \qquad (2)$$

Since $\bar{a} = 0.6\alpha$, the second equation may be rewritten

$$F \times 0.6 - 40 \times 0.45 = 4.8\bar{a}$$

Combine this with equation (1) to get

$$\bar{a} = 0.37 \text{ m/s}^2 \qquad \text{and} \qquad F = 7.14 \text{ N} \qquad \text{(to the left as assumed)}$$

16.6. The coefficient of friction between a plane and a homogeneous sphere of weight 16.1 lb is 0.10. Determine the angular acceleration of the sphere and the linear acceleration of its mass center.

SOLUTION

Draw a free-body diagram showing the frictional force F as unknown. See Fig. 16-9. At the beginning, it is not known whether or not the sphere will slip; therefore determine F to see whether or not it is greater than μN, that is, $0.10N$. If it is greater, this means that not enough friction is available and both rolling and sliding occur.

The equations of motion are

$$\sum F = m\bar{a} \qquad \text{or} \qquad 16.1 \sin 30° - F = \frac{16.1}{32.2}\bar{a} \qquad\qquad (1)$$

$$\sum \bar{M} = \bar{I}\alpha \qquad \text{or} \qquad F \times \frac{1}{2} = \frac{2}{5}\left(\frac{16.1}{32.2}\right)\left(\frac{1}{2}\right)^2\alpha \qquad (2)$$

Note that α is assumed counterclockwise; hence, $\bar{a}$ must be positive down the plane. Also, $\bar{a} = r\alpha = \frac{1}{2}\alpha$.

From (1), $8.05 - F = \frac{1}{2}\bar{a}$; from (2), $F = \frac{1}{5}\bar{a}$. Hence, $\bar{a} = 11.5 \text{ ft/s}^2$ and $F = 2.30 \text{ lb}$.

By inspection $N = 16.1 \cos 30° = 13.9 \text{ lb}$. Maximum friction available is $\mu N = 0.10(13.9) = 1.39 \text{ lb}$. But the F to prevent sliding (or cause rolling) is 2.30 lb. This means that the problem must be reworked using the maximum friction available, namely 1.39 lb instead of F. The relation $\bar{a} = r\alpha$ will no longer hold. Incidentally, we are also assuming that the coefficients of static and kinetic friction are equal.

Draw a new free-body diagram (see Fig. 16-10). The equations of motion are

$$16.1 \sin 30° - 1.39 = \left(\frac{16.1}{32.2}\right)\bar{a} \qquad\qquad (3)$$

$$1.39 \times \frac{1}{2} = \frac{2}{5}\left(\frac{16.1}{32.2}\right)\left(\frac{1}{2}\right)^2\alpha \qquad\qquad (4)$$

Hence, $\bar{a} = 13.3 \text{ ft/s}^2$ and $\alpha = 13.9 \text{ rad/s}^2$. The sphere will both roll and slip.

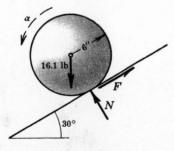

Fig. 16-9

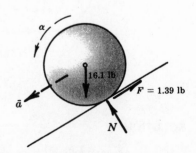

Fig. 16-10

16.7. The cord passes over a massless and frictionless pulley, as shown in Fig. 16-11(a), carrying a mass M_1 at one end and wrapped around a cylinder of mass M_2 that rolls on a horizontal plane. What is the acceleration of the mass M_1?

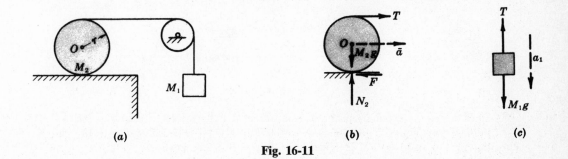

(a)   (b)   (c)

Fig. 16-11

SOLUTION

The free-body diagrams are shown in Fig. 16-1(b) and (c). Note that the magnitude of the acceleration a_1 of M_1 does not equal the magnitude of the acceleration $\bar{a}$ of the center of gravity of the cylinder.

Equations (1) and (2) apply to the cylinder, and equation (3) applies to the mass M_1:

$$\sum F_h = m\bar{a} \quad \text{or} \quad T - F = M_2\bar{a} \tag{1}$$

$$\sum M_O = \bar{I}\alpha \quad \text{or} \quad (F + T)r = \tfrac{1}{2}M_2 r^2 \alpha \tag{2}$$

$$\sum F_v = ma_1 \quad \text{or} \quad M_1 g - T = M_1 a_1 \tag{3}$$

Substitute $\bar{a}/r$ for α in (2) and divide through by r to obtain

$$F + T = \tfrac{1}{2}M_2\bar{a} \tag{4}$$

Add equation (4) to equation (1) to obtain

$$T = \tfrac{3}{4}M_2\bar{a} \tag{5}$$

This may be solved simultaneously with equation (3) if the relation begween $\bar{a}$ and a_1 is obtained. The horizontal component of the acceleration of the top point of the cylinder is equal to the sum of the acceleration $\bar{a}$ of the center and the product $r\alpha$. In this case, assuming pure rolling, $r\alpha$ is equal to $\bar{a}$ by kinematic considerations. Hence, the acceleration of the top point, which is the same as the acceleration a_1 of the mass M_1, is $\bar{a} + r\alpha = 2\bar{a}$.

From (5), $T = \tfrac{3}{4}M_2(\tfrac{1}{2}a_1)$. Substituting into ($3$) and solving, $a_1 = M_1 g/(M_1 + \tfrac{3}{8}M_2)$.

Mathcad

16.8. A wheel with a groove cut in it as shown in Fig. 16-12(a) is pulled up a rail inclined 30° with the horizontal by a rope passing over a pulley and supporting an 80-lb weight. The wheel weighs 100 lb and has a moment of inertia $\bar{I}$ equal to 4 slug-ft². How long will it take the mass center to attain a speed of 20 ft/s starting from rest?

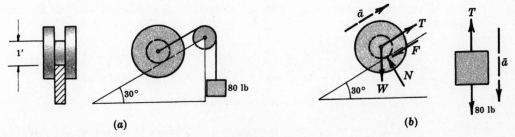

(a)   (b)

Fig. 16-12

SOLUTION

The free-body diagrams for the wheel and the weight are shown in Fig. 16-12(b). Note that tension T in the rope is not 80 lb, because the weight is accelerating.

The necessary equations for the weight and the wheel are

$$\sum F_{\text{weight}} = 80 - T = \frac{80}{32.2}\bar{a} \tag{1}$$

$$\sum F_{\text{wheel}} = T - F - 100\sin 30° = \frac{100}{32.2}\bar{a} \tag{2}$$

$$\sum \bar{M} = F \times \tfrac{1}{2} = 4\alpha \tag{3}$$

Since $\bar{a} = r\alpha = \tfrac{1}{2}\alpha$, equation (3) may be written $F \times \tfrac{1}{2} = 4 \times 2\bar{a} = 8\bar{a}$ or $F = 16\bar{a}$.
Putting $F = 16\bar{a}$ in (2) gives $T - 16\bar{a} - 100 \times 0.500 = (100/32.2)\bar{a}$ or $T = 19.1\bar{a} + 50$.
Putting $T = 19.1\bar{a} + 50$ in (1) gives $80 - 19.1\bar{a} - 50 = 2.48\bar{a}$ or $\bar{a} = 1.39$ ft/s².
To find the time required to attain a speed of 20 ft/s from rest, apply the kinematics equation:

$$v = v_0 + \bar{a}t \qquad 20 = 0 + 1.39t \qquad t = 14.4\text{ s}$$

16.9. Assuming the pulley in Fig. 16-13 to be massless and frictionless, determine the smallest coefficient of static friction that will cause the cylinder to roll. For the cylinder, $M = 70$ kg and $k_O = 400$ mm. Refer to Fig. 16-13(a).

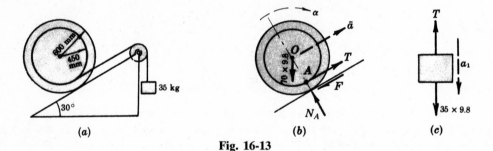

Fig. 16-13

SOLUTION

Draw free-body diagrams of the cylinder and weight as shown in Fig. 16-13(b) and (c). Assume that the cylinder is rolling up the plane. Since pure rolling is called for, the relation $\bar{a} = r\alpha$ holds. (Acceleration $\mathbf{a}_1$ has the same magnitude as the component of the absolute acceleration of point A parallel to the plane.) By kinematic considerations, $\mathbf{a}_A = \mathbf{a}_{A/O} + \bar{\mathbf{a}}$. Dealing only with components parallel to the plane, the component of $a_{A/O} = 0A \times \alpha = 0.45\alpha$ down the plane since α is clockwise. The value of $\bar{a}$ is 0.6α up the plane. Hence, $a_1 = 0.15\alpha$ up the plane.

The equations of motion may now be written. The moment of inertia for the cylinder is $\bar{I} = mk_O^2 = 11.2$ kg · m².

$$\sum F = ma_1 \qquad \text{or} \qquad 35 \times 9.8 - T = 35a_1 \tag{1}$$

$$\sum F = m\bar{a} \qquad \text{or} \qquad T - F - 70 \times 9.8\sin 30° = 70\bar{a} \tag{2}$$

$$\sum \bar{M} = \bar{I}\alpha \qquad \text{or} \qquad F \times 0.6 - T \times 0.45 = 11.2\alpha \tag{3}$$

But $\bar{a} = 0.6\alpha$ and $a_1 = 0.15\alpha$, as has been shown. Eliminate F between (2) and (3) to obtain $0.25T - 686 \times 0.5 = 60.67\alpha$. Solve this simultaneously with (1) to find $\alpha = -4.15$ rad/s² and $T = 365$ N.

Hence, $F = 196$ N. By inspection $N_A = 70 \times 9.8 \cos 30° = 594$ N. Thus, the required coefficient of friction $\mu = 196/594 = 0.33$.

Note that in this problem the cylinder will roll down the plane and the 35-kg mass will ascend.

16.10. In Fig. 16-14, the homogeneous 200-kg sphere rolls without slipping on a horizontal surface. Determine the acceleration of the mass center and the friction needed.

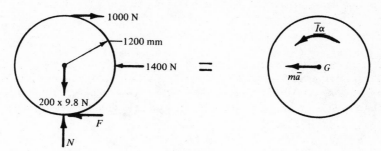

Fig. 16-14

SOLUTION

Assume that the sphere rolls to the left and that the friction force F is also to the left. By inspection, $N = 200 \times 9.8 = 1960$ N.

The equations of motion are

$$\sum F = m\bar{a} \qquad \text{or} \qquad F + 1400 - 1000 = 200\bar{a}$$

$$\sum \bar{M} = \bar{I}\alpha \qquad \text{or} \qquad -F \times 1.2 - 1000 \times 1.2 = \tfrac{2}{5}(200)(1.2)^2\alpha$$

Using $\bar{a} = 1.2\alpha$, these equations yield $\bar{a} = -2.14$ m/s^2 and $F = -828$ N.

Thus, the sphere will roll to the right and the friction is to the right, instead of the direction originally assumed.

16.11. Assume that the disk A in Fig. 16-15 rolls without slipping. Determine the tensions in the ropes and the acceleration of the mass center of disk A.

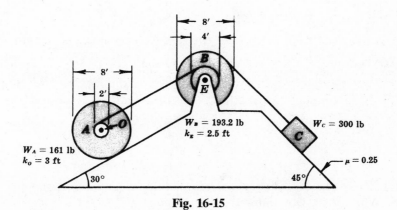

Fig. 16-15

SOLUTION

$I_A = mk_O^2 = (161/32.2)(3)^2 = 45$ lb-s^2-ft, $I_B = (193.2/32.2)(2.5)^2 = 37.5$ lb-s^2-ft.

Draw free-body diagrams of A, B, and C. See Fig. 16-16.

Assume in the diagrams that C moves down the plane and that A rolls up the plane. The following kinematic ideas are needed to solve the problem. First, $\alpha_A = \frac{1}{4}\bar{a}$. Second, the acceleration of rope T_1 equals that component of the absolute acceleration of point D (on disk A) that is parallel to the plane. This component has a magnitude that is the sum of the magnitude of the acceleration relative to the mass center $(1 \times \alpha_A)$ and the magnitude of the acceleration of the mass center $(\bar{a})$. Therefore the magnitude of the absolute acceleration of T_1 is $\frac{5}{4}\bar{a}$. Since this is also the magnitude of the acceleration of a point 2 ft from the center of B, $\alpha_B = \frac{5}{4}\bar{a}/2 = \frac{5}{8}\bar{a}$. The magnitude of the acceleration of T_2 is equal to that of $a_C(\frac{5}{8}\bar{a} \times 4 = \frac{5}{2}\bar{a})$.

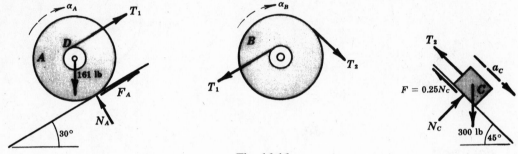

Fig. 16-16

The equations of motion for A are

$$\sum F = m\bar{a} \qquad \text{or} \qquad T_1 + F_A - 161 \sin 30° = \frac{161}{32.2}\bar{a} \tag{1}$$

$$\sum \bar{M} = \bar{I}\alpha_A \qquad \text{or} \qquad T_1 \times 1 - F_A \times 4 = 45 \times \frac{1}{4}\bar{a} \tag{2}$$

The equation of motion for B is

$$\sum \bar{M} = \bar{I}\alpha_B \qquad \text{or} \qquad T_2 \times 4 - T_1 \times 2 = 37.5 \times \frac{5}{8}\bar{a} \tag{3}$$

The equations of motion for C paralell and perpendicular to the plane are

$$\sum F_\| = ma_C \qquad \text{or} \qquad 300 \times 0.707 - 0.25N - T_2 = \frac{300}{32.2}\left(\frac{5}{2}\right)\bar{a} \tag{4}$$

$$\sum F_\perp = 0 \qquad \text{or} \qquad N - 300 \times 0.707 = 0 \tag{5}$$

Solve equation (5) for N; substitute this into equation (4). Solve equation (4) for T_2 in terms of $\bar{a}$. Eliminate F_A between equations (1) and (2) to find T_1 in terms of $\bar{a}$. Substitute these values of T_1 and T_2 into equation (3) to find $\bar{a}$.

These values are $N = 212$ lb, $T_2 = 159 - 23.3\bar{a}$, $T_1 = 64.4 + 6.25\bar{a}$.

Substitute into equation (3) to obtain $4(159 - 23.3\bar{a}) - 2(64.4 + 6.25\bar{a}) = 23.4\bar{a}$.

Hence, $\bar{a} = 3.93$ ft/s^2, $T_1 = 89.0$ lb, $T_2 = 67.4$ lb.

16.12. A solid homogeneous cylinder weighing 644 lb rolls without slipping on the inclined rails shown in Fig. 16-17. Determine the mass center acceleration.

SOLUTION

Draw the free-body diagram assuming that the cylinder rolls up the plane. Note that $\bar{a} = r\alpha = 1.5\alpha$ in this case.

The equations of motion are

$$\sum F = m\bar{a} \qquad \text{or} \qquad F - 644 \sin 30° + 30 \cos 30° = \frac{644}{32.2}\bar{a} \qquad\qquad (1)$$

$$\sum \bar{M} = \bar{I}\alpha \qquad \text{or} \qquad -F \times 1.5 + 30 \times 2 = \frac{1}{2}\left(\frac{644}{32.2}\right)(2)^2\left(\frac{\bar{a}}{1.5}\right) \qquad (2)$$

Solve simultaneously to obtain $\bar{a} = -6.78$ ft/s². This indicates that the cylinder will roll down the plane.

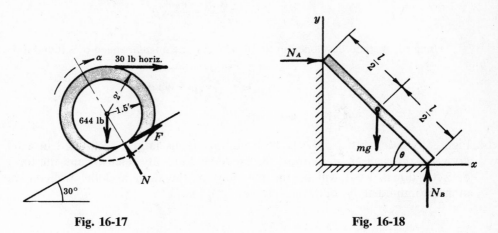

Fig. 16-17   Fig. 16-18

16.13. A uniform bar of length l and mass m rests on smooth surfaces as shown in Fig. 16-18. Describe its motion. Choose the x axis along the horizontal surface and the y axis along the vertical surface.

SOLUTION

The free-body diagram shows only normal reactions of the surfaces on the bar, because friction is negligible.
The equations of motion are

$$\sum F_x = N_A = m\bar{a}_x \qquad\qquad (1)$$

$$\sum F_y = N_B - mg = m\bar{a}_y \qquad\qquad (2)$$

$$\sum \bar{M} = N_A(\tfrac{1}{2}l \sin \theta) - N_B(\tfrac{1}{2}l \cos \theta) = \bar{I}\alpha \qquad\qquad (3)$$

Note that $\bar{I}$ for the rod about its mass center is $\frac{1}{12}ml^2$, and that α is the second derivative of θ with respect to time.

Since the three equations contain five unknowns, two more equations in the unknowns are required to obtain a solution. The geometry of the figure indicates that

$$\bar{x} = \tfrac{1}{2}l \cos \theta \qquad \text{and} \qquad \bar{y} = \tfrac{1}{2}l \sin \theta$$

Then
$$\frac{d\bar{x}}{dt} = \frac{d\bar{x}}{d\theta}\frac{d\theta}{dt} = -\tfrac{1}{2}l \sin \theta \frac{d\theta}{dt} \qquad \text{and} \qquad \frac{d\bar{y}}{dt} = \frac{d\bar{y}}{d\theta}\frac{d\theta}{dt} = \tfrac{1}{2}l \cos \theta \frac{d\theta}{dt}$$

Differentiate again to obtain

$$\bar{a}_x = \frac{d^2\bar{x}}{dt^2} = -\tfrac{1}{2}l\left(\cos \theta \frac{d\theta}{dt}\right)\frac{d\theta}{dt} - \tfrac{1}{2}l \sin \theta \frac{d^2\theta}{dt^2}$$

and
$$\bar{a}_y = \frac{d^2\bar{y}}{dt^2} = \tfrac{1}{2}l\left(-\sin \theta \frac{d\theta}{dt}\right)\frac{d\theta}{dt} + \tfrac{1}{2}l \cos \theta \frac{d^2\theta}{dt^2}$$

These values will be substituted for $\bar{a}_x$ and $\bar{a}_y$ in the original equations. However, it is well to substitue first the values $N_A = m\bar{a}_x$ and $N_B = m\bar{a}_y$ from equations (1) and (2), respectively, into equation (3). Thus, equation (3) becomes, after dividing all terms by $\frac{1}{2}mgl$,

$$\frac{\sin\theta}{g}\bar{a}_x - \frac{\cos\theta}{g}\bar{a}_y = \cos\theta + \frac{l}{6g}\frac{d^2\theta}{dt^2}$$

Next substitute the values of $\bar{a}_x$ and $\bar{a}_y$ just determined and simplify to obtain

$$\frac{d^2\theta}{dt^2} = -\frac{3g}{2l}\cos\theta$$

Using the method developed in Problem 16.39, the angular speed ω is found to be

$$\omega = \sqrt{\frac{3g}{l}(1 - \sin\theta)}$$

16.14. Figure 16-19(a) shows a 12-ft, 20-lb homogeneous ladder restrained in a 60° position. The floor and the wall are smooth. If the ladder is suddenly released and the top is given a speed of 3 ft/s down, what will be the reactions of the floor and the wall on the ladder at the moment immediately following release?

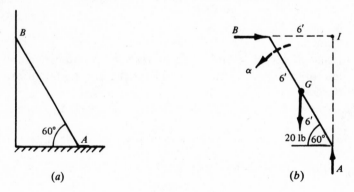

(a)   (b)

Fig. 16-19

SOLUTION

The free-body diagram in Fig. 16-19(b) shows the angular acceleration α assumed counterclockwise. Using the instant center I, the angular velocity is $v_B/IB = 3.6 = 0.5$ rad/s.

The acceleration of the mass center G will be written referring to the acceleration of both A and B:

$$a_G = (a_{G/A})_t + (a_{G/A})_n + a_A \qquad (1)$$

$$a_G = (a_{G/B})_t + (a_{G/B})_n + a_B \qquad (2)$$

The quantities on the right-hand sides of equations (1) and (2) have the following information:

(a) $(a_{G/A})_t$ is 6α down to left at 30° with horizontal;

(b) $(a_{G/A})_n$ is $6(0.5)^2 = 1.5$ down to right at 60° with horizontal;

(c) a_A is horizontal;

(d) $(a_{G/B})_t$ is 6α up to right at 30° with horizontal;

(e) $(a_{G/B})_n$ is $6(0.5)^2 = 1.5$ up to left at 60° with horizontal;

(f) a_B is vertical.

From equation (1), where a_A has no vertical component, we sum vertically to get

(g) $(a_G)_y$ with components $6\alpha \sin 30°$ down and $1.5 \cos 30°$ down or $(3\alpha + 1.3)$ down.

From equation (2), where a_B has no horizontal component, we sum horizontally to get

(h) $(a_G)_x$ with components $6\alpha \cos 30°$ right and $1.5 \cos 60°$ left or $(5.2\alpha - 0.75)$ right.
The equations of motion are

$$\sum F_x = m(a_G)_x \qquad \text{or} \qquad +B = \frac{20}{g}(5.2\alpha - 0.75) \tag{3}$$

$$\sum F_y = m(a_G)_y \qquad \text{or} \qquad 20 - A = \frac{20}{g}(3\alpha + 1.3) \tag{4}$$

$$\sum M_G = I_G\alpha \qquad \text{or} \qquad A \times 3 - B \times 3 = \frac{1}{12}\left(\frac{20}{g}\right)(12)^2\alpha \tag{5}$$

From these, we find $B = 3.2\alpha - 0.466$ and $A = 19.2 - 1.86\alpha$. Substitute these values into equation (5) and solve for $\alpha = 2.6$ rad/s² counterclockwise.

Hence, $A = 14.4$ lb up and $B = 7.93$ lb to the right.

16.15. A homogeneous right circular cone is precariously balanced on its apex in unstable equlibrium on a smooth horizontal plane as shown in Fig. 16-20(a). If it is disturbed, what is the path of its center of mass G?

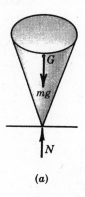

(a)

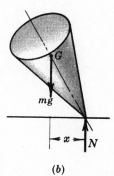

(b)

Fig. 16-20

SOLUTION

Draw a free-body diagram to indicate the cone in any position while falling [see Fig. 16-20(b)]. Note that there is no horizontal force present (no friction assumed). Then the sum of the horizontal forces is zero. But the sum of the horizontal forces equals the product of the cone's mass m and the horizontal acceleration $\bar{a}_x$ of the mass center. Since there is a mass, the product $m\bar{a}_x$ can be zero only if $\bar{a}_x$ is zero.

If the initial horizontal speed is zero and the value of $\bar{a}_x$ is zero, the center of mass G can only have motion in a vertical line passing through the point where the apex of the cone was in its original undisturbaed position.

16.16. Two homogeneous cylindrical disks are rigidly connected to an axle as shown in Fig. 16-21. Each disk has a mass of 7.25 kg and is 900 mm in diameter. The axle is 200 mm in diameter and has a mass of 9 kg. A string wrapped as shown exerts a force of 45 N at the middle of the axle. Analyze the motion.

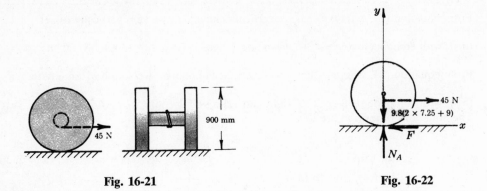

Fig. 16-21   Fig. 16-22

SOLUTION

A free-body diagram is shown indicating only a side view (see Fig. 16-22). Friction F is assumed acting to the left. The equations of motion are

$$\sum F_x = 45 - F = m\bar{a}_x \tag{1}$$

$$\sum F_y = N_A - 9.8 \times 23.5 = m\bar{a}_y = 0 \tag{2}$$

$$\sum \bar{M} = 0.45F - 45(0.1) = \bar{I}\alpha \tag{3}$$

The moment of inertia $\bar{I}$ is the sum of the moments of inertia of the two disks and the axle:

$$\bar{I} = 2(\tfrac{1}{2}m_{disk}r_{disk}^2) + \tfrac{1}{2}m_{axle}r_{axle}^2 = 7.25(0.45)^2 + \tfrac{1}{2}(9)(0.1)^2 = 1.513 \text{ kg} \cdot \text{m}^2$$

Substituting these values, equations (1), (2), and (3) become

$$45 - F = 23.5\bar{a}_x \tag{1'}$$

$$N_A - 230 = 0 \tag{2'}$$

$$0.45F - 4.5 = 1.513\alpha \tag{3'}$$

Since $\alpha = \bar{a}_x/0.45$ where the distance from the center of rolling to the surface on which the rolling occurs is 0.45 m, equation (3') may be written

$$0.45F - 4.5 = 1.513\left(\frac{\bar{a}_x}{0.45}\right) \qquad \text{or} \qquad F - 10 = 7.47\bar{a}_x$$

Add this equation to (1') to obtain $\bar{a}_x = 1.13 \text{ m/s}^2$ to the right.

16.17. A homogeneous sphere and a homogeneous cylinder roll, without slipping, simultaneously from rest at the top of an inclined plane to the bottom. Which reaches the bottom first?

SOLUTION

It is likely from the wording of the problem that the radii of the solids have no influence. Let the subscripts s and c refer to the sphere and cylinder, respectively.

The free-body diagram shown in Fig. 16-23 refers to either solid.

The equations of motion are listed below. The moments of inertia for the sphere and cylinder are, respectively, $\frac{2}{5}m_s r_s^2$ and $\frac{1}{2}m_c r_c^2$.

Sphere		Cylinder	
$\sum F_x = m_s g \sin\theta - F_s = m_s(\bar{a}_x)_s$	*(1)*	$\sum F_x = m_c g \sin\theta - F_c = m_c(\bar{a}_x)_c$	*(4)*
$\sum F_y = N_s - m_s g \cos\theta = 0$	*(2)*	$\sum F_y = N_c - m_c g \cos\theta = 0$	*(5)*
$\sum \bar{M} = F_s r_s = \frac{2}{5}m_s r_s^2 \alpha_s$	*(3)*	$\sum \bar{M} = F_c r_c = \frac{1}{2}m_c r_c^2 \alpha_c$	*(6)*
Substitute $r_s\alpha_s = (\bar{a}_s)_s$ into equation *(3)* and obtain		Substitute $r_c\alpha_c = (\bar{a}_s)_c$ into *(6)* and obtain	
$F_s = \frac{2}{5}m_s(\bar{a}_x)_s$	*(3′)*	$F_c = \frac{1}{2}m_c(\bar{a}_x)_c$	*(6′)*
Substitute this value into *(1)* to obtain		Substitute this value into *(4)* to obtain	
$(\bar{a}_x)_s = \frac{5}{7}g \sin\theta$	*(1′)*	$(\bar{a}_x)_c = \frac{2}{3}g \sin\theta$	*(4′)*

Thus, the sphere, having the larger acceleration, will reach the bottom first or the object with the least mass moment of inertia.

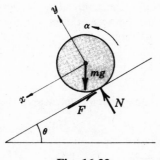

Fig. 16-23

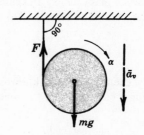

Fig. 16-24

16.18. A cylinder 100 mm in diameter has a cord wrapped around it at the midsection. Attach the free end of the cord to a fixed support and allow the cylinder to fall. Analyze the motion.

SOLUTION

Two equations—one summing the forces in the vertical direction assuming downward as positive and the other summing moments about the mass center assuming clockwise moments as positive—are useful here. They are

$$\sum F_v = mg - F = m\bar{a}_v \qquad (1)$$
$$\sum \bar{M} = Fr = \bar{I}\alpha \qquad (2)$$

Refer to Fig. 16-24. F is the tension in the cord. The moment of inertia $\bar{I} = \frac{1}{2}mr^2$, where $r = 0.05$ m. Also, $\alpha = \bar{a}_v/r = 20\bar{a}_v$, since the motion is equivalent to rolling the cylinder on the cord. Equation *(2)* becomes

$$0.05F = \frac{1}{2}m(0.05)^2(20\bar{a}_v) \qquad \text{or} \qquad F = \frac{1}{2}m\bar{a}_v$$

Substitute this value of F into equation *(1)* to obtain $\bar{a}_v = 2g/3 = 6.53$ m/s^2.

The yo-yo works on this principle. It is designed so that $\bar{a}_v$ is much smaller than g. This occurs if the cylinder is grooved so that F is applied at a smaller radius than that of the cylinder.

16.19. A solid sphere and a thin hoop of equal masses m and radii R are harnessed together by a rigging and are free to roll without slipping down the inclined plane shown in Fig. 16-25(a). Neglecting the mass of the rigging, determine the force in it. Assume frictionless bearings.

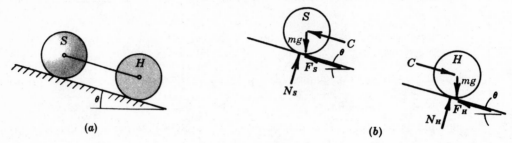

Fig. 16-25

SOLUTION

Draw free-body diagrams assuming that C is the compression in the rigging [see Fig. 16-25(b)]. If the sign should be negative, this merely indicates that C is tension.

Let the subscripts S and H indicate sphere and hoop, respectively. The acceleration $\bar{a}$ is the same for either mass center; and since the radii are equal, the angular acceleration α for each is the same. Sum forces parallel to the plane to obtain the following two equations:

$$mg \sin \theta - F_S - C = m\bar{a} \qquad (1)$$

$$mg \sin \theta - F_H + C = m\bar{a} \qquad (2)$$

Since there are four unknowns (F_S, F_B, C, $\bar{a}$), two more equations are necessary. These are obtained by taking moments about the mass centers:

$$F_S \times r = I_S \alpha = \tfrac{2}{5} mr^2 \alpha \qquad (3)$$

$$F_B \times r = I_H \alpha = mr^2 \alpha \qquad (4)$$

The relation $\bar{a} = r\alpha$ holds for either equation. From (3) and (4), $F_S = \tfrac{2}{5} m\bar{a}$ and $F_H = m\bar{a}$. Substitute these values into equations (1) and (2); then add the resulting equations to eliminate C and obtain $\bar{a} = \tfrac{10}{17} g \sin \theta$. Next solve for $C = \tfrac{3}{17} mg \sin \theta$ (compression).

16.20. At what height above the table should a billiard ball of radius r be hit with a cue held horizontally so that the ball will start moving with no friction between it and the table? Refer to Fig. 16-26.

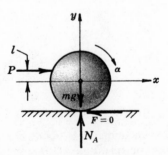

Fig. 16-26

SOLUTION

The equations of motion are

$$\sum F_x = P = m\bar{a}_x \tag{1}$$

$$\sum F_y = N_A - mg = 0 \tag{2}$$

$$\sum \bar{M} = Pl = \tfrac{2}{5}mr^2\alpha \tag{3}$$

Since $\alpha = \bar{a}_x/r$, equation (3) reduces to $P = 2mr\bar{a}_x/5l$.
Substitute this value of P into equation (1) to obtain $l = \tfrac{2}{5}r$.
In other words, the proper distance above the table is $r + \tfrac{2}{5}r = \tfrac{7}{5}r$.

The same procedure is used to find the proper height of the cushion on a billiard table so that the ball rebounds without causing any friction on the table. In this case the force P of the cushion takes the place of the force of the cue. Of course, the answer is the same.

16.21. A solid homogeneous cylinder weighing 16.1 lb rolls without slipping on the inside of a curved fixed surface. In the phase shown, the speed of the mass center of the cylinder is 9 ft/s down to the right. What is the reaction of the surface on the cylinder?

SOLUTION

Draw a free-body diagram showing the reaction components F and N. See Fig. 16-27(a). Figure 16-27(b) illustrates any position of the cylinder to study the kinematic relationship necessary to solve the problem. Note that as the cylinder rolls up to the left, ϕ and θ increase as indicated. Hence, assume that the angular acceleration of the cylinder is positive in the counterclockwise direction and thus that $\bar{a}_t$ is positive to the upper left.

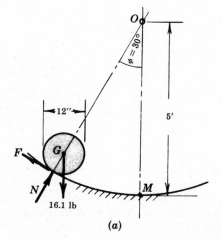

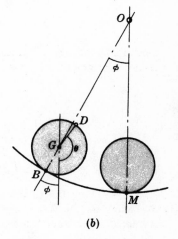

(a)   (b)

Fig. 16-27

The equations of motion are

$$\sum F_n = m\bar{a}_n \quad \text{or} \quad N - 16.1\cos 30° = \frac{16.1}{32.2}\bar{a}_n \tag{1}$$

$$\sum F_t = m\bar{a}_t \quad \text{or} \quad -16.1\sin 30 + F = \frac{16.1}{32.2}\bar{a}_t \tag{2}$$

$$\sum \bar{M} = \bar{I}\alpha \quad \text{or} \quad -F \times \frac{1}{2} = \frac{1}{2}\left(\frac{16.1}{32.2}\right)\left(\frac{1}{2}\right)^2\alpha \tag{3}$$

where $\bar{a}_n$ and $\bar{a}_t$ are the magnitudes of the components of the acceleration of the mass center G and α is the magnitude of the angular acceleration of the cylinder about G. The mass center G can be considered as rotating about the center of curvature O. Hence, $\bar{a}_n = v^2/OG = 81/4.5 = 18 \text{ ft/s}^2$. The tangential magnitude $\bar{a}_t$ is as yet unknown.

To express α in terms of $\bar{a}_t$, refer to Fig. 16-27(b), where D is the point on the cylinder originally in contact with M. Since pure rolling is assumed, the arc BM in the curved surface must equal the arc BD on the cylinder (otherwise slip occurs). Let r = radius of the cylinder and R = radius of surface.

Then $r(\phi + \theta) = R\phi$ or $\theta = (R/r - 1)\phi$. This relationship also holds for the time derivatives of θ and ϕ. The second derivatives of θ with respect to time is the magnitude of the angular acceleration α. The second derivative of ϕ with respect to time may be found in terms of $\bar{a}_t$, since center G is considered to rotate about O; that is, $\bar{a}_t = (R - r)(d^2\phi/dt^2)$. Hence,

$$\alpha = \frac{d^2\theta}{dt^2} = \frac{R-r}{r}\frac{d^2\phi}{dt^2} = \frac{R-r}{r}\frac{\bar{a}_t}{R-r} = \frac{\bar{a}_t}{r}$$

In this case, $R = 5$ ft and $r = \frac{1}{2}$ ft; therefore $\alpha = 2\bar{a}_t$ and the equations of motion simplify to

$$N - 13.9 = 9 \tag{1'}$$

$$-8.05 + F = \tfrac{1}{2}\bar{a}_t \tag{2'}$$

$$-F = \tfrac{1}{4}\bar{a}_t \tag{3'}$$

from which $F = 2.68$ lb and $N = 22.9$ lb.

16.22. A disk of mass m and radius of gyration k has an angular speed ω_0 clockwise when set on a horizontal floor. See Fig. 16-28(a). If the coefficient of friction between the disk and the floor is μ, derive an expression for the time at which skidding stops and rolling occurs.

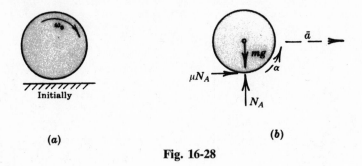

Initially

(a)   (b)

Fig. 16-28

SOLUTION

A free-body diagram shows the friction force μN_A, the normal force N_A, and the gravitational force mg acting on the disk. These will cause a mass center acceleration $\bar{a}$ to the right and an angular acceleration counterclockwise as shown in Fig. 16-28(b).

The equations of motion are

$$\sum F_h = m\bar{a} \quad \text{or} \quad \mu N_A = m\bar{a} \tag{1}$$

$$\sum F_v = 0 \quad \text{or} \quad N_A = mg \tag{2}$$

$$\sum \bar{M} = \bar{I}\alpha \quad \text{or} \quad \mu N_A r = mk^2\alpha \tag{3}$$

Substituting $N_A = mg$ into equation (1) and then into equation (3), we obtain

$$\bar{a} = \mu g \tag{4}$$

$$\alpha = \frac{\mu g r}{k^2} \tag{5}$$

The speed $\bar{v}$ at any time t can be found from equation (4) to be

$$\bar{v} = \bar{v}_0 + \mu g t = \mu g t \tag{6}$$

since $\bar{v}_0 = 0$. Also, the angular speed ω at any time t can be found from (5) to be

$$\omega = \omega_0 - \frac{\mu g r t}{k^2} \tag{7}$$

where it is assumed that clockwise is positive.

When skidding stops and rolling occurs, $\bar{v} = r\omega$. Hence, we multiply (7) by r to obtain $\bar{v}$, which is also given by (6). This yields the required time t':

$$\mu g t' = r\omega_0 - \frac{\mu g r^2 t'}{k^2} \qquad \text{or} \qquad t' = \frac{r\omega_0}{\mu g(1 + r^2/k^2)} \tag{8}$$

16.23. A uniform cylinder of radius R is on a platform that is subjected to a constant horizontal acceleration of magnitude a. Assuming no slip, determine a_0, the magnitude of the acceleration of the mass center of the cylinder.

SOLUTION

The free-body diagram is shown in Fig. 16-29 with all forces acting on the cylinder.

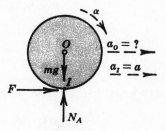

Fig. 16-29

The angular acceleration of the cylinder is assumed clockwise. Point I is the instant center between the cylinder and the platform, and therefore has an acceleration to the right equal to that of the platform.

From kinematic concepts, we can write

$$\mathbf{a}_O = \mathbf{a}_{O/I} + \mathbf{a}_I$$

Summing in the horizontal direction, this equation yields (to the right is positive)

$$a_0 = R\alpha + a \tag{1}$$

The equations of motion are next applied to the cylinder:

$$\sum F_h = ma_O \qquad \text{or} \qquad F = ma_O \tag{2}$$

$$\sum M_O = I_O\alpha \qquad \text{or} \qquad -FR = \tfrac{1}{2}mR^2\alpha \tag{3}$$

The sign on the moment of the frictional force F is negative because the moment is opposite to the assumed α.

Substituting from equation (1) into (2), we find $F = m(R\alpha + a)$. Now put this value of F into equation (3) to obtain $R\alpha = -\tfrac{2}{3}a$. Thus, by equation (1), the acceleration of the mass center is $a_0 = -\tfrac{2}{3}a + a = \tfrac{1}{3}a$.

16.24. The homogeneous bar ABC in Fig. 16-30(a) is 3000 mm long and has a mass of 20 kg. It is pinned to the ground at A and to the homogeneous bar CD, which is 2000 mm long with a mass of 10 kg. Determine the angular acceleration of each bar immediately after a restraining wire is cut. Assume no friction as D starts moving to the right.

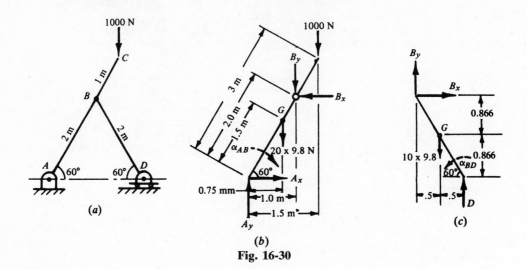

(a)

(b)

(c)

Fig. 16-30

SOLUTION

Figures 16-30(b) and (c) are free-body diagrams of the two bars at the instant of release. The acceleration of B is the same on either bar, and thus we write

$$a_B = (a_{B/D})_n + (a_{B/D})_t + a_D \qquad (1)$$

The acceleration of B as a point on ABC is $2\alpha_{AB}$, and since it is perpendicular to the bar, it is directed to the right and down at an angle of 30° with the horizontal.

The acceleration $(a_{B/D})_n$ is zero because there is no angular velocity at the instant of release. The acceleration $(a_{B/D})_t$ is to the left and down at an angle of 30°, and equals $2\alpha_{BD}$. Finally, note that a_D can only be horizontal. In equation (1), we sum the vertical components of the accelerations with down positive and obtain

$$2\alpha_{AB} \sin 30° = 2\alpha_{BD} \sin 30° + 0$$

Since this shows $\alpha_{AB} = \alpha_{BD}$, we shall use α for both quantities.

Summing the horizontal components of the accelerations in equation (1), with positive to the right being used, we obtain

$$2\alpha_{AB} \cos 30° = -2\alpha_{BD} \cos 30° + a_D$$

Writing α for both angular accelerations, we now have $a_D = 3.46\alpha$ to the right.

For the bar BD, we can write for its mass center G the following acceleration:

$$a_G = (a_{G/D})_t + (a_{G/D})_n + a_D \qquad (2)$$

As before, $(a_{G/D})_n$ is zero. The $(a_{G/D})_t = (1)\alpha$ is directed to left and down at an angle of 30°. Of course, we have already shown $a_D = 3.46\alpha$ directed to the right.

Sum components horizontally in equation (2) to get

$$(a_G)_x = 3.46\alpha - (1\alpha)(0.866) = 2.6\alpha \qquad \text{(to the right)}$$

sum components vertically to obtain

$$(a_G)_y = (1\alpha)(0.5) \qquad \text{or} \qquad 0.5\alpha \qquad \text{(down)}$$

In Fig. 16-30(c), sum horizontally, using to the right as positive, and sum vertically, using down as positive. This yields

$$B_x = m(a_G)_x = 10(2.6\alpha) = 26\alpha \qquad \text{(to the right)} \tag{3}$$

$$-B_y + 10 \times 9.8 - D = 10(0.5\alpha) \tag{4}$$

Next sum moments about G, using counterclockwise as positive. This yields

$$-B_x \times 0.866 - B_y \times 0.5 + D \times 0.5 = \tfrac{1}{12}(10)(2)^2\alpha \tag{5}$$

Equation (4) yields

$$D = -B_y + 98 - 5\alpha$$

Using this value of D and that of B_x from equation (3) in equation (5) to get

$$-0.866(26\alpha) - B_y(0.5) + 0.5(-B_y + 98 - 5\alpha) = 3.33\alpha \tag{6}$$

From equation (6),

$$B_y = 49 - 28.3\alpha$$

Sum moments about A for the free-body diagram in Fig. 16-30(b) to obtain

$$\sum M_A = (20 \times 9.8)(0.75) + 1000(1.5) + B_y \times 1 - B_x \times 1.732 = \tfrac{1}{3}(20)(3)^2\alpha \tag{7}$$

Solve equations (6) and (7) simultaneously to find

$$\alpha = 12.7 \text{ rad/s}^2$$

Translation

16.25. A 50-lb homogeneous door is supported on rollers A and B resting on a horizontal track as shown in Fig. 16-31. A constant force P of 10 lb is applied. What will be the velocity of the door 5 s after starting from rest? What are the reactions of the rollers? Assume that roller friction is negligible.

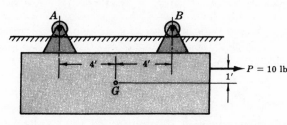

Fig. 16-31

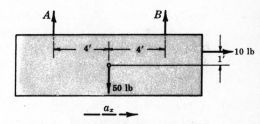

Fig. 16-32

SOLUTION

The acceleration a_x is assumed to be in the x direction in the free-body diagram shown in Fig. 16-32.

Apply the equations $\sum F_x = ma_x$, $\sum F_y = ma_y$, and $\sum \bar{M} = 0$ to obtain the following;

$$\sum F_x = 10 = \frac{50}{32.2} a_x \tag{1}$$

$$\sum F_y = A + B - 50 = \frac{50}{32.2} 0 \tag{2}$$

$$\sum \bar{M} = B \times 4 - A \times 4 - 10 \times 1 = 0 \tag{3}$$

From equation (1), $a_x = 6.44$ ft/s^2. Since a_x is constant, the formula involving the known quantities v_0, a, t, and the unknown v should be used:

$$v = v_0 + at = 0 + (6.44 \text{ ft/s}^2)(5 \text{ s}) = 32.2 \text{ ft/s}$$

Solve equations (2) and (3) simultaneously to obtain $A = 23.7$ lb and $B = 26.3$ lb.

16.26. Figure 16-33(a) shows a homogeneous bar AB that is 2000 mm long and has a mass of 2 kg. The ends are constrained to slide in smooth horizontal guides. Determine the acceleration of the bar and the normal forces at A and B under the action of the horizontal 20-N force.

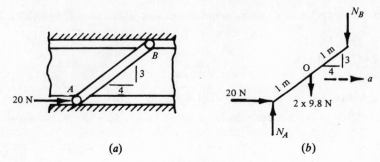

(a)   (b)

Fig. 16-33

SOLUTION

The free-body diagram in Fig. 16-33(b) shows the given force, the normal forces at A and B, and the gravitational force 2×9.8 N. The acceleration is shown acting to the right. (There is no angular acceleration.)

Assume that forces to the right are positive, forces up are positive, and counterclockwise moments are positive. The equations of motion are

$$\sum F_x = ma \qquad \text{or} \qquad 20 = 2a \tag{1}$$

$$\sum F_y = 0 \qquad \text{or} \qquad N_A - N_B = 19.6 = 0 \tag{2}$$

$$\sum \bar{M}_O = 0 \qquad \text{or} \qquad 20(1)(3/5) - N_A(1)(4/5) - N_B(1)(4/5) = 0 \tag{3}$$

From equation (1), $a = 10 \text{ m/s}^2$. Equations (2) and (3) solved simultaneously yield $N_A = 17.3$ N up and $N_B = -2.3$ N up. (The reactions are not in the directions assumed to be correct. A minus sign means that the reaction acts opposite to the direction shown, i.e., N_B acts up.)

16.27. A freight car starts from rest at the top of a 1.5 percent grade 1 mi long. Assuming a resistance of 8 lb/ton, how far along the level track at the bottom of the grade will it roll before coming to rest?

SOLUTION

On the incline the forces acting parallel to the track are the component of the weight $0.015W$ and the resistance of $(8/2000)W = 0.004W$ lb. The first accelerates the car, but the latter decelerates it.

The equation of motion is $\sum F_\parallel = 0.015W - 0.004W = (W/g)a$.

The resulting acceleration is $a = 0.011g$. The velocity at the bottom of the grade is found from

$$v^2 = v_0^2 + 2as = 0 + 2(0.011g)(5280)$$

To determine the distance s along the level, apply the same formulas, noting that the only horizontal force acting is a constant decelerating one (because of train resistance), that is, $-0.004W$.

As before, $\sum F_\parallel = -0.004W = (W/g)a$; hence, $a = -0.004g$. Also, v_0^2 now becomes the value of v^2 found above. Thus,

$$v^2 = v_0^2 + 2as \qquad \text{or} \qquad 0 = 2(0.011g)(5280) - 2(0.004gs)$$

from which $s = 14,500$ ft along the level.

16.28. A block A rests on a cart B as shown in Fig. 16-34. The coefficient of friction between the cart and the block is 0.30. If A has a mass of 70 kg, investigate the maximum acceleration the cart may have.

SOLUTION

The two possibilities to consider are (a) the acceleration that will cause the block to slide off the cart and (b) the acceleration that will cause tipping.

In case (a), the friction is a maximum and is equal to the product of the coefficient of friction and the normal force between the cart and the block.

In case (b), the necessary friction may be (1) less than maximum, (2) maximum, or (3) greater than maximum. If the friction necessary to cause tipping is greater than maximum, naturally it cannot be obtained, and therefore the block will slide at the value of the acceleration calculated in (a).

The free-body diagram for case (a) is shown in Fig. 16-35. Note that the normal force N_A is applied at some unknown point in this representation. The friction is shown acting to the right, since the cart is being accelerated to the right and pulls the block with it because of friction. Only two equations are necessary:

$$\sum F_h = ma_h; \qquad \text{that is,} \qquad F = ma_h \tag{1}$$

$$\sum F_v = ma_v = 0; \qquad \text{that is,} \qquad N_A - 686 = 0 \tag{2}$$

From equation (2), $N_A = 686$ N.
Equation (1) becomes $0.30 \times 686 = 70a_h$ or $a_h = 2.94$ m/s^2.
Accelerations above 2.94 m/s^2 will cause the block to slide.

The free-body diagram for the case of tipping is shown in Fig. 16-36. In this case motion impends about the back edge of the block. Therefore the normal and frictional forces are shown there.

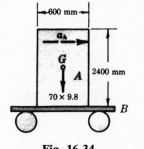

Fig. 16-34

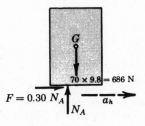

Fig. 16-35

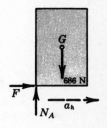

Fig. 16-36

Note that F is not placed equal to $0.30N_A$. The limiting value of friction is used only if slipping impends, as in case (a). The equations are

$$\sum F_h = ma_h \tag{3}$$

$$\sum F_v = ma_v = 0 \tag{4}$$

$$\sum \bar{M} = 0 \tag{5}$$

These become

$$F = 70a_h \tag{3'}$$

$$N_A - 686 = 0 \tag{4'}$$

$$-N_A \times 0{\cdot}3 + F \times 1{\cdot}2 = 0 \tag{5'}$$

From ($4'$), $N_A = 686$ N. Substituting into ($5'$), $F = 172$ N.
Substituting $F = 172$ N into ($3'$), $a_h = 2.46$ m/s^2.

Thus at an acceleration of 2.46 m/s^2 the block tends to tip. Since at this acceleration the force of friction is 172 N, which is less than the limiting value (0.30×686 N), the tipping will occur before sliding can exist.

Or, looking at the problem slightly differently, an acceleration of 2.46 m/s^2 will cause tipping, whereas an acceleration of 2.94 m/s² will cause sliding. Since the acceleration 2.46 m/s² is reached first, tipping will occur before sliding.

It can be deduced directly from equation (5') that tipping will occur first, since the ratio of F to N_A in this case is 0.25, which is less than the maximum ratio of 0.30.

16.29. A 60-kg operator is standing on a spring scale in a 600-kg elevator that is accelerating 6 m/s² up. What is the scale reading in kilograms and what is the tension in the cables?

SOLUTION

Figure 16-37 shows the free-body diagram of the 60-kg elevator operator, with the gravitational force acting down and the scale force acting up. Because the acceleration is up, we shall assume that direction is positive. The equation of motion for the operator is

$$\sum F_v = P - 60 \times 9.8 = ma = 60 \times 6$$

From this, $P = 948$ N and the scale reading will be $948/9.8 = 96.7$ kg. Thus, there is an apparent increase in the mass of the operator of $96.7 - 60 = 36.7$ kg.

Fig. 16-37   **Fig. 16-38**

To determine the tension T in the cable, draw the free-body diagram as shown in Fig. 16-38. If we call the up direction positive, the equation of motion for the elevator and the operator is

$$\sum F_v = ma \quad \text{or} \quad T - 660 \times 9.8 = 660 \times 6$$

From this equation, $T = 10,400$ N.

16.30. Block B is accelerated along the horizontal plane by means of a mass A attached to it by a flexible, inextensible, massless rope passing over a smooth pulley as shown in Fig. 16-39. Assume that the coefficient of friction between B, which has a mass of 2 kg, and the plane is 0.20. Discuss the motion if A has a mass of 1.2 kg.

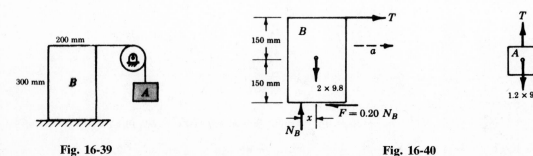

Fig. 16-39   **Fig. 16-40**

SOLUTION

Draw the free-body diagrams of A and B (see Fig. 16-40). Note that the force N_B is shown acting at a distance x from the vertical line through the mass center of block B. Assume the acceleration of block B is to the right as shown.

Sum the vertical forces acting on block A (assume down is positive to agree with the arrow on the acceleration vector) to obtain

$$1.2 \times 9.8 - T = 1.2a \qquad (1)$$

By inspection of the diagram for block B, $N_B = 2 \times 9.8 = 19.6 \text{ N}$.
Sum the horizontal forces on block B (to the right is positive) to obtain

$$T - 0.20N_B = 2a \qquad (2)$$

Equations (1) and (2) yield $a = 2.45 \text{ m/s}^2$.

The sum of the moments of the force about the mass center of B is equal to zero; thus, with counterclockwise moments being positive, the equation is

$$-T \times 0.15 - N_B \times x - F \times 0.15 = 0 \qquad (3)$$

Substituting the values of T, F, and N_B into equation (3), the value of $x = -97.5 \text{ mm}$. The minus sign indicates that the force N_B is acting to the right of the mass center instead of to the left as assumed.

16.31. Refer to Fig. 16-41. The stand A is accelerated to the left at 2.5 m/s^2. But B is pinned at P, and its top rests against the smooth vertical surface at H. A and B have masses of 30 kg and 7 kg, respectively. Determine the horizontal force F and the horizontal push H on the top of the bar.

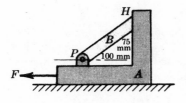

Fig. 16-41

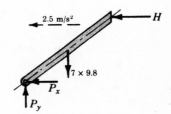

Fig. 16-42

SOLUTION

Taking the entire body as a free body, sum the forces horizontally (to the left is positive) to obtain

$$F = Ma = (30 + 7)(2.5) = 92.5 \text{ N}$$

Next draw the free-body diagram or bar B as shown in Fig. 16-42. The equations of motion are

$$\sum F_h = Ma_h \qquad \text{or} \qquad P_x + H = 7(2.5) \qquad (1)$$

$$\sum F_v = Ma_v \qquad \text{or} \qquad P_y - 7 \times 9.8 = 0 \qquad (2)$$

$$\sum \bar{M} = 0 \qquad \text{or} \qquad (0.037\,5)H - (0.037\,5)P_x - (0.05)P_y = 0 \qquad (3)$$

From equation (1), $P_x = 17.5 - H$

From equation (2), $P_y = 68.6$

Substituting these in equation (3) gives

$$(0.037\,5)H - (0.037\,5)(17.5) + (0.037\,5)H - (0.05)(68.6) = 0$$

Hence, $H = 54.5 \text{ N}$.

16.32. Both bodies A and B in Fig. 16-43 have a mass of 20 kg. A small strip is nailed to B to prevent A from sliding while P accelerates the system to the left on the smooth plane. Determine the maximum value of P without causing A to tip. If the system was moving initially with velocity 3 m/s is to the right, what will be its velocity after moving 4 m?

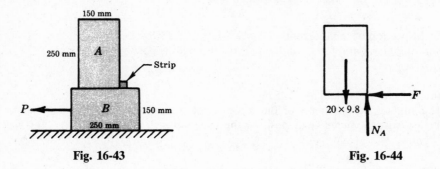

Fig. 16-43   Fig. 16-44

SOLUTION

Draw a free-body diagram of A (see Fig. 16-44). The normal force N_A is shown acting up at the extreme right, since the block is about to tip. The equations of motion are

$$\sum F_h = Ma_h \qquad \text{or} \qquad F = 20a \qquad\qquad (1)$$

$$\sum F_v = Ma_v = 0 \qquad \text{or} \qquad N_A - 20 \times 9.8 = 0 \qquad\qquad (2)$$

$$\sum \bar{M} = 0 \qquad \text{or} \qquad N_A \times 0.075 - F \times 0.125 = 0 \qquad\qquad (3)$$

Substitute $N_A = 20 \times 9.8$ from equation (2) into equation (3) to obtain $F = 118$ N. Substitute into equation (1) to find $a = 5.9$ m/s^2.

To determine the velocity after traversing 4 m, first consider the distance moved to the right before the velocity becomes zero. Applying the kinematics equation,

$$v^2 = v_0^2 + 2as \qquad 0 = (+3)^2 - 2(5.9)s \qquad s = 0.76 \text{ m}$$

The remainder of the 4 m will be to the left, i.e., 3.24 m. Using the same kinematics equation and assuming the positive direction to the left, $v^2 = 0 + 2 \times 5.9 \times 3.24$. Hence, the final velocity is to the left and equals 6.18 m/s.

16.33. Block A with mass 30 kg is located on top of a block B with mass 45 kg as shown in Fig. 16-45. The coefficient of friction between the blocks is $\frac{1}{3}$. The coefficient of friction between block B and the horizontal plane is $\frac{1}{10}$. Determine the maximum value of P that will not cause block A to slide or tip.

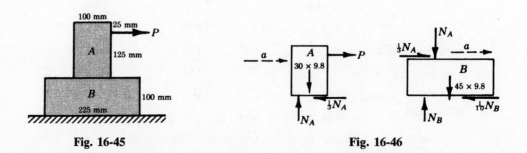

Fig. 16-45   Fig. 16-46

SOLUTION

First determine the force P to cause A to slip on B. Draw the free-body using friction as $\frac{1}{3}N_A$ between blocks. See Fig. 16-46.

The location of the reaction N_A is unknown, but is also immaterial in this solution. By inspection, its value is $30 \times 9.8 = 294$ N. Likewise by inspection, the reaction $N_B = 294 + 45 \times 9.8 = 735$ N.

The other equation of motion for body A is

$$\sum F_h = ma_h \qquad \text{or} \qquad P - \tfrac{1}{3}N_A = 30a \tag{1}$$

For body B, the needed equation of motion is

$$\sum F_h = ma_h \qquad \text{or} \qquad \tfrac{1}{3}N_A - \tfrac{1}{10}N_B = 45a \tag{2}$$

These equations yield $a = 0.544$ m/s^2 and $P = 114$ N (for slip of A on B).

Next draw free-body diagrams to determine the force P to cause tipping (see Fig. 16-47). Now the normal component N_A of the reaction must be drawn at the lower right corner. The frictional component F of the reaction is unknown.

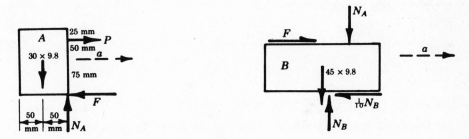

Fig. 16-47

By inspection, $N_A = 30 \times 9.8 = 294$ N, $N_B = 735$ N, and the friction at the bottom is $\frac{1}{10}$ of $735 = 73.5$ N.

For block A, the equations of motion are

$$\sum F = ma_h \qquad \text{or} \qquad P - F = 30a \tag{1}$$

$$\sum \bar{M} = 0 \qquad \text{or} \qquad -P \times 0.05 + N_A \times 0.050 - F \times 0.075 = 0 \tag{2}$$

For block B, the needed equation is

$$\sum F_h = ma_h \qquad \text{or} \qquad F - 73.5 = 45a \tag{3}$$

Solve equations (1), (2), and (3) simultaneously to obtain $P = 132$ N.

The smaller of the two values of P, i.e., 114 N, is the maximum value that will not cause block A to slide or tip.

16.34. Find the minimum time for the lift truck shown in Fig. 16-48 to attain its rated speed of 1.5 m/s without causing the six cartons, 1500 mm high, to tip. Each carton is 250 mm high with a base 400 mm on each side. Assume coefficient friction present so that sliding will not occur.

SOLUTION

The free-body diagram of the cartons on the truck indicates motion to the right. Thus, the reversed effective force $M\bar{a}$ is applied to the left through the mass center to hold the system in "equilibrium for study purposes." Tipping will tend to occur about the left lower edge.

Moments about that edge yield

$$M \times 9.8 \times 0.2 = M\bar{a} \times 0.75$$

The maximum acceleration is thus $\bar{a} = 2.61$ m/s^2. Since $v = v_0 + at$, minimum time $t = 0.57$ s.

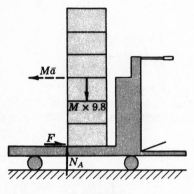

Fig. 16-48

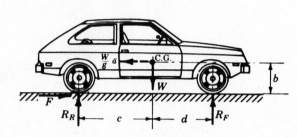

Fig. 16-49

16.35. The car shown in Fig. 16-49 has a rear-wheel drive.

(a) What force must the rear wheels exert on the ground to cause an acceleration $\bar{a}$ to the right?

(b) Neglecting the rotational inertia effect of the wheels, what is the maximum acceleration possible for a given coefficient of friction μ between the tires and the pavement?

(c) Assuming sufficient friction to be available, what is the acceleration required to cause the car to start to tip backward?

SOLUTION

(a) Apply the reversed effective force $(W/g)\bar{a}$ to hold the car in "equilibrium for study purposes." Summing forces horizontally, $\sum F_h = F - (W/g)\bar{a} = 0$. From this equation, $F = (W/g)\bar{a}$.

(b) to find the maximum acceleration possible, solve for R_R. The greatest friction may be determined from $F = \mu R_R$.

Taking moments about R_F to obtain

$$-R_R \times (c + d) + W \times d + \frac{W}{g}\bar{a} \times b = 0$$

From this equation,

$$R_R = \frac{Wd}{c + d} + \frac{Wb\bar{a}}{g(c + d)}$$

Hence,

$$F = \mu R_R = \mu \frac{Wd}{c + d} + \mu \frac{Wb\bar{a}}{g(c + d)}$$

Substitute this value of F into $F = (W/g)\bar{a}$ and then solve for $\bar{a} = \mu\, dg/(c + d - \mu b)$.

Notice that with this type of solution (using the equations of motion as shown), moments may be taken about a point other than the mass center.

(c) In this case the reaction R_F on the front wheels is zero. Sum moments about the line of contact of the rear wheels with the ground to obtain $\sum M_{R_R} = 0 = (W/g)\bar{a} \times b - W \times c$. Hence, the acceleration to cause tipping is $\bar{a} = cg/b$.

16.36. The van shown in Fig. 16-50 has a front-wheel drive. It can accelerate from 0 to 60 mi/h in 13.8 s. Determine the vertical reactions (R_F and R_R) on the front and rear wheels under this acceleration and also the necessary frictional force between the front wheels and the road. The van weighs 3385 lb. Its mass center G is located 46 in behind the front-wheel contact with the road and is 31 in above the road.

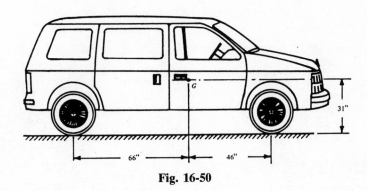

Fig. 16-50

SOLUTION

To find $\bar{a}$, use the equation $v = v_0 + \bar{a}t$ or

$$60 \times 5280/3600 = 0 + \bar{a} \times 13.8$$

Hence, $$\bar{a} = 6.38 \text{ ft/s}^2$$

Summation of the horizontal forces yields the friction force:

$$\sum F_h = F = (3385/g) \times 6.38 \qquad F = 671 \text{ lb}$$

Summation of the vertical forces and summation of the moments about G yields simultaneous equations in R_F and R_R:

$$\sum F_v = R_R + R_F - 3385 = 0$$
$$\sum \bar{M} = -66R_R + 46R_F + 671 \times 31 = 0$$
$$R_F = 1809 \text{ lb} \qquad R_R = 1576 \text{ lb}$$

Rotation

16.37. The homogeneous bar in Fig. 16-51 is on a smooth horizontal table. It is pivoted at its left end about a vertical axis through O. A horizontal force F is applied perpendicular to the bar at its free end. Note that the mass of the bar is m, its length is l, and its angular speed is zero when the force is applied. Determine the acceleration α and the reaction at the axis through O.

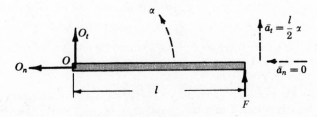

Fig. 16-51

SOLUTION

The equations of motion are

$$\sum F_n = m\bar{r}\omega^2 = 0 \qquad \text{or} \qquad O_n = 0$$

$$\sum F_t = m\bar{r}\alpha \qquad \text{or} \qquad O_t + F = m\frac{l}{2}\alpha$$

$$\sum M_O = I_O\alpha \qquad \text{or} \qquad Fl = \tfrac{1}{3}ml^2\alpha$$

It can readily be determined that $\alpha = 3F/ml$ and $O_t = \tfrac{1}{2}F$.

16.38. Solve Problem 16.37, but assume that the horizontal force F is applied through the center of percussion.

SOLUTION

The arm for F is the distance q to the center of percussion. Thus,

$$q = \frac{k_O^2}{\bar{r}} = \frac{I_O}{m\bar{r}} = \frac{\tfrac{1}{3}ml^2}{\tfrac{1}{2}ml} = \tfrac{2}{3}l$$

The equations of motion then become

$$\sum F_t = m\bar{r}\alpha \qquad \text{or} \qquad O_n + F = \tfrac{1}{2}ml\alpha$$

$$\sum M_O = I_O\alpha \qquad \text{or} \qquad F(\tfrac{2}{3}l) = (\tfrac{1}{3}ml^2)\alpha$$

The solution becomes $\alpha = 2F/ml$ and $O_t = 0$.

This means that the tangential component of the reaction is zero if the force passes through the center of percussion. Batters know that a ball that hits their bat about two-thirds of its length from their hands does not sting. This is because O_t equals zero or nearly so.

16.39. A bar pivoted about a horizontal axis through its lower end is allowed to fall from a vertical position. Describe the motion.

SOLUTION

Assume that the bar is of length l and mass m. The free-body diagram in Fig. 16-52 shows the gravitational force mg acting down and the reactions R_t and R_n at the point of support. The bar is assumed to be at a position θ from the vertical rest position.

The bar is rotating about a fixed axis under the action of an unbalanced force system, the equations of motion being $\sum F_n = m\bar{r}\omega^2$, $\sum F_t = m\bar{r}\alpha$, and $\sum M_O = I_O\alpha$. However, in describing the motion, it will be sufficient to determine the magnitudes of the angular velocity and acceleration as functions of displacement θ. Hence, the only equation needed is $\sum M_O = I_O\alpha$.

The only external force with a moment about O is the gravitational force mg. Hence, taking clockwise moments as positive, $mgd = I_O\alpha$.

But $d = \tfrac{1}{2}l \sin\theta$, $I_O = \tfrac{1}{3}ml^2$, and $\alpha = d^2\theta/dt^2$. Hence,

$$mg(\tfrac{1}{2}l \sin\theta) = \tfrac{1}{3}ml^2\frac{d^2\theta}{dt^2} \qquad \text{or} \qquad \frac{d^2\theta}{dt^2} = \frac{3g}{2l}\sin\theta$$

This differential equation may be solved by noting that

$$\frac{d^2\theta}{dt^2} = \frac{d}{dt}\left(\frac{d\theta}{dt}\right) = \frac{d\omega}{dt} = \frac{d\omega}{d\theta}\frac{d\theta}{dt} = \frac{d\omega}{d\theta}\omega$$

The original equation $d^2\theta/dt^2 = (3g/2l)\sin\theta$ is now written as

$$\omega\frac{d\omega}{d\theta} = \frac{3g}{2l}\sin\theta$$

Multiply both sides of the equation by $d\theta$ to obtain

$$\omega\,d\omega = \frac{3g}{2l}\sin\theta\,d\theta$$

Integration yields $\frac{1}{2}\omega^2 = -(3g/l)\cos\theta + C$, where C is the constant of integration. To evaluate C, note that $\omega = 0$ when $\theta = 0$. Hence, $0 = -(3g/2l)(1) + C$ or $C = 3g/2l$. Then

$$\tfrac{1}{2}\omega^2 = -\left(\frac{3g}{2l}\right)\cos\theta + \frac{3g}{2l} \quad\text{and}\quad \omega = \sqrt{\frac{3g}{l}(1 - \cos\theta)}$$

This indicates a method of finding ω in terms of θ and, of course, $\alpha = (3g/2l)\sin\theta$.

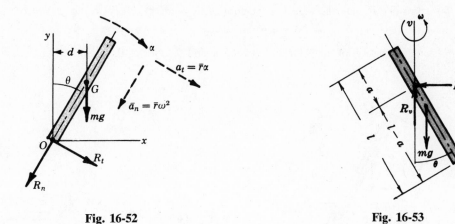

Fig. 16-52   **Fig. 16-53**

16.40. A uniform bar of length l and mass m is rotating at a constant angular velocity **ω** about a vertical axis through a point at a distance a from one end. For the phase shown in Fig. 16-53, when the bar is passing through the plane of the paper, determine the horizontal and vertical components of the reaction of the support on the bar.

SOLUTION

Assume that a is less than half the length. The gravitational force is mg, and the horizontal and vertical components of the bearing reactions are shown acting on the rod.

Note that, since the angular speed ω is constant, the magnitude α of the angular acceleration (about the vertical axis) is zero. Hence, there are no forces acting that have moments about the vertical axis. This conclusion should be evident from the scalar equation $\sum M_v = I_v\alpha = I_v(0)$.

Note also that the equations of motion ($\sum F_n = m\bar{r}\omega^2$, $\sum F_t = m\bar{r}\alpha$, and $\sum M_O = I_O\alpha$) do *not* apply to the body as a whole, since the rod is rotating about an axis that is *not* perpendicular to the plane of symmetry of the rod.

Consider the bar to be composed of a series of thin pieces, each moving with an angular velocity ω about the vertical axis and at a distance ρ from the axis.

From Fig. 16-54,

$$\rho = z\sin\theta \quad\text{and}\quad dm = \frac{dz}{l}m$$

Each differential mass is moving on a circular path (radius ρ). Hence, a normal acceleration a_n exists toward the axis along ρ. A normal force dF accompanies this acceleration a_n ($dF = dm \times a_n$). The sum of all such normal forces (all horizontal) must be the horizontal bearing reaction R_h on the bar. Since in the phase shown the longer length of the bar is to the right of the axis, the normal forces directed to the left will be greater than those acting to the right. Hence, the reaction R_h that supplies these normal forces acts to the left.

The normal force on the particle dm shown in the figure acts toward the axis, i.e., to the left, which means in a negative direction. Hence,

$$dF = -dm\,\rho\omega^2 = -\frac{dz}{l}m(z\sin\theta)\omega^2$$

and the total force is

$$F = \int dF = \int_{-a}^{l-a} -\frac{dz}{l} m(z \sin \theta)\omega^2$$

Since θ and ω^2 may be removed from within the integral sign:

$$F = R_h = -\frac{m\omega^2 \sin \theta}{l} \int_{-a}^{l-a} z \, dz = -m\omega^2(\tfrac{1}{2}l - a) \sin \theta$$

To determine R_v, use the fact that the sum of the vertical forces must equal zero:

$$\sum F_v = 0 = R_v - mg \qquad \text{or} \qquad R_v = mg$$

For an explanation of the relation between ω and θ, refer to Problem 16.42.

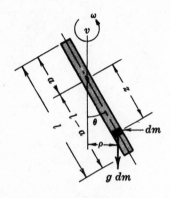

Fig. 16-54

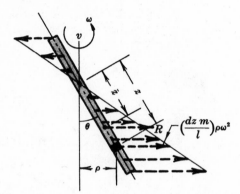

Fig. 16-55

16.41. In Problem 16.40, determine the location of the reversed effective forces to hold the bar in "equilibrium for study purposes only."

SOLUTION

Since the rod is rotating about an axis that is not perpendicular to a plane of symmetry, we take elements of mass of the rod that may be considered as rotating in a plane perpendicular to the axis of rotation.

In Fig. 16-55, the reversed effective normal forces

$$dm \, \rho\omega^2 = \frac{dz}{l} m\omega^2 \rho$$

for the masses of rod of length dz are shown acting away from the vertical axis. Also shown is the single force R at a distance $\bar{z}$ from the support. This force must equal the sum of the individual forces for the elements and must also have the same moment as they have about the pivot point. Hence,

$$R = \int_{-a}^{l-a} \frac{dz}{l} m\omega^2 \rho = \int_{-a}^{l-a} \frac{dz}{l} m\omega^2 z \sin \theta = \frac{m\omega^2 \sin \theta}{2}(l - 2a)$$

This result, of course, is equivalent in magnitude to that derived in Problem 16-40.

Next take moments of the individual inertia forces about the pivot point and equate them to the moment of R about the pivot point. In all cases, the forces are horizontal and the moment arms are vertical (therefore equal to the product of the z distance and $\cos \theta$).

$$R\bar{z} \cos \theta = \int_{-a}^{l-a} z \cos \theta \frac{dz}{l} m\omega^2 \rho$$

Substituting for R its value just derived and for ρ its value $z \sin \theta$, the equation becomes

$$\frac{m\omega^2 \sin \theta}{2}(l - 2a)\bar{z} \cos \theta = \frac{\cos \theta \, m\omega^2 \sin \theta}{l} \int_{-a}^{l-a} z^2 \, dz$$

When $\theta \neq 0$ or $90°$,

$$\frac{\bar{z}}{2}(l - 2a) = \frac{1}{3l}[(l - a)^3 - (-a)^3] \qquad \text{or} \qquad \bar{z} = \frac{2}{3}\left(\frac{l^2 - 3la + 3a^2}{l - 2a}\right)$$

Of course, if the bar is turning about an axis through its end, the above equation for $\bar{z}$ reduces to $\frac{2}{3}l$, since a will then be zero. Under this condition, the force R is

$$R = \tfrac{1}{2}m\omega^2 l \sin \theta$$

16.42. Rework Problem 16.40 using the inertia force method.

SOLUTION

In Problem 16.41 it was indicated that the reversed effective force ($\frac{1}{2}ml\omega^2 \sin \theta$) for a bar of length l rotating about its end should be applied at a distance two-thirds of the length of the bar. Apply reversed effective forces to each of the two parts of the bar as shown in Fig. 16-56 and solve as a problem in statics.

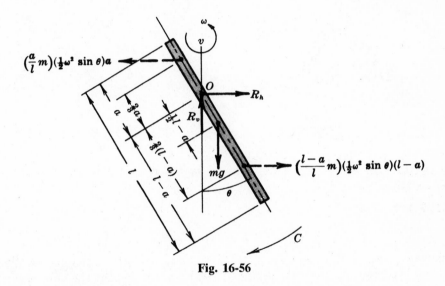

Fig. 16-56

Note that a couple C is applied to the system at the pivot point because it is not evident by inspection that the sum of moments about the pivot point of the two reversed effective forces and the weight will be zero.

The equations of equilibrium are

$$\sum F_h = 0 = +R_h - \left(\frac{a}{l}m\right)(\tfrac{1}{2}\omega^2 \sin \theta)a + \left(\frac{l-a}{l}m\right)(\tfrac{1}{2}\omega^2 \sin \theta)(l - a) \qquad (1)$$

$$\sum F_v = 0 = -mg + R_v \qquad (2)$$

$$\sum M_O = 0 = \left[\left(\frac{l-a}{l}m\right)(\tfrac{1}{2}\omega^2 \sin \theta)(l - a)\right][\tfrac{2}{3}(l - a) \cos \theta]$$

$$+ \left[\left(\frac{a}{l}m\right)(\tfrac{1}{2}\omega^2 \sin \theta)a\right](\tfrac{2}{3}a \cos \theta) - mg(\tfrac{1}{2}l - a) \sin \theta + C \qquad (3)$$

From equation (1), $R_h = -\frac{1}{2}\omega^2 m(l - 2a) \sin \theta$.

This checks, of course, with the result in Problem 16.40. The minus sign indicates that R_h was assumed in the wrong direction, and therefore actually acts to the left.

From equation (3), $C = m \sin \theta[g(\frac{1}{2}l - a) - \frac{1}{3}\omega^2(l^2 - 3la + 3a^2) \cos \theta]$.

The significance of the couple C may now be appreciated. To hold the bar at a desired angle θ

when it is rotating with a given angular speed ω requires a couple C of the magnitude just derived. However, if the couple C is not available at the pivot, then the rod will seek and maintain a definite angle θ for a given angular speed ω. This angle θ may be determined by setting the expression for C equal to zero. Then, since $\sin \theta \neq 0$,

$$\cos \theta = \frac{3g(\frac{1}{2}l - a)}{(l^2 - 3la + a^2)\omega^2}$$

Once the value of θ has been determined, the value of R_h may be found by substituting the value of $\sin \theta$ in its equation.

16.43. A cylinder of weight 161 lb (chosen because mass in slugs is then 5.00) rotates from rest in frictionless bearings under the action of a weight of 16.1 lb carried by a rope wrapped around the cylinder as shown in Fig. 16.57(a). If the diameter is 36ft, what will be the angular speed of the cylinder 2 s after motion starts?

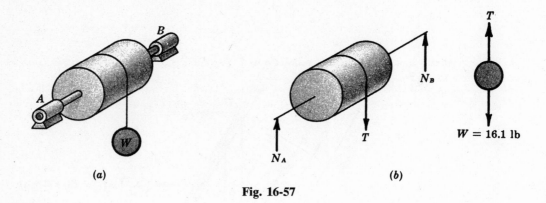

Fig. 16-57

SOLUTION

Free-body diagrams of the cylinder and the weight are shown in Fig. 16-57(b). Note that the tension T in the rope is common to both diagrams.

To determine the angular speed ω after 2 s, it is necessary to find the magnitude α of the angular acceleration. This is done by the principles of the present chapter. The only equation necessary for the cylinder is the moment equation:

$$\sum M_{AB} = \bar{I}_{AB}\alpha$$

The subscript AB means *with respect to the axis AB*. Then

$$T \times r = \tfrac{1}{2}mr^2\alpha \qquad \text{or} \qquad T \times 1.5 = \frac{1}{2}\left(\frac{161}{32.2}\right)(1.5)^2\alpha \tag{1}$$

Since both tension T and angular acceleration α are unknown, another equation is required. Write the equation of motion of the weight W by summing forces in the vertical direction:

$$\sum F_v = ma_v \qquad \text{or} \qquad 16.1 \text{ lb} - T = \frac{16.1}{32.2}a_v \tag{2}$$

Substituting $a_v = r\alpha = 1.5\alpha$ into equation (2),

$$16.1 - T = \frac{16.1}{32.2}(1.5\alpha) \tag{2'}$$

Solve (1) for $T = 3.75\alpha$ and substitute into (2') to obtain $\alpha = 3.58 \text{ rad/s}^2$.

To find ω after 2s: $\omega = \omega_0 + at = 0 + (3.58)(2) = 7.16 \text{ rad/s}$.

16.44. Study the motion of the compound pendulum shown in Fig. 16-58.

SOLUTION

The compound pendulum differs from the simple pendulum in which only one small particle is considered. Here we are concerned with many particles with various linear velocities and accelerations. The system rotates about an axis perpendicular to the plane of the paper but not a centroidal axis.

Assume the pendulum to be moving counterclockwise; that is, θ is positive in that direction. Then its weight has a moment about the axis of rotation that tends to retard motion. Hence, the equation obtained by taking moments about the axis of rotation is

$$\sum M_O = I_O \alpha$$

and since the horizontal moment arm for the weight (force) mg is $\bar{r} \sin \theta$, this is written

$$\sum M_O = -(mg)(\bar{r} \sin \theta) = I_O \frac{d^2\theta}{dt^2} \quad \text{or} \quad \frac{d^2\theta}{dt^2} = -\frac{g}{I_O/m\bar{r}} \sin \theta$$

This is the same type of equation as derived for the simple pendulum of length l, that is,

$$\frac{d^2\theta}{dt^2} = -\frac{g}{l} \sin \theta$$

Thus, it is seen that the compound pendulum has the same period as a simple pendulum whose length l is equal to $I_O/m\bar{r}$.

This may be stated somewhat differently if instead of I_O there is substituted its value in terms of the radius of gyration k_O of the compound pendulum about the axis of rotation.

Since $I_O = mk_O^2$, the length l of the equivalent simple pendulum may be written

$$l = \frac{I_O}{m\bar{r}} = \frac{mk_O^2}{m\bar{r}} = \frac{k_O^2}{\bar{r}}$$

The compound pendulum behaves as a simple pendulum with its mass concentrated at a point $k_O^2\bar{r}$ from the axis of rotation. This value $k_O^2/\bar{r}$ occurs frequently in problems of rotation.

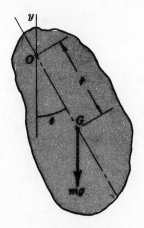

Fig. 16-58

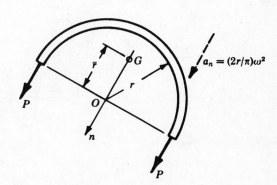

Fig. 16-59

16.45. Find the stress in the rim of a flywheel rotating with constant angular speed ω.

SOLUTION

Draw a free-body diagram of one-half the rim as shown in Fig. 16-59. The n axis is shown with sense from G to O, where G is the mass center of the thin semicircular rim. Note that $OG = 2r/\pi$.

The tensile forces P represent the pull of the other half of the rim on the free body shown. Since the wheel is rotating with constant angular speed, there is no angular acceleration; thus, no tangential effective force ($m\bar{r}\alpha$) need be shown.

Let δ be the mass density of the material. Only one equation of motion is necessary:

$$\sum F_n = m\bar{a}_n \qquad \text{or} \qquad 2P = m\frac{2r}{\pi}\omega^2$$

But the unit tensile stress is the force P (on one side) divided by the cross-sectional area A on which or through which it acts, or $\sigma = P/A = mr\omega^2/\pi A$.

Next express the mass m of the half rim in terms of the rim cross section A, half rim length πr, and mass density δ:

$$m = A\pi r\delta$$

This is, of course, based on the assumption that the rim thickness is small compared with the rim diameter. Substituting this expression for m in the equation for stress, we have

$$\sigma = \delta r^2\omega^2$$

Since the rim speed v is equal to $r\omega$, the formula for stress may also be written

$$\sigma = \delta v^2$$

where in U.S. Customary units, σ is the unit stress in lb/ft^2, δ is the mass density in slugs/ft^3, and v is the rim speed in ft/s.

In SI units, σ is the unit stress in N/m^2, δ is the mass density in kg/m^3, and v is the rim speed in m/s.

16.46. It is necessary to bank rail road tracks on a curve in such a way that the outer rail is above the inner rail. Since the roadbed is usually at some elevation above sea level, it is customary to call the vertical distance of the outer rail above the inner rail not elevation but the superelevation e. The greater the superelevation e, the faster a train may travel around the curve. Determine the value of e in terms of the speed v of the train, the radius r of the curve and the angle of bank. Refer to Fig. 16-60(a).

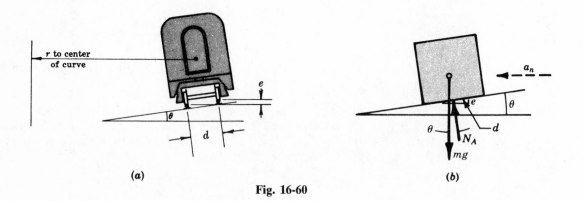

(a)   (b)

Fig. 16-60

SOLUTION

This is an example of rotation of a mass that can be considered a particle turning about the center of the curve.

Consider the curve to be properly banked for a definite speed v. This means that the train has no tendency to slide in or out along the radius. Hence, the side pressure of either rail against the flange of the wheel is zero. The only reaction on the car is the push, perpendicular to the roadbed, of the tracks on the wheels.

The free-body diagram in Fig. 16-60(b) shows the gravitational force mg and a single reaction N_A

of the tracks. Note that the acceleration a_n is directed to the left toward the center of the curve. Its magnitude is $r\omega^2$. The equations of motion are

$$\sum F_n = mr\omega^2 \qquad \text{or} \qquad N_A \sin\theta = mr\omega^2$$
$$\sum F_v = 0 \qquad \text{or} \qquad N_A \cos\theta = mg$$

Dividing, $\tan\theta = r\omega^2/g$.

Since θ is usually a small angle, $\tan\theta$ is approximately equal to $\sin\theta$ (up to about 6°); therefore the above equation may be written

$$\sin\theta \approx \tan\theta = \frac{r\omega^2}{g} = \frac{r^2\omega^2}{gr}$$

But from the figure, $\sin\theta = e/d$. Hence, $e/d = r^2\omega^2/gr = v^2/gr$ or $e = dv^2/gr$.

16.47. In Problem 16.46, calculate the superelevation for a 2000-ft curve for a speed of 120 mi/h. Assume that the gage $d = 4$ ft $8\frac{1}{2}$ in.

SOLUTION

$$e = \frac{dv^2}{gr} = \frac{(4.71\ \text{ft})(176\ \text{ft/s})^2}{(32.2\ \text{ft/s}^2)(2000\ \text{ft})} = 2.27\ \text{ft} = 27.2\ \text{in}$$

16.48. In Fig. 16-61(a) weight A is accelerating down 5 ft/s². It is connected by a weightless, flexible, inextensible rope passing over a smooth drum to a homogeneous cylinder B of weight 161 lb. The cylinder is acted upon by a moment $C = 50$ lb-ft counterclockwise. Detemrine the weight of A and the components of the reaction at O on the cylinder B.

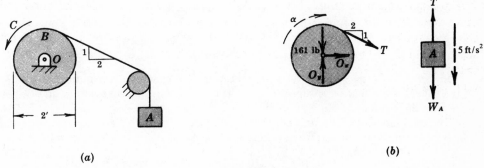

Fig. 16-61

SOLUTION

Draw free-body diagrams of the cylinder and the weight A as shown in Fig. 16-61(b).

The magnitude a_t of the tangential acceleration of a point on the rim is 5 ft/s². Hence, the magnitude α of the angular acceleration is $a_t/r = 5$ rad/s². Since the cylinder rotates about an axis of symmetry, the three equations of motion for the cylinder are

$$\sum \bar{M} = \bar{I}\alpha \qquad \text{or} \qquad T \times 1 - 50 = \frac{1}{2}\left(\frac{161}{32.2}\right)(1)^2(5) \tag{1}$$

$$\sum F_x = 0 \qquad \text{or} \qquad O_x + T\left(\frac{2}{\sqrt{5}}\right) = 0 \tag{2}$$

$$\sum F_y = 0 \qquad \text{or} \qquad O_y - 161 - T\left(\frac{1}{\sqrt{5}}\right) = 0 \tag{3}$$

The equations of motion for weight A is

$$W_A - T = \frac{W_A}{g}(5) \tag{4}$$

Equation (1) yields $T = 62.5$ lb. hence, from equation (4), $W_A = 74.0$ lb.
Solve equations (2) and (3) to obtain $O_x = -55.9$ lb and $O_y = +189$ lb.

16.49. A homogeneous sphere having a mass of 100 kg is attached to a slender rod having a mass of
20 kg. In the horizontal position shown in Fig. 16.62(a), the angular speed of the system is
8 rad/s. Determine the magnitude of the angular acceleration of the system and the reaction
at O on the rod.

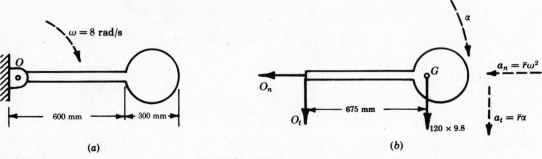

(a)   (b)

Fig. 16-62

SOLUTION

Locate the centroid of the system with reference to the pin O:

$$\bar{r} = \frac{20(300) + 100(750)}{120} = 675 \text{ mm}$$

Figure 16-62(b) shows the three external forces O_n, O_t, and the gravitational force through G acting
on the system. Note that O_n is shown to the left to agree with the sense of $\bar{a}_n$, and O_t is shown down to
agree with the sense of $\bar{a}_t$.

The total moment of inertia I equals the moment of inertia of the rod about its end ($\frac{1}{3}ml^2$) plus the
transferred moment of inertia of the sphere ($\frac{2}{5}mr^2 + md^2$), where $d = 750$ mm $= 0.75$ m.

$$I = \tfrac{1}{3}20(0.6)^2 + \tfrac{2}{5}100(0.15)^2 + 100(0.75)^2 = 59.55 \text{ kg} \cdot \text{m}^2$$

To find the angular acceleration, use the moment equation

$$\sum M_O = I\alpha$$

or $120 \times 9.8 \times 0.675 = 59.55\alpha$

Hence, $\alpha = 13.3 \text{ rad/s}^2$

Sum forces horizontally (along the n axis) to obtain

$$O_n = m\bar{r}\omega^2 = 120(0.675)(8)^2 = 5180 \text{ N} \qquad \text{(to the left)}$$

Sum forces vertically (along the t axis) to obtain

$$O_t + 120 \times 9.8 = m\bar{r}\alpha = 120 \times 0.675 \times 13.3$$

or $O_t = -98.7 \text{ N} \qquad \text{(therefore up)}$

16.50. An eccentric cylinder used in a vibrator weights 40 lb and rotates about an axis 2 in from its geometric center and perpendicular to the top view shown in Fig. 16-63. If the magnitudes ω and α of its angular velocity and angular acceleration are, respectively, 10 rad/s and 2 rad/s² in the phase shown, determine the reaction of the vertical shaft on the cylinder.

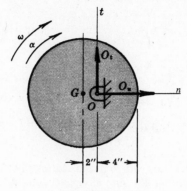

Fig. 16-63

SOLUTION

Choose n and t axes as shown. The distance from O to the mass center G is $\bar{r} = \frac{1}{6}$ ft.
The equations of motion are

$$\sum F_n = m\bar{r}\omega^2 \qquad \text{or} \qquad O_n = \frac{40}{32.2}\left(\frac{1}{6}\right)(10)^2 \tag{1}$$

$$\sum F_t = m\bar{r}\alpha \qquad \text{or} \qquad O_t = \frac{40}{32.2}\left(\frac{1}{6}\right)(2) \tag{2}$$

$$\sum M_O = I_O\alpha = \left[\frac{1}{2}\left(\frac{40}{32.2}\right)\left(\frac{1}{2}\right)^2 + \frac{40}{32.2}\left(\frac{1}{6}\right)^2\right] \times 2 \tag{3}$$

Here the reactions O_n and O_t of the shaft are the external forces acting *on* the cylinder.

Note that I_O is found by the transfer formula. The necessary couple $I_O\alpha$ equals 0.38 lb-ft applied clockwise to the cylinder by the shaft.

The reaction components are $O_n = 20.7$ lb to the right and $O_t = 0.41$ lb up.

16.51. The 6-lb sphere in Fig. 16-64(a) moves in a circular and horizontal path under the actions of the 8-ft weightless bar AB and the very light cord BC. When the speed of B is 10 ft/s, determine the forces in the two supporting members.

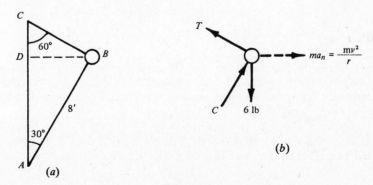

Fig. 16-64

SOLUTION

The free-body diagram in Fig. 16-64(b) shows the reversed effective force applied to hold the sphere in "equilibrium for study purposes." From trigonometry, $DB = \bar{r} = 4$ ft. Hence, the reversed effective force is

$$\frac{6}{g}\frac{(10)^2}{4} = 4.66 \text{ lb}$$

The equations become

$$\sum F_h = 0 = C \cos 60° - T \cos 30° + 4.66$$

$$\sum F_v = 0 = C \cos 30° + T \cos 60° - 6$$

The solutions are $C = 2.87$ lb and $T = 7.04$ lb.

In this particular problem, with BC and AB at right angles, a neat solution would be setting the sum of the forces along AB and the sum along BC equal to zero. In each equation, only one unknown is involved.

16.52. In Fig. 16-65(a) the bar AB is held in a vertical position by the weightless cord BC as the system rotates about the vertical y–y axis. The pin at a is smooth and the bar AB weighs 32.2 lb. If the breaking strength of the cord is 120 lb, how fast can the system rotate without breaking the cord?

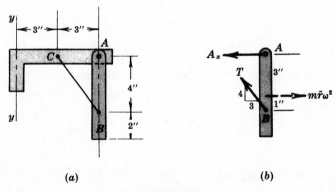

(a)   (b)

Fig. 16-65

SOLUTION

Draw a free-body diagram of the bar AB, omitting the weight and the vertical component A_y for simplicity. See Fig. 16-65(b). The reversed effective force is placed at the center of the bar. This is true only because each horizontal element or slice of the bar is the same distance from y–y as any other element.

Take moments about A to obtain $+m\bar{r}\omega^2 \times 3 - T \times \frac{3}{5} \times 4 = 0$. Note that only the horizontal component of T has a moment about A. Substitute $m = 1$ slug, $T = 120$ lb, and $\bar{r} = 6/12$ ft to obtain $\omega = 13.8$ rad/s or 132 rpm, the value beyond which the cord will break.

16.53. The compound pulley system shown in Fig. 16-66(a) has a mass of 30 kg and a radius of gyration of 450 mm. Determine the tension in each cord and the angular acceleration of the pulleys when the masses are released.

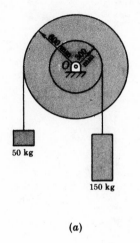

(a)

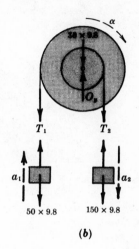

(b)

Fig. 16-66

SOLUTION A

Draw free-body diagrams of the three components of the system as in Fig. 16-66(b). Note that $a_1 = 0.6\alpha$ and $a_2 = 0.3\alpha$.

The equations of motion are

$$\sum F = T_1 - 50 \times 9.8 = 50a_1 = 30\alpha \tag{1}$$

$$\sum F = 150 \times 9.8 - T_2 = 150a_2 = 45\alpha \tag{2}$$

$$\sum \bar{M} = \bar{I}\alpha \quad \text{or} \quad T_2 \times 0.3 - T_1 \times 0.6 = 30(0.45)^2\alpha = 6.08\alpha \tag{3}$$

The solution is $\alpha = 3.9 \text{ rad/s}^2$. From this the tensions are

$$T_1 = 490 + 30(3.91) = 607 \text{ N} \quad \text{and} \quad T_2 = 1470 - 45(3.91) = 1290 \text{ N}$$

SOLUTION B

If the reversed effective (inertia) forces and moments are applied to the system to hold it in "equilibrium for study purposes only," the method of virtual work can be applied for a solution (see Problems 11.6 and 11.7).

Apply the inertia forces as follows: $50(0.6\alpha)$ down through the center of the 50-kg mass, $150(0.3\alpha)$ up through the center of the 150-kg mass, and a counterclockwise moment of 6.08α to the puley. Since the system is now in "equilibrium," we can give the system a virtual displacement $\delta\theta$ clockwise. The work done by the external forces *only* for this virtual displacement is zero. Note the virtual displacements are $\delta\theta$ for the pulleys, $(0.6) \delta\theta$ for the 50-kg mass, and $(0.3) \delta\theta$ for the 150-kg mass. Hence,

$$\delta U = [-50 \times 9.8 - 50(0.6)\alpha](0.6) \delta\theta + [150 \times 9.8 - 150(0.3)\alpha](0.3) \delta\theta + (-6.08\alpha) \delta\theta = 0$$

The solution after dividing each term of $\delta\theta$ is $\alpha = 3.9 \text{ rad/s}^2$. The convenience in finding the angular acceleration is noted, but to find the tension in either cord, a free-body diagram must be drawn.

16.54. In Fig. 16-67(a), the weight C of 28 lb is moving down with a velocity of 16 ft/s. The moment of inertia of drum B is 12 ft-lb-s², and it rotates in frictionless bearings. If the coefficient of friction between the brake A and the drum is 0.40, what force P is necessary to stop the system in 2 s? What is the reaction at D on the rod?

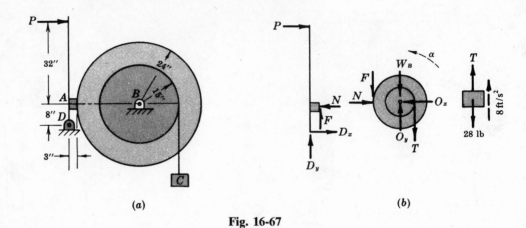

(a)   (b)

Fig. 16-67

SOLUTION

Draw the free-body diagrams of the bar, the drum, and the weight as shown in Fig. 16-67(b). Next determine the magnitude of the acceleration of the weight knowing the initial speed is 16 ft/s down, the final speed is zero, and the time is 2 s. The value of acceleration is 8 ft/s² up.

A vertical summation of forces acting on weight C gives $T - 28 = (28/g)(8)$ or $T = 35.0$ lb.

To find the friction F, sum moments of forces about O of the drum. The angular acceleration of the drum is counterclockwise (the system is slowing down). Its magnitude may be determined because the linear acceleration of a point 15 in from O is 8 ft/s². Hence, $\alpha = 8/(15/12) = 6.40$ rad/s². The moment equation is $F(2) - T(1.25) = I\alpha$ or $F(2) - 35.0(1.25) = 12(6.40)$, from which $F = 60.3$ lb.

But $F = \mu N$. Hence the necessary normal force $N = 60.3/0.40 = 151$ lb.

Summing moments about D of all forces acting on the bar, $-P(40) + N(8) + F(3) = 0$ or $P = 34.7$ lb.

Sum forces horizontally on the bar to obtain $34.7 - 151 + D_x = 0$ or $D_x = 116$ lb to the right.

Summing forces vertically, $D_y + 60.3 = 0$ or $D_y = 60.3$ lb down.

16.55. An 8-kg ball A is mounted on a horizontal bar attached to a vertical shaft as shown in Fig. 16-68(a). Neglecting the mass of the bar and shaft, what are the reactions at B and C when the system is rotating at a constant speed of 90 rpm?

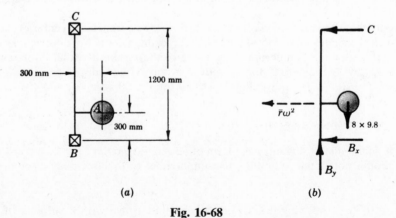

(a)   (b)

Fig. 16-68

SOLUTION

The free-body diagram in Fig. 16-68(b) shows the forces C, B_x, B_y, and the gravitational force $8 \times 9.8 = 78.4$ N. The value of the angular speed is $\omega = 90 \times 2\pi/60 = 9.42$ rad/s.

The equations of motion are

$$\sum F_x = m\bar{r}\omega^2 \qquad \text{or} \qquad B_x + C = 8(0.3)(9.42)^2 = 213$$

$$\sum F_y = 0 \qquad \text{or} \qquad B_y - 78.4 = 0$$

$$\sum \bar{M} = 0 \qquad \text{or} \qquad C \times 0.9 - B_x \times 0.3 - B_y \times 0.3 = 0$$

The solutions are $B_x = 140$ N, $B_y = 78.4$ N, and $C = 72.9$ N, all acting as shown in the free-body diagram.

Inertia-Force Method

16.56. Solve Problem 16.55 using the inertia-force method.

SOLUTION

Figure 16-69 shows the free-body diagram, with the reversed effective force indicated. The equivalent equilibrium equations are then

$$m\bar{r}\omega^2 = 8(0.3)(9.42)^2 = 213 \text{ N}$$

$$\sum M_B = 1.2C - 78.4(0.3) - 213(0.3) = 0 \qquad C = 72.9 \text{ N}$$

$$\sum F_x = -B_x - C + 213 = 0 \qquad B_x = 140 \text{ N}$$

$$\sum F_y = B_y - 78.4 = 0 \qquad B_y = 78.4 \text{ N}$$

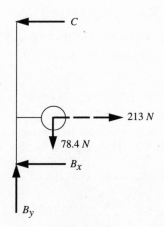

Fig. 16-69

16.57. Solve Problem 16.33 using the inertia-force method.

SOLUTION

Figure 16-70 shows the free-body diagrams of A and B, with the reversed effective forces indicated. Writing the equivalent equilibrium equations, we have

for A, $\qquad \sum F_h = P - 30a - \frac{1}{3}N_A = 0$ $\qquad\qquad\qquad$ (1)

for B, $\qquad \sum F_h = -45a + \frac{1}{3}N_A - \frac{1}{10}N_B = 0$ $\qquad\qquad\qquad$ (2)

But $N_A = 294$ N and $N_B = 735$ N.

From equation (2), $a = 0.544 \text{ m/s}^2$
Substituting in equation (1), $P = 114 \text{ N}$

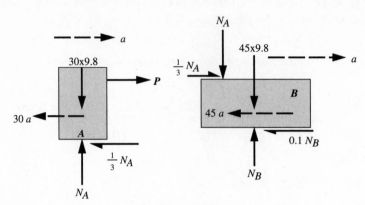

Fig. 16-70

16.58. Find the reactions on the rollers in Problem 16.25 using the inertia-force method. The height of the door is 5 ft.

SOLUTION

Figure 16-71 shows the free body diagram of the door, with the inertia force indicated.
The equations for the equivalent equilibrium problem can be written as

$$\sum F_x = 10 - (50/32.2)a = 0 \tag{1}$$

$$\sum M_A = 8B + (1.5)10 - (2.5)(50/32.2)a - (4)50 = 0 \tag{2}$$

$$\sum F_y = A + B - 50 = 0 \tag{3}$$

Hence, from (1), $a = 6.44 \text{ ft/s}^2$
from (2), $B = 26.3 \text{ lb}$
from (3), $A = 23.7 \text{ lb}$

It can be seen that, without D'Alembert's Principle, it would require a solution of simultaneous equations to determine A and B. This is generally the advantage to be gained from the inertia-force method.

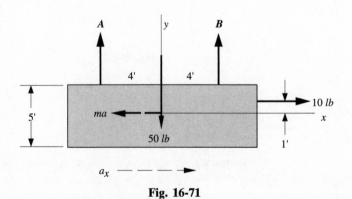

Fig. 16-71

16.59. Solve Problem 16.37 using the inertia-force method.

SOLUTION

Figure 16-72 shows the free-body diagram, with the reversed effective force and reversed effective couple indicated.

The equivalent equilibrium equations are

$$\sum F_n = O_n = 0$$
$$\sum M_O = Fl - m(\tfrac{1}{2}l)\alpha(\tfrac{1}{2}l) - \tfrac{1}{12}ml^2\alpha = 0$$
$$\sum F_t = O_t + F - m(\tfrac{1}{2}l)\alpha = 0$$

Solving these equations yields

$$\alpha = 3F/ml$$
$$O_t = \tfrac{1}{2}F$$
$$O_n = 0$$

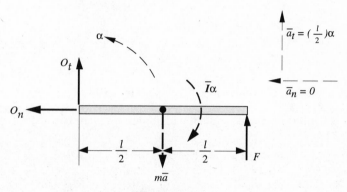

Fig. 16-72

16.60. Find the angular acceleration in Problem 16.49 using the inertia-force method.

SOLUTION

Figure 16-73 shows the free-body diagram, with the reversed effective forces and couples indicated. Writing the moment equation with respect to O yields

$$\sum M_O = -20g\bar{r}_1 + m_1\bar{r}_1^2\alpha + \bar{I}_1\alpha - 100g\bar{r}_2 + m_2\bar{r}_2^2\alpha + \bar{I}_2\alpha = 0$$
$$\sum M_O = -20(9.8)(0.3) + 20(0.3)^2\alpha + (20/12)(0.6)^2\alpha$$
$$- 100(9.8)(0.75) + 100(0.75)^2\alpha + \tfrac{2}{5}(100)(0.15)^2\alpha = 0$$

from which $\alpha = 13.3 \text{ rad/s}^2$

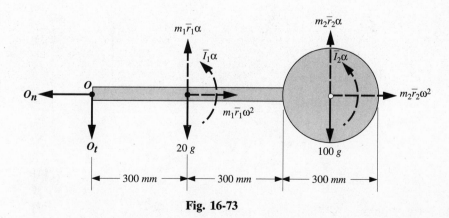

Fig. 16-73

16.61. Solve Problem 16.2 using the inertia-force method.

SOLUTION

Figure 16-74 shows the free-body diagram, with the reversed effective force and couple indicated. Now, $m = 600/32.2 = 18.6$ slugs and $\bar{I} = \frac{1}{2}mr^2$, $\bar{I} = \frac{1}{2}(18.6)(\frac{15}{12})^2 = 14.5$ slug-ft^2

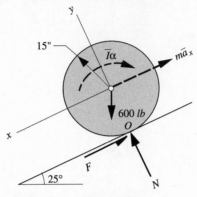

Fig. 16-74

The equations for the equivalent equilibrium problem become

$$\sum M_O = r(W \sin 25°) - r\frac{W}{g}\bar{a}_x - \bar{I}\alpha = 0$$

$$\sum F_x = W \sin 25° - F - \frac{W}{g}\bar{a}_x = 0$$

Or $$\sum M_O = (\tfrac{15}{12})600 \sin 25° - (\tfrac{15}{12})18.6\bar{a}_x - 14.5\alpha = 0$$

where $$\bar{a}_x = (\tfrac{15}{12})\alpha$$

Solving yields $$\bar{a}_x = 9.09 \text{ ft/s}^2$$

$$\sum F_x = 600 \sin 25° - F - 18.6(9.09) = 0 \qquad F = 84.3 \text{ lb}$$

16.62. A 7-kg cylinder is supported by a string wrapped around it and attached to the ceiling. Find the force in the string using the inertia-force method. See Fig. 16-75(a).

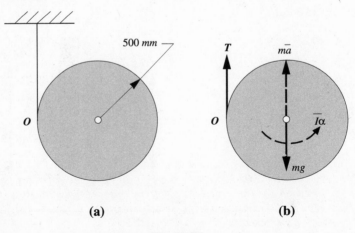

(a)   (b)

Fig. 16-75

SOLUTION

Figure 16-75(*b*) shows the free-body diagram with the reversed effective force and couple indicated. Now, $m = 7$ kg, $\bar{I} = \frac{1}{2}mr^2 = (\frac{1}{2})7(0.5)^2 = 0.875$ kg $\cdot$ m^2 and $\bar{a} = 0.5\alpha$.

The equations of the equivalent equilibrium problem become

$$\sum M_O = -mgr + m\bar{a}r + \bar{I}\alpha = 0$$

$$\sum \bar{M} = -rT + \bar{I}\alpha = 0$$

Or

$$\sum M_O = -(7)(9.8)(0.5) + 7\bar{a}(0.5) + 0.875\alpha = 0$$

$$\sum \bar{M} = -0.5T + 0.875\alpha = 0$$

Solving for α and T gives

$$\alpha = 13.1 \text{ rad/s}^2$$

$$T = 22.9 \text{ N}$$

Supplementary Problems

16.63   A wheel weighing 150 lb is 10 ft in diameter and rolls down a 45° plane. If there is no slippage, determine the angular acceleration of the wheel.   *Ans.*   3.04 rad/s^2

16.64.   A 180-kg cylinder is 200 mm in diameter and rests on horizontal rails perpendicular to its geometric axis. A force P of 550 N is applied tangentially to the right at the bottom of the cylinder. Assuming that the coefficients of static and kinetic friction are 0.29 and 0.25, respectively, what is the motion? Refer to Problem 16.4.
Ans.   Cylinder slides to right and rotates counterclockwise; $\bar{a} = 0.6$ m/s^2; $\alpha = 12.1$ rad/s^2

16.65.   In Fig. 16-76, a cylinder of weight W and radius of gyration k has a rope wrapped around a groove of radius r. Determine the acceleration of the mass center if pure rolling is assumed. Force P is horizontal, as is the plane.   *Ans.* $\bar{a} = PgR(R - r)/W(R^2 + k^2)$

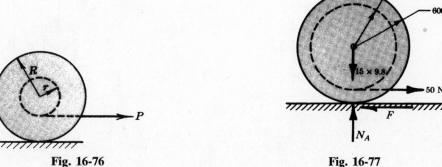

Fig. 16-76   Fig. 16-77

16.66.   A homogeneous cylinder having a mass of 15 kg has a narrow peripheral slot cut in it as shown in Fig. 16-77. A force of 50 N is exerted on a string wrapped in the slot. If the cylinder rolls without slipping, determine the acceleration of its mass center and the frictional force F. Neglect the effect of the slot.   *Ans.* $a = 0.556$ m/s^2, $F = 41.7$ N to the left

16.67.   A 50-kg cylindrical wheel 1200 mm in diameter is pulled up a 20° plane by a cord wrapped around its circumference. If the cord passes over a smooth pulley at the top of the plane and supports a hanging mass of 90 kg, determine the angular acceleration of the wheel. Assume that the cord pulls at the top of the wheel and is parallel to the plane.   *Ans.*   6.1 rad/s^2

16.68.   What is the tension in the cable in the system shown in Fig. 16-78. Assume that the pulleys are weightless and frictionless. the plane is smooth.   *Ans.* $T = 5.16$ lb

16.69.   In Problem 16.68 assume that the lower pulley is a homogeneous cylinder 1 ft in diameter and weighing 4 lb. What is the tension in the cable parallel to the plane?   *Ans.* $T = 7.79$ lb

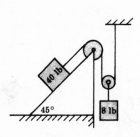

Fig. 16-78

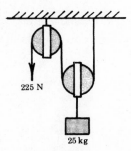

Fig. 16-79

16.70.   A constant force of 225 N is applied as shown in Fig. 16-79. How long will it take the 25-kg block to reach a speed of 1.2 m/s starting from rest? Assume that the pulleys are massless and frictionless. *Ans.* $t = 0.145$ s

16.71.   In the system shown in Fig. 16-80, the pulleys are to be considered weightless and frictionless. The masses are 1, 2, 3, and 4 kg. Determine the acceleration of each mass and the tension in the top cord. *Ans.* $a_1 = 9.02$ m/s^2 up, $a_2 = 0.39$ m/s^2 down, $a_3 = 3.53$ m/s^2 down, $a_4 = 5.10$ m/s^2 down, $T = 75.3$ N

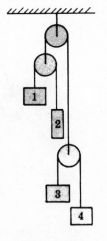

Fig. 16-80

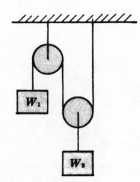

Fig. 16-81

16.72.   In Fig. 16-81 assume that the weight W_2 is moving down. If the pulleys are "weightless" and frictionless, determine the tension in the rope when the weights are released from rest. *Ans.* $I = 3W_1W_2/(4W_1 + W_2)$

16.73.   A horizontal force of 180 N is applied tangentially at the top of a 45-kg cylinder having a diameter of

1500 mm. If there is no slippage at the ground, determine the linear acceleration of the center of the cylinder.   *Ans.*   5.33 m/s² horizontally

16.74. A thin hoop of weight W rolls horizontally under the action of a horizontal force P applied at the top as shown in Fig. 16-82. Express the mass center acceleration $\bar{a}$ in terms of P, W, and radius R. Also show that the frictional force of the plane on the hoop is zero.   *Ans.* $\bar{a} = Pg/W$

16.75. In Problem 16.11, what must be the weight of C so that the acceleration of the disk A is 6 ft/s² up the plane?   *Ans.*   1340 lb

16.76. A 2-kg uniform sphere of radius 80 mm has a rope wrapped around it with one end attached to the ceiling as shown in Fig. 16-83. If the sphere is released from rest, determine the tension in the rope, the angular acceleration of the sphere, and the speed of the mass center 2 s after release.
Ans. $T = 5.6$ N, $\alpha = 87.5$ rad/s² clockwise, $\bar{v} = 14.0$ m/s down.

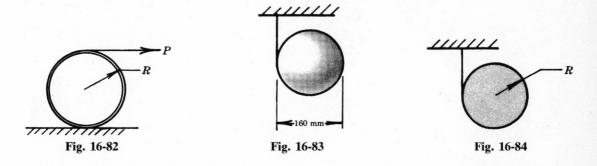

Fig. 16-82   **Fig. 16-83**   **Fig. 16-84**

16.77. A cylinder of weight W and radius of gyration k is released from rest in the position shown in Fig. 16-84. If the radius of the cylinder is R, determine the tension in the cord and the acceleration of the mass center.   *Ans.* $T = W/(1 + R^2/k^2)$, $\bar{a} = g/(1 + k^2/R^2)$

16.78. Figure 16-85 shows a 40-lb weight attached to a cord that wraps around a 30-lb cylinder. The cylinder rolls without slipping up the 30° plane. Determine the angular acceleration of the cylinder.
Ans.   0.98 rad/s²

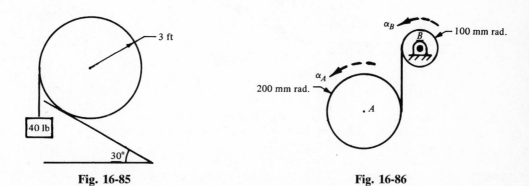

Fig. 16-85   **Fig. 16-86**

16.79. In Fig. 16-86, a 2-kg cylinder B with a radius of 100 mm is free to turn in frictionless bearings. A rope is wrapped around B and is then wrapped around the 4-kg cylinder with radius 200 mm. Determine the angular accelerations of A and B immediately after the cylinders are released with zero velocity.
Ans. $\alpha_A = 14$ rad/s², $\alpha_B = 56$ rad/s²

16.80. In Fig. 16-87, a homogeneous cylinder and a homogeneous sphere of equal masses $m = 8$ kg and equal

radii R are harnessed together by a light frame and are free to roll without slipping down the plane inclined 28° with the horizontal. Determine the force in the frame. Assume that the bearings are frictionless.   *Ans.*   1.27 N

16.81.   A homogeneous ball of diameter 3 ft is spinning at 60 rpm about a horizontal centroidal axis when it is lowered onto a plane for which the coefficient of sliding friction is 0.30. Determine the time it takes before rolling begins.   *Ans.* $t = 0.279$ s

16.82.   A homogeneous cylinder with a mass of 3 kg and of diameter 120 mm is rotating 8 rad/s about a horizontal axis when dropped on a horizontal plane for which the coefficient of friction is 0.25. How far will the center travel before rolling begins and skidding stops?   *Ans.* $d = 5.2$ mm

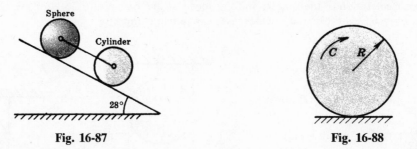

Fig. 16-87   **Fig. 16-88**

16.83.   A homogeneoous cylinder of mass m and radius R is at rest on a horizontal plane when a couple C is applied as shown in Fig. 16-88. Determine the magnitude of the coefficient of friction between the wheel and the plane so that rolling will occur.   *Ans.* $\mu \geq \frac{2}{3}C/mgR$

16.84.   In Fig. 16-89, the thin-walled cylinder is on a horizontal platform that has an acceleration a. Determine the acceleration of the center assuming rolling without sliding.   *Ans.* $\bar{a} = 0.5a$

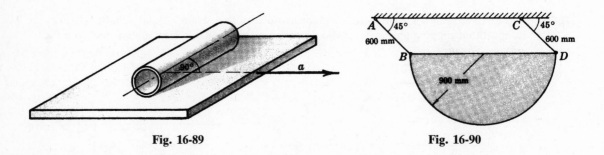

Fig. 16-89   **Fig. 16-90**

16.85.   In Problem 16.84, assume that the coefficient of friction is $\mu = 0.35$. Determine the maximum acceleration the platform may have without slip between the cylinder and platform.
Ans. $a = 6.86$ m/s² or 22.5 ft/s²

16.86.   The 30-kg semicircular homogeneous plate of 900-mm radius is released from rest in the position shown in Fig. 16-90. What are the forces in the two cords for this position?
Ans. $T_{AB} = 148$ N, $T_{CD} = 59.8$ N

16.87.   In a loop-the-loop 20 ft in diameter, a car weighing 500 lb leaves the platform 30 ft above the bottom of the loop. What is the normal force of the track on the car at the top of the loop? Assume that the center of gravity of the car is 10 ft from the center of the loop.   *Ans.*   500 lb

16.88.   A disk of weight W has its center of mass G at a distance e from the geometric center O. The disk is rolling on the horizontal plane with constant angular velocity ω. Determine the normal force and frictional force of the floor on the disk when it is in the position shown in Fig. 16-91.

Ans. $F = (W/g)e\omega^2 \cos \theta$ to the left, $N = W + (W/g)e\omega^2 \sin \theta$ up

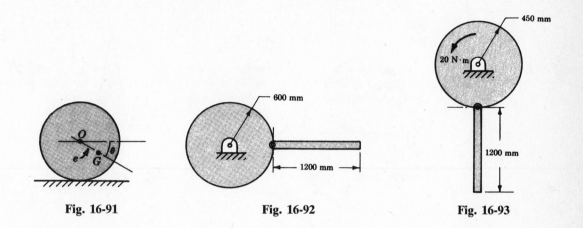

Fig. 16-91   Fig. 16-92   Fig. 16-93

16.89. The disk in Fig. 16-92 has a mass of 50 kg, a radius of 600 mm, and a radius of gyration relative to its mass center of 450 mm. The uniform slender bar is 1200 mm long and has a mass of 18 kg. Its left end is pinned to the disk. It is initially held in the horizontal position. Determine the angular accelerations of the disk and bar immediately after release from rest.

Ans. $\alpha_{\text{disk}} = 2.25$ rad/s² clockwise, $\alpha_{\text{bar}} = 10.5$ rad/s² clockwise

16.90. A 20-kg drum is homogeneous and has a radius of 450 mm. A 10-kg bar is homogeneous and is 1200 mm long. Its upper end is pinned to the bottom of the drum as shown in Fig. 16.93. Assuming that a couple with moment of 20 N · m is applied to the drum, determine the angular accelerations of the drum and the bar.   *Ans.* $\alpha_{\text{drum}} = 7.9$ rad/s² counterclockwise, $\alpha_{\text{bar}} = 4.44$ rad/s² clockwise

Translation

16.91. A homogeneous door weighing 200 lb is free to roll on a horizontal track on frictionless wheels at A and B as shown in Fig. 16.94. What horizontal force P will reduce the vertical reaction at A to zero? What will be the acceleration of the door under the action of this force P? What will be the reaction at B?

Ans. $P = 450$ lb to the left, $a = 72.5$ ft/s² to the left, $B = 200$ lb up

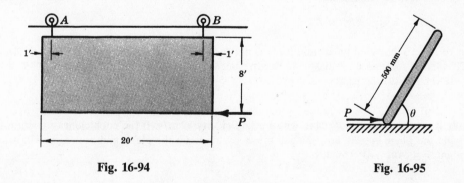

Fig. 16-94   Fig. 16-95

16.92. Figure 16-95 shows a homogeneous slender bar with a mass of 4 kg and a length of 500 mm being pushed along the smooth horizontal surface by a horizontal force $P = 60$ N. Determine the angle θ for translation. What is the accompanying acceleration?   *Ans.* $\theta = 33.2°$, $a = 15$ m/s²

16.93.   A cylinder having a mass of 10 kg with a diameter of 1.2 m is pushed to the right without rotation and with acceleration 2 m/s^2. See Fig. 16-96. Determine the magnitude and location of the force P if the coefficient of friction is 0.20.   *Ans.* $P = 39.6 \text{ N}$ to the right, $h = 0.3 \text{ m}$ above surface

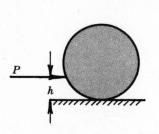

Fig. 16-96

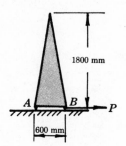

Fig. 16-97

16.94.   A homogeneous lamina with a face that is an isosceles triangle is pulled to the right with an acceleration of 1.2 m/s^2. See Fig. 16-97. Assuming the supports at A and B are frictionless, determine the force P and the reactions at A and B if the mass is 2.5 kg.
Ans. $P = 3 \text{ N}$ to the right, $A = 15.2 \text{ N}$ up, $B = 9.25 \text{ N}$ up

16.95.   A homogeneous sphere of weight W lb and radius R is acted upon by a horizontal force P as shown in Fig. 16-98. The coefficient of sliding friction between the plane and sphere is μ. Where must the force P be applied in order that the sphere skids without rotating? What is the acceleration?
Ans. $d = R(1 - \mu W/P)$, $a = g(P/W - \mu)$ to the right

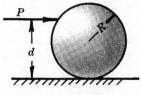

Fig. 16-98

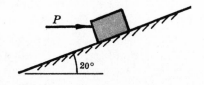

Fig. 16-99

16.96.   Refer to Fig. 16-99. What horizontal force is needed to give the 50-kg block an acceleration of 3 m/s^2 up the 20° plane? Assume a coefficient of friction between the block and the plane of 0.25.
Ans. $P = 507 \text{ N}$ to the right

16.97.   A body is projected up a 30° plane with initial velocity 12 m/s. If the coefficient of friction between the body and the plane is 0.20, how far will the body move up the plane and how long will it take to reach this point?   *Ans.*   10.9 m, 1.82 s

16.98.   A water slide is inclined 35° with the horizontal. Assuming no friction, how long will it take a child starting from a rest position to slide 15 ft?   *Ans.* $t = 1.27 \text{ s}$

16.99.   An object weighing W lb starts from rest at the top of a plane that has a slope of 50° with the

horizontal. After traveling 20 ft down the plane, the object slides 30 ft across the horizontal floor and comes to rest. Determine the coefficient of friction between the object and the surfaces.
Ans. $\mu = 0.357$

16.100. A dish slides 500 mm on a level table before coming to rest. If the coefficient of friction between the dish and the table is 0.12, what was the time of travel?   *Ans.* $t = 0.92$ s

16.101. A 9.3-lb block is prevented from slipping by means of a strip as shown in Fig. 16-100. What is the greatest acceleration the cart may have without tipping the block?   *Ans.* $a = 12.1$ ft/s^2 to the left

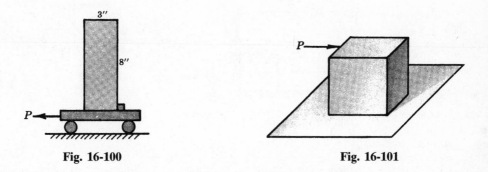

Fig. 16-100   **Fig. 16-101**

16.102. The homogeneous cube in Fig. 16-101 is moved along the horizontal plane by a horizontal force *P*. If the coefficient of friction is 0.2, what is the maximum acceleration the block can have if it slides but is on the verge of tipping?   *Ans.* $a = 5.88$ m/s^2 to the right

16.103. A cylindrical shaft 6 ft in diameter and 35 in high stands on end in a flat car that has a constant velocity of 35 mi/h. If the car is brought to rest with uniform deceleration in a distance of 40 ft, will the cylinder tip?   *Ans.*   No

16.104. An 8-kg mass is being lowered by a rope that can safely support 60 N. What is the minimum acceleration the mass can have under these conditions?   *Ans.*   2.3 m/s^2 down

16.105. An elevator requires 2 s from rest to acquire a downward velocity of 10 ft/s. Assuming uniform acceleration, what is the force of the floor during this time on an operator whose normal weight is 150 lb?   *Ans.*   127 lb

16.106. An elevator having a mass of 500 kg is ascending with an acceleration of 4 m/s^2. The mass of the operator is 65 kg. Assuming that the operator is standing on spring scales during the ascension, what is the scale reading? What is the tension in the cables?   *Ans.*   91.5 kg, 7800 N

16.107. A body with a mass of 12 kg at rest is on a spring scale in an elevator. What should be the acceleration (magnitude and direction) in order for the scale to indicate a mass of 10 kg?
Ans. $a = 1.63$ m/s^2 down

16.108. A 10-lb weight hangs from a spring balance in an elevator that is accelerating upward at 8.05 ft/s^2. Determine the spring balance reading.   *Ans.*   12.5 lb

16.109. An elevator having a mass of 700 kg is designed for a maximum acceleration of 2 m/s^2. What is the maximum load that may be placed in the elevator if the allowable load in the supporting cable is 19 kN? *Ans.* $M = 910 \text{ kg}$

16.110. A horizontal cord connects the 10- and 20-kg masses shown in Fig. 16-102. There is no friction between the masses and the horizontal plane. If a 4-N force is applied horizontally, determine the acceleration of the masses and the tension in the cord.   *Ans.* $a = 0.133 \text{ m/s}^2$, $T = 2.67 \text{ N}$

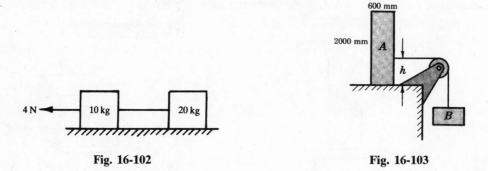

Fig. 16-102   Fig. 16-103

16.111. A 5-kg block A is pulled to the right under the action of a falling 5-kg mass B as shown in Fig. 16-103. If the coefficient of friction between the block A and the plane is 0.5, determine the acceleration of B. Also determine the maximum value of h if block A is not to tip.
Ans. $a = 2.45 \text{ m/s}^2$ down, $h = 730 \text{ mm}$ above plane

16.112. In Fig. 16-104, the 30-lb weight is on a horizontal floor. The coefficient of friction between the weight and the floor is 0.2. The pulleys shown are assumed to be massless and frictionless. Determine the acceleration of the system and the tensions in the cords.
Ans. $a = 3.33 \text{ ft/s}^2$, $T_{AB} = 8.83 \text{ lb}$, $T_{BC} = 17.9 \text{ lb}$

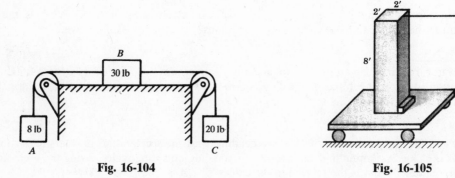

Fig. 16-104   Fig. 16-105

16.113. A 1000-lb block stands on a truck as shown in Fig. 16-105. The weight W falls, accelerating the block and the 300-lb truck. A strip nailed to the platform prevents sliding. What is the maximum acceleration the system may have without tipping the block? Find W. Assume rolling of the truck and neglect the inertia effects of the wheels.   *Ans.* $a = 5.03 \text{ ft/s}^2$ to the right, $W = 241 \text{ lb}$

16.114. In Fig. 16-106, the homogeneous bar AB has a mass of 5 kg and is fastened by a frictionless pin at A

and rests against the smooth vertical part of the cart. The cart has a mass of 50 kg and is pulled to the right by a horizontal force P. At what value of P will the bar exert no force on the cart at B?
Ans. $P = 642$ N

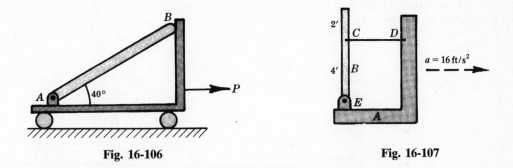

Fig. 16-106   Fig. 16-107

16.115. The platform A accelerates to the right on a horizontal plane as shown in Fig. 16-107. The bar B is held in the vertical position by the horizontal cord CD. Determine the tension in the cord and the pin reactions at E. The bar is uniform and weighs 18 lb.
Ans. $T = 6.71$ lb, $E_y = 18.0$ lb up, $E_x = 2.24$ lb to the right

16.116. In Fig. 16-108, a block is moving to the right with an acceleration of 1.5 m/s^2. A homogeneous bar 1 m long with a mass of 1 kg is suspended as shown by means of a frictionless pin at A. What is the angle θ?   *Ans.* $\theta = 8.7°$

16.117. A chain hangs freely from the rear of a truck. If the truck accelerates to the left 3 m/s^2, what will be the approximate angle between the chain and the horizontal?   *Ans.* $\theta = 73°$

16.118. A helicopter carrying an 80-ft transmission-line tower by means of a sling accelerates 6 ft/s^2 horizontally. What angle does the centerline of the tower make with the vertical?   *Ans.* $\theta = 10.5°$

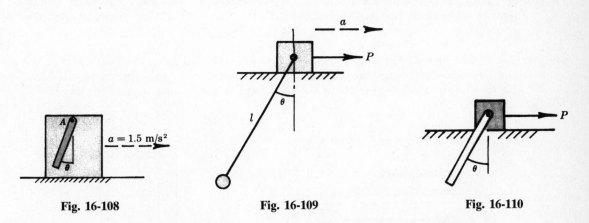

Fig. 16-108   Fig. 16-109   Fig. 16-110

16.119. A small block moves with an acceleration a along the horizontal surface shown in Fig. 16-109. The bob at the end of the string maintains a constant angle θ with the vertical. Show that θ is a measure of the acceleration a in this simple accelerometer.   *Ans.* $a = g \tan \theta$

16.120. In Fig. 16-110, the 50-kg block is subjected to an acceleration of 3 m/s^2 horizontally to the right under the action of a horizontal force P. The homogeneous slender bar has a mass of 8 kg and is 1 m long. Assuming a frictionless surface, determine P and θ. What are the horizontal and vertical components of the pin reaction on the bar?   *Ans.* $P = 174$ N, $\theta = 17°$, $V = 78.4$ N, $H = 24.0$ N

16.121. In the preceding problem, what are the values of P and θ if the coefficient of friction between the block and the plane is 0.25?   *Ans.* $P = 316\,\text{N}$, $\theta = 17°$

16.122. A box weighing 1000 lb is shown on a truck in Fig. 16-111. The coefficient of friction between the block and the truck is 0.30. Determine the forward acceleration at which the box tips or slides. Assume homogeneity.   *Ans.*   Tips at $a = 8.05\,\text{ft/s}^2$ to the right

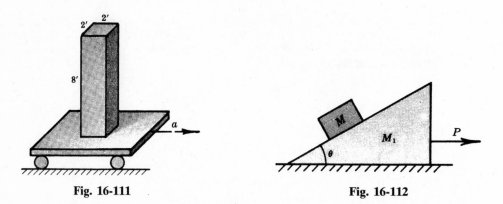

Fig. 16-111   **Fig. 16-112**

16.123. A block of mass M is on a surface (the coefficient of friction is μ) that is inclined at angle θ with the horizontal as shown in Fig. 16-112. This surface is part of a triangular block of mass M_1. A horizontal force P causes the system to have an acceleration a to the right. What value of P will cause the top block to move relative to the surface? Assume no friction on the bottom surface. What is the acceleration?

Ans. $a = \dfrac{g(\mu - \tan \theta)}{1 + \mu \tan \theta}$, $\quad P = \dfrac{[(M_1 + M_2)g](\mu - \tan \theta)}{1 + \mu \tan \theta}$

16.124. A truck is traveling along a level road at a constant speed. Its body has a mass m and contains a box of mass m. Show that if the box drops off the truck, the truck body will experience an upward acceleration of mg/M.

16.125. The 2500-lb automobile in Fig. 16-113 is brought to rest from a speed of 60 mi/h. If the car is equipped with four-wheel brakes and the coefficient of friction between the tires and the road is 0.6, what will be the time required to bring the car to rest? What will be the distance covered in that time?
Ans.   4.56 s, 200 ft

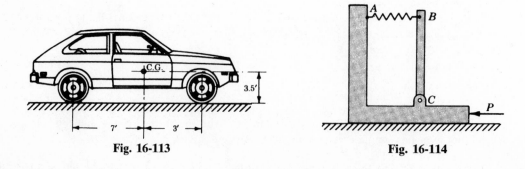

Fig. 16-113   **Fig. 16-114**

16.126. In Problem 16.125, determine the vertical reaction on the two front tires and on the two rear tires.
    Ans. $R_F = 2275\,\text{lb}$, $R_R = 225\,\text{lb}$

16.127. The bar BC in Fig. 16-114 is vertical when the system is at rest. The bar is 3 m long and has a mass of 10 kg. It is noted that the bar is rotated 5° from the vertical when the cart to which it is fastened is accelerating. The spring constant is 100 N/m. Assuming the spring remains horizontal, determine the acceleration. The cart is on a smooth horizontal surface.   *Ans.* $a = 4.37$ m/s^2

Rotation

16.128. The 6-m homogeneous bar of mass M shown in Fig. 16-115 falls from its vertical rest position. Assuming no friction at the pivot, determine the angular velocity of the bar at the time when the tangential component of the acceleration of the unpivoted end is equal to g, the acceleration due to gravity. *Ans.* $\theta = 0.73$ rad, $\dot{\theta} = 1.12$ rad/s

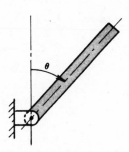

Fig. 16-115

16.129. A homogeneous cylinder 600 mm in diameter with a mass of 40 kg has an angular acceleration of 2 rad/s^2 about its geometric axis. What torque is acting to produce this acceleration? *Ans.*   3.6 N · m

16.130. A homogeneous disk 6 ft in diameter weighs 800 lb and is made to revolve about its geometric axis by a force of 80 lb applied tangentially to its circumference. Determine the angular acceleration of the disk. *Ans.*   2.15 rad/s^2

16.131. A 90-lb solid cylindrical disk 3 ft in diameter is rotating 60 rpm about a central axis perpendicular to a diameter. What constant tangential force must be applied to the rim of the disk to bring it to rest in 2 min?   *Ans.*   0.11 lb

16.132. A 200-kg mass hangs vertically downward from the end of a massless cord that is wrapped around a cylinder 900 mm in diameter. The mass descends 8 m in 4 s. What is the mass of the cylinder? *Ans.*   3520 kg

16.133. A homogeneous bar 10 ft long weighs 30 lb and is suspended vertically from a pivot point located at one end. The bar is struck a horizontal blow of 90 lb at a point 2 ft below the pivot point. Determine the horizontal reaction at the pivot. What is the angular acceleration of the bar due to the blow? *Ans.* $R_h = 63$ lb, $\alpha = 5.8$ rad/s^2

16.134. A slender bar 1200 mm long has a mass of 3.6 kg. It is hanging vertically at rest when struck by a horizontal force $P = 12$ N at the lower end as shown in Fig. 16-116. Determine (a) the angular acceleration of the bar and (b) the horizontal and vertical components of the pin reaction O on the bar. *Ans.* $\alpha = 8.33$ rad/s^2 counterclockwise, $O_y = 35.3$ N up, $O_x = 6$ N to right

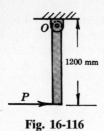

Fig. 16-116

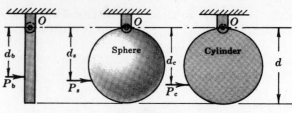

Fig. 16-117

16.135. In the preceding problem, where should the 12-N force be applied so that the horizontal component of the pin reaction will be zero?   *Ans.*   800 mm below O

16.136. At what point should the horizontal force P be applied to the homogeneous bar, the homogeneous cylinder, and the homogeneous sphere so that the horizontal component of the pin reaction at O is zero? See Fig. 16-117.   *Ans.* $d_b = \frac{2}{3}d$, $d_c = \frac{3}{4}d$, $d_s = \frac{7}{10}d$

16.137. The coefficient of friction between the horizontal floor and a runner's shoes is 0.5. Find the radius of the smallest circular path around which the runner can travel at a constant speed of 16 ft/s without slipping. *Ans.* $r = 15.9$ ft

16.138. A train weighing 100,000 lb has its center of gravity 5.5 ft above the rails. At a speed of 30 mi/h, it rounds an unbanked curve having a radius of 2500 ft. If the centerlines of the rails are 4 ft $8\frac{1}{2}$ in apart, find the vertical force on the outer rail.   *Ans.*   52,800 lb

16.139. Determine the angle of banking for a roadway to allow an automobile speed of 100 km/h on a curve of 90-m radius so that there will be no side thrust on the wheels.   *Ans.*   41.2°

16.140. A highway curve with a radius of 600 m is to be banked for a speed of 80 km/h. What should be the angle of bank so that at the design speed there will be no frictional force necessary on the tire in a direction perpendicular to the plane of rolling?   *Ans.* $\theta = 0.084$ rad or 4.8°

16.141. A 0.06-lb bead slides from rest down a frictionless wire, bent as shown in Fig. 16-118. Determine the normal force of the wire on the bead at point A. Next find the normal forces at points B and C with the bead on the arc of the circle.
Ans. $N_A = 0.036$ lb, $\theta_x = \tan^{-1} 0.75 = 36.9°$; $N_B = 0.276$ lb, $\theta_x = 36.9°$; $N_C = 0.348$ lb up

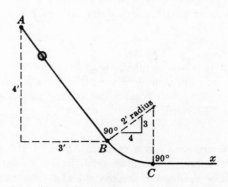

Fig. 16-118

16.142. A flywheel with radius of gyration k and mass m is acted upon by a constant torque $M = C$. What will be the angular speed after the flywheel has rotated θ rad from rest?   *Ans.* $\omega = (1/k)\sqrt{2C\theta/m}$

16.143. A uniform disk of diameter 6 ft and weighing 34 lb is released from rest when O and G are on a horizontal line as shown in Fig. 16-119. What will be the angular velocity when G is vertically below O? Determine the reactions at O at that time.   *Ans.* $\omega = 3.78$ rad/s, $O_n = 79.3$ lb, $O_t = 0$

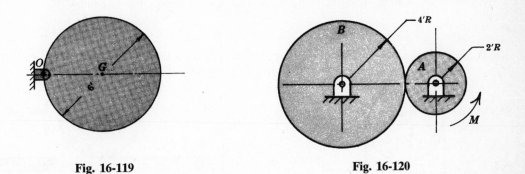

Fig. 16-119   Fig. 16-120

16.144. If the complete object of Problem 16.49 is used as a pendulum, determine the frequency of the motion. *Ans.*   0.581Hz

16.145. Refer to Fig. 16-120. A moment M of 45 lb-ft is applied to the uniform disk A, which then drives uniform disk B without slip occurring between the two disks. What is the angular acceleration of each disk? Disk A weighs 64.4 lb and disk B weighs 128.8 lb.
Ans. $\alpha_A = 3.75$ rad/s^2 counterclockwise, $\alpha_B = 1.88$ rad/s^2 clockwise

16.146. A massless bar 4 m long rotates in a horizontal plane about its center. From each end of the rod is suspended a 4-kg mass on a 600-mm cord. When the system is revolving at $\frac{2}{3}$ rev/s, determine the angle between the cords and the vertical.
Ans. $(2 + 0.6 \sin \theta)/\tan \theta = 0.559$ and $\theta = 77.8°$ by computer solution

16.147. What is the angular acceleration of the pulley shown in Fig. 16-121 turning under the action of the two masses?   *Ans.* $\alpha = 0.666$ rad/s^2 clockwise

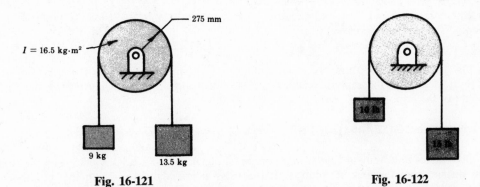

Fig. 16-121   Fig. 16-122

16.148. In Fig. 16-122, the simple Atwood machine has weights of 10 and 15 lb measured on earth with $g = 32.2$ ft/s^2. If the machine is moved to the moon, where the acceleration of gravity is 0.16 of the earth's g, what will be the tension in the cord when the masses are released from rest?
Ans. $T = 1.29$ lb

16.149. Two masses are connected by a light inextensible string that passes over a pulley rotating in frictionless bearings. See Fig. 16-123. At a place where $g = 9.8 \text{ m/s}^2$, the masses of A, B, and the pulley are 14, 9, and 5 kg, respectively. The radius of gyration of the pulley is 400 mm. Determine the acceleration of the masses if the system is released from rest at a place where $g = 4.9 \text{ m/s}^2$.   *Ans.* $a = 0.935 \text{ m/s}^2$

16.150. In Fig. 16-124, the disk of weight 76 lb and radius of gyration 2.34 ft is rotating at 600 rpm. What force P must be applied to the braking mechanism to stop the disk in 25 s? Use a coefficient of friction between the disk and the horizontal member equal to 0.30.   *Ans.* $P = 17.8$ lb

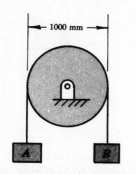

Fig. 16-123

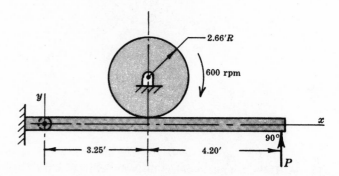

Fig. 16-124

16.151. A thin triangular lamina of weight 20 lb is rotating at 30 rpm about a horizontal axis in two frictionless bearings at A and B as shown in Fig. 16-125. When the lamina is vertical as shown, determine the bearing reactions at A and B.   *Ans.* $A = 8.9$ lb up, $B = 10.7$ lb up

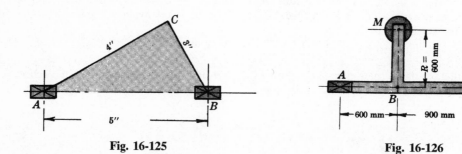

Fig. 16-125

Fig. 16-126

16.152. In Fig. 16-126, $AB = 600$ mm, $BC = 900$ mm, $M = 50$ kg, and $R = 600$ mm. If the system is rotating at $1\frac{2}{3}$ rev/s, determine the support reactions due to "centrifugal" force at journals A and C. Neglect the weight of the arm.   *Ans.* $A = 1970$ N, $B = 1320$ N

16.153. Solve Problem 16.152 if the speed is $16\frac{2}{3}$ rev/s.   *Ans.* $A = 197$ kN, $B = 132$ kN

16.154. A disk rotates uniformly at $\frac{1}{4}$ rev/s in a horizontal plane. A 40-kg mass is placed on the disk at a point 2 m distant from the axis of rotation. If the mass is just about to slip in this position, what is the coefficient of friction between the mass and the disk?   *Ans.* $\mu = 0.5$

16.155. A 2-lb block is held stationary on a stop 18 inches from the center and a horizontal turntable by a cord that passes down through a hole in the shaft as shown in Fig. 16-127. What tension in the cord is necessary to keep the block 18 in from the center as the block and turntable rotate around the center at 30 rad/s? Assume that there is no friction. *Ans.* $T = 83.8$ lb

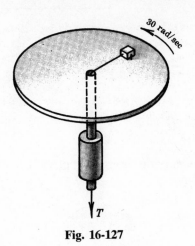

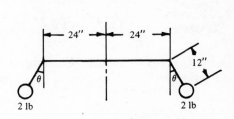

Fig. 16-127   **Fig. 16-128**

16.156. The horizontal member in Fig. 16-128 is 48 in long and is rotating at 20 rpm about a vertical axis through its midpoint. Each 2-lb ball is suspended by a cord that is 12 in long. What angle θ will each cord make with the vertical? Use a trial-and-error solution.   *Ans.* $\theta = 17.4°$

Do the following Problems using the inertia-force method.

16.157. Solve Problem 16.78.

16.158. Solve Problem 16.79.

16.159. Solve Problem 16.90.

16.160. Solve Problem 16.93.

16.161. Solve Problem 16.101.

16.162. Solve Problem 16.111.

16.163. Solve Problem 16.115.

16.164. Solve Problem 16.145.

16.165. Solve Problem 16.150.

Chapter 17

Work and Energy

17.1 WORK

Work U done by a force $\mathbf{F}$ acting on a particle moving along any path is defined as the line integral from position P_1 at time t_1 to position P_2 at time t_2;

$$U = \int_C \mathbf{F} \cdot d\mathbf{r}$$

where $d\mathbf{r}$ is the infinitesimal change in the position vector $\mathbf{r}$.

The expression for work may also be written (see Fig. 17-1) as

$$U = \int_{s_1}^{s_2} F_t \, ds$$

where s_1, s_2 = respective distances of the particle from reference point P_0 at the beginning and end of the motion

F_t = magnitude of the tangential component of the force $\mathbf{F}$ as indicated in Fig. 17-1

ds = infinitesimal change in position of the particle along the path

Then, since

$$\mathbf{F} = F_x\mathbf{i} + F_y\mathbf{j} + F_z\mathbf{k} \qquad \text{and} \qquad d\mathbf{r} = dx\,\mathbf{i} + dy\,\mathbf{j} + dz\,\mathbf{k}$$

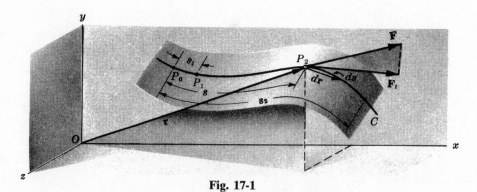

Fig. 17-1

the above line integral can be written as

$$U = \int_C (F_x\mathbf{i} + F_y\mathbf{j} + F_z\mathbf{k}) \cdot (dx\,\mathbf{i} + dy\,\mathbf{j} + dz\,\mathbf{k}) = \int_C (F_x\,dx + F_y\,dy + F_z\,dz)$$

This can be converted to a time integral as follows;

$$U = \int_{t_1}^{t_2} \left(F_x\frac{dx}{dt} + F_y\frac{dy}{dt} + F_z\frac{dz}{dt} \right) dt$$

17.2 SPECIAL CASES

Special cases arise as follows:

1. Force constant in magnitude and with its direction along a straight line:

$$U = Fs$$

where U = work done
        F = constant force
        s = displacement during the motion along the straight line

2. Force constant in magnitude but at a constant angle with a straight line displacement:

$$U = Fs \cos \theta$$

where U = work done
        F = constant force
        s = displacement during the motion along the straight line
        θ = angle between the action line of the force and the displacement

3. Forces constituting a couple (rotation occurs):

$$U = \int_{\theta_1}^{\theta_2} M \, d\theta$$

where M = couple
           $d\theta$ = differential angular displacement
           θ_1, θ_2 = initial and final angular displacements

Work is positive if the force acts in the direction of motion. Work is negative if the force acts opposite to the direction of motion.

The work done on a rigid body by two or more concurrent forces during a displacement is equivalent to the work done by the resultant of the force system during the same displacement.

Work is a scalar quantity. In the U.S. Customary system, the unit is the foot-pound (ft-lb). In SI, the unit is the newton-meter (N · m), which is called the joule (J). (1 N · m = 1 J.)

17.3 POWER

Power, which is the rate of doing work, equals $dU/dt = \mathbf{F} \cdot d\mathbf{r}/dt = \mathbf{F} \cdot \mathbf{v}$. In the case of a couple, this would be Power $= dU/dt = \mathbf{M} \cdot \boldsymbol{\omega}$.

(a) In the U.S. Customary system, the unit of power is the foot-pound per second (ft-lb/s). This unit is small; thus, a larger unit, the horsepower, which equals 550 ft-lb/s, is used.

The formulas used for horsepower are

$$\text{hp} = \frac{\mathbf{F} \cdot \mathbf{v}}{550} \quad \text{and} \quad \text{hp} = \frac{\mathbf{M} \cdot \boldsymbol{\omega}}{550}$$

where $\mathbf{F}$ = force in pounds
        $\mathbf{v}$ = velocity in ft/s
        $\mathbf{M}$ = moment of couple in lb-ft
        $\boldsymbol{\omega}$ = angular velocity in rad/s

(b) In SI, the unit of power is the joule per second (J/s). This is, of course, the watt (1 J/s = 1 W).

17.4 EFFICIENCY

Efficiency is equal to work output divided by work input for the same period of time. It is also expressed as power output divided by power input. The work output is less than the work input by the work dissipated (usually in overcoming friction).

17.5 KINETIC ENERGY OF A PARTICLE

Kinetic energy T of a particle with mass m and moving with speed v is defined as $\frac{1}{2}mv^2$.
In the U.S. Customary system, the unit is the foot-pound (ft-lb).
In SI, the unit is the joule (J).

17.6 WORK-ENERGY RELATIONS FOR A PARTICLE

Work done on a particle by all forces is equal to the change in the kinetic energy T of the particle.

Proof:

$$U = \int_{r_1}^{r_2} \mathbf{F} \cdot d\mathbf{r} = \int_{r_1}^{r_2} m\ddot{\mathbf{r}} \cdot d\mathbf{r} = \int_{t_1}^{t_2} m\ddot{\mathbf{r}} \cdot \frac{d\mathbf{r}}{dt} \, dt$$

$$= m \int_{t_1}^{t_2} (\ddot{\mathbf{r}} \cdot \dot{\mathbf{r}}) \, dt = \frac{1}{2} m \int_{t_1}^{t_2} \frac{d}{dt} (\dot{\mathbf{r}})^2 \, dt$$

$$= \frac{1}{2} m [v^2]_{v_1}^{v_2} = \frac{1}{2} m v_2^2 - \frac{1}{2} m v_1^2 = T_2 - T_1$$

where U = work done
           m = mass of particle
           v_1, v_2 = initial and final speeds, respectively, at P_1 and P_2
           T_1, T_2 = initial and final kinetic energy, respectively, at P_1 and P_2

17.7 KINETIC ENERGY T OF A RIGID BODY IN TRANSLATION

The kinetic energy T of a rigid body in translation is $T = \frac{1}{2}mv^2$.

Proof: All particles of the rigid body have the same velocity $\mathbf{v}$. Then, if m_i = mass of the ith particle,

$$T = \sum_{i=1}^{n} \frac{1}{2} m_i v^2 = \frac{1}{2} v^2 \sum_{i=1}^{n} m_i = \frac{1}{2} m v^2$$

where m = mass of the entire body
         v = speed of the body

17.8 KINETIC ENERGY T OF A RIGID BODY IN ROTATION

The kinetic energy T of a rigid body in rotation is $T = \frac{1}{2} I_0 \omega^2$.

Proof: Let $\mathbf{r}_i$ be the radius vector of the ith particle and $\dot{\mathbf{r}}_i$ its velocity as shown in Fig. 17-2; then

$$T = \sum_{i=1}^{n} \frac{1}{2} m_i (\dot{\mathbf{r}}_i)^2 = \sum_{i=1}^{n} \frac{1}{2} m_i (\boldsymbol{\omega} \times \mathbf{r}_i)^2$$

$$= \sum_{i=1}^{n} \frac{1}{2} m_i \omega^2 r_i^2 = \frac{1}{2} \omega^2 \sum_{i=1}^{n} m_i r_i^2 = \frac{1}{2} I_O \omega^2$$

where I_O = mass moment of inertia of the entire body about the axis of rotation
         ω = angular speed

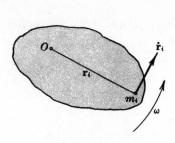

Fig. 17-2

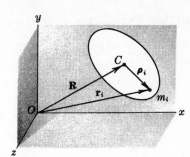

Fig. 17-3

17.9   KINETIC ENERGY T OF A BODY IN PLANE MOTION

The kinetic energy T of a body in plane motion is $T = \frac{1}{2}m\bar{v}^2 + \frac{1}{2}\bar{I}\omega^2$.

Proof: Study the motion of the ith particle, choosing the mass center as the base point. Refer to Fig. 17-3. Then

$$\mathbf{r}_i = \mathbf{R} + \boldsymbol{\rho}_i \qquad \text{and} \qquad \dot{\mathbf{r}}_i = \dot{\mathbf{R}} + \dot{\boldsymbol{\rho}}_i$$

$$\dot{\mathbf{r}}_i^2 = (\dot{\mathbf{R}} + \dot{\boldsymbol{\rho}}_i) \cdot (\dot{\mathbf{R}} + \dot{\boldsymbol{\rho}}_i) = \dot{\mathbf{R}}^2 + 2\dot{\mathbf{R}} \cdot \dot{\boldsymbol{\rho}}_i + \dot{\boldsymbol{\rho}}_i^2$$

Then
$$T = \sum_{i=1}^{n} \frac{1}{2}m_i\dot{\mathbf{r}}_i^2 = \frac{1}{2}\sum_{i=1}^{n} m_i\dot{\mathbf{R}}^2 + \sum_{i=1}^{n} m_i\dot{\mathbf{R}} \cdot \dot{\boldsymbol{\rho}}_i + \frac{1}{2}\sum_{i=1}^{n} m_i\dot{\boldsymbol{\rho}}_i^2$$

But $\dot{\mathbf{R}}^2 = \bar{v}^2$, $\dot{\boldsymbol{\rho}}_i = \boldsymbol{\omega} \times \boldsymbol{\rho}_i$, and $m_i\dot{\mathbf{R}} \cdot \dot{\boldsymbol{\rho}}_i = m_i\dot{\mathbf{R}} \cdot (\boldsymbol{\omega} \times \boldsymbol{\rho}_i) = \dot{\mathbf{R}} \cdot (\boldsymbol{\omega} \times m_i\boldsymbol{\rho}_i)$. Thus

$$T = \frac{1}{2}\bar{v}^2 \sum_{i=1}^{n} m_i + \dot{\mathbf{R}} \cdot \left(\boldsymbol{\omega} \times \sum_{i=1}^{n} m_i\boldsymbol{\rho}_i \right) + \frac{1}{2}\sum_{i=1}^{n} m_i\omega^2\boldsymbol{\rho}_i^2$$

Since $\sum_{i=1}^{n} m_i\boldsymbol{\rho}_i = 0$ when $\boldsymbol{\rho}_i$ is drawn from an axis through the mass center,

$$T = \frac{1}{2}m\bar{v}^2 + 0 + \frac{1}{2}\bar{I}\omega^2$$

where $\bar{v}$ = linear speed of the mass center
  ω = angular speed
  $\bar{I}$ = moment of inertia about an axis through the mass center parallel to the z axis

17.10   POTENTIAL ENERGY

A conservative force is one that does the same amount of work regardless of the path taken by its point of application. (See Problem 17-1).

The potential energy of a body is measured by the work done against conservative forces acting on the body in bringing the body from some reference or datum position to the position in question. The potential energy may be defined as the negative of the work done *by* the conservative force acting on the body in bringing it from the datum position to the position in question. The selection of the datum position is arbitrary—usually for convenience.

17.11   WORK-ENERGY RELATIONS FOR A RIGID BODY

The principle of work and energy states that the work done by the external forces acting on a rigid body during a displacement is equal to the change in kinetic energy of the body during the same displacement.

17.12 LAW OF CONSERVATION OF ENERGY

The law of conservation of energy states that if a particle (or body) is acted upon by a conservative force system, the sum of the kinetic energy and potential energy is a constant.

Proof: Let P and Q be respectively the initial and final positions of the particle. The work done by a conservative force $\mathbf{F}$ as the particle moves from P to Q has been defined as $U = \int_P^Q \mathbf{F} \cdot d\mathbf{r}$. But, in terms of our standard position S with radius vector r_s, we can write

$$U = \int_P^S \mathbf{F} \cdot d\mathbf{r} + \int_S^Q \mathbf{F} \cdot d\mathbf{r} = \int_P^S \mathbf{F} \cdot d\mathbf{r} - \int_Q^S \mathbf{F} \cdot d\mathbf{r}$$

But potential energy $V_P = \int_P^S \mathbf{F} \cdot d\mathbf{r}$ and $V_Q = \int_Q^S \mathbf{F} \cdot d\mathbf{r}$. Hence,

$$U = V_P - V_Q$$

We have also shown in terms of kinetic energy that

$$U = T_Q - T_P$$

Thus, $$V_P - V_Q = T_Q - T_P \qquad \text{or} \qquad V_P + T_P = V_Q + T_Q$$

Solved Problems

17.1. Determine the potential energy V of a body of mass m at a height h above a datum plane (assumed as a standard position). Assume that h is small enough that the gravitational force on m does not vary.

SOLUTION

The potential energy V is the negative of the work done against the only force acting on the body as it traverses any smooth path from S (datum) to P. This force is the gravitational force mg, which is assumed constant throughout this travel. Refer to Fig. 17-4(a).

Since $\mathbf{F}$ is constant, we can write

$$V = \int_C - \mathbf{F} \cdot d\mathbf{r} = -\mathbf{F} \cdot \int_{r_S}^{r_P} d\mathbf{r} = -\mathbf{F} \cdot (\mathbf{r}_p - \mathbf{r}_S)$$

But Fig. 17-4(b) indicates that

$$-\mathbf{F} \cdot (\mathbf{r}_p - \mathbf{r}_S) = \mathbf{F} \cdot (\mathbf{r}_S - \mathbf{r}_p) = mg\, |\mathbf{r}_S - \mathbf{r}_p| \cos \theta = mgh$$

Thus, the potential energy V of a body of mass m at a height h above an arbitrary datum plane is *mgh*.

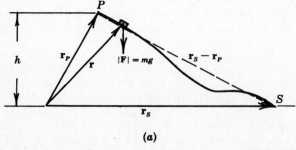

(a)

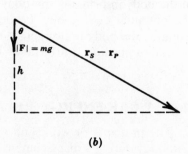

(b)

Fig. 17-4

17.2. A body on a frictionless horizontal plane is attached to a spring with modulus k lb/in. Refer to Fig. 17-5(a) for the top view. When the body is at S, a distance l from the wall, there is no force in the spring ($kl = 0$). This will be considered the standard position. When the body is at P, a distance $l + x$ from the wall, determine the potential energy V_P of the system (actually of the spring, because there is no change in the potential energy of the body, which is always on the horizontal plane).

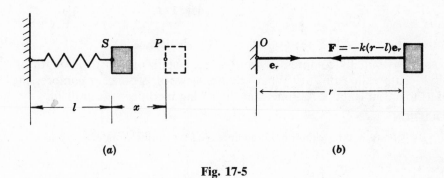

(a)   (b)

Fig. 17-5

SOLUTION

Figure 17-5(b) shows the body at a distance r from the wall as the origin. The force acting on the body in this position is proportional to the deformation of the spring from the standard unstressed length; thus, $\mathbf{F} = -k(r - l)\mathbf{e}_r$. The work done by this force for a differential change $d\mathbf{r}$ (where $d\mathbf{r} = dr\,\mathbf{e}_r$) is $-k(r - l)\mathbf{e}_r \cdot dr\,\mathbf{e}_r = -k(r - l)\,dr$, since $\mathbf{e}_r \cdot \mathbf{e}_r = 1$. The potential energy is

$$V_P = -\int_{l}^{l+x} [-k(r - l)]\,dr = \tfrac{1}{2}kx^2$$

The body possesses this potential energy because of the pull of the spring, which varies with the deformation. The pull of the spring (its internal force) always acts along the spring. Thus, the potential energy for a given deformation remains the same even if the spring is moved sideways on the smooth horizontal surface, provided the distance from O does not change. This sideways motion could be conceived as the sum of infinitesimal displacements perpendicular to the spring during which forces act only along the spring (zero work is done perpendicular to the spring).

In Problem 17.1, the given body possesses potential energy because of the attraction of the earth considered constant for relatively small changes of height h. Problem 17.3 introduces an attractive force that varies inversely with the square of the distance.

17.3. Determine the potential energy of a body attracted by a force that is inversely proportional to the square of the distance r from the source O of the force.

SOLUTION

The standard position S of the body will be chosen at infinity. (*Note:* Some students might have tried the source O as the standard position, but V is defined in terms of the work done in moving from the standard position; and since the force tends to become infinite as the body approaches O, the choice of O is seen to be unwise.) The body is shown in Fig. 17-6 at a distance r from the origin. The force acting on the body in this position is $\mathbf{F} = -(C/r^2)\mathbf{e}_r$. As in Problem 17.2, the potential energy is

$$V_P = \int_{\infty}^{r} -\left(\frac{C}{r^2}\right)\mathbf{e}_r \cdot (-dr\,\mathbf{e}_r) = \int_{\infty}^{r} +\left(\frac{C}{r^2}\right)dr = -C\left[\frac{1}{r}\right]_{\infty}^{r} = -\frac{C}{r}$$

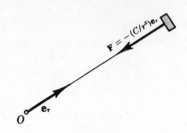

Fig. 17-6

17.4. A spring is initially compressed from a free length of 8 in to a length of 6 in, making a net initial compression of 2 in. What additional work is done in compressing it 3 in more (5 in total) to a 3-in length? Assume that the spring modulus $k = 20$ lb/in.

SOLUTION

By definition, the work done in compressing the spring from 2 in to 5 in is

$$U = \int_{S_1}^{S_2} F_t\, ds = \int_2^5 20s\, ds = 210 \text{ in-lb}$$

The same value may be obtained by using Problem 17.2 to determine the work done in compressing the spring first 2 in and then 5 in. Note that both expressions are for work done from the unstressed or zero position. Their difference is the required value;

$$U = \int_0^5 20s\, ds - \int_0^2 20s\, ds = 250 - 40 = 210 \text{ in-lb}$$

17.5. Solve Problem 17.4 graphically.

SOLUTION

To solve graphically, plot the varying force $F = 20s$ against the displacement as shown in Fig 17-7.

Note that the area included between the 2- and 5-in displacements is in terms of inches horizontally and pounds vertically.

The shaded area is therefore equal to the work done, or equal to the average height in pounds times change in displacement in inches:

$$U = \tfrac{1}{2}(40 \text{ lb} + 100 \text{ lb})(3 \text{ in}) = 210 \text{ in-lb}$$

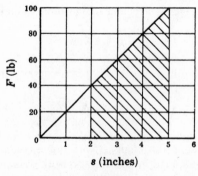

Fig. 17-7

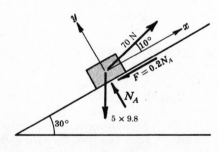

Fig. 17-8

17.6. Determine the total work done on a 5-kg body that is pulled 6 m up a rough plane inclined 30° with the horizontal, as shown in Fig. 17-8. Assume that the coefficient of friction $\mu = 0.20$.

SOLUTION

From the free-body diagram, N_A may be found by summing forces perpendicular to the plane:

$$\sum F_y = 0 = -5 \times 9.8 \cos 30° + N_A + 70 \sin 10°$$

solving, $N_A = 30.3$ N.

The work done by each force is determined next. The sign is positive if the force acts in the direction of motion of the body.

Force	Sign of work done	Magnitude of work done	Result, N · m
70 N	+	$(70 \cos 10°)(6)$	+413.6
5×9.8	−	$(5 \times 9.8 \sin 30°)(6)$	−147
N_A	no work, since N_A acts $\perp$ to direction of motion		0
F	−	$(0.20 \times 30.3)(6)$	−36.4

Therefore the total work done = $+413.6 - 147 - 36.4 = 230$ N · m.

Now find the resultant of all the forces and then determine the work done by the resultant. Since only the x component of the resultant will do work, determine only that:

$$R_x = \sum F_x = +70 \cos 10° - 0.20(30.3) - 5 \times 9.8 \sin 30° = 38.4$$

Since 38.4 N is a constant force, the total work done = $R_x(6) = 38.4(6) = 230$ N · m.

17.7. Find the work done in rolling a 20-kg wheel a distance 1.5 m up a plane inclined 30° with the horizontal as shown in Fig. 17-9. Assume a coefficient of friction of 0.25.

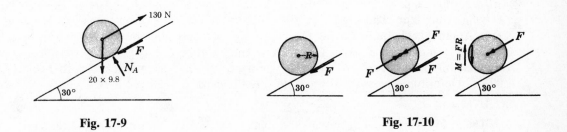

Fig. 17-9   **Fig. 17-10**

SOLUTION

The normal force N_A does no work because it has no component in the direction of motion.

The force of friction F also does no work, but for a different reason. Its point of contact moves with the wheel, and since there is no relative motion (no slipping involved) between the frictional force and the wheel, the work done by the force of friction is zero. This is not the case in the preceding problem, where the frictional force did negative work, but of course the body slipped along the plane.

Another way of attacking the friction work in this problem would be to replace the single force F with an equal parallel force F in the same direction through the center and a couple as shown in Fig. 17-10. (Neglect all other forces for the time being.)

The work done by the equivalent system is equal to the sum of the work done by the force F and the couple M (magnitude FR). Let us determine the work done for one revolution, i.e., for 2π rad. The center moves a distance $2\pi R$ during one revolution. Hence, the work done by F is $-F(2\pi R)$. The work done by the couple aids the wheel in moving up the hill and equals $+M\theta$ or $2\pi FR$. The sum of these two values is zero work.

Therefore, of the given system, the only forces doing work are the 130-N force and the component of the gravitational force along the plane (refer to Fig. 17-9). The component of the gravitational force is $20 \times 9.8 \sin 30°$.

$$\text{Work done} = +130(1.5) - (20 \times 9.8 \sin 30°)(1.5) = 48 \, \text{N} \cdot \text{m}$$

17.8. Work is expressed as $U = \int (\sum F_x) \, dx$, where $\sum F_x$ is the sum of the forces in the x direction. Show that the work of the expanding gas in the simple engine shown in Fig. 17-11 is $U = \int p \, dV$, where p is the gas pressure in lb/ft^2 and v is the volume in ft^3.

SOLUTION

The force F acting on the piston is pA, where A is the area of the piston in square feet. Also, the change in volume $dV = A \, dx$. Hence,

$$U = \int \left(\sum F_x \right) dx = \int (pA) \frac{dV}{A} = \int p \, dV$$

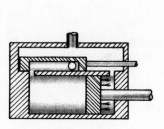

Fig. 17-11

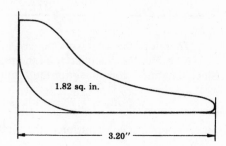

1.82 sq. in.

3.20''

Fig. 17-12

17.9. Figure 17-12 illustrates an indicator card taken on the head end of a steam engine. The horizontal line is proportional to the stroke of the engine and therefore to the volume of the steam in cubic feet. The height of the card indicates the internal pressure in either psi or lb/ft^2. Assume that the length of the card is 3.20 in and that the engine is 6 in by 8 in. The indicator spring scale is 100. This means that a pressure of 100 psi produces a vertical displacement on the card of 1 in. The area of the card as measured with a planimeter is $1.82 \, \text{in}^2$. Determine the work represented by the card.

SOLUTION

The 3.20 in horizontally represents the volume contained in the engine. Since the bore is 6 in and the stroke 8 in, this volume is

$$V = Al = \tfrac{1}{4}\pi d^2 l = \tfrac{1}{4}\pi (6/12)^2 (8/12) \, \text{ft}^3 = 0.131 \, \text{ft}^3$$

One horizontal inch represents $(0.131 \, \text{ft}^3)/3.20 = 0.041 \, \text{ft}^3$.
One vertical inch represents $(100 \, \text{lb/in}^2)(144 \, \text{in}^2/\text{ft}^2) = 14{,}400 \, \text{lb/ft}^2$.
Then $1.82 \, \text{in}^2$ of card area represents $1.82(14{,}400)(0.041) = 1075 \, \text{ft-lb}$ of work.

17.10. At a certain instant during acceleration along a level track, the drawbar pull of an electric locomotive is 100 kN. What power is being developed if the speed of the train is 90 km/h?

SOLUTION

$$\text{Power} = (100\,000 \, \text{N})\left(\frac{90\,000 \, \text{m}}{3600 \, \text{s}}\right) = 2\,500\,000 \, \text{N} \cdot \text{m/s} = 2.5 \times 10^6 \, \text{J/s}$$

$$= 2.5 \, \text{MJ/s or } 2.5 \, \text{MW}$$

17.11. A belt wrapped around a pulley 600 mm in diameter has a tension of 800 N on the tight side and a tension of 180 N on the slack side. What power is being transmitted if the pulley is rotating at 200 rpm?

SOLUTION

The torque M is found by taking the algebraic sum of the moments of the tensile forces in the belt about the center of the pulley:

Torque $M = 800 \times 0.3 - 180 \times 0.3 = 186$ N $\cdot$ m.

Power $= M(\text{N} \cdot \text{m}) \cdot \omega(\text{rad/s}) = (186 \text{ N} \cdot \text{m})(200 \times 2\pi/60 \text{ rad/s}) = 3.9$ kW.

17.12. The force P acting on the Prony brake shown in Fig. 17-13 is 25.6 lb. The moment M transmitted by the shaft of the engine to the brake drum gives it a counterclockwise angular velocity of 600 rpm. Determine the power dissipated by the Prony brake. Neglect the weight of the brake.

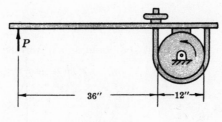

Fig. 17-13

SOLUTION

The moment M being transmitted by the shaft equals the moment of the force P about the drum center; that is, $(25.6 \text{ lb})(42/12 \text{ ft}) = 89.6$ lb$\cdot$-ft.

$$\text{Hp} = \frac{M\omega}{550} = \frac{89.6 \text{ lb-ft} \times (2\pi \times 600/60) \text{ rad/s}}{550 \text{ ft-lb/s}} = 10.2$$

Thus, the power being transmitted by the shaft is 10.2 hp.

17.13. The power measured by a Prony brake attached to an engine flywheel is 3.8 brake horsepower (bhp). The indicated horsepower (ihp), as measured by means of indicator cards, is 4.1. What is the efficiency of the engine?

SOLUTION

$$\text{Efficiency} = \frac{\text{power output}}{\text{power input}} = \frac{3.8 \text{ bhp}}{4.1 \text{ ihp}} = 93\%$$

17.14. A block of mass m is projected with initial speed v_0 along a horizontal plane. If it covers a distance s before coming to rest, what is the coefficient of friction, assuming that the force of friction is proportional to the normal force?

SOLUTION

The initial kinetic energy is $T_1 = \frac{1}{2}mv_0^2$. Its final kinetic energy is zero.

The normal force is mg. Hence, the friction is μmg and does an amount of work equal to $-\mu mgs$.

$U = T_2 - T_1$ or $-\mu mgs = -\frac{1}{2}mv_0^2$, from which $\mu = v_0^2/2gs$.

17.15. The strength of a magnetic field is given by $F = -3/x$, where F is in pounds and x is the distance from the magnet in feet. A disk weighing 4 oz is placed 6 ft from the magnet on a smooth horizontal plane. What will be the speed of the disk when it is 3 ft from the magnet?

SOLUTION

Using the work–energy relation $U = \Delta T$, we have

$$U = \int F\, dx = \int_6^3 -\frac{3}{x}\, dx = [-3 \ln x]_6^3 = 2.08 \text{ ft-lb}$$

$$\Delta T = \frac{1}{2}\left(\frac{4/16}{g}\right)v^2 - 0 = 2.08 \text{ ft-lb}$$

Hence, $v = 23.1$ ft/s

17.16. A slender rod 2 m long and having a mass of 4 kg increases its speed about a vertical axis through one end from 20 to 50 rpm in 10 revolutions. Find the constant moment M required to do this.

SOLUTION

The moment of inertia I_O of the rod about an axis through one end is

$$I_O = \tfrac{1}{3}ml^2 = \tfrac{1}{3}(4)(2)^2 = 5.33 \text{ kg} \cdot \text{m}^2$$

$\omega_1 = 2\pi(20/60) = 2.09$ rad/s, $\omega_2 = 2\pi(50/60) = 5.23$ rad/s, and $\theta = 2\pi(10) = 62.8$ rad.

Work done = changes in kinetic energy of rotating body

$$M\theta = \tfrac{1}{2}I_O(\omega_2^2 - \omega_1^2)$$
$$M(62.8) = \tfrac{1}{2}(5.33)(\overline{5.23}^2 - \overline{2.09}^2) \qquad \text{or} \qquad M = 0.975 \text{ N} \cdot \text{m}$$

17.17. A slender rod having a mass m and length l is pinned at one end to a horizontal plane. The rod, initially in a vertical position, is allowed to fall. See Fig. 17-14. What will be its angular speed when it strikes the floor?

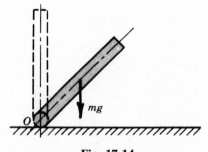

Fig. 17-14

SOLUTION

The only force doing work is the gravitational force mg, assumed concentrated at the center of gravity, which falls a total vertical distance $\tfrac{1}{2}l$. The work done by gravity is thus $\tfrac{1}{2}mgl$.

The kinetic energy is changed from $T_1 = 0$ to $T_2 = \tfrac{1}{2}I_O\omega^2 = \tfrac{1}{2}(\tfrac{1}{3}ml^2)\omega^2$. Then

$$U = T_2 - T_1 \qquad \text{or} \qquad \tfrac{1}{2}mgl = \tfrac{1}{6}ml^2\omega^2$$

from which $\omega = \sqrt{\dfrac{3g}{l}}$

17.18. A car and four wheels weigh 1800 lb. Each wheel weighs 200 lb and is 2.50 ft in diameter. The car, moving at 15 mi/h, coasts to rest on a level track in 2 mi. What is the rolling resistance F?

SOLUTION

The initial kinetic energy T_1 is the kinetic energy (of translation) of the 1000-lb car and the kinetic energy (both rotation and translation) of the four wheels. The final kinetic energy $T_2 = 0$. The work done by the rolling resistance F equals $F(2)(5280)$ ft-lb. Also, 15 mi/h = 22 ft/s. Then

$$U = T_2 - T_1$$
$$10{,}560F = 0 - \tfrac{1}{2}(1000/32.2)(22)^2 - 4(\tfrac{1}{2})(200/32.2)[(22)^2 + \tfrac{1}{2}r^2\omega^2]$$

Substitute $\omega^2 r^2 = v^2 = (22)^2$ into the expression to obtain $F = -1.57$ lb.

17.19. A 100-lb slender rod 4 ft long falls from rest in the horizontal line to the 45° position shown in Fig 17-15. In that phase, what are the bearing reactions at O on the rod?

SOLUTION

Choose n and t axes along and perpendicular to the bar. apply reversed effective forces through the center of percussion P to hold the bar in "equilibrium for study purposes."

To determine the angular acceleration α in the phase shown, use the equation $\sum M_O = 0$ or $-W(2)(0.707) + m\bar{r}\alpha(k_0^2/\bar{r}) = 0$. The distance to P is $k_0^2/\bar{r} = I_O/m\bar{r} = \tfrac{8}{3}$ ft. The equation then becomes $-100(2)(0.707) + (100/32.2)(2\alpha)(\tfrac{8}{3}) = 0$, whence $\alpha = 8.53$ rad/s².

Summing forces along the t axis, $O_t + m\bar{r}\alpha - 100\cos 45° = 0$ or $O_t = 17.7$ lb in direction assumed.

To determine O_n, it is first necessary to find ω. Use work–energy methods. The only force doing work is the weight, whose center G has fallen $2(0.707) = 1.414$ ft. Hence, work = $+100(1.414) = 141.4$ ft-lb.

The kinetic energy change is the final value $\tfrac{1}{2}I_O\omega^2 = \tfrac{1}{2}[\tfrac{1}{3}(100/32.2)(4)^2]\omega^2 = 8.28\omega^2$.

Since U = change in T, we have $141.4 = 8.28\omega^2$ or $\omega^2 = 17.1$.

Summing the forces along the bar,

$$-O_n + 100(0.707) + (100/32.2)(2)(17.1) = 0 \qquad \text{or} \qquad O_n = 177 \text{ lb}$$

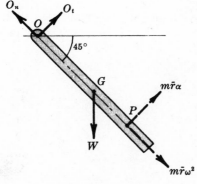

Fig. 17-15

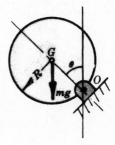

Fig. 17-16

17.20. A solid homogeneous cylinder of radius R and mass m rotates freely from its initial rest position (G vertically above O) about a fixed horizontal axis perpendicular to the plane of the paper. What is the value of its angular speed in the position θ? Refer to Fig. 17-16.

SOLUTION

By inspection, G falls a vertical distance $R - R\cos\theta$. The work done on the cylinder by gravity is thus equal to $mgR(1 - \cos\theta)$.

The kinetic energy in its rest position is $T_1 = 0$. The kinetic energy in the θ position is $T_2 = \frac{1}{2}I_0\omega^2$. By the transfer theorem for moments of inertia,

$$I_O = \bar{I} + mR^2 = \frac{1}{2}mR^2 + mR^2 = \frac{3}{2}mR^2$$

Then $\qquad U = T_2 - T_1 \qquad mgR(1 - \cos\theta) = \frac{3}{4}mR^2\omega^2 \qquad \omega = \sqrt{\dfrac{4g(1 - \cos\theta)}{3R}}$

17.21. a sphere, rolling with an initial velocity of 9 m/s, starts up a plane inclined 30° with the horizontal as shown in Fig 17-17. How far will roll up the plane?

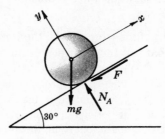

Fig. 17-17

SOLUTION

The initial kinetic energy T_1 decreases to $T_2 = 0$ at the top of the travel. The only force that does work is the component (negative) of the weight W along the plane.

Work done $= -(mg\sin30°)x$, where x is the required distance.

The initial kinetic energy T_1 for the body in plane motion is $T_1 = \frac{1}{2}m\bar{v}_1^2 + \frac{1}{2}\bar{I}\omega_1^2$.

Since here $\bar{I} = \frac{2}{5}mR^2$ and $v_1 = \omega_1 R$, we have $T_1 = \frac{1}{2}m\bar{v}_1^2 + \frac{1}{5}m\bar{v}_1^2 = \frac{7}{10}m(9)^2$. Then

$$U = T_2 - T_1 \qquad -(mg\sin30°)x = 0 - \tfrac{7}{10}m(9)^2 \qquad x = 11.6 \text{ in}$$

17.22. Fig. 17-18(a) shows a 322-lb homogeneous cylinder that rolls from rest without slipping on a horizontal plane under the action of the horizontal 12-lb force. Determine the angular velocity of the cylinder after it has rotated 90°. The diameter is 3.2 ft.

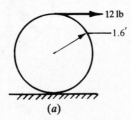

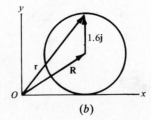

Fig. 17-18

SOLUTION

In the free-body diagram, Fig. 17-18(b), the only force doing work is the 12-lb force. Using an (x, y) set of axes as shown, the radius vector for the top point is $\mathbf{r} = \mathbf{R} + 1.6\mathbf{j}$.

When the cylinder rotates through a differential angular displacement $d\phi$, we can write

$$d\mathbf{r} = d\mathbf{R} + 1.6(d\phi)\mathbf{i}$$

Also note that $d\mathbf{R} = 1.6(d\phi)\mathbf{i}$. The differential work done by the 12-lb force is now written

$$dU = 12\mathbf{i} \cdot d\mathbf{r} = 12\mathbf{i} \cdot (1.6 \, d\phi \, \mathbf{i} + 1.6 \, d\phi \, \mathbf{i})$$

$$= 38.4 \, d\phi$$

and
$$U = \int_0^{\pi/2} 38.4 \, d\phi = 60.3 \text{ ft-lb}$$

This result can also be obtained by replacing the 12-lb force at the top with a horizontal 12-lb force through the center *and* a clockwise couple equal to $12 \times 1.6 = 19.2$ lb-ft.

The work done by the force through the center and the couple is

$$U = 12 \times 1.6 \times \frac{\pi}{2} + 19.2\left(\frac{\pi}{2}\right) = 60.3 \text{ ft-lb}$$

Next determine the kinetic energy of the cylinder as

$$\tfrac{1}{2}m\bar{v}^2 + \tfrac{1}{2}\bar{I}\omega^2 = \frac{1}{2}\left(\frac{322}{g}\right)\left(\frac{\omega}{1.6}\right)^2 + \frac{1}{2}\left[\frac{1}{2}\left(\frac{322}{g}\right)(1.6)^2\right]\omega^2$$

Equating this to the work done, we find

$$\omega = 2.69 \text{ rad/s}$$

17.23. A solid homogeneous cylinder rolls without slipping on the horizontal plane. The cylinder has a mass of 90 kg and is at rest in the phase shown in Fig. 17-19. The modulus of the spring is 450 N/m and its unstretched length is 600 mm. What will be the angular speed of the cylinder when the spring has moved the center 500 mm to the right?

SOLUTION

The cylinder has initial kinetic energy $T_1 = 0$ and final $T_2 = \tfrac{1}{2}m\bar{v}_2^2 + \tfrac{1}{2}\bar{I}\omega_2^2$. Substituting $\bar{v}_2 = r\omega_2 = 0.15\omega_2$ and $\bar{I} = \tfrac{1}{2}mr^2 = \tfrac{1}{2}(90)(0.15)^2 = 1.01 \text{ kg} \cdot \text{m}^2$, we obtain $T_2 = 1.52\omega_2^2$.

Since the point of application of the frictional force moves with the cylinder (and hence friction does no work), the forces form a conservative system. The conservation of energy law will be used here. The system possesses potential energy because of the spring configuration (see Problem 17.2).

The initial and final lengths of the spring are $s_1 = \sqrt{(0.6)^2 + (0.9)^2} = 1.08$ m and $s_2 = \sqrt{(0.6)^2 + (0.4)^2} = 0.72$ m. Then

$$V_1 = \tfrac{1}{2}k(1.08 - 0.6)^2 = \tfrac{1}{2}(450)(0.48)^2 = 51.8 \text{ N} \cdot \text{m} \qquad V_2 = \tfrac{1}{2}k(0.72 - 0.6)^2 = 3.24 \text{ N} \cdot \text{m}$$

Finally, from the conservation of energy law, we can write

$$T_1 + V_1 = T_2 + V_2 \qquad 0 + 51.8 = 1.52\omega_2^2 + 3.24 \qquad \omega_2 = 5.65 \text{ rad/s}$$

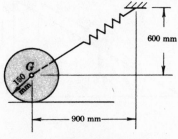

Fig. 17-19

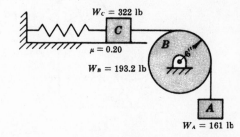

Fig. 17-20

17.24. In the phase shown in Fig. 17-20 the block A is moving down 5 ft/s. The cylinder B is considered a homogeneous solid and moves in frictionless bearings. The spring is originally compressed 6 in and has a modulus of 6 lb/ft. What will be the speed v of A after dropping 4 ft?

SOLUTION

Consider the system consisting of bodies A, B, and C as a unit in determining kinetic energy and work done. The initial kinetic energy T_1 is the sum

$$T_1 = \frac{1}{2}\left(\frac{W_C}{g}\right)v_1^2 + \frac{1}{2}\bar{I}_B\omega_1^2 + \frac{1}{2}\left(\frac{W_A}{g}\right)v_1^2 = 225 \text{ ft-lb}$$

where $\bar{I}_B = \frac{1}{2}(W_B/g)(\frac{1}{2})^2$, and $\omega_1 = v_1/r = 5/\frac{1}{2} = 10$ rad/s, since no slip of rope on cylinder is assumed. The final kinetic energy T_2 in terms of the required speed v is

$$T_2 = \frac{1}{2}\left(\frac{W_C}{g}\right)v^2 + \frac{1}{2}\bar{I}_B(2v)^2 + \frac{1}{2}\left(\frac{W_A}{g}\right)v^2 = 9.0v^2$$

Work is done on the system as follows: W_A does positive work; friction on C (0.20×322) does negative work; for the first 6 in, or $\frac{1}{2}$ ft, the spring does positive work, but after that it is being stretched from the neutral position and does negative work. Thus,

$$\text{Work } U = +161(4) - 64.4(4) + \int_0^{1/2} ks\, ds - \int_0^{3.5} ks\, ds = +26.4 \text{ ft-lb}$$

Then $U = T_2 - T_1$, $26.4 = 9.0v^2 - 225$, and $v = 5.29$ ft/s.

17.25. A cylinder is pulled up a plane by the tension in a rope that passes over a frictionless pulley and is attached to a 70-kg mass as shown in Fig. 17-21. The 45-kg cylinder has radius 600 mm. The cylinder moves from rest up a distance of 5 m. What will be its speed?

SOLUTION

The initial kinetic energy of the cylinder and mass is $T_1 = 0$. The final kinetic energy of the system is

$$T_2 = T_c + T_m = (\tfrac{1}{2}m_c\bar{v}^2 + \tfrac{1}{2}\bar{I}_c\omega^2) + (\tfrac{1}{2}m_m\bar{v}^2) = 68.75\bar{v}^2$$

After substituting, $m_c = 45$, $m_m = 70$, $\bar{I}_c = \frac{1}{2}m_cR^2$, and $R^2\omega^2 = \bar{v}^2$.

The initial potential energy for the cylinder will be assumed zero, and it will gain potential energy $(9.8 \times 45)(5 \sin 50°)$. The initial potential energy of the 70-kg mass will be assumed zero, and it will lose potential energy $(9.8 \times 70)(5)$. Thus,

$$T_1 + V_1 = T_2 + V_2 \qquad 0 + 0 = 68.75\bar{v}^2 + (9.8 \times 45)(5 \sin 50°) - (9.8 \times 70)(5) \qquad \bar{v} = 5.03 \text{ m/s}$$

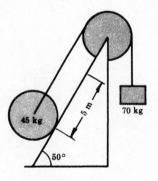

Fig. 17-21

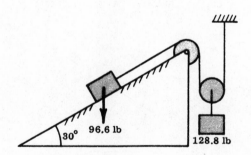

Fig. 17-22

19.26. The 96.6-lb block shown in Fig. 17-22 rests on a smooth plane. It is connected by a flexible inextensible cord that passes around weightless, frictionless pulleys to a support. The 128.8-lb weight is attached as shown. After the system is released from rest, in what distance will the block on the plane attain a speed of 8 ft/s?

SOLUTION

The 128.8-lb block travels half the distance that the 96.6-lb block does, and its speed is half the speed of the 96.6-lb block.

The initial kinetic energy of the system is $T_1 = 0$. Its final kinetic energy is

$$T_2 = \tfrac{1}{2}(3)(8)^2 + \tfrac{1}{2}(4)(8/2)^2 = 128 \text{ ft-lb}$$

Work is done by the component of the weight along the plane and by the 128.8-lb weight. Assuming motion up the plane, the work done is

$$U = -(96.6 \sin 30°)s + 128.8\,\frac{s}{2} = 16.1s$$

$$16.1s = 128 \quad \text{or} \quad s = 7.95 \text{ ft} \quad \text{(up the plane)}$$

17.27. A flexible chain of length l and mass a kg/m is released when it has a free overhang of c meters. What will be its speed when leaving the smooth table?

SOLUTION

The free overhang of mass ac will fall a distance $l - c$; the work done by gravity on this part will be $gac(l - c)$. The part on the table will fall an average distance $\tfrac{1}{2}(l - c)$; the work done by gravity on this part is $ga(l - c) \times \tfrac{1}{2}(l - c)$. The total work equals the final kinetic energy $\tfrac{1}{2}mv^2 = \tfrac{1}{2}alv^2$. Then

$$gac(l - c) + \tfrac{1}{2}ga(l - c)^2 = \tfrac{1}{2}alv^2, \quad \text{from which} \quad v = \sqrt{\frac{g(l^2 - c^2)}{l}}$$

17.28. Determine the work done in winding up a homogeneous cable that hangs from a horizontal drum if its free length is 6 m and it has a mass of 50 kg.

SOLUTION

Figure 17-23 shows the cable in its original phase. In analyzing the problem, note that any element dx is acted upon by a gravitational force $9.8(50/6)\,dx = 81.7\,dx$.

Assume that the element dx is at a distance x from the free end. This element is then raised a distance $(6 - x)$ m. The work done on it is the product of the gravitational force and the distance raised.

The total work is the integral of the work done on a differential element. Note that x varies from zero to 6 m.

$$\text{Work} = \int_0^6 81.7(6 - x)\,dx = 1470 \text{ N} \cdot \text{m}$$

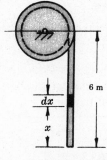

Fig. 17-23

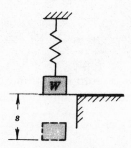

Fig. 17-24

17.29. Refer to Fig. 17-24. If weight W hangs freely, it stretches the spring a distance c. Show that if the weight (held so that the spring is unstressed) is suddenly released, the spring stretches a distance $2c$ before the weight starts to return upward.

SOLUTION

The kinetic energy of the weight at the top (initial position) and the bottom is zero. Hence, the total work done on the weight must be zero. But the total work done is the positive work of gravity. (Ws) offset by the negative work of the spring ($\frac{1}{2}ks^2$) on the weight. Therefore

$$Ws - \tfrac{1}{2}ks^2 = 0$$

But in the freely hanging position, W is balanced by the spring force kc. Putting $W = kc$ in the above equation, we obtain $s = 2c$.

The tension in the spring for this type of loading is double the weight.

17.30. In the spring gun shown in Fig. 17-25, ball W rests against the compressed spring of constant k. Its initial compression is x_0. What will be the speed of the ball leaving the gun? Assume the spring is unstressed when the bearing plate is at the end of the gun.

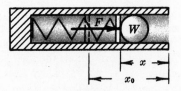

Fig. 17-25

SOLUTION

The ball is shown is any position x from the end of the gun. The force doing work is the spring force $F = kx$.

The work done $= \int_0^{x_0} kx\,dx = \frac{1}{2}kx_0^2$. This is equal to the gain in kinetic energy of the ball, which is $\frac{1}{2}(W/g)v^2$. Hence, $v = x_0\sqrt{kg/W}$.

17.31. Two balls are connected by a spring whose unstressed length is 450 mm and whose modulus is 0.044 N/mm. The balls are pushed together (compressing the spring) until they are 150 mm apart. They are then released on a smooth horizontal table. What work is done on the balls in returning them to their original distance apart?

SOLUTION

The work done on the balls by the varying spring force is equal to the work done in compressing the spring from 450 to 150 mm; that is, $U = \int_0^{300} kx\,dx = [\frac{1}{2}kx^2]_0^{300} = \frac{1}{2}(0.044)(300)^2 = 1.98\text{ N}\cdot\text{m}$.

17.32. A rope is wrapped around a 10-kg solid homogeneous cylinder as shown in Fig. 17-26. Find the speed of its center G after it has moved 1.2 m down from rest.

SOLUTION

The only force doing work is the gravitational force $10 \times 9.8 = 98$ N.
Initial kinetic energy is zero. Final kinetic energy is

$$\tfrac{1}{2}m\bar{v}^2 + \tfrac{1}{2}\bar{I}\omega^2 = \tfrac{1}{2}m\bar{v}^2 + \tfrac{1}{2}(\tfrac{1}{2}mr^2)\left(\frac{\bar{v}^2}{r^2}\right) = \tfrac{3}{4}m\bar{v}^2 = 7.5\bar{v}^2$$

Then $U = 98 \times 1.2 = 7.5\bar{v}_1^2$ and $\bar{v} = 3.96$ m/s

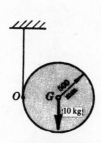

Fig. 17-26

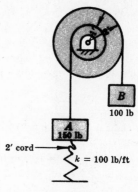

Fig. 17-27

17.33. In Fig. 17-27, block A initially rests on the spring, to which it is connected by a 2-ft inextensible cord, which becomes taut after the system is released. What will be the stretch of the spring to bring the system to rest? The cylinder may be considered homogeneous; it weighs 161 lb and rotates in frictionless bearings.

SOLUTION

First determine the kinetic energy of the system before the spring comes into play, i.e., while the weight A rises 2.0 ft. the work done equals $100s_B - 150s_A$.

The weight B drops 4.0 ft while A rises 2.0 ft. Hence, the kinetic energy when the spring action starts is the work done, or $T_1 = 100(4.0) - 150(2.0) = 100$ ft-lb. The final kinetic energy of the system is $T_2 = 0$.

The work done on the system (A and B) by gravity and by the spring as it stretches a distance x ft is

$$U = -100(2x) - 150(x) - \tfrac{1}{2}kx^2 = 50x - 50x^2$$

Since $U = T_2 - T_1$, where $50x - 50x^2 = 0 - 100$, $x^2 - x - 2 = 0$, and $x = 2.0$ ft.

17.34. In Fig. 17-28, what weight of B will cause the cylinder having an $I = 100$ slug-ft^2 to attain a speed of 4 rad/s after rotating counterclockwise 6 rad from rest?

SOLUTION

The work done in the counterclockwise direction is

$$U = [W_B(2) - 64.4(\tfrac{1}{2})]\theta = 12W_B - 193$$

The system has initial kinetic energy $T_1 = 0$ and final kinetic energy

$$T_2 = \tfrac{1}{2}m_Bv_B^2 + \tfrac{1}{2}I\omega^2 + \tfrac{1}{2}m_Cv_C^2$$

But $v_B = 2\omega$, $v_C = \tfrac{1}{2}\omega$, and $\omega = 4$ rad/s; hence,

$$T_2 = \frac{1}{2}\left(\frac{W_B}{32.2}\right)(8)^2 + \tfrac{1}{2}(100)(4)^2 + \tfrac{1}{2}(2)(2)^2 = 0.994W_B + 804$$

Then $U = T_2 - T_1$, $12W_B - 193 = 0.994W_B + 804$, and $W_B = 90.6$ lb.

17.35. In Fig. 17-29, block A weighs 96.6 lb and block B weighs 128.8 lb. The drum has a moment of inertia $\bar{I} = 12$ slug-ft^2. Through what distance will A fall before it reaches a speed of 6 ft/s?

SOLUTION

Work done $= 96.6s_A - 128.8s_B = 53.7s_A$, since $s_B = \tfrac{1}{3}s_A$.

The system has initial kinetic energy $T_1 = 0$ and final kinetic energy $T_2 = \tfrac{1}{2}m_Av_A^2 + \tfrac{1}{2}\bar{I}\omega^2 + \tfrac{1}{2}m_Bv_B^2 = 86$ ft-lb, since $v_A = 6$ ft/s, $v_B = 2$ ft/s, and $\omega = v_A/r = 6/3 = 2$ rad/s.

Then $53.7s_A = 86$ or $s = 1.60$ ft.

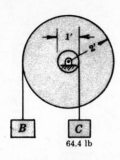

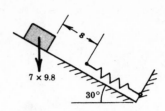

| **Fig. 17-28** | **Fig. 17-29** | **Fig. 17-30** |

17.36. The 7-kg block shown in Fig. 17-30 is released from rest and slides a distance s down the inclined plane. It strikes the spring, which it compresses 75 mm before motion impends up the plane. Assuming that the coefficient of friction is 0.25 and that the spring constant $k = 2.8 \, \text{N/mm}$, determine the value of s.

SOLUTION

The initial kinetic energy and the final kinetic energy (when the block has moved $s + 0.075$ m) are zero. Hence, the work done by friction, gravity, and the spring must be zero.

The normal reaction $= 9.8 \times 7 \cos 30° = 59.4 \, \text{N}$, does no work.

The friction $= 0.25 \times 59.4 = 14.85 \, \text{N}$. The component of the gravitational force along the plane $= 9.8 \times 7 \sin 30° = 34.3 \, \text{N}$. Each of these forces does work for $s + 0.075$ m; frictional work is negative, the other is positive.

The work of the spring is negative and equals $\frac{1}{2}k(0.075)^2 = 7.88 \, \text{N} \cdot \text{m}$.

$$U = (34.3 - 14.85)(s + 0.075) - 7.88 \quad \text{and} \quad s = 330 \, \text{mm}$$

17.37. A 5-kg mass drops 2 m upon a spring whose modulus is 10 N/mm. What will be the speed of the block when the spring is deformed 100 mm?

SOLUTION

The mass drops $(2 + 0.1)$ m $= 2.1$ m. The work done by gravity is $9.8 \times 5 \times 2.1 = 102.9 \, \text{N} \cdot \text{m}$. The work done by the spring on the mass is negative and equals $\frac{1}{2}kx^2 = \frac{1}{2}(10\,000 \, \text{N/m})(0.1 \, \text{m})^2 = 50 \, \text{N} \cdot \text{m}$.

The kinetic energy of the block increases from zero to $\frac{1}{2}mv^2 = \frac{1}{2}(5)v^2 = 2.5v^2$.

$$U = \text{change of kinetic energy} \quad 102.9 - 50 = 2.5v^2 \quad v = 4.6 \, \text{m/s}$$

17.38. A weight dropped from rest through 6 ft on a spring whose modulus is 20 lb/in causes a maximum shortening in the spring of 8 in. What is the value of the weight?

SOLUTION

Work done by gravity $= W(72 + 8) = 80W$ in-lb.

Work done by the spring $= -\frac{1}{2}kx^2 = -\frac{1}{2}(20)(8^2) = -640$ in-lb.

Since the block starts from rest and ends at rest, the change in kinetic energy is zero. Then $U = 0$, $80W - 640 = 0$, and $W = 8 \, \text{lb}$.

17.39. Determine the speed of escape, i.e., the initial speed, that must be given to a particle on the earth's surface to project it to an infinite height.

SOLUTION

The particle of weight W is shown in Fig 17-31 at a distance x from the center of the earth of radius R. The earth's attraction F is known to be inversely proportional to the square of the distance x; that is, $F = -C/x^2$.

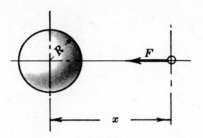

Fig. 17-31

To determine C, note that the attraction on the earth's surface is the weight W. Thus, $-W = -C/R^2$, $C = WR^2$, and hence $F = -WR^2/x^2$.

The work done in going from $x = R$ to $x = \infty$ is

$$\int_R^\infty F\,dx = \int_R^\infty -\left(\frac{WR^2}{x^2}\right)dx = WR^2\left[\frac{1}{x}\right]_R^\infty = -WR$$

This work done equals the change in kinetic energy. $T_1 = \frac{1}{2}(W/g)v_0^2$ and $T_2 = 0$ (since $v = 0$ when x becomes infinite). Hence,

$$-WR = -\frac{1}{2}\left(\frac{W}{g}\right)v_0^2 \qquad \text{or} \qquad v_0 = \sqrt{2gR}$$

Assuming the diameter of the earth to be 7900 mi, the required speed of escape is calculated to be 6.93 mi/s.

17.40. Determine the spring constant k in Fig. 17-32 such that the slender rod AB just reaches the vertical down position when it is released from rest in the horizontal position. The spring is stretched 1 in in the position shown. The weight of the bar is 8 lb.

SOLUTION

The unstretched length of the spring is 19 in. In the vertical position, the spring is stretched 28 in − 19 in = 9 in. The work done by the spring is

$$-\tfrac{1}{2}k(s_2^2 - s_1^2) = -\tfrac{1}{2}k\left[\left(\frac{9}{12}\right)^2 - \left(\frac{1}{12}\right)^2\right] = -0.278k \text{ ft-lb}$$

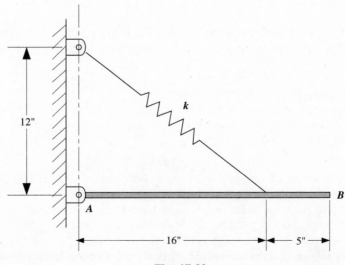

Fig. 17-32

The work done by the gravity force is

$$mgh = (8)\left(\frac{10.5}{12}\right) = 7 \text{ ft-lb}$$

Since the bar starts from rest and ends at rest, the change in the kinetic energy is zero. Hence,

$$U = T_2 - T_1 \qquad -0.278k + 7 = 0$$

and $k = 25.2$ lb/ft.

17.41. Refer to Fig. 17-33. A 100-lb cylinder of radius 1 ft rolls without slipping under the action of an 80-lb force. A spring is attached to a cord that is wound around the cylinder. What is the speed of the center of the cylinder after it has moved 6 in? The spring is unstretched when the 80-lb force is applied.

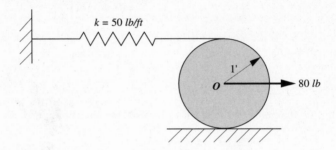

Fig. 17-33

SOLUTION

Since the cylinder rolls without slipping, the spring becomes stretched 12 in when the center of the cylinder moves 6 in to the right. Noting that the friction force and normal force under the cylinder do no work (see Problem 17.7), the work is

$$U = -\tfrac{1}{2}50(s_2^2 - s_1^2) + Fs = -\tfrac{1}{2}50\left(\left(\frac{12}{12}\right)^2 - 0\right) + 80\left(\frac{6}{12}\right) = 15 \text{ ft-lb}$$

The initial kinetic energy is zero. Hence, the change in kinetic energy, where $v_0 = 1\omega$ for no slip, is

$$\Delta T = T_2 - T_1 = \tfrac{1}{2}mv_0^2 + \tfrac{1}{2}I_0\omega^2 = \frac{1}{2}\frac{100}{g}v_0^2 + \frac{1}{2}\left(\frac{1}{2} \times \frac{100}{g} \times 1^2\right)\left(\frac{v_0}{1}\right)^2$$

Hence, $U = T_2 - T_1,$ $15 = \frac{3}{4}\left(\frac{100}{g}\right)v_0^2,$ $v_0 = 2.54$ ft/s

Supplementary Problems

17.42. A cylindrical well is 2 m in diameter and 12 m deep. If there is 3 m of water in the bottom of the well, determine the work done in pumping all this water to the surface.   *Ans.*   970 kN · m

17.43. Referring to Problem 17.42, what work must be done by a pump that is 60-percent effiicient?
Ans.   1620 kN · m

17.44. A person who can push 60 lb wishes to roll a barrel weighing 200 lb into a truck that is 3 ft above the ground. How long a board must be used and how much work will the person do in getting the barrel into the truck?   *Ans.*   10 ft, 600 ft-lb

17.45. A 10-lb block slides 4 ft on a horizontal surface. (*a*) If the coefficient of friction is 0.3 what work is done by the block on the surface? (*b*) What work is done by the surface on the block?
Ans.   (*a*) $U = 0$, (*b*) $U = 12$ ft-lb

17.46. The 8-kg block is acted upon by a 100-N force as shown in Fig. 17-34. If the coefficient of sliding friction is 0.30, determine the work done by all forces as the block moves 4 m to the right.
Ans. $U = 241$ N · m

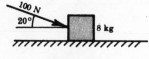

Fig. 17-34

17.47. A 10-lb block slides 6 ft down a plane inclined 40° with the horizontal. Determine the work done by all forces acting on the block. The coefficient of sliding friction is 0.40.   *Ans.* $U = 20.2$ ft-lb

17.48. A particle moves along the path $x = 2t$, $y = t^3$, where t is in seconds and distances are in feet. What work is done in the interval from $t = 0$ to $t = 3$ s by a force whose components are $F_x = 2 + t$ and $F_y = 2t^2$? Forces are in pounds.   *Ans.* $U = 313$ ft-lb

17.49. A 2-oz bead is raised slowly along a frictionless wire from A to B as shown in Fig. 17-35. What work is done?   *Ans.* $U = 0.5$ ft-lb

17.50. A freestanding crane has a horizontal boom 250 ft long that is 400 ft above the ground. If the crane is slowly lifting 8000 lb of concrete up a distance of 300 ft, what work is done? During this lifting, what is the moment that tends to overturn the crane?   *Ans.* $U = 2.4 \times 10^6$ ft-lb, $M = 2.0 \times 10^6$ ft-lb

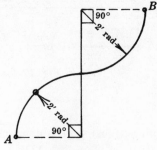

Fig. 17-35

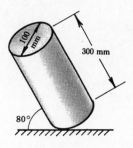

Fig. 17-36

17.51. The 20-kg solid cylinder shown in Fig. 17-36 is released from rest. Determine the work done by the earth's pull when the bottom hits the floor.   *Ans.* $U = 1.25$ N · m

17.52. The empty cylindrical tank A in Fig. 17-37 is filled with water from the cubical tank B. Water has a density of 1000 kg/m^3. Assume that B was filled at the beginning. How much work is done?
Ans. $U = 176$ kN · m

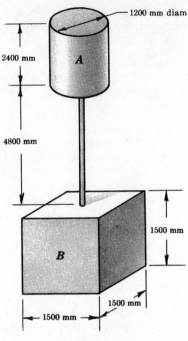

Fig. 17-37

17.53. The 16-kg mass in Fig. 17-38 drops 2.5 m. It is attached by a light rope to a drum that rotates in frictionless bearings. A constant torque $M = 80\,N \cdot m$ is supplied to the drum. What work is done on the system?   *Ans.* $U = 59\,N \cdot m$.

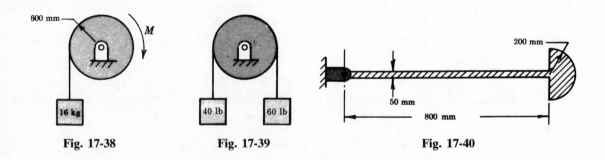

Fig. 17-38   **Fig. 17-39**   **Fig. 17-40**

17.54. The drum shown in Fig. 17-39 rotates in frictionless bearings. What work is done on the system when the 60-lb weight falls 3 ft?   *Ans.* $U = 60$ ft-lb

17.55. The homogeneous object shown in Fig. 17-40 is 25 mm thick and has a mass density of $7840\,kg/m^3$. It falls from a horizontal to a vertical position. What work is done on the object?   *Ans.* $U = 138\,N \cdot m$

17.56. A couple $M = 2\theta^3 - \theta$ is applied to a shaft that rotates from $\theta = 0°$ to $\theta = 90°$. Determine the work done if M is in $N \cdot m$.   *Ans.* $U = 1.81\,N \cdot m$

17.57. A force of 30 N will stretch an elastic cord 250 mm. If the force required to stretch the cord varies directly as the deformation, what is the work done in stretching the cord 1500 mm?   *Ans.* $135\,N \cdot m$

17.58.   A force of 200 lb is required to compress a spring through a distance of 4 in. If the force required to compress the spring varies directly with its deformation, how much work is done in compressing it through 9 in?   *Ans.*   2025 in-lb

17.59.   At a certain instant during acceleration along a level track, the drawbar pull of a locomotive is 90 kN. What power is being developed if the speed of the train is 60 km/h?   *Ans.*   1.5 MW

17.60.   An automobile weighing 2500 lb climbs a 10-percent grade at a uniform rate of 15 mi/h. If the resistance is 20 lb/ton, what horsepower is the car developing?   *Ans.*   11.0 hp

17.61.   A steam engine raises an 1800-kg mass vertically at the rate of 9 m/s. What is the power of the engine assuming an efficiency of 70%   *Ans.*   227 kW

17.62.   The power measured by a Prony brake attached to an engine flywheel is 6.3 brake horsepower (bhp). The indicated horsepower (ihp), as measured by means of indicator cards, is 7.1. What is the efficiency of the engine?   *Ans.*   Eff = 89%

17.63.   What horsepower is required to raise a 100-lb weight to a height of 8 ft in 4 s?   *Ans.*   0.36 hp

17.64.   A 50-kg homogeneous cylinder 1200 mm in diameter is rotating at 100 rpm. A torque of 30 N · m is needed to keep this speed constant (overcoming friction). What power is required?   *Ans.*   314 W

17.65.   A 6-in-diameter pulley is rotating at 2000 rpm. The belt driving it has tensions of 1 and 3 lb in the slack and tight sides, respectively. What horsepower is being delivered to the pulley?   *Ans.*   0.19 hp

17.66.   A 100-g body falls 1500 mm to the surface of the earth. What is its kinetic energy as it hits the ground?   *Ans.* $T = 1.47$ N · m

17.67.   A 50-kg body starts from rest and is pulled along the ground by a horizontal force of 300 N. If the kinetic coefficient of friction is 0.1 and the force acts for a distance of 2 m and then ceases to act, determine the distance required for the body to come to rest.   *Ans.*   10.2 m

17.68.   A bullet enters a 2-in plank with a speed of 2000 ft/s and leaves with a speed of 800 ft/s. Determine the greatest thickness of plank that could be penetrated by the bullet.   *Ans.*   2.38 in

17.69.   If the block in Problem 17.6 starts from rest, what is its speed after traversing the 6 m?
Ans.   9.6 m/s

17.70.   A 2-kg block slides down a plane inclined 50° with the horizontal. The coefficient of friction between the block and the plane is 0.25. Determine the speed of the block after it has moved 4 m along the plane starting with a velocity of 2 m/s.   *Ans.* $v = 7.17$ m/s

17.71.   A block weighing 80 lb is acted upon by a 30-lb horizontal force. The coefficient of friction is 0.25. Determine the speed of the block after it has moved 20 ft from rest. Refer to Fig. 17-41.
Ans. $v = 12.7$ ft/s to the right

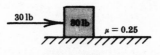

Fig. 17-41

17.72.   Determine the kinetic energy possessed by a 100-kg disk that is 500 mm in diameter, 75 mm thick, and rotating at 100 rpm about its center.   *Ans.*   171 N · m

17.73.   A 100-lb 6-in-diameter sphere rotates at 120 rpm about an axis 16 in from its center. What is the kinetic energy of rotation?   *Ans.*   442 ft-lb

17.74. A homogeneous cylinder weighing 20 lb and with a radius of 8 in is moving with a mass center speed of 6 ft/s. What is its kinetic energy?   *Ans.* $T = 16.8$ ft-lb

17.75. A 4-kg sphere has a radius of 1000 mm and a radius of gyration of 600 mm. It is rolling on a horizontal plane with angular speed 3 rad/s. What is the kinetic energy of the sphere?   *Ans.* $T = 24.5$ N · m

17.76. In Fig. 17-42, the angular velocity of crank *CD* is 1.5 rad/s clockwise. The weights of the slender bars are as follows: *AB* is 3 lb, *BC* is 5 lb, and *CD* is 4 lb. Determine the kinetic energy of the system. *Ans.*   4.09 ft-lb

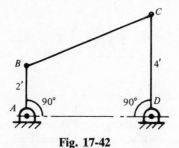

Fig. 17-42   **Fig. 17-43**

17.77. The slender rods *AB* and *BC* in Fig. 17-43 weigh 10 and 6 lb, respectively. *AB* rotates with an angular velocity of 8 rad/s clockwise, while *BC* has an angular velocity of 6 rad/s counterclockwise. Determine the kinetic energy of the system.   *Ans.*   243 ft-lb

17.78. A 900-kg solid cylindrical flywheel is 1200 mm in diameter. If the axle is 150 mm in diameter and the coefficient of journal friction is 0.15, find the time required for the flywheel to coast to rest from a speed of 500 rpm.   *Ans.*   85.5 s

17.79. An electric motor has a rotor weighing 20 lb with radius of gyration $k = 1.83$ in. A frictional torque of 8 oz-in is present. How many revolutions will the rotor make while coming to rest from a speed of 1800 rpm?   *Ans.* $\theta = 979$ rev

17.80. A drum rotating 20 rpm is lifting a 1-ton cage connected to it by a light inextensible cable as shown in Fig. 17-44. If power is cut off, how high will the cage rise before coming to rest? Assume frictionless bearings.   *Ans.* $h = 0.36$ ft

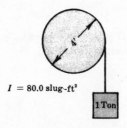

Fig. 17-44

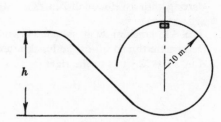

Fig. 17-45

17.81. A car travels down the smooth incline and then moves inside the loop shown in Fig. 17-45. Determine the least value of *h* so that the car will remain in contact with the track.   *Ans.* $h = 25$ m

17.82. a 4-kg homogeneous slender bar is 1 m long. It pivots about one end. When released from rest in the horizontal position, it falls under the action of gravity and a constant retarding torque of 5 N · m. What will be its angualr speed as it passes through its lowest position?   *Ans.* $\omega = 4.19$ rad/s

17.83. The 30-kg drum in Fig. 17-46 has a radius of gyration $k = 800$ mm. Assume no friction and determine the speed of the drum after it has made one revolution starting from rest.   *Ans.* $\omega = 3.35$ rad/s

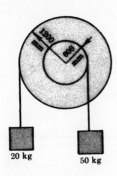

20 kg   50 kg

Fig. 17-46

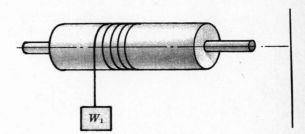

W_1

Fig. 17-47

17.84. A well bucket of weight W_1 is attached to a windlass of weight W_2 and radius of gyration k. See Fig. 17-47. If r is the radius of the windlass, how long will it take the bucket to drop a distance s from rest to the water level? Neglect bearing friction and the weight of the rope.
Ans. $t = \sqrt{2s(1 + W_2 k^2 / W_t r^2)/g}$

17.85. A homogeneous bar of length L is pivoted about a point a distance a from one end as shown in Fig. 17-48. If the bar is released from rest in the 30° position, what will be the angular speed when the bar is vertical?

Ans. $\omega^2 = \dfrac{0.402g(L - 2a)}{L^2 - 3La + 3a^2}$

17.86. A 200-lb sphere, 2 ft in diameter, rolls from rest down a 25° plane for a distance of 100 ft. What is its kinetic energy at the end of the 100 ft?   *Ans.*   8450 ft-lb

17.87. In Problem 17.86, what is the speed of the center of the sphere after it has traveled the 100 ft?
Ans.   44.2 ft/s

17.88. A homogeneous sphere rolls a distance s down a plane inclined at angle θ with the horizontal. What is the speed of the sphere if it starts from rest?   *Ans.* $v = 6.78\sqrt{s \sin \theta}$

Mathcad

17.89. A car has a 900-kg body, four 20-kg wheels, and a 70-kg driver. The wheels are 700 mm in diameter and have a radius of gyration $k = 300$ mm. What will be the speed of the car if it moves from rest 300 m down a 5-percent grade?   *Ans.*   60 km/h

17.90. The 8-lb homogeneous sphere shown in Fig. 17-49 has a string wrapped around a slot as shown. What will be the speed of the center if it falls 3 ft from the rest position?   *Ans.* $v = 11.8$ ft/s

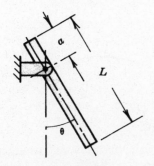

Fig. 17-48

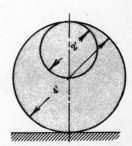

Fig. 17-49

Fig. 17-50

17.91. A 5-lb homogeneous disk 2 ft in diameter is attached rigidly to a 20-lb homogeneous disk 4 ft in diameter. If the assembly is released from rest as shown in Fig. 17-50, what will be the angular speed when the small disk is at the bottom of its travel?   *Ans.* $\omega = 2.25$ rad/s

17.92. In Fig. 17-51, A has a mass of 7 kg and B has a mass of 4 kg. If B falls 400 mm from rest, determine its speed (a) if no friction exists and (b) if the coefficient of friction between A and the horizontal plane is 0.20.   *Ans.* (a) $v = 1.69$ m/s, (b) $v = 1.36$ m/s

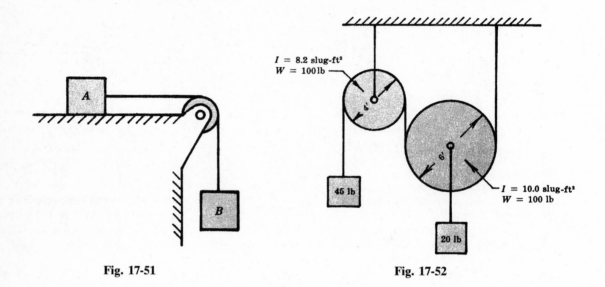

Fig. 17-51	**Fig. 17-52**

17.93. In the system shown in Fig. 17-52, all ropes are vertical. The 45-lb weight rises 1.6 ft from rest. What will be its speed?   *Ans.* $v = 3.21$ ft/s

17.94. A 36-Mg freight car moving 8 km/h horizontally hits a bumper with a spring constant of 1750 N/mm. What will be the maximum compression of the spring?   *Ans.* $d = 320$ mm

17.95. A 16-lb weight slides 6 in from rest down the 25° plane shown in Fig. 17-53. It hits a spring whose modulus is 10 lb/in. The coefficient of kinetic friction is 0.20. Determine the maximum compression of the spring.   *Ans.* $d = 2.57$ in

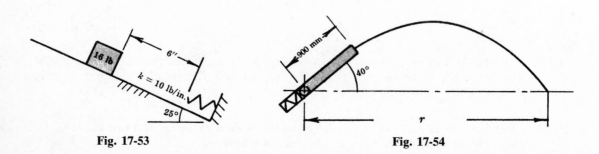

Fig. 17-53	**Fig. 17-54**

17.96. A spring compressed 75 mm and with a modulus $k = 5$ N/mm is used to propel a 50-g mass from the frictionless tube shown in Fig. 17-54. Determine the horizontal distance r at which the mass will be at the same height as it was initially. Neglect air resistance.   *Ans.* $r = 55.4$ m

17.97. (*a*) In Fig. 17-55(*a*), the 2-lb weight W slides from rest at A along a frictionless rod bent into a quarter circle. The spring with modulus $k = 1.2$ lb/ft has an unstretched length of 18 in. Determine the speed of W at B. (*b*) If the path is elliptical, as in Fig. 17-55(*b*), what is the speed at B?
Ans.   (*a*) $v = 11.3$ ft/s, (*b*) $v = 9.49$ ft/s

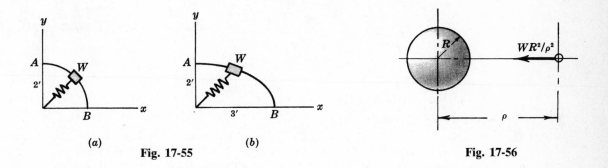

(*a*)   (*b*)
Fig. 17-55   **Fig. 17-56**

17.98. The pull of the earth on a particle varies inversely with the square of the distance from the center of the earth. If W is the weight of the particle on the earth's surface (radius R) then the pull at distance ρ is WR^2/ρ^2. (*a*) What work must be done against this gravitational pull to move the particle from the earth's surface to a distance x from the earth's center? (*b*) To infinity? See Fig. 17-56.
Ans.   (*a*) $U = WR - WR^2/x$, (*b*) $U = WR$

17.99. The wheel shown in Fig. 17-57 weighs 322 lb and has a radius of gyration of 1.20 ft with respect to its center of mass G. In the initial phase shown, the velocity of G is 6 ft/s down the plane, and the spring is stretched 0.50 ft. If the spring modulis is 80 lb/ft, what will be total stretch of the spring?
Ans.   3.78 ft

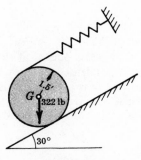

Fig. 17-57

17.100. In Problem 17.40, the bar is started, when it is vertically up and the spring is unstretched, with an angular velocity of 2 rad/s clockwise. What will be the angular velocity when the bar rotates through 180°?   *Ans.* $\omega = 7.63$ rad/s

17.101. Using the result of Problem 17.100, determine the moment on the bar at A that will bring the bar to rest when it is horizontal to the left.   *Ans.* $M = 4.74$ lb-ft

17.102. The circular disk in Fig. 17-58 has a mass of 3 kg and the slender bar AB has a mass of 8 kg. The initial angular velocity of the disk is 6 rad/s as shown. What is the value of the moment M that will stop the disk when it has rotated 90° counterclockwise? Assume that the roller at B is massless.
Ans. 0.54 N · m counterclockwise

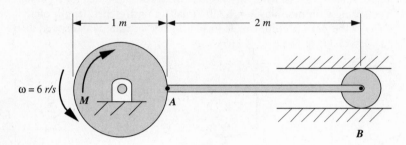

Fig. 17-58

17.103. The plank AB in Fig. 17-59 has a mass of 7 kg. The rollers D and E are each of mass 5 kg and are 0.5 m in diameter. The plank is released from rest with roller D under the end A and roller E under the mass center C. Assuming no slip, determine the velocity of the plank when roller E is under end B.
Ans. 7.11 m/s

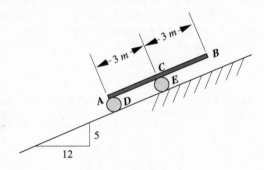

Fig. 17-59

17.104. In Fig. 17-60, the 5-kg slender rod AB is pinned at A to a 3-kg uniform disk. The rod rests on the horizontal plane at B. If the system is released from rest in the position shown, what will be the velocity of the center of the disk when the bar is horizontal? The disk rolls without slipping, and the friction under the bar at B can be neglected.   *Ans.* $v_0 = 1.81$ m/s to the right

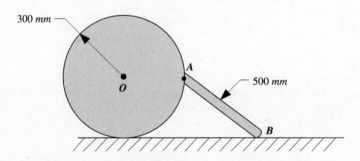

Fig. 17-60

17.105. In Fig. 17-61, gear A drives gear B under the action of a clockwise moment $M = 250$ lb-in. If the two gears are at rest when the moment is applied, what is the angular velocity of gear B when gear A has turned through four revolutions? Assume the gears A and B to be disks of weight 8 lb and 32 lb, respectively.   *Ans.* $\omega_B = 27.4$ rad/s

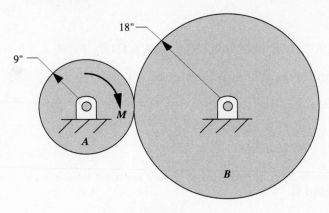

Fig. 17-61

Chapter 18

Impulse and Momentum

18.1 LINEAR IMPULSE-MOMENTUM RELATION FOR A PARTICLE

The linear momentum of a particle was defined in Section 13.1 as

$$\mathbf{G} = m\mathbf{v}$$

where m = mass of the particle
   $\mathbf{v}$ = velocity of the particle

The vector sum of the external forces acting on a particle is equal to the time rate of change of the linear momentum $\mathbf{G}$:

$$\sum \mathbf{F} = \frac{d(m\mathbf{v})}{dt} = \frac{d\mathbf{G}}{dt} = \dot{\mathbf{G}} \tag{1}$$

Integration over the time interval during which of the velocity of the particle changes from $\mathbf{v}_1$ to $\mathbf{v}_2$ results in the expression

$$\int_{t_1}^{t_2} \sum \mathbf{F} \, dt = \int_{\mathbf{G}_1}^{\mathbf{G}_2} d\mathbf{G} = \mathbf{G}_2 - \mathbf{G}_1 = m\mathbf{v}_2 - m\mathbf{v}_1 \tag{2}$$

The left-hand side of equation (2) is called the linear impulse $\mathbf{I}$ of the resultant force $(\sum \mathbf{F})$ during the time interval from t_1 to t_2. Thus, the linear impulse is equal to the change in the linear momentum during this time interval.

18.2 LINEAR IMPULSE-MOMENTUM RELATION FOR AN ASSEMBLAGE OF PARTICLES

The vector sum of the external forces acting on an assemblage of n particles equals the time rate of change of the linear momentum of a mass m that is equal to the sum of the masses of the n particles and that possesses a velocity equal to that of the mass center of the n particles:

$$\sum \mathbf{F} = \frac{d(m\bar{\mathbf{v}})}{dt} = \frac{d\bar{\mathbf{G}}}{dt} \tag{3}$$

where $\sum \mathbf{F}$ = sum of the external forces acting on the group of particles
   $m = \sum_{i=1}^{n} m_i$ = mass of all n particles
   $\bar{\mathbf{v}}$ = velocity of the mass center of the group of n particles

As in the case of one particle, the above equation may be integrated as

$$\int_{t_1}^{t_2} \sum \mathbf{F} \, dt = \int_{\bar{\mathbf{v}}_1}^{\bar{\mathbf{v}}_2} d(m\bar{\mathbf{v}}) = m\bar{\mathbf{v}}_2 - m\bar{\mathbf{v}}_1 = \Delta\bar{\mathbf{G}} \tag{4}$$

This states that the linear impulse $\mathbf{I}$ of all the forces acting in the stated time interval is equal to the change in linear momentum of a mass m, as stated in the first sentence of this section. Note that the vector $\Delta\bar{\mathbf{G}}$ does not, in general, pass through the mass center of the assemblage of particles.

18.3 MOMENT OF MOMENTUM H_O

The moment of momentum $\mathbf{H}_O$ (also called angular momentum) is the moment about any point O of the linear momentum vector $\mathbf{G}$. In Fig. 18-1, O can be any point, fixed or moving. Thus,

$$\mathbf{H}_O = \boldsymbol{\rho} \times \mathbf{G} = \boldsymbol{\rho} \times (m\mathbf{v}) \tag{5}$$

where $\boldsymbol{\rho}$ = radius vector of particle P relative to O
     $\mathbf{v}$ = absolute velocity of P (tangent to the path)

The sum of the moments about a *fixed point* O of the external forces acting on a *particle* is equal to the time rate of change of the moment of momentum $\mathbf{H}_O$; that is,

$$\sum \mathbf{M}_O = \frac{d\mathbf{H}_O}{dt} = \dot{\mathbf{H}}_O \tag{6}$$

Proof: From equation (5), $\mathbf{H}_O = \mathbf{r} \times (m\mathbf{v})$, where $\mathbf{r}$ is the radius vector in a Newtonian frame of particle m, and $\mathbf{v}$ = absolute velocity of the particle. Taking the time derivative,

$$\frac{d\mathbf{H}_O}{dt} = \dot{\mathbf{r}} \times (m\mathbf{v}) + \mathbf{r} \times (m\dot{\mathbf{v}})$$

Since $\dot{\mathbf{r}} = \mathbf{v}$, $\mathbf{v} \times \mathbf{v} = 0$, $m\dot{\mathbf{v}} = m\mathbf{a} = \sum \mathbf{F}$, and $\mathbf{r} \times (m\mathbf{a}) = \mathbf{r} \times (\sum \mathbf{F}) = \sum \mathbf{M}_O$, we obtain $d\mathbf{H}_O/dt = \sum \mathbf{M}_O$. Equation (6) may be integrated as follows:

$$\int_{t_1}^{t_2} \sum \mathbf{M}_O \, dt = \int_{\mathbf{H}_1}^{\mathbf{H}_2} d\mathbf{H}_O = \mathbf{H}_2 - \mathbf{H}_1 = \mathbf{r} \times (m\mathbf{v}_2 - m\mathbf{v}_1) \tag{7}$$

The integral on the left is the *angular impulse* acting throughout the time interval t_1 to t_2, and the right-hand side of (7) is the change that occurs in angular momentum during this time interval.

Equation (6) can be applied as follows to an assemblage of particles. The sum of the moments about point O of the external forces acting on an *assemblage* of n particles is equal to the time rate of change of the moment of momentum $\mathbf{H}_O$ about this point O, only if (a) point O is at rest or (b) the mass center of the n particles is at rest or (c) the velocities of O and the mass center are parallel (certainly true if O is the mass center). See Problem 18.1, which proves equation (6) holds for an assemblage of particles.

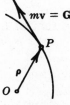

Fig. 18-1

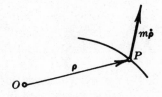

Fig. 18-2

18.4 MOMENT OF RELATIVE MOMENTUM H'_O

The moment of relative momentum $\mathbf{H}'_O$ is the moment about any point O of the product of the mass of the particle and the time rate of change of the radius vector $\boldsymbol{\rho}$ of the particle relative to O (see Fig. 18-2):

$$\mathbf{H}'_O = \boldsymbol{\rho} \times (m\dot{\boldsymbol{\rho}}) \tag{8}$$

where $\boldsymbol{\rho}$ = radius vector of particle P relative to O
     $\dot{\boldsymbol{\rho}}$ = time rate of change of $\boldsymbol{\rho}$

The sum of the moments about O of the external forces acting on an assemblage of n particles is equal to the time rate of change of the moment of relative momentum $\mathbf{H}'_O$ about this point O, that is

$$\sum \mathbf{M}_O = \frac{d\mathbf{H}'_O}{dt} \tag{9}$$

only if (a) O is the mass center of the n particles or (b) O is a point of constant velocity (or at rest) or (c) O is a point with an acceleration vector that passes through the mass center. See Problems 18.2 and 18.3.

18.5 CORRESPONDING SCALAR EQUATIONS

For a body in translation (all particles have the same velocity), equation (3) can be replaced by the scalar equations

$$\sum (\text{Imp})_x = \Delta G_x = m(v''_x - v'_x) \tag{10}$$

$$\sum (\text{Imp})_y = \Delta G_y = m(v''_y - v'_y) \tag{11}$$

where $\sum (\text{Imp})_x$, $\sum (\text{Imp})_y$ = linear impulses of external forces in the x and y directions
$\qquad\qquad m$ = mass of the body
$\qquad\qquad v''_x, v''_y$ = final velocities of the body in the x and y directions
$\qquad\qquad v'_x, v'_y$ = initial velocities of the body in the x and y directions

For a body in rotation about a fixed axis, the above equations become

$$\sum (\text{Ang Imp})_O = \Delta H_O = I_O(\omega'' - \omega') \tag{12}$$

where $\sum (\text{Ang Imp})_O$ = angular impulse of external forces about the axis of rotation through O
$\qquad\qquad I_O$ = moment of inertia of the body about the axis of the rotation
$\qquad\qquad \omega''$ = final angular velocity of the body
$\qquad\qquad \omega'$ = initial angular velocity of the body

For proof see Problem 18.4.
For a body in plane motion, the above equations become

$$\sum (\text{Imp})_x = \Delta G_x = m(\bar{v}''_x - \bar{v}'_x) \tag{13}$$

$$\sum (\text{Imp})_y = \Delta G_y = m(\bar{v}''_y - \bar{v}'_y) \tag{14}$$

$$\sum (\text{Ang Imp})_G = \Delta \tilde{H} = \bar{I}(\omega'' - \omega') \tag{15}$$

where $\sum (\text{Imp})_x$, $\sum (\text{Imp})_y$ = linear impulses of external forces in the x and y directions
$\qquad\qquad m$ = mass of the body
$\qquad\qquad \bar{v}''_x, \bar{v}''_y$ = final velocities of the mass center in the x and y directions
$\qquad\qquad \bar{v}'_x, \bar{v}'_y$ = initial velocities of the mass center in the x and y directions
$\qquad\quad \sum (\text{Ang Imp})_G$ = angular impulse of the external forces about the axis through the
$\qquad\qquad\qquad$ mass center G
$\qquad\qquad \bar{I}$ = moment of inertia of the body about the mass center G
$\qquad\qquad \omega''$ = final angular velocity of the body
$\qquad\qquad \omega'$ = initial angular velocity of the body

For proof, see Problem 18.5.
Alternatively, for a body in general plane motion, if the axis of angular momentum is not the center of mass, the scalar angular momentum becomes

$$H_O = I_O\omega + m\bar{x}v_{Oy} - m\bar{y}v_{Ox}$$

where H_O = angular momentum about an axis through O
   I_O = moment of inertia about an axis through O
   ω = angular velocity of the body
   m = mass of the body
   $\bar{x}, \bar{y}$ = coordinates of the mass center
   v_{Ox}, v_{Oy} = components of the velocity of the axis O

This formulation of the angular momentum is of particular value in problems involving eccentric collisions.

18.6   UNITS

Unit	U.S. Customary	SI
Mass	slug = lb-s^2/ft	kg
Linear impulse	lb-s	N $\cdot$ s
Linear momentum	slug-ft per s = lb-s	kg $\cdot$ m/s = N $\cdot$ s
Angular impulse	lb-s-ft	N $\cdot$ m $\cdot$ s
Angular momentum	(slug-ft^2)(rad/s) = lb-s-ft	(kg $\cdot$ m^2)(rad/s) = N $\cdot$ m $\cdot$ s

18.7   CONSERVATION OF LINEAR MOMENTUM

Conservation of linear momentum in a given direction occurs if the sum of the external forces in that direction is zero. This follows because there is then no linear impulse in that direction, and hence no change in linear momentum can occur.

18.8   CONSERVATION OF ANGULAR MOMENTUM

Conservation of angular momentum about an axis occurs if the sum of the moments of the external forces about that axis is zero. This follows because there is then no angular impulse about that axis, and hence no change in angular momentum can occur.

18.9   IMPACT

Impact covers the cases where the time intervals during which the forces act are quite small and usually indeterminate. The surfaces of two colliding bodies have a common normal, which is the line of impact.

(a)   Direct impact occurs if the initial velocities of the two colliding bodies are along the line of impact.

(b)   Direct central impact occurs if the mass centers in (a) are also along the line of impact.

(c)   Direct eccentric impact occurs if the initial velocities are parallel to the normal to the striking surfaces but are not collinear.

(d)   Oblique impact occurs if the initial velocities are not along the line of impact.

In direct central impact of the two bodies, the coefficient of restitution is the ratio of the relative velocity of separation of the two bodies to their relative velocity of approach. Thus,

$$e = \frac{v_2 - v_1}{u_1 - u_2} = -\frac{v_2 - v_1}{u_2 - u_1}$$

where e = coefficient of restitution

u_1, u_2 = velocities of bodies 1 and 2, respectively, before impact ($u_1 > u_2$ for collision to occur if both are moving in the same direction)

v_1, v_2 = velocities of bodies 1 and 2, respectively, after impact

Note: When the impact is oblique, the *normal* components of the velocities are used in the above formula.

Since during impact the same force acts on each body (equal and opposite reaction), the sum of the momenta before impact must equal the sum of the momenta after impact; i.e., momentum is conserved. This relation is expressed as

$$m_1 u_1 + m_2 u_2 = m_1 v_1 + m_2 v_2$$

18.10 VARIABLE MASS

Suppose at time t a mass m is moving along a straight line with an absolute speed v (see Fig. 18-3). Further suppose a mass dm immediately in front of mass m is moving along the same straight line with absolute speed u. If the mass m absorbs the mass dm in a time interval dt then the combined mass ($m + dm$) will move with speed ($v + dv$).

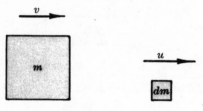

Fig. 18-3

The momentum of the system at time t is ($mv + dm\,u$). The momentum of the system at time ($t + dt$) is ($m + dm$)($v + dv$). The change in momentum is then

$$dG = (m + dm)(v + dv) - (mv + dm\,u)$$

$$= mv + m\,dv + dm\,v + dm\,dv - mv - dm\,u$$

Since the magnitude of ($dm\,dv$) is of second order, we shall drop the term. Dividing by dt,

$$\frac{dG}{dt} = m\frac{dv}{dt} + \frac{dm}{dt}(v - u)$$

If the mass is released (decrease in mass) then dm/dt will be negative.

The above formula was shown for straight line motion, but is of a more general nature.

Since the sum of the forces acting equals the time rate of change of momentum,

$$\sum F = \frac{dG}{dt} = m\frac{dv}{dt} + \frac{dm}{dt}(v - u) \qquad (16)$$

Solved Problems

18.1. Given an assemblage of n particles of masses $m_1, m_2, m_3, \ldots, m_n$, show that the sum of the moments about point O of the external forces equals the time rate of change of the moment of momentum of the group of particles about this point O only if (a) point O is at rest or (b) the mass center of the n particles is at rest or (c) the velocities of O and the mass center are parallel (which is certainly true if O is the mass center).

SOLUTION

Figure 18-4 shows the ith particle of the group with mass m_i. The given point O has a position vector $\mathbf{r}_O$ relative to a Newtonian frame of reference: O' is fixed. The position vector of P, relative to the fixed frame, is $\mathbf{r}_P$. Let $\mathbf{H}_O$ be the angular momentum relative to O of all the particles of which m_i is representative. Thus,

$$\mathbf{H}_O = \sum_{i=1}^n \boldsymbol{\rho}_i \times (m_i \mathbf{v}_i) \tag{a}$$

Taking the time derivative of equation (a),

$$\dot{\mathbf{H}}_O = \sum_{i=1}^n \dot{\boldsymbol{\rho}}_i \times (m_i \mathbf{v}_i) + \sum_{i=1}^n \boldsymbol{\rho}_i + \frac{d}{dt}(m_i \mathbf{v}_i) \tag{b}$$

From the figure, $\mathbf{r}_P = \mathbf{r}_O + \boldsymbol{\rho}_i$ and hence $\dot{\mathbf{r}}_P = \dot{\mathbf{r}}_O + \dot{\boldsymbol{\rho}}_i$. Substituting for $\dot{\boldsymbol{\rho}}_i$ into (b),

$$\dot{\mathbf{H}}_O = \sum_{i=1}^n (\dot{\mathbf{r}}_P - \dot{\mathbf{r}}_O) \times (m_i \mathbf{v}_i) + \sum_{i=1}^n \boldsymbol{\rho}_i \times (m_i \dot{\mathbf{v}}_i) \tag{c}$$

Expand the first term on the right of (c) into

$$\sum_{i=1}^n \dot{\mathbf{r}}_P \times (m_i \mathbf{v}_i) - \sum_{i=1}^n \dot{\mathbf{r}}_O \times (m_i \mathbf{v}_i) \tag{d}$$

Since $\dot{\mathbf{r}}_P$ is the absolute velocity $\mathbf{v}_i$ of P, the first term of (d) is zero. Also, since $\dot{\mathbf{r}}_O$ does not change during the summation and is therefore independent of i, the second term of (d) can be written

$$\dot{\mathbf{r}}_O \times \sum_{i=1}^n (m_i \mathbf{v}_i) \qquad \text{or} \qquad \dot{\mathbf{r}}_O \times (m\bar{\mathbf{v}})$$

where $\bar{\mathbf{v}}$ is the velocity of the mass center.

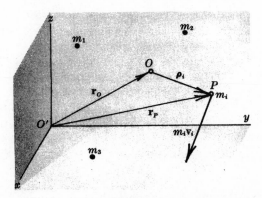

Fig. 18-4

The last term of equation (c) is equal to $\sum \mathbf{M}_O$, the sum of the moments of the external forces on all the particles. Thus, (c) may now be written as

$$\dot{\mathbf{H}}_O = -\dot{\mathbf{r}}_O \times (m\bar{\mathbf{v}}) + \sum \mathbf{M}_O \tag{e}$$

Equation (e) indicates that the time rate of change of the moment of momentum about O equals $\sum \mathbf{M}_O$ only if $-\dot{\mathbf{r}}_O \times (m\bar{\mathbf{v}})$ is zero. This occurs when (1) O is fixed, that is, $\dot{\mathbf{r}}_O = 0$; (2) $\bar{\mathbf{v}} = 0$; or (3) $\dot{\mathbf{r}}_O$ and $\bar{\mathbf{v}}$ are parallel (cross product of parallel vectors is zero). If O is the mass center then $\dot{\mathbf{r}}_O = \bar{\mathbf{v}}$ and $\dot{\mathbf{r}}_O \times m\bar{\mathbf{v}} = 0$.

18.2.   Given an assemblage of n particles of masses $m_1, m_2, m_3, \ldots, m_n$, show that the sum of the moments about O of the external forces equals the time rate of change of the moment of *relative* momentum of the group of particles about this point O only if (a) point O is the mass center of the n particles or (b) point O has constant velocity (or is at rest) or (c) point O has an acceleration vector that passes through the mass center.

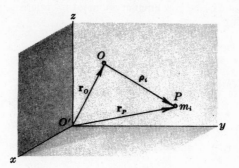

Fig. 18-5

SOLUTION

Figure 18-5 shows the ith particle of the group with mass m_i. The given point O has a position vector $\mathbf{r}_O$ relative to a Newtonian (inertial) frame of reference, i.e., O' is fixed. The position vector of P relative to the fixed frame is $\mathbf{r}_P$. Let $\mathbf{H}'_O$ be the moment of relative momentum with respect to O of all the particles of which m_i is representative. Then, from equation (8),

$$\mathbf{H}'_O = \sum_{i=1}^{n} \boldsymbol{\rho}_i \times (m_i \dot{\boldsymbol{\rho}}_i) \qquad\qquad (a)$$

Taking the time derivative of equation (a),

$$\dot{\mathbf{H}}'_O = \sum_{i=1}^{n} \dot{\boldsymbol{\rho}}_i \times (m_i \dot{\boldsymbol{\rho}}_i) + \sum_{i=1}^{n} \boldsymbol{\rho}_i \times (m_i \ddot{\boldsymbol{\rho}}_i) \qquad\qquad (b)$$

From the figure, $\mathbf{r}_P = \mathbf{r}_O + \boldsymbol{\rho}_i$ and hence $\ddot{\mathbf{r}}_P = \ddot{\mathbf{r}}_O + \ddot{\boldsymbol{\rho}}_i$. Substituting for $\ddot{\boldsymbol{\rho}}_i$ in (b) and noting that the first term on the right of (b) is zero ($\dot{\boldsymbol{\rho}}_i \times \dot{\boldsymbol{\rho}}_i = 0$), we have

$$\dot{\mathbf{H}}'_O = \sum_{i=1}^{n} \boldsymbol{\rho}_i \times (m_i \ddot{\mathbf{r}}_P) - \sum_{i=1}^{n} \boldsymbol{\rho}_i \times (m_i \ddot{\mathbf{r}}_O) \qquad\qquad (c)$$

The last term in (c) may be written

$$\left(\sum_{i=1}^{n} m_i \boldsymbol{\rho}_i \right) \times \ddot{\mathbf{r}}_O$$

since $\ddot{\mathbf{r}}_O$ does not change as the sum is taken over the n particles. Also,

$$\sum_{i=1}^{n} \boldsymbol{\rho}_i \times (m_i \ddot{\mathbf{r}}_P) = \sum \mathbf{M}_O \qquad \text{and} \qquad \sum_{i=1}^{n} m_i \boldsymbol{\rho}_i = m \bar{\boldsymbol{\rho}}$$

where $\bar{\boldsymbol{\rho}}$ is the position vector relative to O of the mass center. Then

$$\dot{\mathbf{H}}'_O = \sum \mathbf{M}_O - m \bar{\boldsymbol{\rho}} \times \ddot{\mathbf{r}}_O \qquad\qquad (d)$$

The last term in (d) is zero if (1) O is the mass center ($\bar{\boldsymbol{\rho}} = 0$); (2) O has constant velocity ($\ddot{\mathbf{r}}_O = 0$); or (3) O has an acceleration $\ddot{\mathbf{r}}_O$ passing through the mass center, i.e., along $\bar{\boldsymbol{\rho}}$ (the cross product of parallel vectors is zero).

18.3. Four equal masses m are spaced at the quarter points of a thin massless rim of radius R. Show that the moment of momentum relative to the mass center O using absolute velocities is the same as that obtained by using the relative velocities of the masses to the mass center O.

SOLUTION

Let $v =$ speed of the mass center as the rim rolls to the right.

Figure 18-6(a) shows the rim with the masses m_2 and m_4 in a vertical line. In Fig. 18-6(b), the linear momentum **G** of each mass is shown using the absolute velocities of each mass. Thus, $v_1 = \sqrt{2}\,v$ at 315°, while $v_3 = \sqrt{2}\,v$ at 45°. Of course, $v_2 = 2v$, and v_4 is that of the instant center with no absolute velocity.

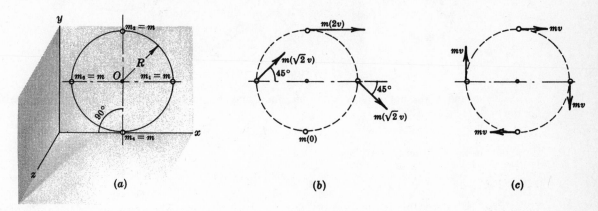

Fig. 18-6

The moment of each linear momentum is in the negative z direction. Thus, summing in 1, 2, 3, and 4 order, we can write

$$\mathbf{H}_O = -[m_1(\sqrt{2}v)(\tfrac{1}{2}\sqrt{2}R) + m_2(2v)R + m_3(\sqrt{2}v)(\tfrac{1}{2}\sqrt{2}R) + 0]\mathbf{k} = -(4mvR)\mathbf{k}$$

Figure 18-6(c) illustrates the velocity of each mass relative to the center multiplied by the mass. The moments about O of these relative momentum vectors are

$$\mathbf{H}_O = -(4mvR)\mathbf{k} \qquad \text{(as before)}$$

18.4. For a body rotating about a fixed axis that is through O and perpendicular to the plane of the paper, show that the sum of the angular impulses of the external forces about the fixed axis is equal to the change in $I_O\boldsymbol{\omega}$, where $I_O\boldsymbol{\omega}$ is the angular momentum $\mathbf{H}_O$ of the entire body.

SOLUTION

In Fig. 18-7, dm represents any differential mass with position vector $\boldsymbol{\rho}$ in the plane of the paper. The angular momentum of dm is $\boldsymbol{\rho} \times (dm\ \mathbf{v})$. The angular momentum $\mathbf{H}_O$ for the entire body is

$$\mathbf{H}_O = \int \boldsymbol{\rho} \times (dm\ \mathbf{v})$$

But $\mathbf{v} = \boldsymbol{\omega} \times \boldsymbol{\rho}$ is in the plane of the paper and has magnitude $\rho\omega$ because the vectors $\boldsymbol{\omega}$ and $\boldsymbol{\rho}$ are at right angles. Also, $\boldsymbol{\rho} \times (\boldsymbol{\omega} \times \boldsymbol{\rho})$ is directed out of the paper and has magnitude $\rho^2\omega$ because $\boldsymbol{\rho}$ and $\mathbf{v}$ are at right angles. Thus,

$$\mathbf{H}_O = \int \rho^2 \omega\, dm\ \mathbf{k}$$

where $\mathbf{k}$ is the unit vector perpendicular to the paper and directed toward the reader.

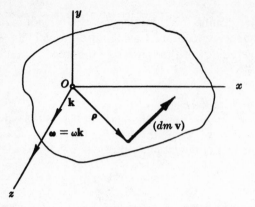

Fig. 18-7

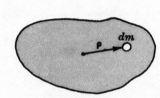

Fig. 18-8

Since ω and $\mathbf{k}$ do not change with dm, they may be taken outside the integral sign. Also, $\int \rho^2 \, dm = I_O$. Hence,

$$\mathbf{H}_O = I_O \omega \mathbf{k}$$

Then, using equation (6), we have

$$\sum \mathbf{M}_O = \frac{d\mathbf{H}_O}{dt} = \frac{d(I_O \omega)}{dt} \mathbf{k}$$

Since the moments of forces and $\boldsymbol{\omega}$ are in the $\mathbf{k}$ direction, the equation can be written in scalar form as

$$\int \left(\sum M_O \, dt \right) = \Delta H_O = I_O(\omega'' - \omega')$$

where ω' and ω'' are the initial and final angular speeds, and $\int \left(\sum M_O \, dt \right) = \sum (\text{Ang Imp})_O$.

18.5. For a rigid body of mass m in plane motion (assume the plane of the paper is the plane of motion), show that equations (13), (14), and (15) at the beginning of this chapter are true. Refer to Fig. 18-8.

SOLUTION

Since a rigid body is an assemblage of particles that remain at constant distances from each other, equation (4) of this chapter applies. Thus

$$\sum \mathbf{F} = \frac{d(m\bar{\mathbf{v}})}{dt} \qquad \text{or} \qquad \int_{t_1}^{t_2} \sum \mathbf{F} \, dt = \Delta \bar{\mathbf{G}} = m\bar{\mathbf{v}}_2 - m\bar{\mathbf{v}}_1$$

This vector equation is equivalent to the two scalar equations

$$\sum (\text{Imp})_x = m(\bar{v}_x'' - \bar{v}_x') \tag{13}$$

$$\sum (\text{Imp})_y = m(\bar{v}_y'' - \bar{v}_y') \tag{14}$$

To derive equation (15), make use of equation (9). As indicated in Problem 18.2, the mass center is one of the points which can be selected in order to apply equation (9). Then $\mathbf{H}_O'$ becomes $\bar{\mathbf{H}}'$, the moment of relative momentum of the rigid body about the mass center, and

$$\bar{\mathbf{H}}' = \int \boldsymbol{\rho} \times (dm \, \dot{\boldsymbol{\rho}})$$

Because of the rigid body constraint, the vector $\boldsymbol{\rho}$ (from the mass center to the element of mass dm) can change only in direction but not in magnitude. Hence, $\dot{\boldsymbol{\rho}}$ is perpendicular to $\boldsymbol{\rho}$ (it is in the plane of the paper) and of magnitude $\rho\omega$. Then $\mathbf{\rho} \times (dm\,\dot{\mathbf{\rho}})$ is perpendicular to the plane of the paper and of magnitude $\rho^2\omega$. Thus, the vector equation may be replaced by the scalar equation

$$\bar{H}' = \omega \int \rho^2\, dm = \bar{I}\omega$$

and equation (9) becomes

$$\sum M_G = \dot{\bar{H}}' = \frac{d(\bar{I}\omega)}{dt}$$

which can now be expressed as

$$(\text{Ang Imp})_G = \int \sum M_G\, dt = \Delta\bar{H} = \bar{I}(\omega'' - \omega') \qquad (15)$$

18.6. A thin rim of mass m and radius R is rolling without slipping on a horizontal plane as shown in Fig. 18-9(a). A horizontal force of magnitude P is applied at the top. Show that the sum of the external forces about the mass center G, when equated to the time rate of change of the relative momentum about G, yields the same result as obtained by equating the sum of the moments of the external forces about the instant center A to the time rate of change of the relative momentum about A.

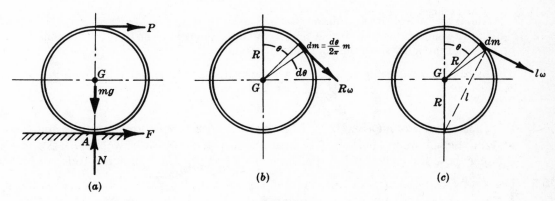

Fig. 18-9

SOLUTION

The free-body diagram in Fig. 18-9(a) shows the normal force N, the friction force F, the applied force P, and the weight mg concentrated at the mass center G.

Scalar equations will be used in this discussion.

(a) Figure 18-9(b) shows a differential mass dm of the rim at an angle θ with the vertical. The mass dm is that part of the rim subtended by the angle $d\theta$; hence, $dm = m\,d\theta/2\pi$. The speed of dm relative to the mass center G is $R\omega$ as shown. The moment of relative momentum about G of dm is thus $dm\,R^2\omega$. The moment of relative momentum of the entire rim is

$$H'_G = \int_0^{2\pi} mR^2\omega\,\frac{d\theta}{2\pi} = mR^2\omega$$

The sum of the moments of the external forces about G (considering clockwise positive) is

$$\sum M_G = PR - FR$$

But for any group of particles (in this case the rim with mass center speed $v = r\omega$),

$$\sum F_x = \dot{G}_x \qquad \text{or} \qquad P + F = \frac{d}{dt}(mR\omega)$$

From this,

$$F = -P + \frac{d}{dt}(mR\omega)$$

The equation $\sum M_G = \dot{H}_G'$ yields

$$PR - \left[-P + \frac{d}{dt}(mR\omega) \right] R = \frac{d}{dt}(mR^2\omega) \qquad \text{or} \qquad 2PR = \frac{d}{dt}(2mR^2\omega)$$

(b)   To use the instant center A as the center, let l be the distance from A to the mass dm as shown in Fig. 18-9(c). The speed of dm relative to A (which is at rest) is $l\omega$. It is perpendicular to the line l as shown. The line has length

$$l = \sqrt{R^2 + R^2 + 2RR\cos\theta}$$

The moment of relative momentum of dm about A is $dm\,(2R^2 + 2R^2\cos\theta)\omega$. Using $dm = m\,d\theta/2\pi$,

$$H_A' = \int_0^{2\pi} mR^2\omega(2 + 2\cos\theta)\frac{d\theta}{2\pi} = 2mR^2\omega$$

The sum of the moments of the external forces about A is

$$\sum M_A = 2PR = \dot{H}_A' = \frac{d}{dt}(2mR^2\omega)$$

This is the same result as determined in part (a).

It is interesting to note that since $2mR^2$ is the moment of inertia I_A of the thin rim about the instant center, the above equation can be written

$$\sum M_A = \frac{d}{dt}(I_A\omega) = I_A\alpha$$

18.7.   If the mass center of an assemblage of particles is at rest, equation (6) is true and any point O may be used. For two particles of equal mass m mounted on a weightless rim of radius R rotating about the center of the rim, show that $\sum M_O = (d/dt)(2mR^2\omega)$. Refer to Fig. 18-10.

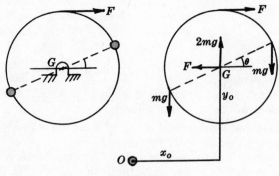

Fig. 18-10

SOLUTION

The bearing reactions are F to the left and $2mg$ up. Summing moments about any point O, $\sum M_O = FR$ and is independent of the moment center O:

$$H_O = m(R\omega\cos\theta)(x_O + R\cos\theta) + m(R\omega\sin\theta)(y_O + R\sin\theta)$$

$$- m(R\omega\cos\theta)(x_O - R\cos\theta) - m(R\omega\sin\theta)(y_O - R\sin\theta) = 2mR^2\omega$$

Hence, $FR = (d/dt)(2mR^2\omega)$, the same equation as when moments are taken relative to mass center G.

18.8. A 100-lb object is pushed for 5 s by a horizontal force F over a horizontal plane where the coefficient of friction is 0.2. During this time interval, the speed changes from 5 to 10 ft/s. Determine the value of the force F.

SOLUTION

The normal force equals the weight of 100 lb. The frictional force opposing motion is $0.2 \times 100 = 20$ lb. The linear impulse in the horizontal direction is equal to the change in linear momentum. Hence,

$$\int_0^5 (F - 20)\, dt = \frac{100}{g}(10) - \frac{100}{g}(5)$$

Solving,

$$F = 23.1 \text{ lb}$$

18.9. A 10-kg block slides from rest down a plane inclined 30° with the horizontal. Assuming a coefficient of kinetic friction between the block and the plane of 0.3, what will be the speed of the block at the end of 5 s?

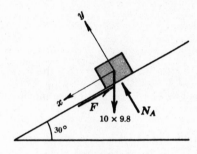

Fig. 18-11

SOLUTION

This problem may, of course, be solved by previous methods. However, with time as one of the quantities given, the impulse–momentum method is the simplest. This is a problem of translation.

Draw a free-body diagram indicating all external forces acting on the block (see Fig. 18-11). Since only motion along the plane is considered, one impulse–momentum equation will suffice.

$$\sum (\text{Imp})_x = \Delta G_x$$

or

$$\left(\sum F_x\right)(t) = m(v_x'' - v_x')$$

where t = elapsed time
  m = mass of block
  v_x'' = final speed
  v_x' = initial speed

Assuming the downward direction as positive, the equation becomes

$$(+98 \sin 30° - \mu N_A)(5) = 10(v_x'' - 0) \qquad\qquad (A)$$

To determine N_A, it is only necessary to sum forces perpendicular to the plane and equate to zero, because no motion in this y direction occurs:

$$\sum F_y = 0 = N_A - 98 \cos 30° \qquad \text{or} \qquad N_A = 84.9 \text{ N}$$

Substituting in (A) for N_A, we obtain $v_x'' = 11.8$ m/s

18.10. An 80-lb block rests on a horizontal plane. It is acted upon by a horizontal force that varies from zero according to the law $F = 20t$. If the force acts for 5 s, what is the speed of the block? The coefficient of static friction is 0.25, and the coefficient of kinetic friction is 0.20.

SOLUTION

The force F increases until it reaches the limiting value of static friction; i.e., $F = 20t = 0.25 \times 80$. Thus, at $t = 1$ s, the block begins to move with friction now equal to $0.2 \times 80 = 16$ lb, which remains constant.

After $t = 1$ s, the linear impulse horizontally equals the change in linear momentum horizontally. Thus,

$$\sum (\text{Imp})_x = \Delta G_x \quad \text{or} \quad \int_1^5 F\, dt - \int_1^5 16\, dt = \frac{80}{32.2}(v - 0)$$

Using $F = 20t$, we obtain $v = 70.8$ ft/s.

18.11. A 1.5-kg block starts from rest on a smooth horizontal plane under the action of a horizontal force F that varies according to the equation $F = 3t - 5t^2$. Determine the maximum speed.

SOLUTION

$$\sum (\text{Imp})_x = \Delta G_x \quad \text{or} \quad \int_0^t F\, dt = \int_0^t (3t - 5t^2)\, dt = 1.5(v - 0)$$

from which $v = t^2 - 1.11t^3$. To find the maximum value of v, determine the value of t at which $dv/dt = 2t - 3.33t^2 = 0$. This gives $t = 0.6$ s, and hence maximum $v = 0.12$ m/s.

18.12. The 40-kg block shown in Fig. 18-12 is moving up initially with a speed of 2.5 m/s. What constant value of P will result in an upward speed of 5 m/s in 12 s? Assume that the pulleys are frictionless and that the coefficient of friction between the blocks and the plane is 0.10.

SOLUTION

Solve the problem by summing impulses and momenta along the line of travel of the system. For example, the force P and a component of the gravitational force on the 10-kg block have positive impulses in the line of travel of the 10-kg block, while the friction acting on the 10-kg block has a negative impulse. The cord tension acts on both the 10- and 15-kg masses in opposite directions; therefore its linear impulse along the line of travel is zero. Proceeding in this way, the impulse–momentum equation becomes

$$\sum (\text{Imp}) = \Delta G$$

$$[P + 9.8 \times 10 \sin 45° - 0.10(9.8 \times 10 \cos 45°) - 0.10(9.8 \times 15) - 9.8 \times 40](12) \text{ N} \cdot \text{s}$$

$$= (10 + 15 + 40)(5 - 2.5) \text{ N} \cdot \text{s}$$

from which $P = 358$ N.

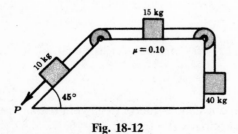

Fig. 18-12

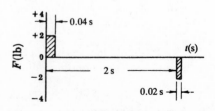

Fig. 18-13

18.13. A 4-lb weight is at rest in a smooth horizontal slot. It is struck a 2-lb blow that lasts for 0.04 s. Two seconds after the start of the first blow, a second blow of -2 lb is delivered and lasts for 0.02 s. What will be the speed of the body after 3 s?

SOLUTION

The forces in this problem may be represented graphically in a plot against time as shown in Fig. 18-13.

The linear impulse for any length of time larger than 2.02 s is the algebraic sum of the two areas; i.e., $+2(0.04) - 2(0.02) = 0.04$ lb-s. Then

$$\sum \text{Imp} = \Delta G \qquad \text{or} \quad 0.04 = \left(\frac{4}{32.2}\right)(v - 0)$$

or
$$v = 0.322 \text{ ft/s}$$

18.14. In Fig. 18-14, the mercury in the left column of the manometer is falling at the rate of 1 in/s. The left column is 18 in long and the right column is 22 in long. What is the vertical momentum of the mercury? The manometer is made of quarter-inch glass (inside diameter). Mercury weighs 850 lb/ft³.

SOLUTION

Neglecting the bend in the tube, it is evident that the momentum of the left column is down while that of the right column is up. The net upward momentum is that of a 4-in, $\frac{1}{4}$-in-diameter column moving $\frac{1}{12}$ ft/s.

$$\text{Momentum} = \left[\frac{\frac{1}{4}\pi(\frac{1}{4})^2 4}{(12)^3} \text{ ft}^3\right]\left(\frac{850}{32.2} \text{ slug/ft}^3\right)(\tfrac{1}{12} \text{ ft/s}) = +0.00025 \text{ lb-s}$$

18.15. A flywheel of mass 2000 kg and radius of gyration 1200 mm rotates about a fixed center O from rest to an angular speed of 120 rpm in 200 s. What moment M is necessary?

SOLUTION

$$\sum (\text{Ang Imp})_O = \Delta H_O = I_O(\omega_2 - \omega_1)$$
$$M(200 \text{ s}) = [2000(1.2)^2 \text{ kg} \cdot \text{m}^2][(240\pi/60 - 0) \text{ rad/s}] \qquad M = 181 \text{ N} \cdot \text{m}$$

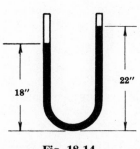

Fig. 18-14

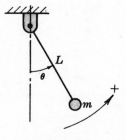

Fig. 18-15

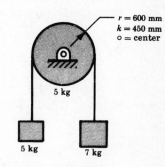

Fig. 18-16

18.16. A pendulum consists of a bob of mass m and a slender rod of negligible mass. See Fig. 18-15. Show that the differential equation of the motion is $\ddot{\theta} + (g/L)\sin\theta = 0$.

SOLUTION

The free-body diagram shows the pendulum displaced an angle θ from the vertical position. The angular momentum $\mathbf{H}_O$ of the bob relative to the support is $I\dot\theta\mathbf{k}$, where $\mathbf{k}$ is the unit vector perpendicular to the page, with the arrow pointing toward the reader.

The only force with a moment is the gravitational force acting on the bob; the moment is clockwise or negative. Then

$$\sum \mathbf{M}_O = \frac{d\mathbf{H}_O}{dt} \qquad \text{or} \qquad -mg(L\sin\theta)\mathbf{k} = \frac{d}{dt}(I\dot\theta\mathbf{k})$$

Since $I = mL^2$ for the bob, we obtain the scalar equation $\ddot\theta + (g/L)\sin\theta = 0$.

18.17. In Fig. 18-16, a massless rope carries two masses of 5 and 7 kg when hanging on a pulley of mass 5 kg, radius 600 mm, and radius of gyration 450 mm. How long will it take to change the speed of the masses from 3 to 6 m/s?

SOLUTION (A)

Draw a free-body diagram.

Let T_1 and T_2 be the tensions in the rope supporting the 5-kg and 7-kg masses, respectively.

The useful impulse–momentum equations are as follows, where equation (*1*) applies to the 5-kg mass, equation (*2*) applies to the 7-kg mass, and equation (*3*) applies to the pulley:

$$(T_1 - 5 \times 9.8)t = 5(6 - 3) \tag{1}$$
$$(7 \times 9.8 - T_2)t = 7(6 - 3) \tag{2}$$
$$(T_2 - T_1)(0.6)t = \bar{I}(\omega_{\text{final}} - \omega_{\text{initial}}) \tag{3}$$

In (*3*), substitute $\bar{I} = mk^2 = 5(0.45)^2 = 1.01$ kg $\cdot$ m^2, $\omega_{\text{final}} = 6/0.6 = 10$ rad/s, and $\omega_{\text{initial}} = 3/0.6 = 5$ rad/s. Solve equations (*1*), (*2*), and (*3*) simultaneously to obtain $t = 2.27$ s.

SOLUTION (B)

Realizing that the moment about the center of the pulley of the linear momentum of either hanging mass is the angular momentum of the mass, a system approach may be used to solve this problem. Thus,

$$\text{(Initial Ang Mom)}_0 + \text{(Ang Imp)}_0 = \text{(Final Ang Mom)}_0 \tag{4}$$

The tensions do not occur in the system impulse expression because they occur in pairs and thus cancel each other. Also, only the two gravitational forces on the two hanging masses have angular impulses about the center of the pulley. The equation becomes

$$(I_0\omega_{\text{initial}} + 5v_{\text{initial}} \times 0.6 + 7v_{\text{initial}} \times 0.6) + (7 \times 9.8t - 5 \times 9.8t)0.6 = (I_0\omega_{\text{final}} + 5v_{\text{final}} \times 0.6 + 7v_{\text{final}} \times 0.6) \tag{5}$$

or $\quad (1.01 \times 5 + 5 \times 3 \times 0.6 + 7 \times 3 \times 0.6) + 11.76t = (1.01 \times 10 + 5 \times 6 \times 0.6 + 7 \times 6 \times 0.6) \tag{6}$

Hence, $t = 2.27$ s.

18.18. Refer to Fig. 18-17. Determine the mass of B necessary to cause the 50-kg mass A to change its speed from 4 to 8 m/s in 6 s. Assume that the drum rotates in frictionless bearings.

SOLUTION

Apply the system solution, noting that the initial angular momentum about the drum center O plus the angular impulses about O of all the external forces is equal to the final angular momentum about O. Note that $N_A = 50 \times 9.8$ and hence $F = 0.25 \times 50 \times 9.8 = 123$ N. The angular speeds of the drum are $4/0.8$ and $8/0.8$, respectively (5 rad/s and 10 rad/s).

$$[30(5) + m_B(4)(0.8) + 50(4)(0.8)] + [9.8m_B(0.8) - 123 \times 0.8]6 = [30(10) + m_B(8)(0.8) + 50(8)(0.8)]$$

The solution is $m_B = 20.5$ kg.

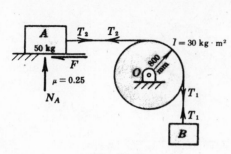

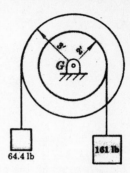

Fig. 18-17	Fig. 18-18

18.19. The drum shown in Fig. 18-18 consists of two homogeneous cylinders integrally connected. The smaller cylinder weighs 250 lb and the larger weighs 500 lb. How much time will elapse before the drum changes its speed from 100 to 300 rpm?

SOLUTION

The moment of inertia of the drum equals

$$\frac{1}{2}\left(\frac{500}{g}\right)(3)^2 + \frac{1}{2}\left(\frac{250}{g}\right)(2)^2 = 85.4 \text{ slug-ft}^2$$

The linear impulse of the 161-lb weight has a moment about G that is clockwise and has a magnitude of $2(161)(t) = 322t$ lb-s-ft. The linear impulse of the 64.4-lb weight has a counterclockwise moment about G and equals $-3(64.4)t = -193t$ lb-s-ft. The total angular impulse is the sum of the two values, or $129t$ lb-s-ft. This total angular impulse acting on the system is equal to the total change in angular momentum of the system with reference to G.

The change in angular momentum of the drum equals

$$I(\omega_2 - \omega_1) = 85.4\left(300 \times \frac{2\pi}{60} - 100 \times \frac{2\pi}{60}\right)$$
$$= 1790 \text{ lb-s-ft}$$

The linear momentum of each weight is multiplied by its arm to determine the moment of its momentum (angular momentum) about G. The 161-lb weight is moving down, and thus its linear momentum has a clockwise (positive) moment of momentum about G. This change in angular momentum equals

$$2\left[\left(\frac{161}{32.2}\right)\left(2 \times \frac{600\pi}{60} - 2 \times \frac{200\pi}{60}\right)\right] = 419 \text{ lb-s-ft}$$

Similarly, the change in linear momentum of the 64.4-lb weight is up, and thus the weight has a positive (clockwise) moment about G. Its value is

$$3\left[\left(\frac{64.4}{32.2}\right)\left(3 \times \frac{600\pi}{60} - 3 \times \frac{200\pi}{60}\right)\right] = 377 \text{ lb-s-ft}$$

The total angular impulse is equal to the total change in angular momentum of the system, or

$$129t = 1790 + 419 + 377$$

which yields $t = 20.0$ s

18.20. Refer to Fig. 18-19. A cylinder of radius r, mass m, and moment of inertia $\bar{I}$ (about the center) rolls from rest down a plane inclined at an angle θ with the horizontal. What is the speed of its center at time t from rest?

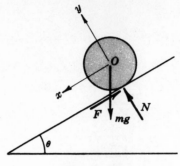

Fig. 18-19

SOLUTION

Draw a free-body diagram showing all external forces acting on the cylinder. Do not make the mistake of assuming that the friction force F is equal to the product of coefficient of friction and the normal force N.

Since the cylinder is in plane motion, the impulse–momentum equations for this type of motion are

$$\sum (\text{Imp})_x = \Delta G_x \qquad (1)$$

and
$$\sum (\text{Ang Imp})_O = \Delta H_O \qquad (2)$$

Note that F is the only external force with a moment about the center O. Equations (1) and (2) become

$$(mg \sin \theta - F)t = m(\bar{v} - 0) \qquad (3)$$

$$Frt = \bar{I}(\omega - 0) \qquad (4)$$

In these equations, the friction force F and the speed $\bar{v}$ are unknown. ($\omega = \bar{v}/r$, since no slipping is assumed.)

From (4), $F = (\bar{I}\omega)/(rt)$. Substitute this into (3), with $\omega = \bar{v}/r$, to obtain

$$\bar{v} = \frac{mg \sin \theta}{m + \bar{I}/r^2} t$$

18.21. A solid homogeneous cylinder 3 ft in diameter weighing 300 lb is rolled up a 20° incline by a force of 250 lb applied parallel to the plane. Assuming no slipping, determine the speed of the cylinder after 6 s if the initial speed is zero. Refer to Fig. 18-20.

SOLUTION

Draw the free-body diagram of the cylinder, assuming that the frictional force F acts up the plane. The impulse–momentum equations of plane motion are

$$\sum (\text{Imp})_\parallel = \Delta G_\parallel \qquad \text{or} \qquad (250 + F - 300 \sin 20)6 = \frac{300}{g} \bar{v} \qquad (1)$$

$$\sum (\text{Ang Imp})_G = \bar{I}\omega \qquad \text{or} \qquad \left[250\left(\frac{3}{2}\right) - F\left(\frac{3}{2}\right)\right]6 = \frac{1}{2}\left(\frac{300}{g}\right)\left(\frac{3}{2}\right)^2\left(\frac{\bar{v}}{3/2}\right) \qquad (2)$$

from which the speed $\bar{v}$ after 6 s is 171 ft/s, parallel to the plane.

18.22. Refer to Fig. 18-21. A solid homogeneous cylinder starts up the 30° inclined plane with a mass center speed of 20 ft/s. If it rolls freely to rest, how long does it take to reach its highest point?

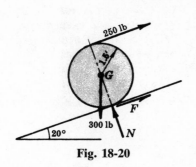

Fig. 18-20

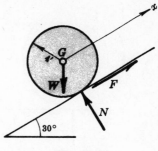

Fig. 18-21

SOLUTION

The initial speed $\bar{v}$ of the mass center G is 20 ft/s. The initial angular velocity $\omega = 20/4 = 5$ rad/s. The final speed is zero. The free-body diagram shows the forces acting on the cylinder.

Apply the two impulse–momentum equations of plane motion:

$$\sum (\text{Imp})_x = \Delta G_x \qquad \text{or} \qquad (F - W \sin 30°)t = \frac{W}{g}(0 - 20) \qquad (1)$$

$$\sum (\text{Ang Imp})_G = \Delta H_G \qquad \text{or} \qquad (-4F)t = \frac{1}{2}\left(\frac{W}{g}\right)(4^2)(0 - 5) \qquad (2)$$

from which $t = 1.86$ s.

The solution may also be found using the instant center as the reference point. The one equation needed is

$$\sum (\text{Ang Imp})_I = \Delta H_I \qquad \text{or} \qquad -(4 \sin 30)Wt = \frac{3}{2}\left(\frac{W}{g}\right)(4)^2(0 - 5)$$

from which $t = 1.86$ s.

18.23. In Fig. 18-22, the 75-kg wheel has a radius of gyration with respect to G of 900 mm. The pulley is massless and runs in frictionless bearings. The wheel is rolling initially 10 rad/s counterclockwise. How long will it take until it is rolling 6 rad/s clockwise?

SOLUTION

In the free-body diagram, friction is assumed to act to the left. By kinematic relations, the initial speed of G is $v_G = -1.2(10) = -12$ m/s, that is, to the left.

Its final speed is $+1.2(6) = +7.2$ m/s, to the right. To find the speeds of the 30-kg mass, determine the initial and final velocities of point B on the wheel: $\mathbf{v}_B = \mathbf{v}_{B/G} + \mathbf{v}_G$. Hence, the initial speed of B is $-0.8\omega - 12 = -0.8(10) - 12 = -20$ m/s. The final speed of B is, by simlar reasoning, $+12$ m/s. Thus, the initial speed of the 30-kg mass is 20 m/s up and its final speed is 12 m/s down.

In the equations, the following sign conventions are used. For the wheel, to the right is positive and clockwise is positive. For the 30-kg mass in equation (3), down is positive.

$$\sum (\text{Imp})_h = \Delta G_h \qquad \text{or} \qquad (T - F)t = 75[+7.2 - (-12)] \qquad (1)$$

$$\sum (\text{Ang Imp})_G = \Delta H_G \qquad \text{or} \qquad (0.8T + 1.2F)t = 75(0.9)^2[+6 - (-10)] \qquad (2)$$

$$\sum (\text{Imp})_v = \Delta G_v \qquad \text{or} \qquad (9.8 \times 30 - T)t = 30[+12 - (-20)] \qquad (3)$$

The solution is $t = 7.86$ s.

18.24. Refer to Fig. 18-23. A solid homogeneous 8-in-diameter cylinder weighing 96.6 lb is spinning 36 rad/s clockwise about its horizontal geometric axis when it is dropped suddenly onto a horizontal plane. Assuming a coefficient of friction is 0.15, determine the speed $\bar{v}$ of the mass center G when pure rolling begins. Through what distance has the mass center moved before this occurs?

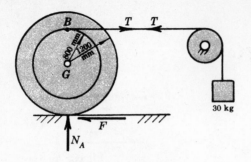

Fig. 18-22   Fig. 18-23

SOLUTION

Draw the free-body diagram when contact with the plane is first established. The same force system will act on the cylinder until pure rolling is attained. The frictional force F acts (1) to decrease the angular speed and (2) to increase the mass center speed from zero to its value $\bar{v} = r\omega$ when pure rolling obtains.

Here $N = W = 96.6$ lb, $F = 0.15N = 14.5$ lb, $m = 3$ slugs, and $\bar{I} = \frac{1}{2}mr^2 = \frac{1}{2}(3)(\frac{1}{3})^2 = \frac{1}{6}$ slug-ft^2.

The impulse–momentum equations are

$$\sum (\text{Imp})_h = \Delta G_h \qquad \text{or} \qquad Ft = m(\bar{v} - 0) \tag{1}$$

$$\sum (\text{Ang Imp})_G = \Delta H_G \qquad \text{or} \qquad -Frt = \bar{I}(\omega - 36) \tag{2}$$

In the second equation, assume that the clockwise direction is positive. The equations become $14.5t = 3\bar{v}$ and $-14.5(\frac{1}{3})t = \frac{1}{6}(\omega - 36)$, where $\omega = \bar{v}/r = 3\bar{v}$. The solution is thus $\bar{v} = 4.0$ ft/s, $t = 0.83$ s. Then $s = \frac{1}{2}(\bar{v} + 0)t = 1.66$ ft.

18.25. A 2-in-diameter stream of water moving 80 ft/s horizontally strikes a flat vertical plate as shown in Fig. 18-24. After striking, the water moves parallel to the plate. What is the force exerted on the plate by the stream of water?

SOLUTION

Consider all the particles of water in time interval Δt. The total mass m of these particles is

$$m = Av(\Delta t)\delta = \frac{1}{4}\pi\left(\frac{2}{12}\right)^2 (80)(\Delta t)\left(\frac{62.4}{32.2}\right)$$

$$= 3.39(\Delta t) \text{ slugs}$$

where A = area of the cross-section of the stream in ft^2
        v = speed of stream in ft/s
        δ = density of water in slugs/ft^3

Let P = force of plate on the mass m of water. Then

$$\sum F_x(\Delta t) = \Delta G_x = m(v''_x - v'_x)$$

or $\qquad\qquad P(\Delta t) = 3.39(\Delta t)(0 - 80)$

from which $P = -271$ lb. The force of the water on the plate is $+271$ lb, that is, to the right.

18.26. Rework Problem 18.25, but assume that the plate is moving to the right with a speed of 20 ft/s.

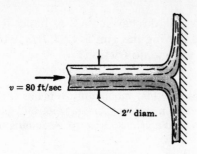

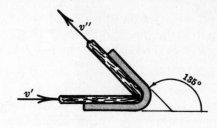

| Fig. 18-24 | Fig. 18-25 |

SOLUTION

Again consider the mass of all particles of water in time Δt. The speed of the wate relative to the plate is $v = (80 - 20) = 60$ ft/s. Then $m = Av(\Delta t)\delta = 2.54(\Delta t)$, and, using the linear impulse–momentum equation in the x direction, we obtain

$$P(\Delta t) = 2.54(\Delta t)(20 - 80)$$

where the final speed of water is that of the plate. Solving, the force of the plate on the water is $P = -153$ lb. Of course, the force of the water on the plate is $+153$ lb, that is, to the right.

18.27. A stream of water with cross-sectional area 2000 mm² and moving 10 m/s horizontally strikes a fixed blade curved as shown in Fig. 18-25. Assuming that the speed of the water relative to the blade is constant (no friction is considered), determine the horizontal and vertical components of the force of the blade on the stream of water.

SOLUTION

Note that the final velocity $\mathbf{v}''$ has the same magnitude v as the initial velocity $\mathbf{v}'$ but a different direction. The mass m of all particles of water in time interval Δt is

$$m = Av(\Delta t)\delta = (2000 \times 10^{-6}\ \text{m}^2)(10\ \text{m/s})(\Delta t)(1000\ \text{kg/m}^3) = 20(\Delta t)$$

Using the linear impulse–momentum equation for the x and y directions,

$$\sum F_x(\Delta t) = \Delta G_x = m(v_x'' - v_x') \qquad \text{or} \qquad P_x(\Delta t) = (20\Delta t)(-10\cos 45° - 10)$$
$$\sum F_y(\Delta t) = \Delta G_y = m(v_y'' - v_y') \qquad \text{or} \qquad P_y(\Delta t) = (20\Delta t)(+10\sin 45° - 0)$$

where P_x and P_y are, respectively, the x and y components of the blade force. Solving, $P_x = -340$ N (to the left on the water) and $P_y = +140$ N (up on the water).

18.28. A tank weighing 150 lb rests on platform scales. A vertical jet empties water into the tank with a speed of 20 ft/s. Its cross-sectional area is 0.20 in². What will be the scale reading 1 min later?

SOLUTION

The water from the jet exerts a continuuous force F on the bottom of the tank and thus to the scale. From the impulse–momentum equation, considering the water at any time t as a free body, we may write $\sum (\text{Imp})_v = \Delta G_v$ or $Ft = m(0 - 20)$, where $m = Avt\delta$. Then

$$Ft = \frac{0.20}{144}(20)(t)\left(\frac{62.4}{32.2}\right)(0 - 20) \qquad \text{or} \qquad F = -1.1\ \text{lb}$$

The negative sign indicates that an upward force (from the scale) is required to stop the water.

At $t = 60$ s, the tank holds $(0.20/144)(20)(60)(62.4) = 104$ lb of water.

Thus, the scale reading after 1 min will be $(1.1 + 104 + 150) = 255$ lb.

18.29. A 130-lb person sitting in a 150-lb iceboat fires a gun, discharging a 2-oz bullet horizontally aft (along the fore-and-aft line). If the pullet speed leaving the gun is 1200 ft/s, what will be the speed of the boat after the bullet is fired, if we assume no friction?

SOLUTION

Since the impulse of the entire system is zero, the total momentum remains zero. This is true because the action on the bullet is equal to the reaction on the person, gun, and boat. The initial momentum is zero. The momentum after the bullet is fired must also be zero. Assuming the bullet speed to be positive, we can write

$$\frac{W_{\text{boat}} + W_{\text{person}}}{g} v + \frac{W_{\text{bullet}}}{g} 1200 = 0$$

Hence,

$$\frac{150 + 130}{g} v + \frac{2/16}{g} 1200 = 0$$

This yields $v = -0.54$ ft s, with the minus sign indicating that the boat and person move in the opposite direction from that of the bullet.

18.30. A 60-g bullet was fired horizontally into a 50-kg sandbag suspended on a rope 900 mm long as shown in Fig 18-26. It was calculated from the observed angle θ that the bag with the bullet embedded in it swung to a height of 30 mm. What was the speed of the bullet as it entered the bag?

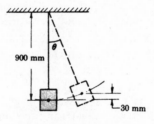

Fig. 18-26

SOLUTION

Let

v_1 = speed of bullet before impact

v_2 = speed of (bag + bullet) after impact = $\sqrt{2gh} = \sqrt{2(9.8)(0.03)} = 0.767$ m/s

Momentum of system before impact = momentum of system after impact

$$(0.06 \text{ kg})v_2 + 0 = (0.06 + 50) \text{ kg} \times 0.767 \text{ m/s}$$

from which $v_1 = 640$ m/s.

18.31. As another example of conservation of linear momentum, consider the recoil of guns.

ANALYSIS

Initially the projectile is at rest in the gun. The charge explodes, pushing the projectile from the gun and at the same time pushing back with the same force on the gun. Since no external forces act during this explosion, the sum of the momenta of the projectile forward and the gun backward must be the same as the initial momentum of the system, i.e., zero. Thus,

$$m_p v_p + m_g v_g = 0$$

where m_p, m_g = masses of projectile and gun, respectively

v_p, v_g = speeds of projectile and gun, respectively, immediately after the explosion

Solving for speed of recoil, $v_g = -(m_p/m_g)v_p$. The minus sign indicates that the gun moves in the opposite direction to the projectile.

The greater the mass of the gun, the less will be the speed of recoil. Hence, the energy to be absorbed by the recoil spring or other devices will be correspondingly less. Of course, the weight is limited by the need for mobility.

18.32. A 60-g bullet moving with a speed of 500 m/s strikes a 5-kg block moving in the same direction with a speed of 30 m/s. What is the resultant speed v of the bullet and the block, assuming the bullet to be embedded in the block?

SOLUTION

Momentum of system before impact = momentum of system after impact

$$(0.06 \text{ kg})(500 \text{ m/s}) + (5 \text{ kg})(30 \text{ m/s}) = (0.06 + 5) \text{ kg} \times v$$

$$v = 35.6 \text{ m/s}$$

18.33. A spring normally 6 in long is connected to the two weights shown in Fig. 18-27 and compressed 2 in. If the system is released on a smooth horizontal plane, what will be the speed of each block when the spring is again its normal length? The spring constant is 12 lb/in.

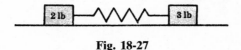

Fig. 18-27

SOLUTION

Since the same spring force acts on the two weights, but in opposite directions, the total linear impulse on the system of both weights is zero. Hence, the momentum of the two weights is constant, i.e., zero, and $(2/g)v_2 + (3/g)v_3 = 0$, where the subscripts refer to the 2- and 3-lb weights. From this equation, $v_2 = -(3/2)v_3$ at all times.

Another equation involving v_2 and v_3 is necessary. The work done by the spring in expanding to its original length (by virtue of its potential energy) is equal to the change in kinetic energy of both weights. As the spring expands a distance x from its compressed position, its compressive force $= 12(2 - x)$ lb.

Hence, the total work done $= \int_0^2 12(2 - x)\, dx = 24$ in-lb or 2 ft-lb. Equating this to the final kinetic energy (initial is zero), $2 = \frac{1}{2}(2/g)v_2^2 + \frac{1}{2}(3/g)v_3^2$. Solve simultaneously with the previous equation to obtain $v_3 = 4.14$ ft/s (to the right) and $v_2 = -6.22$ ft/s (to the left).

18.34. Figure 18-28 shows a remote-controlled 10-lb object at a distance of 4 ft from the center of a horizontal "weightless" turntable that is turning 2 rad/s about a vertical axis. An opposing moment $M = 2t$ lb-ft is applied to the shaft. Determine (a) how long it will take for the turntable to reach a speed of 1.5 rad/s and (b) how far the object must be moved, if the moment is removed, in order to bring the turntable back to a speed of 2 rad/s.

SOLUTION

(a) The angular momentum of the object is the product of its linear momentum mv and its radial distance r. The angular impulse equals the change in angular momentum. Hence, using $v = r\omega$, we have

$$\int_0^t -2t\, dt = \frac{10}{g}(4 \times 1.5)(4) - \frac{10}{g}(4 \times 2)(4)$$

Hence, $t = 1.58$ s

(*b*) To solve the second part of the problem, use conservation of angular momentum to determine the necessary radius *r* at which to place the object:

$$\frac{10}{g}(4)(1.5)(4) = \frac{10}{g}(4)(2)(r)$$

and thus $r = 3.46$ ft, and the object must be moved from 4 to 3.46 ft, or 0.54 ft toward the center.

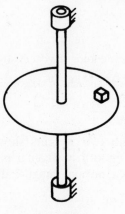

Fig. 18-28

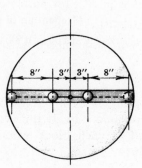

Fig. 18-29

18.35. Two small spheres, each weighing 2.5 lb, are connected by a light string as shown in Fig. 18-29. The horizontal platform is rotating under no external moments at 36 rad/s when the string breaks. Assuming no friction between the spheres and the groove in which they ride, determine the angular speed of the system when the spheres hit the outer stops. The moment of inertia *I* for the disk is 0.4 slug-ft^2.

SOLUTION

The initial moment of inertia I_i is that of the disk and the two spheres at distance 3 in from the center. Hence,

$$I_i = 0.4 + 2\left(\frac{2.5}{32.2}\right)\left(\frac{3}{12}\right)^2 = 0.41 \text{ slug-ft}^2$$

The initial angular momentum is $I_i\omega_i$ or 0.41(36). The final moment of inertia I_f is

$$I_f = 0.4 + 2\left(\frac{2.5}{32.2}\right)\left(\frac{11}{12}\right)^2 = 0.531 \text{ slug-ft}^2$$

Since no external moments act on the system, there is conservation of angular momentum; then

$$I_f\omega_f = I_i\omega_i \qquad 0.531\omega_f = 0.41(36) \qquad \text{and} \qquad \omega_f = 27.8 \text{ rad/s}$$

18.36. Solve the following impact equations for unknowns v_1 and v_2:

$$e = \frac{v_2 - v_1}{u_1 - u_2} \qquad (1)$$

$$m_1 u_1 + m_2 u_2 = m_1 v_1 + m_2 v_2 \qquad (2)$$

where   *e* = coefficient of restitution
u_1, u_2 = speed of bodies 1 and 2, respectively, before impact
v_1, v_2 = speeds of bodies 1 and 2, respectively, after impact
m_1, m_2 = masses of bodies 1 and 2, respectively

SOLUTION

Multiply (*1*) by $(u_1 - u_2)m_1$ to obtain (*3*) $em_1(u_1 - u_2) = m_1v_2 - m_1v_1$. add (*2*) and (*3*) to get $m_1u_1 + m_2u_2 + em_1(u_1 - u_2) = (m_2 + m_1)v_2$. Then

$$v_2 = \frac{m_1u_1(1 + e) + u_2(m_2 - em_1)}{m_2 + m_1}$$

Similarly,

$$v_1 = \frac{m_2u_2(1 + e) + u_1(m_1 - em_2)}{m_2 + m_1}$$

18.37. Discuss impact for purely elastic bodies ($e = 1$).

SOLUTION

For elastic impact, substitute $e = 1$ in the equations for v_2 and v_1 derived in Problem 18.36:

$$v_1 = \frac{2m_2u_2 + u_1(m_1 - m_2)}{m_2 + m_1} \tag{1}$$

$$v_2 = \frac{2m_1u_1 + u_2(m_2 - m_1)}{m_2 + m_1} \tag{2}$$

Of special interest is the case when $m_1 = m_2 = m$. Then the equations become

$$v_1 = \frac{2mu_2 + 0}{m + m} = u_2 \tag{1'}$$

$$v_2 = \frac{2mu_1 + 0}{m + m} = u_1 \tag{2'}$$

The above equations explain what takes place when a moving coin strikes a stationary coin on a smooth surface. The final speed of the moving coin will be the initial speed of the other coin (in this case zero), while the final speed of the formerly stationary coin will be the initial speed of the moving coin. In other words, the moving coin stops and the stationary one assumes its speed. The same thing is true of a straight row of coins. The first one (farthest away from the moving coin) moves away while the others remain stationary.

18.38. What happens during inelastic impact?

SOLUTION

This is the case when one body absorbs the other or clings to it. Common sense indicates that they have a common final speed. Note that if $e = 0$ is substituted into the equations for v_1 and v_2 in Problem 18.36, this is actually so:

$$v_1 = \frac{m_2u_2 + m_1u_1}{m_2 + m_1} \qquad v_2 = \frac{m_1u_1 + m_2u_2}{m_2 + m_1} \qquad \text{or} \qquad v_1 = v_2$$

18.39. Two equal billiard balls meet centrally with sppeds of 6 and -8 ft/s. What will be their final speeds after impact if the coefficient of restitution is assumed to be 0.8?

SOLUTION

Let subscript 1 refer to the 6-ft/s ball and subscript 2 to the -8-ft/s ball. The masses are equal ($m_1 = m_2 = m$).

Hence, $u_1 = 6$ ft/s and $u_2 = -8$ ft/s. The problem is to determine v_1 and v_2. Work directly from the following fundamental equations rather than applying the results of Problem 18.36:

$$e = \frac{v_2 - v_1}{u_1 - u_2} \qquad 0.8 = \frac{v_2 - v_1}{6 - (-8)} \qquad \text{or} \qquad v_2 - v_1 = 11.2 \text{ ft/s} \tag{1}$$

$$m_1u_1 + m_2u_1 = m_1v_1 + m_2v_2 \qquad m(+6) + m(-8) = mv_1 + mv_2 \qquad \text{or} \qquad v_1 + v_2 = -2 \text{ ft/s} \tag{2}$$

Solve the two equations simultaneously to get $v_1 = -6.6$ ft/s and $v_2 = 4.6$ ft/s.

18.40. A ball rebounds vertically from a horizontal floor to a height of 20 m. On the next rebound, it reaches a height of 14 m. What is the coefficient of restitution between the ball and the floor?

SOLUTION

Let subscript 1 refer to the ball. The initial and final speeds of the floor, u_2 and v_2, are zero since it is assumed that the floor remains stationary during impact.

The second time the ball hits the floor it has fallen from a height of 20 m. Its speed u_1 is

$$u_1 = \sqrt{2gh} = \sqrt{2(9.8 \text{ m/s}^2)(20 \text{ m})} = 19.8 \text{ m/s}$$

By similar reasoning, the speed v_1 of rebound may be found. The ball with speed v_1 rises to a height of 14 m. Hence, its starting speed v_1 is $v_1 = \sqrt{2(9.8)(14)} = 16.6 \text{ m/s}$.

Now apply the following equation of impact, where downward speeds are chosen positive:

$$e = \frac{v_2 - v_1}{u_1 - u_2} = \frac{0 - (-16.6)}{19.8 - 0} = 0.84$$

The value of e could be determined by considering the square roots of the successive heights to which the ball bounced. This is true since the value of e in this case is actually the ratio of the speeds which depend on the square roots of the heights $e = \sqrt{14/20} = 0.84$.

18.41. A 2-lb ball moving horizontally with a speed of 12 ft/s, as shown in Fig. 18-30(*a*), hits the bottom of the rigid, slender 5-lb bar, which pivots about its top. If the coefficient of restitution is 0.7, determine the angular speed of the bar and the linear speed of the ball just after impact.

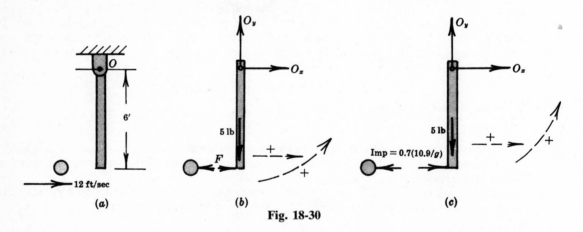

Fig. 18-30

SOLUTION

During the first phase of the impact, from time zero to t_1, the linear impulse acting between the ball and bar will cause the sphere to slow down and the lower end of the bar to speed up to a common speed u. The free-body diagram for this first phase is shown in Fig. 18-30(*b*). The linear impulse of the interacting force F causes the change in linear momentum of the ball as indicated in equation (*1*), where to the right is considered positive:

$$-\int_0^{t_1} F \, dt = \Delta G = \frac{2}{g}(u - 12) \tag{1}$$

This linear impulse has a moment about the pivot point O of the bar, and causes a change in the angular momentum of the bar as indicated in equation (*2*), where counterclockwise is considered positive:

$$+\int_0^{t_1} F(6) \, dt = \Delta H_O = I_O(\omega - 0) \tag{2}$$

where $I_O = \frac{1}{3}mL^2 = \frac{1}{3}(5/g)(6^2) = 60/g$ slug-ft^2, $\omega = u/6$ rad/s. Putting these values into equation (2) and dividing by 6, we have

$$+\int_0^{t_1} F\,dt = \frac{5u}{3g} \qquad\qquad\qquad (3)$$

Adding (1) and (3), we find $u = 6.55$ ft/s. Thus, at the end of this first phase (compression phase) the ball and the lower end of the bar have a common speed of 6.55 ft/s to the right. The bar then has an angular speed $\omega = 6.55/6 = 1.09$ rad/s counterclockwise.

The linear impulse acting during this phase can be calculated from (1) as

$$\int_0^{t_1} F\,dt = -\frac{2}{g}(6.55 - 12) = \frac{10.9}{g}\,\text{lb-s}$$

This value can be checked by substituting $u = 6.55$ ft/s into (3).

In the second phase (restitution phase), the linear impulse acting is only 0.7 of that acting in the compression phase, as shown in Fig. 18-30(c). Proceeding as before, using v as the speed of the ball immediately after impact and ω' as the angular speed of the bar immediately after impact, the equations for the ball and bar are, respectively,

$$-0.7\left(\frac{10.9}{g}\right) = \frac{2}{g}(v - 6.55) \qquad\qquad\qquad (4)$$

$$+0.7\left(\frac{10.9}{g}\right)(6) = \frac{60}{g}(\omega' - 1.09) \qquad\qquad\qquad (5)$$

from which $v = 2.72$ ft/s to the right and $\omega' = 7.73$ rad/s counterclockwise.

18.42. Refer to Fig. 18-31. The 9-kg ball moving down at 3 m/s strikes the 5.5-kg ball moving as shown at 2.5 m/s. The coefficient of restitution $e = 0.80$. Find the speeds v_1 and v_2 after impact.

SOLUTION

To determine the y components of the final speeds, apply the two equations

$$e = \frac{(v_2)_y - (v_1)_y}{(u_1)_y - (u_2)_y} \qquad \text{and} \qquad m_1(v_1)_y + m_2(v_2)_y = m_1(u_1)_y + m_2(u_2)_y$$

Then $\qquad 0.80 = \dfrac{(v_2)_y - (v_1)_y}{(-3) - (1.77)} \qquad$ and $\qquad 9(v_1)_y + 5.5(v_2)_y = 9(-3) + 5.5(+1.77)$

whose solution is $(v_2)_y = -3.56$ m/s, i.e., down; $(v_1)_y = +0.26$ m/s, i.e., up.

In the x direction, the 5.5-kg ball will continue to the right with undiminished speed. Thus, $(v_2)_x = 1.77$ m/s to the right, and $(v_1)_x = 0$.

To summarize, the 9-kg ball will rebound up with a speed of 0.26 m/s, and the 5.5-kg ball will move to the right and down with components of 1.77 m/s and 3.56 m/s, respectively.

18.43. Refer to Fig. 18-32. A body at rest is free to move on a smooth tabletop. It is struck by an impact force F at a distance d from its mass center G. Investigate the motion and locate the instantaneous center of the body in terms of d and its radius of gyration about its mass center.

SOLUTION

Employing the impulse–momentum equations for plane motion,

$$\sum (\text{Imp})_h = \Delta G_h \qquad \text{or} \qquad \int F\,dt = m(\bar{v} - 0) \qquad\qquad (1)$$

$$\sum (\text{Ang Imp})_G = \Delta H_G \qquad \text{or} \qquad \int Fd\,dt = \bar{I}(\omega - 0) \qquad\qquad (2)$$

Since the distance d is constant, it may be removed from under the integral sign. Then $\bar{v} = \int F\,dt/m$ and $\omega = (d/\bar{I})\int F\,dt$. But $\bar{I} = m\bar{k}^2$; hence, $\omega = (d/m\bar{k}^2)\int F\,dt$, where $\bar{k}$ is the radius of gyration of the body with respect to the mass center.

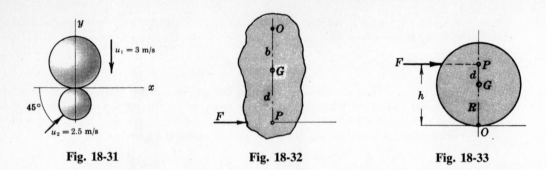

Fig. 18-31   Fig. 18-32   Fig. 18-33

To locate the instantaneous center—a point about which the body tends to pivot only—select any point O at a distance b from the mass center on a line perpendicular to the action line of the impact force F, as shown in the figure. The velocity of O is $\mathbf{v}_O = \mathbf{v}_{O/G} + \bar{\mathbf{v}}$. But $\bar{v}$ has been found in terms of the linear impulse. Also, $v_{O/G} = b\omega$, where ω has just been defined in terms of the linear impulse. Note that for the object shown, $\bar{v}$ is to the right $(+)$ and ω is counterclockwise. Hence, $v_{O/G}$ is to the left $(-)$.

Substituting in the formula,

$$v_O = -b\omega + \bar{v} = \frac{-bd}{m\bar{k}^2}\int F\,dt + \frac{1}{m}\int F\,dt$$

To make O an instantaneous center, v_O must be zero. Set the above expression equal to zero and solve to obtain $b = \bar{k}^2/d$.

This may be viewed slightly differently. Suppose the distance d was chosen originally as $\bar{k}^2/b$. Then point O will remain at rest. When this occurs, the force F is said to be applied through the *center of percussion P*.

18.44. Apply the results of the preceding problem to find the height h above the plane at which a solid cylinder of radius R should be struck by a horizontal impact force F in order to have no sliding at the point of contact O. See Fig. 18-33.

SOLUTION

According to the theory, F should be struck through the center of percussion P in order that the point of contact O be the instantaneous center.

Hence, $h = d + R = \bar{k}^2/R + R$. But, for a cylinder, $\bar{k}^2 = \bar{I}/m = \frac{1}{2}mR^2/m = \frac{1}{2}R^2$. Then $h = \frac{1}{2}R + R = \frac{3}{2}R$.

18.45. The sphere of mass m_1 in Fig. 18-34(a) is moving to the right on a smooth horizontal plane with a velocity u_1 when it strikes (at right angles) a slender bar of mass m_2 that has a mass-center velocity u_2 to the right and an initial angular velocity ω_i clockwise. If the coefficient of restitution is e, set up the equations needed to solve for the final velocity v_1 of the sphere, the final mass-center velocity v_2 of the bar, as well as the final angular velocity ω_f of the bar. The moment of inertia of the bar relative to the mass center G is $\bar{I}$.

SOLUTION

Figure 18-34(b) depicts the conditions immediately after impact. Note that the initial velocity of the point of impact on the bar is given by the kinematic relation $u_2 + d\omega_i$. Similar reasoning shows its final velocity to be $v_2 + dw_f$.

Linear momentum is conserved; hence, calling to the right positive, there results

$$m_1u_1 + m_2u_2 = m_1v_1 + m_1v_2 \qquad (1)$$

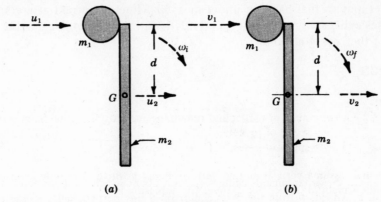

Fig. 18-34

Angular momentum relative to the mass center G is conserved; hence, calling clockwise positive, there results

$$m_1 u_1 d + \bar{I}\omega_1 = m_1 v_1 d + \bar{I}\omega_f \tag{2}$$

A third equation is found by using the definition of e along the impact line:

$$e = -\frac{(v_2 + d\omega_f) - v_1}{(u_2 + d\omega_i) - u_1} \tag{3}$$

The three unknowns v_1, v_2, and ω_f may now be found.

18.46. A box in the shape of a cube with edge 1.5 m is sliding across a floor with a speed of 4 m/s. The leading lower edge of the box strikes an upraised floor tile with a completely inelastic impact. What is the angular velocity of the box immediately after the box hits the tile. The mass of the box is 9 kg.

SOLUTION

Since the impact forces on the box act through the leading lower edge, the angular momentum about that edge is conserved. Because the impact is completely inelastic, the leading edge has zero velocity during impact, and so the motion subsequent to the impact is rotation about O, the leading lower edge of the box. Hence, as shown in Fig. 18-35,

$$H_O = H_O'$$
$$mv(\tfrac{1}{2}s) = I_O\omega$$

But
$$I_O = \bar{I} + md^2 \quad\text{or}\quad I_O = \tfrac{1}{6}ms^2 + m[(\tfrac{1}{2}s)^2 + (\tfrac{1}{2}s)^2]$$
$$I_O = \tfrac{2}{3}ms^2 = \tfrac{2}{3}(9)(1.5)^2 = 13.5 \text{ kg} \cdot \text{m}$$

Substituting in the conservation of momentum equation yields

$$(9)(4)(0.75) = 13.5\omega$$

from which $\omega = 2$ rad/s clockwise.

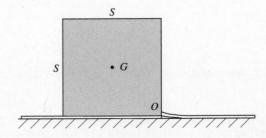

Fig. 18-35

18.47. A rocket and its fuel have an initial mass m_0. Fuel is burned at a constant rate, $dm/dt = C$. The gases exhaust at a constant speed relative to the rocket. Neglecting air resistance, find the speed of the rocket at time t.

SOLUTION

Let

$$m = \text{mass of rocket and remaining fuel at time } t; \text{ that is, } m = m_O - Ct$$
$$v = \text{speed of rocket}$$
$$u = \text{speed of gases}$$

We shall assume vertical motion with up being positive. The only external force to the system at time t is the weight mg which is negative. Our equation is

$$\sum F = m\frac{dv}{dt} - \frac{dm}{dt}(v - u)$$

where $(v - u)$ is the relative speed of the rocket to the gases and is a constant K. Then

$$-mg = m\frac{dv}{dt} - CK \qquad -(m_0 - Ct)g = (m_0 - Ct)\frac{dv}{dt} - CK \qquad \text{or} \qquad \frac{dv}{dt} = -g + \frac{CK}{m_0 - Ct}$$

Integrating the last equation,

$$\int_0^v dv = \int_0^t -g\,dt + \int_0^t \frac{CK}{m_0 - Ct}\,dt$$

or $\qquad v = -gt + [-K \ln (m_0 - Ct)]_0^t = -gt - K \ln (m_0 - Ct) + K \ln m_0 = K \ln \frac{m_0}{m_0 - Ct} - gt$

18.48. An empty rocket has a mass of 2000 kg. It is fired vertically up with a fuel load of 8500 kg. The exit speed of the exhaust gases relative to the nozzles is a constant 2000 m/s. At what rate, in kg/s, must the gases be discharged at the beginning if the desired acceleration is 9.8 m/s² up? Assume that the nozzles exhaust at atmospheric pressure.

SOLUTION

Selecting up as positive, we insert values into the equation

$$\sum F = m\frac{dv}{dt} - \frac{dm}{dt}(v - u)$$

to obtain

$$-10\,500 \times 9.8 = 10\,500(9.8) - \frac{dm}{dt}(-2000)$$

Solving, $\qquad\qquad\qquad\qquad \dfrac{dm}{dt} = -103 \text{ kg/s}$

The negative sign indicates the loss in the mass of the system.

18.49. Study the motion of the gyroscope shown schematically in Fig. 18-36(a) as a spinning wheel and rotor that is attached to point O in such a way that it is free to turn in any direction about O.

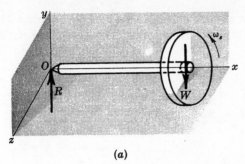

Fig. 18-36

ANALYSIS

Assume that the wheel is spinning with a large angular velocity ω_s about the x axis, which is called the spin axis. Also assume that the system is released; the external forces acting on it are the weight W of the wheel (neglect weight of rotor) and the upward reaction R at O. The weight W exerts a torque on the system about the z axis, which is called the moment axis. It is known by actual experiment that the wheel does not fall but turns about the y axis, which is called the axis of precession.

The explanation for this follows. Before the system is released, the angular momentum of the system is equal to the product of its moment of inertia and its angular velocity ω_s, that is, $\bar{I}\omega_s$. This can be represented by a vector **OB** as shown in Fig. 18-36(b) directed along the x axis to the right. (According to the right-hand rule, the thumb points to the right when the fingers are curled in the direction of ω_s. This is also expressed by saying that a right-handed screw would advance to the right when turned in the direction of ω_s.)

Immediately upon release, there is an external torque **M** acting on the system. It acts clockwise about the z axis when viewed from the positive end of that axis, and its magnitude is equal to the product of the weight and the perpendicular distance from the z axis to the center of mass. For a short time interval dt, the impulse of this torque is **M**$\,dt$. Its units are the same as the units of angular momentum, and it may be represented according to the right-hand rule by a vector along or parallel to the z axis and acting in the negative direction. Figure 18-36(b) illustrates this vector **BC**. But this angular impulse **M**$\,dt$ is the change in the angular momentum of the system. The new angular momentum is the vector sum of **OB** and **BC**. This line **OC** is in the xz plane but at an angle $d\phi$ with the original x axis. Note that the new axis of spin is along OC. Using scalar quantities and substituting $d\phi$ for $\tan d\phi$, there results

$$d\phi = \frac{M\,dt}{\bar{I}\omega_s} \qquad \text{or} \qquad M = \bar{I}\omega_s \frac{d\phi}{dt}$$

But $d\phi/dt$ is the angular speed of the spin axis about the precession axis, say ω_p.

The equation of the gyroscope is $M = \bar{I}\omega_s\omega_p$. Note that the spin axis remains in the xz plane, provided ω_s remains large.

18.50. A rotor weighing 4000 lb and with a radius of gyration 3.00 ft is mounted with its geometric axis along the fore-and-aft line of a ship that is turning 1 rpm counterclockwise as viewed from above. The rotor turns 300 rpm counterclockwise when observed from the rear of the ship. Assuming the center-to-center distance between bearings to be 3.50 ft, determine the front and rear bearing reactions on the rotor.

SOLUTION

Draw a free-body diagram of the rotor with the x axis as the fore-and-aft line of the ship. See Fig. 18-37. The front and rear reactions are R_F and R_R, respectively. According to the right-hand rule, the angular momentum about the x axis ($\bar{I}\omega_s$) is drawn to the left along the x axis. To precess in the given direction about the y axis, the rotor must be acted upon by an angular impulse (**M**$\,dt$) as shown.

According to the right-hand rule, this moment (caused by the reactions) must be counterclockwise when viewed from the positive end of the z axis. The magnitude of this moment is $M = (3.50/2)(R_F - R_R)$.

The values given are $\omega_s = 2\pi(300/60) = 31.4$ rad/s, $\omega_p = 2\pi(1/60) = 0.105$ rad/s, and $\bar{I} = mk^2 = (4000)/(32.2)(3.00)^2 = 1118$ slug-ft^2.

Then $M = \bar{I}\omega_s\omega_p = 3690$ lb-ft and $(R_F - R_R) = M(2/3.50) = 2110$ lb.

A vertical sum yields $R_F + R_R = 4000$ lb. The solution is

$$R_F = 3055 \text{ lb} \qquad \text{and} \qquad R_R = 945 \text{ lb}$$

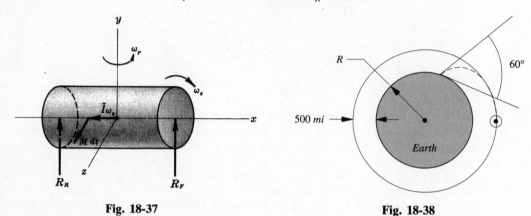

Fig. 18-37   Fig. 18-38

18.51. A satellite in circular orbit about the earth ejects a pod tangent to its path. If the satellite is orbiting 500 mi above the earth, what must be the speed of the pod such that it will strike the earth at an angle of 60° with respect to the ground? Refer to Fig. 18-38

SOLUTION

Since the gravitational force on the pod acts through the center of the earth, the angular momentum of the pod about the earth's center is conserved. So $\Delta H_O = 0$ and $mv_1(R + 500) = mv_2 \cos 60°(R)$, where R is the radius of the earth or 4000 mi. Hence, $v_2 = 2.25v_1$.

Also, from the work–energy equation, $U = \Delta T$,

$$\int_{R+500}^{R} \left(-\frac{GMm}{r^2}\right) dr = \tfrac{1}{2}m(v_2^2 - v_1^2) \qquad \text{where} \qquad GM = 1.41 \times 10^{16} \text{ ft}^3/\text{s}^2$$

$$GM\left(\frac{1}{R} - \frac{1}{R + 500}\right) = \tfrac{1}{2}(v_2^2 - v_1^2) = 2.031v_1^2$$

$$v_1 = 6040 \text{ ft/s}$$

18.52. One end of an elastic band, shown schematically in Fig. 18-39, is attached to a fixed pin. The other end is attached to a $\tfrac{1}{2}$-lb ball. The elastic constant of the band is 4 lb/ft. The elastic is stretched 1 ft and released with a velocity of 5 ft/s perpendicular to the elastic band. The unstretched length of the elastic band is 2 ft, and motion takes place on a smooth horizontal plane. How close to the fixed pin will the ball pass?

SOLUTION

Since the force of the elastic band on the ball acts through point O the angular momentum about O is conserved. Hence, $(mv_0)3 = (mv)d$. Or, $vd = 15$. Furthermore, $U = \Delta T$ or

$$-\tfrac{1}{2}k(s^2 - s_0^2) = \tfrac{1}{2}m(v^2 - v_0^2)$$

$$-\tfrac{1}{2}4(0^2 - 1^2) = \frac{1}{2}\frac{1}{2g}(v^2 - 5^2) \qquad v = 16.77 \text{ ft/s}$$

Substituting in $vd = 15$ yields $d = 0.89$ ft

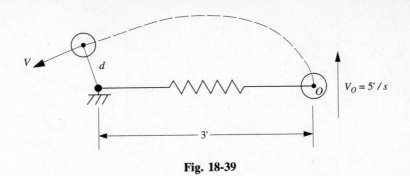

Fig. 18-39

Supplementary Problems

18.53. In Problem 18.6, show that the selection of the top point as the moment center yields $2PR = (d/dt)(4mR^2\omega)$, which does not agree with the results obtained in that problem. [It should not agree, because the top point does not fit the conditions imposed on point O in equation (9).]

18.54. A 50-kg mass falls freely from rest for 4 s. Find its momentum at that time.
 Ans.   1960 kg · m/s = 1960 N · s

18.55. A pile-driver hammer weighing 1000 lb is dropped from a height of 25 ft. If the time required to stop the pile driver is 1/20 s, determine the average force acting during that time.   *Ans.* $F_{av} = 25,000$ lb

18.56. A body is thrown upward with an initial speed of 15 m/s. Find the time required for the body to attain a speed of 6 m/s downward.   *Ans.*   2.14 s

18.57. A 50-lb body is projected up a 30° plane with initial speed of 20 ft/s. If the coefficient of friction is 0.25, determine the time required for the body to have an upward speed of 10 ft/s.   *Ans.*   0.433 s

18.58. A 2000-lb sphere with a diameter of 10 ft is rotating about a diameter at 600 rpm. What angular torque will bring the sphere to rest in 3 min?   *Ans.* $T = 216$ lb-ft

18.59. A block weighing 120 lb rests on a horizontal floor. It is acted upon by a horizontal force that varies from zero according to the law $F = 15t$. If the force acts for 10 s, what is the speed of the block? Assume that the coefficient of static and kinetic friction is 0.25.   *Ans.*   129 ft/s

18.60. The magnitude of a force applied tangentially to the rim of a 1200-mm-diameter pulley varies according to the law $F = 0.03t$. Determine the angular impulse of the force about the axis of rotation for the interval 0 to 35 s. (*Hint:*   Use the integral definition of angular impulse.)   *Ans.*   11.0 N · s · m

18.61. A 300-lb disk rotating at 1000 rpm has a diameter of 3 ft. Find its angular momentum.
 Ans.   1100 lb-s-ft

18.62. A bob of mass m travels in a circular path of radius R on a smooth horizontal plane under the restraint of a string which passes through a hole O in the plane as shown in Fig. 18-40. If the angular speed of the bob is ω_1 when the radius is R, what will be the angular speed if the string is pulled from underneath until the radius of the path is $\frac{1}{2}R$? Find the ratio of the final tension in the string to its initial value.
 Ans.   8

18.63. A 12-kg disk with a radius of gyration equal to 600 mm is subjected to a torque $M = 2t$ N · m. Determine the angular speed of the disk 2 s after it started from rest. Assume frictionless bearings.
 Ans. $\omega = 0.93$ rad/s

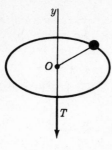

Fig. 18-40

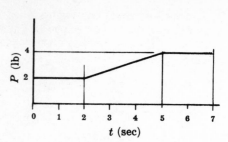

Fig. 18-41

18.64. A 20-lb block is acted upon by a horizontal force P that varies according to the graph shown in Fig. 18-41. The block moves from rest on a smooth horizontal surface. What will be the position and speed after 7 s?   *Ans.* $v = 33.8$ ft/s, $s = 134$ ft

18.65. A 12-lb homogeneous rotor with radius of gyration 1.2 ft comes to rest in 85 s from a speed of 180 rpm. What was the frictional torque that stopped the rotor?   *Ans.* $T = 0.119$ lb-ft

18.66. The two masses shown in Fig. 18-42 are connected by a light inextensible cord that passes over a homogeneous 4-kg cylinder having a diameter of 1200 mm. How long will it take the 7-kg mass to move from rest to a speed of 2 m/s   *Ans.* $t = 1.43$ s

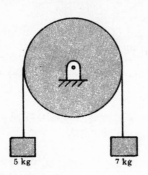

Fig. 18-42

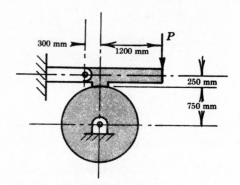

Fig. 18-43

18.67. A 100-kg rotor has a radius of gyration of 700 mm. The rotor is rotating at 6 rad/s counterclockwise when the force P is applied to the brake as shown in Fig. 18-43. What is the magnitude of P if the rotor stops in 92 s? The coefficient of friction between the brake and rotor is 0.4.   *Ans.* $P = 1.42$ N

18.68. A drum rotating at 20 rpm is lifting a 1-ton cage connected to it by a light inextensible cable as shown in Fig. 18-44. If power is cut off, how long will it take the cage to come to rest? Assume frictionless bearings.   *Ans.* $t = 0.172$ s

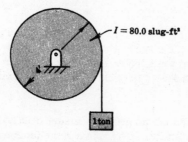

Fig. 18-44

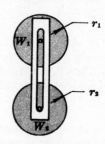

Fig. 18-45

18.69. Two homogeneous disks of weights W_1 and W_2 and radii r_1 and r_2 are free to turn in frictionless bearings in the fixed vertical frame shown in Fig. 18-45. The upper disk W_1 is turning with angular velocity ω clockwise when it is dropped onto the lower disk W_2. Assuming that the coefficient of friction μ is constant, determine the time until slipping stops. What are the angular velocities at that time?

Ans. $\quad t = \dfrac{r_1 \omega}{2\mu g(1 + W_1/W_2)}, \quad \omega_1 = \dfrac{\omega}{1 + W_2/W_1}$ clockwise, $\omega_2 = \dfrac{r_1}{r_2}\left(\dfrac{\omega}{1 + W_2/W_1}\right)$ counterclockwise

18.70. A rotor with a moment of inertia of 320 slug-ft^2 is supported by a shaft that turns in two bearings. The friction in the bearings produces a resisting moment of 250 lb-in. After the power is cut off, the rotor slows to 180 rpm in 150 s. What was the original speed of the rotor?   *Ans.* 273 rpm

18.71. A table fan with a mass m is at rest on a horizontal tabletop. The coefficient of friction between the fan and the top is μ. Let A be the area of the air drawn through the fan and δ the mass density of the air. If v is the velocity of the downstream air, derive an expression for the theoretical maximum velocity before there is slip of the fan on the tabletop.   *Ans.* $v = (\sqrt{\mu mg})/A\delta$

18.72. Water issues at the rate of 60 ft/s from a 4-in-diameter nozzle. Determine the reaction of the nozzle against its supports.   *Ans.* 610 lb

18.73. A jet of water 50 mm in diameter exerts a force of 1200 N on a flat vane perpendicular to the stream. What is the nozzle speed of the jet?   *Ans.* 24.7 m/s

18.74. A jet of water issues from a 25-mm-diameter nozzle at 30 m/s. Determine the total force against a circularly curved vane where the direction of the jet changes 45°. Assume no friction.
Ans. 338 N.

18.75. A jet of water leaves a nozzle at a speed of 30 ft/s. It impinges at right angles against a vertical plate, which is stopped from moving by a horizontal force of 85 lb. What is the diameter of the nozzle?
Ans. 2.99 in

18.76. The force on a curved vane that changes the direction of a jet of water by 45° is 450 lb. If the nozzle speed of the jet is 80 ft/s, calculate the diameter of the nozzle.   *Ans.* 2.95 in

18.77. Refer to Fig. 18-46. A stream of water 2700 mm^2 in cross section and moving horizontally with a speed of 30 m/s splits into two equal parts against the fixed blade. Assuming no friction between the water and the blade, determine the force of the water on the blade.   *Ans.* $P_x = 4150$ N to the right

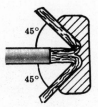

Fig. 18-46

18.78. In the preceding problem, assume that the stream splits so that two-thirds of the water moves to the top portion of the blade and one-third to the lower portion of the blade. Determine the reaction of the water on the blade.   *Ans.* $P_x = 4150$ N to the right, $P_y = 573$ N up

18.79. A nozzle with a 1700-mm^2 cross-sectional area discharges a horizontal stream of water with a speed of 25 m/s against a fixed vertical plate as shown in Fig. 18-47. What is the force against the plate?
Ans. 1060 N to the right

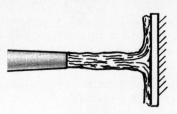

Fig. 18-47

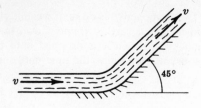

Fig. 18-48

18.80. A stream of water with a 3-in² cross-sectional area moving at 30 ft/s strikes a stationary curved blade as shown in Fig. 18-48. Assuming that friction is negligible, find the horizontal and vertical components of the force of the blade on the stream. *Ans.* $P_x = 10.7$ lb to the left, $P_y = 25.7$ lb up

18.81. A horizontal jet of water issues left to right from a 4-in-diameter nozzle at a rate of 100 ft/s. Determine the force exerted on the curved vane if the direction of the stream is changed through 120°.
Ans. $F_x = 2540$ lb to the left, $F_y = 1460$ lb up

18.82. Two homogeneous sliding disks A and B are mounted on the same shaft. A is a 50-kg disk 1000 mm in diameter, 50 mm thick, and at rest. B is a 100-kg disk 1000 mm in diameter, 100 mm thick, and rotating clockwise at 600 rpm. If the disks are engaged so as to rotate together, what is their common angular speed? *Ans.* 400 rpm

18.83. In Problem 18.82, what is the percent loss is kinetic energy of the system because the disks are coupled together? *Ans.* 33.3%

18.84. A 70-lb child is standing in a 100-lb boat that is initially at rest. If the child jumps horizontally with a speed of 6 ft/s relative to the boat, determine the speed of the boat. *Ans.* 2.47 ft/s

18.85. After a flood, an 18-kg goat finds itself adrift on one end of a 25-kg log 2 m long. As the other end touches shore, the goat maneuvers to that end. When it gets there, how far is it from shore? Assume that the log is at right angles to the shore and the water is almost calm after the storm.
Ans. 0.837 m

18.86. A 0.2-lb bullet is fired from a 14-lb rifle with a muzzle speed of 1000 ft/s. What is the speed of the rifle recoil? *Ans.* 14.3 ft/s

18.87. A 50-Mg gun fires a 500-kg projectile. If the recoil apparatus exerts a constant force of 400 kN and the gun moves back 200 mm, calculate the muzzle speed of the projectile. *Ans.* 179 m/s

18.88. A 600-lb projectile is shot with an initial velocity of 2000 ft/s from a 200,000-lb gun. What is the velocity of recoil of the gun? *Ans.* $v_R = 6$ ft/s backward

18.89. In Fig. 18-49, a series of n identical balls is shown on a smooth horizontal surface. If number 1 moves horizontally with speed u into number 2, which in turn collides with number 3, etc., and if the coefficient of restititution for each impact is e, determine the speed of the nth ball. *Ans.* $v_n = (1 + e)^{n-1} u / 2^{n-1}$

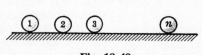

Fig. 18-49

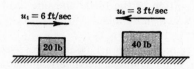

Fig. 18-50

18.90. In Fig. 18-50, the 20-lb block is moving to the right with a speed of 6 ft/s. The 40-lb block is moving to the left with a speed of 3 ft/s. If the coefficient of restitution is assumed to be 0.4, determine the speeds immediately after impact. *Ans.* $v_{20} = 2.4$ ft/s to the left, $v_{40} = 1.2$ ft/s to the right

18.91. A 3-kg ball and a 5-kg ball of the same diameter move on a smooth horizontal plane along a straight line with speeds of $+5$ and -3 m/s, respectively. Determine their speeds after impact if the impact is (a) inelastic, (b) elastic, and (c) such that the coefficient of restitution is 0.4. *Ans.* (a) 0, 0; (b) -5, $+3$; (c) -2, -1.2 m/s

18.92. A 20-kg block A moving 12 m/s horizontally to the right meets a 16-kg block B moving 8 m/s horizontally to the left. If the coefficient of restitution is $e = 0.7$, determine the speeds of A and B immediately after impact. *Ans.* $v_a = 3.11$ m/s to the left, $v_B = 10.9$ m/s to the right

18.93. A 2000-lb car traveling at 30 mi/h overtakes a 1500-lb car traveling at 15 mi/h in the same direction. What is their common speed after coupling? What is the loss in kinetic energy?
Ans. 34.6 ft/s, 6300 ft-lb

18.94. A 30-kg ball (A) moving to the right with a speed of 30 m/s strikes squarely a 10-kg ball (B) that has a speed of 7 m/s in the opposite direction. If the coefficient of restitution is 0.6, determine the speeds of the balls after impact. *Ans.* $v_A = 15.2$ m/s to the right, $v_B = 37.4$ m/s to the right

18.95. A ball falls freely from rest 5 m above a smooth plane angled at 30° to the horizontal. If $e = 0.5$, to what height will the ball rebound? *Ans.* $h = 0.08$ m

18.96. A ball falls 6 m from rest. It hits a horizontal plane and rebounds to a height of 5 m. Determine the coefficient of restitution. *Ans.* $e = 0.91$

18.97. A glass ball is dropped onto a smooth horizontal floor, from which it bounces to a height of 9 m. On the second bounce, it attains a height of 6 m. What is the coefficient of restitution between the glass and the floor? *Ans.* $e = 0.82$

18.98. A 4-lb weight falls 0.5 ft onto a 2-lb platform mounted on springs whose combined $k = 50$ lb-ft. If the impact is fully plastic ($e = 0$), determine the maximum distance the platform moves down from its initial position. See Fig. 18-51. *Ans.* 3.90 in

18.99. In Fig. 18-52, the mass M is moving with speed u when the string becomes taut. What will be the speed of mass m if the coefficient of restitution is e. The masses are on a smooth horizontal plane.
Ans. $v = Mu(1 + e)/(M + m)$

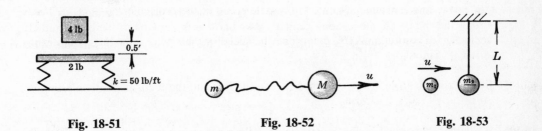

Fig. 18-51   Fig. 18-52   Fig. 18-53

18.100. A mass m_1 moving with speed u strikes a stationary mass m_2 hanging on a string of length L as shown in Fig. 18-53. If the coefficient of restitution is e, determine (a) the speed of each mass immediately after the impact and (b) the height h to which mass m_2 will rise.
Ans. (a) $v_1 = u\dfrac{m_1 - em_2}{m_1 - m_2}$, $v_2 = u\dfrac{m_1(1 + e)}{m_1 + m_2}$; (b) $h = \dfrac{u^2 m_1^2(1 + e)^2}{(m_1 + m_2)^2 2g}$

18.101. A 1-kg ball traverses the frictionless tube shown in Fig. 18-54, falling a vertical distance of 900 mm. It then strikes a 1-kg ball hanging on a 900-mm cord. Determine the height to which the hanging ball will rise (*a*) if the collision is perfectly elastic, (*b*) if the coefficient of restitution is 0.7.
Ans.   (*a*) 900 mm, (*b*) 650 mm

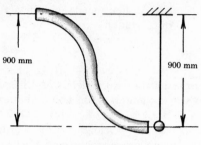

Fig. 18-54

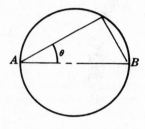

Fig. 18-55

18.102. Refer to Fig. 18-55. A ball thrown from position *A* against a smooth vertical circular wall rebounds and hits position *B* at the other end of the diameter through *A*. Show that the coefficient of restitution is equal to the square of the tangent of angle *θ*.

18.103. A billiard ball moving at 4 m/s strikes a smooth horizontal plane at an angle of 35° as shown in Fig 18-56. If the coefficient of restitution is 0.6, what is the velocity with which the ball rebounds?
Ans. $v = 3.55$ m/s, $\theta = 22.8°$

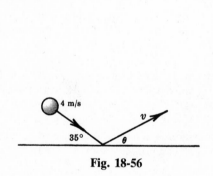

Fig. 18-56

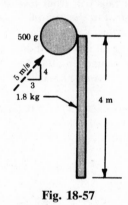

Fig. 18-57

18.104. The 500-g sphere shown in Fig. 18-57 is moving on a smooth horizontal plane with a velocity of 5 m/s. It strikes the end of a homogeneous slender 1.8-kg bar that is 4 m long. If the bar is initially at rest and the coefficient of restitution is 0.6, determine the speed of the sphere immediately after impact.
Ans.   4.06 m/s

18.105. Solve Problem 18.46 if the collision between the box and the tile is perfectly elastic.
Ans. $\omega = 4$ rad/s clockwise

18.106. A slender, horizontal homogeneous bar is falling downward at 10 ft/s when its right-hand end strikes the edge of a table. If the coefficient of restitution between the table and the bar is 0.45, determine the angular velocity of the bar immediately after impact. The bar is 3 ft long and weighs 3 lb.
Ans. $\omega = 7.25$ rad/s counterclockwise

18.107. In Fig. 18-58, a large drum is filled with liquid weighing 50 lb/ft³. The drum and liquid weigh 200 lb and are at rest on a horizontal sheet of ice for which the coefficient of friction is 0.05. If a 4-in plug 3 ft below the surface of the liquid is suddenly removed, will the drum move?   *Ans.*   Yes

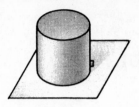

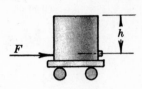

Fig. 18-58   **Fig. 18-59**

18.108. A tank of liquid with a mass density of δ rests on a cart. If h is the height of the fluid above the orifice, what horizontal force F is necessary to hold the tank at rest when the fluid starts issuing from the nozzle? The cross-sectional area of the orifice is A. See Fig. 18-59.   *Ans.* $F = 2Ah\delta g$

18.109. A rocket has a mass of 3000 kg empty. If it is projected vertically up from the earth with a fuel load of 7000 kg, calculate the initial acceleration. Assume that the gases exhaust at 2000 m/s relative to the rocket and that the initial rate of fuel burning is 150 kg/s.   *Ans.* $a = 20.2\ \text{m/s}^2$

18.110. The rotor of a gyroscope is a homogeneous 4-in-diameter cylinder weighing 6 oz. It is mounted horizontally midway between bearings 6 in apart. The rotor is turning at 9000 rpm in a clockwise direction when viewed from the rear. The assembly is turning at 2 rad/s about a vertical axis in a clockwise direction when viewed from above. What are the bearing reactions on the rotor shaft?
Ans. $R_F = 12.8$ oz up, $R_R = 6.8$ oz down

18.111. Refer to Fig. 18-60. A solid 6-in-diameter wheel that is 2 in thick rotates at 6000 rpm. Neglect the mass of the shaft attached to it. Assume a weight density of 480 lb/ft³. Determine the speed of precession.   *Ans.* $\omega_p = 1.10$ rad/s clockwise about the y axis when viewed from above.

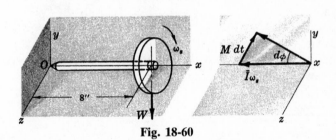

Fig. 18-60

18.112. A space shuttle is to link up with a space laboratory orbiting at a constant altitude of 400 km. At an altitude of 80 km and speed of 4000 m/s, the shuttle engine is shut off. What must be the angle that the velocity makes with the vertical, at shut-off, if the shuttle is to arrive tangent to the orbit of the laboratory?   *Ans.* 56.8°

18.113. A satellite is launched parallel to the surface of the earth at an altitude of 400 mi. The speed of launch is 20,500 mi/h. Determine the maximum altitude reached by the satellite.   *Ans.* 8420 mi

18.114. Repeat Problem 18.113 if the satellite is launched outward at 60° from the vertical.   *Ans.* 1470 mi

Chapter 19

Mechanical Vibrations

19.1 DEFINITIONS

Mechanical vibration of a system possessing masses and elasticity is motion about an equilibrium position which repeats itself in a definite time interval.

The *period* is the time interval for the vibration to repeat itself.

A *cycle* is each repetition of the entire motion completed during the period.

The *frequency* is the number of cycles in a unit of time.

Free vibrations occur in a system not acted upon by periodic external disturbing forces.

Natural frequency is the frequency of a system undergoing free vibrations.

Forced vibrations occur in a system acted upon by periodic external disturbing forces.

Resonance occurs when the frequency of the forced vibrations coincides with or at least approaches the natural frequency of the system.

Transient vibrations disappear with time. Free vibrations are transient in character.

Steady-state vibrations continue to repeat themselves with time. Forced vibrations are examples of steady-state vibrations.

19.2 DEGREES OF FREEDOM

The degrees of freedom of a system depend on the number of variables (coordinates) needed to describe its motion.

For example, in Fig. 19-1(*a*), the motion of the mass on a spring that is assumed to vibrate only in a vertical line can be described with one coordinate, and thus possesses a single degree of freedom. A bar supported by the two springs in Fig. 19-1(*b*) needs two variables as shown, and therefore possesses two degrees of freedom.

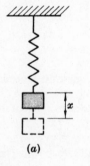

(*a*)

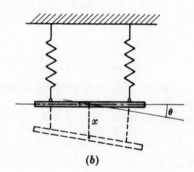

(*b*)

Fig. 19-1

19.3 SIMPLE HARMONIC MOTION

Simple harmonic motion can, as shown in Section 12.2, be represented by a sine or cosine function of time. Thus, $x = X \sin \omega t$ is an equation of simple harmonic motion. It could represent the projection on a diameter of a vector of length X as the tip of the vector X rotates on a circular path with a constant angular velocity ω radians per second. For this motion,

$$x = \text{length of projection}$$
$$X = \text{length of rotating vector}$$
$$\omega = \text{circular frequency in rad/s}$$
$$\tau = 2\pi/\omega = \text{period in seconds}$$
$$f = \omega/2\pi = \text{frequency in cycles per second (Hz)}$$

All bodies vibrate with simple harmonic motion or some combination of simple harmonic motions with different frequencies and amplitudes.

19.4 MULTICOMPONENT SYSTEMS

A multicomponent system is analyzed by replacing it with an equivalent system of masses, springs, and damping devices. The differential equations of this idealized system, when solved, will approximate the desired results. Engineering judgment will suggest proper modifications to fit the actual system.

The problems that follow illustrate free vibrations with and without damping, and forced vibrations with and without damping. Only viscous damping (where the damping force is proportional to the velocity of the body) is considered. However, it is well to point out two other types of damping: (1) Coulomb damping, which is independent of the velocity and arises because of sliding of the body on dry surfaces (its force is thus proportional to the normal force between the body and the surface on which it slides), and (2) solid damping, which occurs as internal friction within the material of the body itself (its force is independent of the frequency and proportional to the maximum stress induced in the body itself).

19.5 UNITS

The following is a table of the U.S. Customary and SI units used in mechanical vibrations problems.

Quantity	U.S. customary	SI
Length	in	m
Time	s	s
Velocity	ft/s or in/s	m/s
Acceleration	ft/s^2 or in/s^2	m/s^2
Mass	slug	kg
Mass moment of inertia	slug-ft^2 or lb-s^2-in	kg · m^2
Force	lb	N
Spring	lb/in	N/mm
Damping constant	lb-s/in	N · s/m

Solved Problems

Free Vibrations—Linear

19.1.　A mass m hangs on a vertical spring whose spring constant or modulus is k. Assuming that the mass of the spring may be neglected, study the motion of the mass if it is released at a distance x_0 below the equilibrium position with an initial velocity v_0 downward.

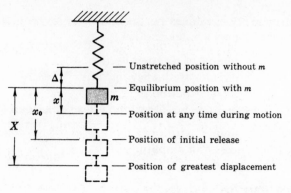

Fig. 19-2

ANALYSIS

In Fig. 19-2, various positions of the mass m are shown. The distances are exaggerated for clarity. The value X is the amplitude of the motion. Of course, the mass will also rise to a height X above the equilibrium position.

The tension in the spring is equal to the product of the spring modulus k and the distance the spring is stretched or compressed from its unstretched position (without m). Draw free-body diagrams of the mass in its equilibrium position and at the position x below equilibrium. Note that the position x of the mass at any time is expressed from its position of static equilibrium.

In Fig. 19-3, no acceleration is shown, since the system is in equilibrium. Hence, $T = k\Delta = mg$. Note that consequently $\Delta = mg/k$.

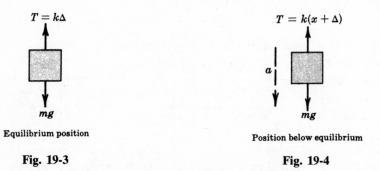

Equilibrium position   Position below equilibrium

Fig. 19-3   **Fig. 19-4**

In Fig. 19-4, assume that displacements below the equilibrium position are positive. Since the direction of the acceleration is unknown, assume it to be positive. A negative sign would then mean it is directed up. Apply Newton's laws to this free-body diagram and obtain

$$\sum F_v = ma \qquad \text{or} \qquad mg - T = ma$$

Substitute $T = k(\Delta + x)$ and $mg = k\Delta$ to get $k\Delta - k\Delta - kx = ma$. Since $a = d^2x/dt^2$, the equation of motion becomes

$$\frac{d^2x}{dt^2} + \frac{k}{m}x = 0$$

It should be apparent that whenever x is positive (below the equilibrium position), the acceleration (which is then negative) is directed up. This means that as the mass moves down to its lowest position, the acceleration is up or the mass is decelerating. Just after reaching the bottom and starting up, the displacement is still positive and the acceleration is still directed oppositely, i.e., up. Therefore the mass will accelerate up to the equilibrium position. Above this position, the displacement is negative and the acceleration is directed down to the equilibrium position. Hence, up to its top point of travel, the mass decelerates, and between the top point and the position of equilibrium it accelerates. The acceleration is always directed toward the equilibrium position.

Assume that the solution of the above second-order differential equation has the form

$$x = A \sin \omega t - B \cos \omega t$$

where ω is the circular frequency in rad/s.

To determine whether or not this is a solution, take the second derivative with respect to time (d^2x/dt^2) and substitute it into the differential equation. Note that

$$\frac{dx}{dt} = A\omega \cos \omega t - B\omega \sin \omega t \qquad \text{and} \qquad \frac{d^2x}{dt^2} = -A\omega^2 \sin \omega t - B\omega^2 \cos \omega t = -\omega^2 x$$

Substitute the value of d^2x/dt^2 into the equation of motion to obtain $-\omega^2 x + (k/m)x = 0$. Hence, ω must equal $\sqrt{k/m}$ if the assumed value of x is to be a solution. Then the solution is thus far

$$x = A \sin \sqrt{\frac{k}{m}}\,t + B \cos \sqrt{\frac{k}{m}}\,t$$

Note that a cycle of motion will be completed at intervals of 2π rad, i.e., when $(\sqrt{k/m})\tau = 2\pi$, where τ is the period or time for one cycle. Hence, $\tau = 2\pi\sqrt{m/k}$. The frequency is the inverse of the period, or

$$f = \frac{1}{\tau} = \frac{1}{2\pi} \sqrt{\frac{k}{m}}$$

As pointed out previously, $\Delta = mg/k$; hence, the above formulas may also be written $\tau = 2\pi\sqrt{\Delta/g}$ and $f = (1/2\pi)\sqrt{g/\Delta}$.

The constants of A and B should be evaluated on the basis of the initial conditions given in the problem. Here it is assumed that at $t = 0$, $x = x_0$, and $v = v_0$. Substitute these values of x and v into the equations for these variables, but be sure also to substitute the time $t = 0$:

$$x_0 = A \sin (\omega \cdot 0) + B \cos (\omega \cdot 0)$$
$$v_0 = A\omega \cos (\omega \cdot 0) - B\omega \sin (\omega \cdot 0)$$

The first equation yields $B = x_0$, and the second equation gives $A = v_0/\omega$.

Hence, the solution is

$$x = \frac{v_0}{\omega} \sin \omega t + x_0 \cos \omega t$$

The solution may be written in another form: $x = X \cos (\omega t - \phi)$, where the amplitude $X = \sqrt{(v_0/\omega)^2 + (x_0)^2}$ and the phase angle $\phi = \tan^{-1} (v_0/x_0\omega)$. Note that $\omega = \sqrt{k/m}$.

This problem is the most commonly used example of free vibrations, which theoretically would continue indefinitely after the mass is set in motion. Since only one variable has been used to describe the motion, the system possesses one degree of freedom. Many additional problems can now be solved by reducing them to this type. In other words, replace the actual elastic suspension by an equivalent spring attached to the vibrating body.

19.2. A 60-lb machine is mounted on 80-lb platform, which in turn is supported by four springs. Assume that each spring carries one-fourth of the load, and determine the period of vibration. Each spring has a spring constant of 18 lb/in.

SOLUTION

Each spring carries 35 lb, and hence the frequency is

$$f = \frac{1}{2\pi} \sqrt{\frac{kg}{W}} = \frac{1}{2\pi} \sqrt{\frac{(18)(386)}{35}} = 2.24 \text{ cps}$$

Note that g is used in in/s² because k was given in lb/in.
The period is $1/f = 0.45$ s.

19.3. Solve Problem 19.1 using the conservation of energy theorem.

SOLUTION

This theorem states that the sum of the potential energy V and kinetic energy T of the system is a constant provided the system is conservative (no friction or damping is assumed at this point in the discussion).

At any distance x below the equilibrium position, the spring tension is $mg + kx$. Hence, the potential energy V_s of the spring is equal numerically to the work done by this force in stretching the spring:

$$V_s = \int_0^x (mg + kx)\, dx = mgx + \tfrac{1}{2}kx^2$$

During this same displacement x, the mass has lost potential energy of the amount mgx. Then the total potential energy of the system is $\tfrac{1}{2}kx^2$.

At a distance x below equilibrium, the kinetic energy T of the mass, which is moving with a velocity dx/dt, is $T = \tfrac{1}{2}m(dx/dt)^2$. Since the spring is assumed to be massless, its kinetic energy is zero. Therefore the kinetic energy of the system is that of the mass only.

The conservation of energy theorem states that $T + V = $ constant, or that $d/dt(T + V) = 0$. Then

$$\frac{d}{dt}\left[\frac{1}{2}m\left(\frac{dx}{dt}\right)^2 + \frac{1}{2}kx^2\right] = 0 \qquad \text{or} \qquad \frac{1}{2}m(2)\left(\frac{dx}{dt}\right)\frac{d^2x}{dt^2} + \frac{1}{2}k2x\frac{dx}{dt} = 0$$

This reduces to the same differential equation obtained in Problem 19.1:

$$\frac{d^2x}{dt^2} + \frac{k}{m}x = 0$$

Hence, the frequency is

$$\sqrt{\frac{k}{m}}\ \text{rad/s} \qquad \text{or} \qquad \frac{1}{2\pi}\sqrt{\frac{k}{m}}\ \text{Hz}$$

Note that in a linear differential equation of the type shown above with the second-order term having a coefficient 1, the circular frequency ω in rad/s equals the square root of the coefficient of the x term. ($\omega = 2\pi f$, where ω is in rad/s and f is in hertz.)

19.4. A small mass m is fastened to a vertical wire that is under tension T as shown in Fig. 19-5. What will be the natural frequency of vibration of the mass if it is displaced laterally a slight distance and then released?

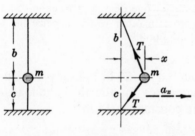

Fig. 19-5

SOLUTION

Assume the mass m at some time during the motion is at a distance x to the right of equilibrium.

In the horizontal direction, this mass is acted upon by the components of the two tensions T shown. For *small* displacements, these x components of the tensions will be Tx/b and Tx/c, both acting to the left, or negative, if we assume x to the right is positive. Hence, the differential equation of motion is

$$-\frac{Tx}{c} - \frac{Tx}{b} = ma_x = m\frac{d^2x}{dt^2} \quad \text{or} \quad \frac{d^2x}{dt^2} + \frac{1}{m}\left(\frac{T}{c} + \frac{T}{b}\right)x = 0$$

Note that this differential equation is entirely similar, except for the coefficient of the x term, to the spring equation. Hence, as indicated in Problem 19.3, the frequency is

$$f = \frac{1}{2\pi}\sqrt{\frac{T}{m}\left(\frac{b+c}{bc}\right)}\,\text{Hz}$$

19.5. A cylinder floats as shown in Fig. 19-6. The cross-sectional area of the cylinder is A and the mass is m. What will be the frequency of oscillation if the cylinder is depressed somewhat and released? The mass density of the liquid is δ. Neglect the damping effects of the liquid as well as the inertia effects of the moving liquid.

SOLUTION

When the cylinder is a distance x below its equilibrium position, it is acted upon by a buoyant force equal in magnitude to the gravitational force on the displaced liquid. This is exactly analogous to the spring in Problem 19.1. Using Newton's laws, the unbalanced force, which is up for a downward displacement, is equated to the product of the mass and its acceleration. Call a downward displacement positive. Then the differential equation of motion is

$$-\delta gAx = m\frac{d^2x}{dt^2} \quad \text{or} \quad \frac{d^2x}{dt^2} + \frac{\delta gA}{m}x = 0$$

Thus, the frequency is

$$f = \frac{1}{2\pi}\sqrt{\frac{\delta gA}{m}}\,\text{Hz}$$

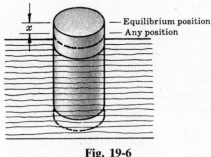

Fig. 19-6

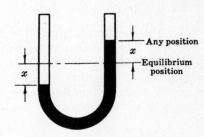

Fig. 19-7

19.6. Liquid of density δ and total length l is used in a manometer as shown in Fig. 19-7. A sudden increase in pressure on one side forces the liquid down. Upon release of the pressure, the liquid oscillates. Neglecting any frictional damping, what will be the frequency of vibration?

SOLUTION

Assume the liquid to be a distance x below the equilibrium position in the left column and, of course, a distance x above the equilibrium position in the right column.

The unbalanced force tending to restore equilibrium is the gravitational force on a column of the liquid $2x$ high. This force is $2xA\delta g$, where A is the area of the cross section of the liquid. The total mass of liquid in motion is $lA\delta$. Using Newton's law,

$$-2xA\delta g = lA\delta \frac{d^2x}{dt^2} \quad \text{or} \quad \frac{d^2x}{dt^2} + \frac{2g}{l}x = 0$$

and

$$f = \frac{1}{2\pi}\sqrt{\frac{2g}{l}} \text{ Hz}$$

Note that f is independent of the density of the liquid and the cross-sectional area. The units for g and l must be in SI units or both be in U.S. Customary units.

19.7. Determine the natural undamped frequency of the system shown in Fig. 19-8. The bar is assumed massless and the lower spring is attached to the bar midway between the points of attachment of the upper springs. The spring constants are as shown.

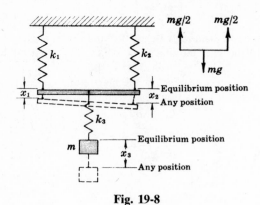

Fig. 19-8

SOLUTION

The total static displacement Δ of the mass m is equal to the stretch x_3 of spring k_3 plus the average of x_1 and x_2, which are the extensions of springs k_1 and k_2, respectively.

This problem may be solved by substituting this value of Δ into the equation $f = (1/2\pi)\sqrt{g/\Delta}$ (refer to Problem 19.1). Note that the entire gravitational force mg is transferred through spring k_3 whereas $\frac{1}{2}mg$ is transferred through each spring k_1 and k_2 to the support. The latter fact can be ascertained by referring to the free-body diagram of the bar shown to the right of the figure. Hence,

$$x_3 = \frac{mg}{k_3} \qquad x_2 = \frac{\frac{1}{2}mg}{k_2} \qquad x_1 = \frac{\frac{1}{2}mg}{k_1}$$

and

$$\Delta = x_3 + \tfrac{1}{2}(x_2 + x_1) = mg\left(\frac{4k_1k_2 + k_1k_3 + k_2k_3}{4k_1k_2k_3}\right)$$

Substituting,

$$f = \frac{1}{2\pi}\sqrt{\frac{4k_1k_2k_3}{m(4k_1k_2 + k_1k_3 + k_2k_3)}}$$

19.8. In Fig. 19-9, the mass m is suspended by means of spring k_2 from the end of a rigid massless beam that is of length l and attached to the frame at its left end. It is also supported in a horizontal position by a spring k_1 attached to the frame as shown. What is the natural frequency f of the system?

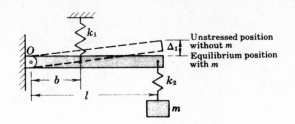
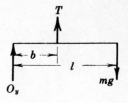

Fig. 19-9

SOLUTION

The horizontal position is the state of equilibrium. Determine the static displacement Δ_1 of the end of the beam from its original position when no load is impressed. In the free-body diagram of the beam in the equilibrium position, note that the tension in spring k_1 must be $T = (l/b)mg$ to hold the beam in equilibrium. (This can be seen by taking moments about O.) Hence, the spring k_1 elongates under this tension by an amount mgl/bk_1. If a point on the beam a distance b from the left end is displaced this amount, mgl/bk_1, the right end is displaced (using similar triangles) by an amount

$$\Delta_1 = \frac{l}{b}\frac{mgl}{bk_1} = \frac{mg}{k_1}\left(\frac{l}{b}\right)^2$$

The total static displacement Δ of m is then equal to the sum of Δ_1 and the elongation of spring k_2 in transmitting the force mg to the beam. Then

$$\Delta = \frac{mg}{k_2} + \left(\frac{l}{b}\right)^2\frac{mg}{k_1} = mg\left(\frac{k_1 + (l/b)^2k_2}{k_1k_2}\right)$$

and

$$f = \frac{1}{2\pi}\sqrt{\frac{g}{\Delta}} = \frac{1}{2\pi}\sqrt{\frac{k_1k_2}{m[k_1 + (l/b)^2k_2]}}\,\text{Hz}$$

Free Vibrations—Angular

19.9. What will be the natural frequency of the system shown in Fig. 19-10(a) for small displacements?

SOLUTION

In the equilibrium position shown in Fig. 19-10(a), the spring has been stretched a distance equal to $b\theta_0$. In any other position during the motion such as that shown in Fig. 19-10(b), the cylinder is displaced through an additional angle θ and the spring is stretched a total distance of $b(\theta + \theta_0)$. Note that the angular displacements are assumed small. Also assume clockwise angular displacements are positive. The acceleration of the mass m is $a = r\alpha$.

The equations of motion for the mass and the cylinder are, respectively,

$$\sum F = ma \quad \text{or} \quad mg - T_2 = mr\alpha \tag{1}$$

$$\sum M_O = J_O\alpha \quad \text{or} \quad T_2r - bk(\theta + \theta_0)b = J_O\alpha \tag{2}$$

where J_O is the centroidal polar moment of inertia. Substitute the value of T_2 from equation (1) into equation (2) to obtain

$$(mg - mr\alpha)r - b^2k\theta - b^2k\theta_0 = J_O\alpha$$

But in the equilibrium position with P directly above O, $\sum M_O = 0$; this means that $(k\theta_0b)b = mgr$. (Note that $T_2 = mg$ when the system is in equilibrium.) The equation of motion for the cylinder becomes

$$(J_O + mr^2)\frac{d^2\theta}{dt^2} + b^2k\theta = 0, \quad \text{from which} \quad f = \frac{1}{2\pi}\sqrt{\frac{b^2k}{J_O + mr^2}}\,\text{Hz}$$

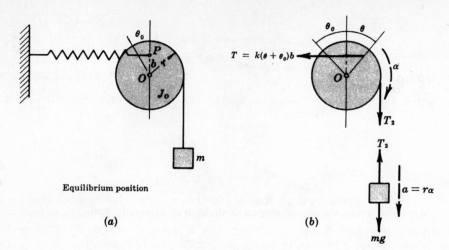

Fig. 19-10

19.10. A circular homogeneous plate with radius R and weight W is supported by three cords of length l, equally spaced in a circle, as shown in Fig. 19-11(a). Determine the natural frequency of oscillations for small displacements of the plate about its vertical centerline.

SOLUTION

The tension in each cord is $\frac{1}{3}W$. If the plate rotates through a *small* angular displacement θ, the bottom of each cord rotates through a distance $r\theta$ as shown in Fig. 19-11(b). The horizontal component of the tension is approximately $\frac{1}{3}W \sin\beta = \frac{1}{3}W(r\theta/l)$. Each of the three tensions supplies a resisting moment about the vertical centerline equal to $\frac{1}{3}W(r\theta/l)r$.

The moment equation about the vertical centerline becomes

$$-3\left(\frac{W}{3}\right)\left(\frac{r\theta}{l}\right)r = \frac{1}{2}\left(\frac{W}{g}\right)(R^2)\frac{d^2\theta}{dt^2}$$

This reduces to

$$\frac{d^2\theta}{dt^2} + \frac{2g}{l}\left(\frac{r^2}{R^2}\right)\theta = 0$$

and

$$f = \frac{1}{2\pi}\left(\frac{r}{R}\right)\sqrt{\frac{2g}{l}} \text{ cps}$$

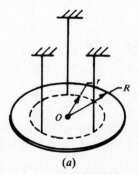

(a)   (b)

Fig. 19-11

19.11. A homogeneous steel disk 200 mm in diameter and 50 mm thick is rigidly attached to a vertical steel wire 2 mm in diameter and 900 mm long. What is the natural frequency of the system?

SOLUTION

When a wire is twisted through an angle θ, there is a restoring torque $K\theta$. The equation of motion for the disk becomes $-K\theta = J_O\alpha$. Hence,

$$\frac{d^2\theta}{dt^2} + \frac{K}{J_O}\theta = 0$$

and

$$f = \frac{1}{2\pi}\sqrt{\frac{K}{J_O}}\,\text{Hz}$$

According to the theory developed in strength of materials, the angle of twist for the wire is

$$\theta = \frac{Tl}{\frac{1}{32}\pi d^4 G}$$

where T = torque in N · m
$\quad\quad l$ = length of wire in meters
$\quad\quad G$ = shear modulus of elasticity (80 GPa for steel)
$\quad\quad d$ = diameter in meters

Hence, the torsional constant is

$$K = \frac{T}{\theta} = \frac{\pi d^4 G}{32l} = \frac{\pi(0.002)^4(80 \times 10^9)}{32(0.9)} = 0.14\,\text{N} \cdot \text{m/rad}$$

The moment of inertia of the disk is

$$J_O = \tfrac{1}{2}mr^2 = \tfrac{1}{2}(\tfrac{1}{4}\pi)(0.2)^2(0.05)(7850)(0.1)^2 = 0.062\,\text{kg} \cdot \text{m}^2$$

Note that the density of steel used was $7850\,\text{kg/m}^3$.

Finally,

$$f = \frac{1}{2\pi}\sqrt{\frac{K}{J_O}} = \frac{1}{2\pi}\sqrt{\frac{0.14}{0.062}} = 0.24\,\text{Hz}$$

19.12. Discuss the motion of two heavy masses with moments of inertia J_1 and J_2 and connected by a shaft of small diameter d, as shown in Fig. 19-12.

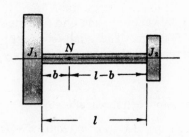

Fig. 19-12

ANALYSIS

If one mass is held and the other rotated and then both released, the system will oscillate. Since no external torques are assumed to act on the system, the *angular momentum* of the system is conserved. Thus, $J_1\omega_1 + J_2\omega_2 = 0$; hence, $\omega_2 = -(J_1/J_2)/\omega_1$.

Note that ω_1 and ω_2 represent the varying angular velocities of J_1 and J_2 during a complete cycle for each mass.

Since the previous equation indicates that the masses always are rotating in opposite directions, there is a section of the shaft that is always at rest. This nodal section can be used to study the motion of the masses, since each can be treated as a torsional pendulum (see Problem 19.11).

The time for one mass to complete a cycle must equal the time for the other to complete a cycle. If they differed, one mass would eventually rotate in the same direction as the other. But the above equation indicates that the masses always rotate in opposite directions. Since the periods are equal, the number of cycles per second for each must be equal. Thus,

$$f = \frac{1}{2\pi}\sqrt{\frac{K_1}{J_1}} = \frac{1}{2\pi}\sqrt{\frac{K_2}{J_2}}$$

where K_1 and K_2 are the torsional spring constants of the parts of the shaft from the nodal section N to each end. Hence, those constants are related by $K_1/K_2 = J_1/J_2$.

For a cylindrical shaft, $K = \pi d^4 G/32l$, where d is its diameter, G is the shearing modulus of elasticity, and l is its length (see Problem 19.11). Assume the nodal section is a distance b from the J_1 mass; then

$$K_1 = \frac{\pi d^4 G}{32b} \quad \text{and} \quad K_2 = \frac{\pi d^4 G}{32(l-b)}$$

Hence, $K_1/K_2 = (l-b)/b = J_1/J_2$, from which $b = J_2 l/(J_1 + J_2)$. This locates the nodal section.

For the left portion,

$$f = \frac{1}{2\pi}\sqrt{\frac{K_1}{J_1}}$$

where

$$K_1 = \frac{\pi d^4 G}{32b} = \frac{\pi d^4 G(J_1 + J_2)}{32l J_2}$$

Then

$$f = \frac{1}{2\pi}\sqrt{\frac{\pi d^4 G(J_1 + J_2)}{32l J_1 J_2}}$$

Since the polar moment of inertia of the area of a circle is $J = \frac{1}{32}\pi d^4$, the above expression may be written

$$f = \frac{1}{2\pi}\sqrt{\frac{JG(J_1 + J_2)}{J_1 J_2 l}}$$

Note that J_1, J_2 refer to the *mass* polar moments of inertia of the cylinders and J refers to the *area* polar moment of inertia of a cross section of the shaft.

19.13. An engine has a 120-kg flywheel at each end of a steel shaft. Assume that each flywheel has a radius of gyration of 300 mm. The shaft connecting the two is 600 mm long and has a diameter of 50 mm. Determine the natural frequency of torsional oscillation.

SOLUTION

The moment of inertia for each flywheel is

$$J_f = mr^2 = 120(0.3)^2 = 10.8 \text{ kg} \cdot \text{m}^2$$

The moment of inertia for the cross-sectional area of the shaft is

$$J = \tfrac{1}{32}\pi(0.05)^4 = 6.14 \times 10^{-7} \text{ m}^4$$

Using the formula from the preceding problem, we have

$$f = \frac{1}{2\pi}\sqrt{\frac{JG(J_1 + J_2)}{l J_1 J_2}} = \frac{1}{2\pi}\sqrt{\frac{6.14 \times 10^{-7}(80 \times 10^9)(10.8 + 10.8)}{0.6(10.8)(10.8)}}$$

$$= 19.6 \text{ Hz}$$

19.14. A 2-in-diameter steel shaft 15 in long is attached at one end to a flywheel weighing 300 lb with a radius of gyration of 6 in and at the other end to a rotor that weighs 100 lb and has a radius of gyration of 4 in. Where is the nodal section and what is the natural frequency of torsional oscillation?

SOLUTION

Moment of inertia of the flywheel is $J_f = (W/g)r^2 = (300/386)(6)^2 = 28.0$ lb-s^2-in.
Moment of inertia of the rotor is $J_r = (W/g)r^2 = (100/386)(4)^2 = 4.15$ lb-s^2-in.
Moment of inertia of the shaft cross section is $J = \frac{1}{32}\pi d^4 = \frac{1}{32}\pi(2)^4 = 1.57$ in^4.

From Problem 19.12, the distance of the nodal section from the flywheel is $b = J_r l/(J_f + J_r) = 1.94$ in.

$$\text{The frequency } f = \frac{1}{2\pi}\sqrt{\frac{JG(J_f + J_r)}{LJ_f J_r}} = \frac{1}{2\pi}\sqrt{\frac{1.57(12 \times 10^6)(28.0 + 4.15)}{15(4.15)(28.0)}} = 94.1 \text{ cps}$$

19.15. Figure 19-13(a) shows a 10-lb homogeneous bar pivoted at its left end. The right end is supported by a weightless cylinder, which is floating as shown in water. The cylinder has a cross-sectional area of 2 in^2. Neglecting the damping effect of the water and the inertia effect of the moving water, determine the frequency of oscillation if the bar is depressed slightly from its horizontal (equilibrium) position.

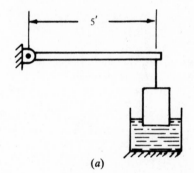

(a)

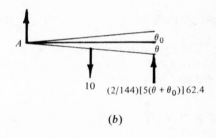

(b)

Fig. 19-13

SOLUTION

In Fig. 19-13(b), the equilibrium position is shown in the horizontal phase. The force at the right end is equal to the weight of the displaced water. The moment equation for any angle θ is

$$\sum M_A = I_A \frac{d^2\theta}{dt^2}$$

or

$$10 \times 2.5 - \left(\frac{2}{144}\right)[5(\theta + \theta_0)]62.4 \times 5 = \frac{1}{3}\left(\frac{10}{g}\right)(5)^2\frac{d^2\theta}{dt^2}$$

However, in the equilibrium phase, the moments about A equal zero, and hence

$$10 \times 2.5 - \left(\frac{2}{144}\right)5\theta_0 \times 62.4 \times 5 = 0$$

The equation of motion is

$$-\left(\frac{2}{144}\right)5\theta \times 62.4 \times 5 = 2.59\frac{d^2\theta}{dt^2}$$

which reduces to

$$\frac{d^2\theta}{dt^2} + 8.37\theta = 0$$

Hence,

$$\omega_n = \sqrt{8.37} = 2.89 \text{ rad/s}$$

19.16. A reed-type tachometer is composed of small cantilever beams with weights attached to their free ends. If the vibration frequency of a disturbing force corresponds to the natural vibration frequency of one of the reeds, it will vibrate. Since each reed is calibrated, it is possible to determine the frequency of the disturbance immediately. What size weight W should be placed on the free end of a spring steel reed 0.05 in thick, 0.20 in wide, and 4.00 in long so that its natural frequency is 50 Hz? See Fig. 19-14.

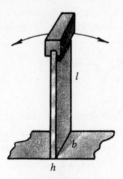

Fig. 19-14

SOLUTION

By the use of theory developed in strength of materials, the static deflection of a cantilever beam due to a concentrated mass m on the free end is $\Delta = mgl^3/3EI$, where E is the tensile (or compressive) modulus of elasticity and I is the moment of inertia of the cross-sectional area about the neutral axis. Hence,

$$f = \frac{1}{2\pi}\sqrt{\frac{3EI}{ml^3}} = \frac{1}{2\pi}\sqrt{\frac{3EIg}{Wl^3}}\text{ Hz}$$

where
$f = 50\text{ cps}$
$g = 386\text{ in/s}^2$
$E = 30 \times 10^6\text{ lb/in}^2$
$l = 4.00\text{ in}$
$I = \frac{1}{12}bh^3 = \frac{1}{12}(0.20)(0.05)^3 = 2.08 \times 10^{-6}\text{ in}^4$

Substituting values in the above equation, we find $W = 0.011\text{ lb}$.

Free Vibrations—Plane Motion

19.17. A homogeneous disk of mass m and radius r rolls without slipping on a horizontal plane. It is attached to a wall by a spring of modulus k. If the disk is displaced to the right and released, derive the differential equation of motion and determine the frequency of oscillation. See Fig. 19-15.

SOLUTION

Figure 19-15(b) shows the free-body diagram of the disk. The equations of motion are

$$\sum F_h = -F - k\bar{x} = m\ddot{\bar{x}}$$

$$\sum M_G = rF = \tfrac{1}{2}mr^2\ddot{\theta}$$

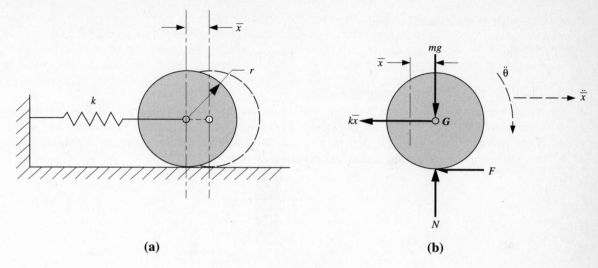

(a)   **(b)**

Fig. 19-15

For a non-slip rolling wheel, $\ddot{x} = r\ddot{\theta}$.

Adding the equations of motion yields

$$-k\bar{x} = m\ddot{\bar{x}} + \tfrac{1}{2}m\ddot{\bar{x}}$$

Or

$$\ddot{\bar{x}} + \frac{2}{3}\left(\frac{k}{m}\right)\bar{x} = 0$$

and

$$f = \frac{1}{2\pi}\sqrt{\frac{2k}{3m}}\,\text{Hz}$$

Alternatively, the conservation of energy theorem can be used.

Consider the center of mass G to be displaced $\bar{x}$ from the equilibrium position. The potential energy V of the spring is given by

$$V = \int k\bar{x}\,dx = \tfrac{1}{2}k\bar{x}^2$$

The kinetic energy of the rolling disk is

$$T = \tfrac{1}{2}m\dot{\bar{x}}^2 + \tfrac{1}{2}I_G\dot{\theta}^2$$

Because energy is conserved, $T + V = $ constant. Or

$$\frac{d(T + V)}{dt} = 0$$

$$\frac{d}{dt}\left[\tfrac{1}{2}m\dot{\bar{x}}^2 + \tfrac{1}{2}I_G\dot{\theta}^2 + \tfrac{1}{2}k(\bar{x})^2\right] = 0$$

$$m\dot{\bar{x}}\ddot{\bar{x}} + I_G\dot{\theta}\ddot{\theta} + k\bar{x}\dot{\bar{x}} = 0$$

But

$$\bar{x} = r\theta \quad \dot{\bar{x}} = r\dot{\theta},\ \ddot{\bar{x}} = r\ddot{\theta} \quad I_G = \tfrac{1}{2}mr^2$$

from which,

$$m\dot{\bar{x}}\ddot{\bar{x}} + \tfrac{1}{2}mr^2\left(\frac{\dot{\bar{x}}}{r}\right)\left(\frac{\ddot{\bar{x}}}{r}\right) + k\bar{x}\dot{\bar{x}} = 0$$

Simplifying yields

$$\ddot{\bar{x}} + \frac{2}{3}\left(\frac{k}{m}\right)\bar{x} = 0$$

The result in equation (1) above

Hence,
$$f = \frac{1}{2\pi} \sqrt{\frac{2k}{3m}} \, \text{Hz}$$

Free Vibrations with Viscous Damping

19.18. A mass m suspended from a spring whose modulus is k is subjected to viscous damping represented in Fig. 19-16 as occurring by means of a dashpot. This damping force is proportional to the velocity; that is, $F = c(dx/dt)$, where c is the damping constant. Discuss the motion as the damping coefficient c varies.

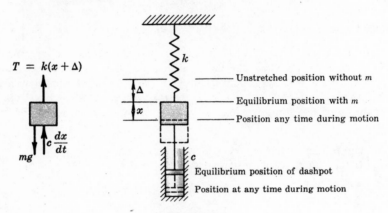

Fig. 19-16

ANALYSIS

Figure 19-16 indicates the essential data. The free-body diagram to the left illustrates all the forces acting on the mass when displaced a distance x below equilibrium and traveling down.

Note, as before, that in the equilibrium position $k\Delta = mg$. Note also that the damping force $c(dx/dt)$ opposes motion. The equation of motion ($\sum F = ma$ with an assumed position down) becomes

$$mg - k(x + \Delta) - c\frac{dx}{dt} = m\frac{d^2x}{dt^2} \quad \text{or} \quad \frac{d^2x}{dt^2} + \frac{c}{m}\frac{dx}{dt} + \frac{k}{m}x = 0$$

Assume a solution of this differential equation in the form $x = Ae^{st}$, where A and s are nonzero constants. Substitute this value into the equation (noting that $dx/dt = Ase^{st}$ and $d^2x/dt^2 = As^2e^{st}$) to obtain

$$As^2e^{st} + A\frac{c}{m}se^{st} + A\frac{k}{m}e^{st} = 0 \quad \text{or} \quad \left(s^2 + \frac{c}{m}s + \frac{k}{m}\right)e^{st} = 0$$

The desired solution must be such that the above equations are zero. Since e^{st} cannot be zero, its coefficient must be zero; that is, $s^2 + (c/m)s + k/m = 0$.

Using the quadratic formula, the two solutions for s are

$$s = \frac{-c}{2m} \pm \sqrt{\left(\frac{c}{2m}\right)^2 - \frac{k}{m}}$$

The general solution is of the form

$$x = Ae^{\left[\frac{-c}{2m} + \sqrt{\left(\frac{c}{2m}\right)^2 - \frac{k}{m}}\right]t} + Be^{\left[\frac{-c}{2m} + \sqrt{\left(\frac{c}{2m}\right)^2 - \frac{k}{m}}\right]t}$$

The radical may be real, imaginary, or zero depending on the magnitude of the damping coefficient c. The value of c that makes the radical zero is called the critical damping coefficient c_c. Its value is obtained by equating the radicand to zero, and is

$$c_c = 2m\sqrt{\frac{k}{m}} = 2m\omega_n$$

Note that ω_n is the undamped natural frequency of the system.

The ratio of the damping coefficient c in any system to the critical damping coefficient c_c is called the damping factor d. Its use simplifies the analysis of the problem.

Multiply $c/2m$ by c_c/c_c and substitute $d = c/c_c$ and $c_c = 2m\omega_n$ to get

$$\frac{c}{2m} = \frac{c}{c_c}\frac{c_c}{2m} = d\frac{2m\omega_n}{2m} = d\omega_n$$

The solution for x obtained above may be written

$$x = Ae^{(-d\omega_n + \sqrt{d^2\omega_n^2 - \omega_n^2})t} + Be^{(-d\omega_n - \sqrt{d^2\omega_n^2 - \omega_n^2})t}$$

$$= Ax^{(-d + \sqrt{d^2 - 1})\omega_n t} + Be^{(-d - \sqrt{d^2 - 1})\omega_n t}$$

Three cases arise, depending on the value of d.

Case A: Large damping $(d > 1)$ means that the radical is real and less than d. Hence, both exponents are negative. The value of x is then equal to the sum of two decreasing exponentials. When $t = 0$, $x = Ax^0 + Be^0 = A + B$. The plot of either part of the solution of such aperiodic motion indicates that the frictional resistance is so large that the mass, after its initial displacement, creeps back to equilibrium without vibrating. See Fig. 19-17. Since there is no period to the motion, it is called aperiodic.

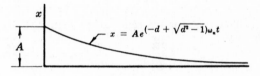

Fig. 19-17

Case B: Light damping $(d < 1)$ means that the radical is imaginary. Using $i = \sqrt{-1}$, the solution may be rewritten

$$x = Ae^{(-d + i\sqrt{1 - d^2})\omega_n t} + Be^{(-d - i\sqrt{1 - d^2})\omega_n t}$$

$$= e^{-d\omega_n t}(Ae^{i\sqrt{1 - d^2}\omega_n t} + Be^{-i\sqrt{1 - d^2}\omega_n t})$$

The term in parentheses may be expressed in terms of a sine or cosine function. When this is done, we obtain $x = Xe^{-d\omega_n t}\sin(\sqrt{1 - d^2}\,\omega_n t + \phi)$ where $X\sin\phi$ = displacement at $t = 0$ and ϕ = phase angle. Note that ω_n, d, X, and ϕ are all constants. A plot of this solution shows the sine curve with its height continuously decreasing because it is multiplied by the factor $e^{-d\omega_n t}$, which decays with time. See Fig. 19-18.

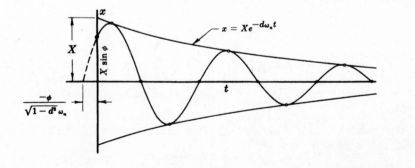

Fig. 19-18

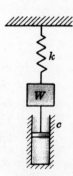

Fig. 19-19

Case C: Critical damping ($d = 1$) means that the solution must be written

$$x = (A + Bt)e^{-\omega_n t}$$

The t term is inserted with B because otherwise only one of the two solutions would be found. This method is developed in a differential equations course.

The graph here is similar to Case A.

The motion is aperiodic, but the time of return to equilibrium is a minimum when the damping is critical.

19.19. In Fig. 19-19, the weight W is suspended from a spring whose constant is 20 lb/in and is connected to a dashpot providing viscous damping. The damping force is 10 lb when the velocity of the dashpot plunger is 20 in/s. The weight of W and the plunger is 12 lb. What will be the frequency of the damped vibrations?

SOLUTION

The damping coefficient $c = \dfrac{10 \text{ lb}}{20 \text{ in/s}} = 0.5 \text{ lb-s/in}$

The natural frequency of the undamped system (let us say that the oil has been removed from the dashpot) is

$$\omega_n = \sqrt{\frac{kg}{W}} = \sqrt{\frac{20(386)}{12}} = 25.4 \text{ rad/s}$$

The critical damping coefficient according to Problem 19-18 is

$$c_c = \frac{2W}{g}\,\omega_n = \frac{2(12)}{386}\,(25.4) = 1.58 \text{ lb-s/in}$$

The damping factor $d = c/c_c = 0.5/1.58 = 0.316$, which is less than 1. According to Problem 19.18, this is light damping and oscillations will be present. The solution is of the form

$$x = Xe^{-d\omega_n t} \sin{(\sqrt{1 - d^2}\,\omega_n t + \phi)}$$

The frequency of the damped vibration ω_d is the coefficient of the time t in the sine term:

$$\omega_d = \sqrt{1 - d^2}\,\omega_n = \sqrt{1 - (0.316)^2}\,(25.4) = 24.1 \text{ rad/s}$$

Note that the period of the damped vibration is $2\pi/24.1 = 0.26$ s, and the period of the undamped system is $2\pi/25.4 = 0.25$ s.

19.20. In Problem 19.19, determine the rate of decay of the oscillations.

SOLUTION

This is conveniently expressed by introducing a new term, the logarithmic decrement δ, which is the natural logarithm of the ratio of any two successive amplitudes one cycle apart:

$$\delta = \ln{\frac{x_1}{x_2}} = \ln{\frac{Xe^{-d\omega_n t} \sin{(\sqrt{1 - d^2}\,\omega_n t + \phi)}}{Xe^{-d\omega_n(t + \tau)} \sin{[\sqrt{1 - d^2}\,\omega_n(t + \tau) + \phi]}}}$$

The numerator indicates a value of x at time t, whereas the denominator gives the value at time $t + \tau$, where τ is the period of motion. Hence, the two amplitudes occur one cycle apart (we neglect the fact that the sine curve is tangent to its envelope $Xe^{-d\omega_n t}$ at a point slightly different from the point of maximum amplitude).

Now $\sin{\sqrt{1 - d^2}\,\omega_n t + \phi)} = \sin{[\sqrt{1 - d^2}\,\omega_n(t + \tau + \phi]}$, since they are evaluated one cycle or 2π rad apart. Hence, the above expression for δ reduces to $\delta = d\omega_n\tau$.

In Problem 19.19, the period $\tau = 2\pi/\omega_d = 2\pi/24.1 = 0.261$ s.

The logarithmic decrement $\delta = d\omega_n\tau = 0.316(25.4)(0.261) = 2.095$.

The ratio of any two successive amplitudes is $e^\delta = e^{2.095} = 8.12$.

19.21. Set up the differential equation of motion for the system shown in Fig. 19-20. Determine the natural frequency of the damped oscillations.

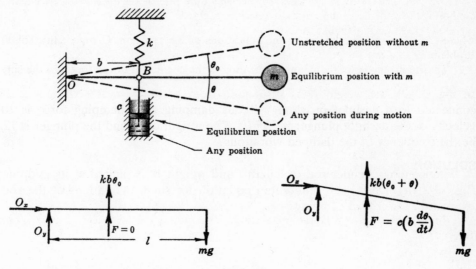

Fig. 19-20

SOLUTION

In linear vibration, we expressed the distance a body moves from equilibrium as a function of time. In rotation such as this, we are interested in expressing the angular displacement of a body in terms of time.

The free-body diagram for any phase during the motion is shown separately. Note that if the arm is moving down, the damping force opposes motion and acts up. It is equal to the product of the damping coefficient c and the velocity of the plunger in the dashpot. This is the velocity of the point B on the rod that is a distance b from O and hence has a linear velocity equal to the product of b and the angular velocity $d\theta/dt$ of the rod.

The spring force is the product of k and the total linear displacement of the spring. This displacement is that of point B, and hence equals $b(\theta_0 + \theta)$. The force is $kb(\theta_0 + \theta)$.

The equilibrium free-body diagram is also shown. Take moments about O to show $kb^2\theta_0 = mgl$. This information will simplify the differential equation of motion.

The sum of moments about O for the free-body diagram for any phase is equated to $I_O\alpha$. But, for a concentrated mass m, $I_O = ml^2$ and $\alpha = d^2\theta/dt^2$. Therefore the equation of motion, $\sum M_O = I_O\alpha$, becomes

$$+mgl - kb^2(\theta_0 + \theta) - cb^2\frac{d\theta}{dt} = ml^2\frac{d^2\theta}{dt^2}$$

But $mgl = kb^2\theta_0$. Then, simplifying,

$$\frac{d^2\theta}{dt^2} + \frac{cb^2}{ml^2}\frac{d\theta}{dt} + \frac{kb^2}{ml^2}\theta = 0$$

To solve, let $\theta = e^{st}$. Then the equation becomes (if e^{st} is a solution)

$$s^2e^{st} + \frac{cb^2}{ml^2}se^{st} + \frac{kb^2}{ml^2}e^{st} = 0 \qquad \text{or} \qquad s = \frac{-cb^2}{2ml^2} \pm \frac{1}{2}\sqrt{\frac{c^2b^4}{m^2l^4} - \frac{4kb^2}{ml^2}}$$

Critical damping occurs when the radicand is zero. Hence,

$$c_c = 2\frac{l}{b}\sqrt{mk}$$

If vibrations occur, the radicand will be negative and the solution will be of the form

$$\theta = Ce^{-(cb^2/2ml^2)t} \sin\left(\sqrt{-\frac{c^2b^4}{4m^2l^4} + \frac{kb^2}{ml^2}}\,t + \phi\right)$$

where C and ϕ are determined by the conditions of the problem. Compare this solution with that of Problem 19.18, Case B.

The frequency ω_d of the damped vibration is the coefficient of the time t in the sine term; i.e.,

$$\omega_d = \sqrt{-\frac{c^2b^4}{4m^2l^4} + \frac{kb^2}{ml^2}} = \frac{b}{l}\sqrt{\frac{k}{m} - \left(\frac{cb}{2ml}\right)^2}\ \text{rad/s}$$

19.22. A homogeneous slender rod of length l and weight W is pinned at its midpoint as shown in Fig. 19-21(a). Derive the differential equation for small oscillations of the rod. What is the expresison for critical damping?

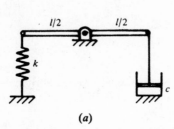

(a)   **Fig. 19-21**   (b)

SOLUTION

The free-body is shown displaced clockwise through a small angle θ in Fig. 19-21(b). The damping force and the spring tension both resist motion as shown. The moment equation about O becomes

$$-\frac{l}{2}\left(k\frac{l}{2}\theta\right) - \frac{l}{2}\left(\frac{l}{2}\frac{d\theta}{dt}c\right) = \frac{1}{12}\left(\frac{W}{g}\right)(l^2)\left(\frac{d^2\theta}{dt^2}\right)$$

This is rewritten as

$$\frac{d^2\theta}{dt^2} + \frac{3gc}{W}\left(\frac{d\theta}{dt}\right) + \frac{3kg}{W}\theta = 0$$

To solve, assume that $\theta = e^{st}$. Hence,

$$\frac{d\theta}{dt} = se^{st} \quad \text{and} \quad \frac{d^2\theta}{dt^2} = s^2e^{st}$$

Substitute to obtain

$$s^2e^{st} + \frac{3gc}{W}se^{st} + \frac{3kg}{W}e^{st} = 0$$

Since e^{st} will not be zero, this equation is satisfied if

$$s^2 + \frac{3gc}{W}s + \frac{3kg}{W} = 0$$

The solution is

$$s = \frac{-3gc}{2W} \pm \frac{1}{2} \sqrt{\left(\frac{3gc}{W}\right)^2 - 4\left(\frac{3kg}{W}\right)}$$

Critical damping occurs when the radicand in the expression for c is zero. Hence,

$$\left(\frac{3gc_c}{W}\right)^2 = \frac{12kg}{W} \qquad \text{or} \qquad c_c = \sqrt{\frac{4kW}{3g}}$$

Forced Vibrations without Damping

19.23. In Fig. 19-22, a mass m is suspended on a spring whose modulus is k. The mass is subjected to a periodic disturbing force $F \cos \omega t$. Discuss the motion.

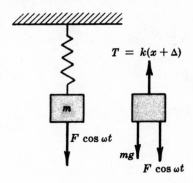

$$T = k(x + \Delta)$$

$$F \cos \omega t$$

$$mg$$

$$F \cos \omega t$$

Fig. 19-22

ANALYSIS

The differential equation now has an additional term when compared with the equation of free vibrations;

$$-kx + F \cos \omega t = m \frac{d^2x}{dt^2}$$

or

$$\frac{d^2x}{dt^2} + \frac{kx}{m} = \frac{F}{m} \cos \omega t$$

According to the theory of differential equations, the solution of this equation consists of the sum of two parts: (1) the solution previously determined in Problem 19.1 for the equation when the right-hand side is set equal to zero (the transient part) and (2) a solution to make

$$\frac{d^2x}{dt^2} + \frac{kx}{m} = \frac{F}{m} \cos \omega t$$

Assume that the second solution, which is called the steady-state solution, is of the form $x = X \cos \omega t$. Then $dx/dt = -X\omega \sin \omega t$ and $d^2x/dt^2 = -X\omega^2 \cos \omega t$.

Substituting,

$$-X\omega^2 \cos \omega t + \frac{k}{m} X \cos \omega t = \frac{F}{m} \cos \omega t$$

Hence, X must be $(F/k)/(1 - \omega^2 m/k)$.

Let Δ_F be the deflection that the force F would impart to the spring if acting on it statically; that is, $\Delta_F = F/k$. Also note that $\omega_n^2 = k/m$, where ω_n is the natural frequency when the disturbing force is absent.

Then X may be written

$$\frac{\Delta_F}{1 - (\omega/\omega_n)^2}$$

For convenience in analysis, let $\omega/\omega_n = r$; then the steady-state solution is

$$x = \frac{\Delta_F}{1 - r^2} \cos \omega t$$

Note that its frequency is the same as the disturbing frequency.

The entire solution is then

$$x = A \sin \sqrt{\frac{k}{m}} t + B \cos \sqrt{\frac{k}{m}} t + \frac{1}{1 - r^2} \Delta_F \cos \omega t$$

The first two terms, representing the free vibrations, are transient in character becausae some damping is always present to cause those vibrations to decay. Hence, study only the solution

$$x = \Delta_F \frac{1}{1 - r^2} \cos \omega t$$

Its maximum value, which occurs when $\cos \omega t = 1$, is $\Delta_F/(1 - r^2)$ and is called the amplitude. The ratio of the amplitude of the steady-state solution to the static deflection Δ_F that F would cause is called the magnification factor. Its value is

$$\frac{\Delta_F/(1 - r^2)}{\Delta_F} = \frac{1}{1 - r^2}$$

Since $r = \omega/\omega_n = f/f_n$, this can be written

$$\frac{1}{1 - (f/f_n)^2}$$

Its value can be positive or negative, depending on whether or not f is less than f_n. When $f = f_n$, resonance occurs and the amplitude is theoretically infinite. Actually, damping, which is always present, holds the amplitude to a finite amount.

A plot of the magnification factor versus the frequency ratio r is shown in Fig. 19-23.

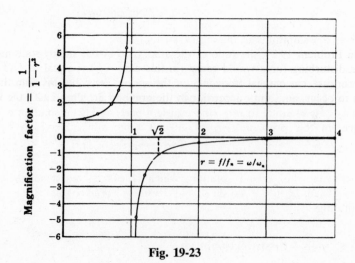

Fig. 19-23

The negative value when $r > 1$ indicates that the force F is directed one way while the displacement x is opposite.

Note that when $r = \sqrt{2}$, the magnification factor is

$$\frac{1}{1 - (\sqrt{2})^2} = -1$$

This means that if the ratio r is made greater than $\sqrt{2}$, the magnitude of the magnification factor will be less than 1. Thus, the disturbing force under these conditions will cause less movement than if it were applied statically.

19.24. A disturbing force of 9 N acts harmonically on a 5-kg mass suspended on a spring whose modulus is 6 N/mm. What will be the amplitude of excursion of the mass if the disturbing frequency is (a) 1 Hz, (b) 5.40 Hz, (c) 50 Hz?

SOLUTION

The natural frequency of the system is

$$f_n = \frac{1}{2\pi} \sqrt{\frac{k}{m}} = \frac{1}{2\pi} \sqrt{\frac{6000}{5}} = 5.51 \text{ Hz}$$

The deflection that the disturbing force would give to the spring if applied statically is $\Delta_F = 9/6 = 1.5$ mm.

(a) The frequency ratio $r = f/f_n = 1/5.51 = 0.185$; hence, the amplitude will be $\Delta_F/(1 - r^2) = 1.5/(1 - 0.0329) = 1.551$ mm.

(b) The ratio $r = 5.40/5.51$; hence, the amplitude will be 37.9 mm.

(c) The ratio $r = 50/5.51$; hence, the amplitude will be -0.018 mm. Note that in this case the amplitude is opposite to the direction in which the force is exerted but is negligible in magnitude.

19.25. A refrigerator unit weighing 60 lb is supported on three springs, each with a modulus of k lb/in. The unit operates at 600 rpm. What should be the value of k if one-twelfth of the disturbing force of the unit is to be transmitted to the supporting box?

SOLUTION

Assume that the disturbing force transmitted is proportional to the amplitude of the motion of the unit. This is logical because the supporting springs transmit forces proportional to their deformation, which equals the amplitude of the motion of the unit.

From Problem 19.23, the ratio of the amplitude of the steady-state motion to the static deflection which the disturbing force would cause (in this case $-\frac{1}{12}$) is equal to $1/(1 - r^2)$. Note that the ratio is negative because the natural frequency of the springs must be less than the disturbing frequency for a reduction to occur, and, thus, according to the graph in Problem 19.23, the magnification factor is below the line.

Hence, in this problem, $-\frac{1}{12} = 1/(1 - r^2)$, from which $r^2 = 13$, $r = f/f_n = \sqrt{13}$, and $f_n = (600 \div 60)/\sqrt{13} = 2.77$ cps. Using the weight W on one spring, f_n is equated to $(1/2\pi)\sqrt{kg/W}$. Then

$$2.77 = \frac{1}{2\pi} \sqrt{\frac{k(386)}{20}}$$

from which $k = 15.7$ lb/in.

Forced Vibrations with Viscous Damping

19.26. In Fig. 19.24(a), a mass m is suspended from a spring whose modulus is k. It is also connected to a dashpot that provides viscous damping. Discuss the motion if the mass is subjected to a harmonic disturbing force $F_0 \cos \omega t$.

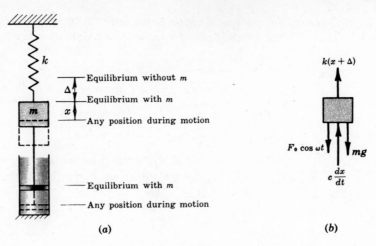

Fig. 19-24

SOLUTION

The free-body diagram in Fig. 19-24(b) shows the forces acting on the mass. Assuming that down is positive, the equation of motion is

$$\sum F = ma$$

or

$$mg + F_0 \cos \omega t - k(x + \Delta) - c\frac{dx}{dt} = m\frac{d^2x}{dt^2}$$

As in previous problems, the free-body diagram for the equilibrium position shows $mg - k\Delta = 0$. Thus, the equation becomes

$$\frac{d^2x}{dt^2} + \frac{c}{m}\frac{dx}{dt} + \frac{k}{m}x = \frac{F_0}{m}\cos \omega t$$

The transient solution will be neglected, as in Problem 19.23. The steady-state solution is

$$x = \frac{F_0 \cos(\omega t - \phi)}{\sqrt{(k - m\omega^2)^2 + (c\omega)^2}} \qquad \text{with} \qquad \tan \phi = \frac{c\omega/k}{1 - m\omega^2/k}$$

Since usually only the amplitude X of the motion is considered, the equation for the amplitude may be written

$$X = \frac{F_0}{\sqrt{(k - m\omega^2)^2 + (c\omega)^2}} = \frac{F_0/k}{\sqrt{(1 - m\omega^2/k)^2 + (c\omega/k)^2}}$$

This is further simplified by noting that F_0/k is the static deflection Δ_{F_0} that the disturbing force would give the spring. Also, $m/k = 1/\omega_n^2$ where ω_n is the natural undamped frequency of the system (rad/s). If d is the ratio of the given damping coefficient c to the critical damping coefficient c_c, the last term in the radicand may be expressed as

$$\frac{c\omega}{k} = \frac{c}{c_c}\frac{c_c\omega}{k} = d\frac{c_c\omega}{k}$$

But the critical damping coefficient (Problem 19.18) is $c_c = 2m\omega_n$. Then the last term of the radicand is

$$d(2m\omega_n)\frac{\omega}{k} = d\left(2\frac{\omega_n\omega}{\omega_n^2}\right) = 2rd$$

where $r = \omega/\omega_n$. Thus,

$$\frac{X}{\Delta_{F_0}} = \frac{1}{\sqrt{(1 - r^2)^2 + (2rd)^2}} \qquad \text{and} \qquad \tan \phi = \frac{2rd}{1 - r^2}$$

The ratio X/Δ_{F_0} is called the magnification factor.

In Fig. 19-25, a graph of the magnification factor versus the frequency ratio shows the peaking that occurs near $r = \omega/\omega_n = 1$ and the influence that damping exerts on the heights of these peaks.

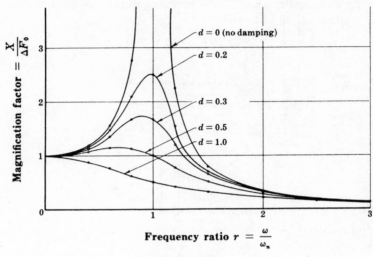

Fig. 19-25

Note that the amplitude peaks in the graph are at a value slightly less than $r = 1$. The exact value may be found by taking the derivative of X/Δ_{F_0} with respect to r and equating the result to zero:

$$\frac{d}{dr}\left(\frac{X}{\Delta_{F_0}}\right) = -\tfrac{1}{2}[(1 - r^2)^2 + (2rd)^2]^{-1/2}[2(1 - r^2)(-2r) + 2(2rd)2d] = 0$$

The radical cannot equal zero; hence, $2(1 - r^2)(-2r) + 2(2rd)2d = 0$ or $r(r^2 + 2d^2 - 1) = 0$.

The solution for the peak amplitude is $r = \sqrt{-2d^2 + 1}$.

In Fig. 19-26, a plot is also shown of the phase angle ϕ for various damping coefficients.

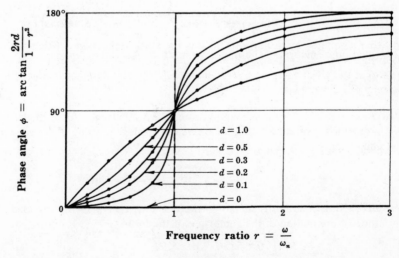

Fig. 19-26

19.27. A small mass m is attached with an eccentricity e to the flywheel of a motor mounted on springs as shown in Fig. 19-27. The modulus of each of the springs is k. The dashpot introduces viscous damping of an amount c. If the total mass of the motor and small mass is M, study the motion of the system under the action of the disturbing force caused by the eccentrically mounted mass m. (This problem illustrates the effect of either reciprocating or rotating unbalance.)

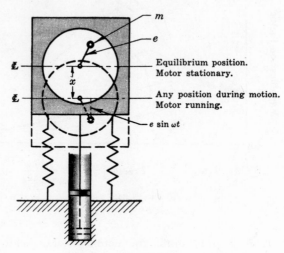

Fig. 19-27

ANALYSIS

In the running position shown, the mass m is below the centerline by an amount $e \sin \omega t$. Hence, if the positive direction is assumed as down, the absolute displacement x_1 of the mass m is the sum of its vertical displacement $(e \sin \omega t)$ relative to the centerline and the absolute displacement x of the centerline; that is, $x_1 = x + e \sin \omega t$.

Let F = unbalanced force exerted by motor on mass m to impart to it an acceleration d^2x_1/dt^2. Then

$$F = m \frac{d^2x_1}{dt^2} = m \frac{d^2(x + e \sin \omega t)}{dt^2} = m \frac{d^2x}{dt^2} - me\omega^2 \sin \omega t$$

The equation of motion of the motor without the small mass m is

$$-F - kx - c \frac{dx}{dt} = (M - m) \frac{d^2x}{dt^2}$$

Note that F is used with a negative sign because this unbalanced force of the mass m on the motor is opposite in direction to that of the motor on m. Substitute the value of F obtained above into the motor equation and simplify to obtain

$$\frac{d^2x}{dt^2} + \frac{c}{M} \frac{dx}{dt} + \frac{k}{M} x = \frac{m}{M} e\omega^2 \sin \omega t$$

This differential equation is similar to the one in Problem 19.26, provided F_0 is replaced by $me\omega^2$. The amplitude of the motion of the centerline of the motor is

$$X = \frac{(m/M)e(\omega/\omega_n)^2}{\sqrt{[1 - (\omega/\omega_n)^2]^2 + (2d\omega/\omega_n)^2}} \quad \text{with} \quad \tan \theta = \frac{2d\omega/\omega_n}{1 - (\omega/\omega_n)^2}$$

where　ω = frequency of disturbance—the motor speed—in rad/s
　　　　ω_n = natural frequency of the spring-supported system in rad/s
　　　　d = damping ratio
　　　　θ = phase angle

At resonance, the above equation reduces to $X = (me/M)/2d$. Also, for very large values of ω/ω_n, the same equation reduces within reasonable limits of accuracy to $X = me/M$.

The distance b from the geometric center O of the motor to the center of mass G of the system, i.e., of the motor plus the mass m, can be determined by taking moments about O. Refer to Fig. 19-28.

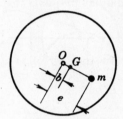

Fig. 19-28

Thus, $bM = em$ or $b = me/M$, which at large values of ω/ω_n is equal to X. Hence, at high motor speeds (ω/ω_n very large), the magnitude of the displacement X of the centerline is equal to the magnitude of b. However, at large values of ω/ω_n, the value of θ, as deduced from the equation

$$\tan \theta = \frac{2d\omega/\omega_n}{1 - (\omega/\omega_n)^2}$$

approaches 180°. Hence, the displacement X equals b but is 180° out of phase.

From the geometry of the figure, the absolute displacement x_G of the center of mass G equals the sum of the absolute displacement of the centerline and the relative displacement of the center of mass to the centerline, or

$$x_G = X \sin (\omega t - 180°) + b \sin \omega t = -b \sin \omega t + b \sin \omega t = 0$$

The center of mass G stands still at high motor speeds.

19.28. A spring-supported mass is subjected to a disturbing force of varying frequency. A resonant amplitude of 12 mm is observed. Also, at very high disturbing frequencies, an almost fixed amplitude of 1.3 mm is observed. What is the damping ratio d of the system?

SOLUTION

At resonance, $X = (me/M)/2d$; and at high frequencies, $X = me/M$. Thus, $me/M = 1.3$ mm. Substituting into the first equation, $12 = 1.3/2d$ or $d = 0.054$.

Supplementary Problems

19.29. A 5-kg mass vibrates with harmonic motion $x = X \sin \omega t$. If the amplitude X is 100 mm and the mass makes 1750 vibrations per minute, find the maximum acceleration of the mass.　　*Ans.*　$a = 3360$ m/s^2

19.30. A cylinder oscillates about a fixed axis with a frequency of 10 cycles per minute. If the motion is harmonic with an amplitude of 0.10 rad, find the maximum acceleration in rad/s².
  Ans. $\alpha = 0.11 \text{ rad/s}^2$

19.31. An instrument weighing 4.4 lb is fastened to four rubber mounts, each of which is rated at 0.125-in deflection per pound loading. What will be the natural frequency of vibration in Hz?
  Ans. $f = 8.44 \text{ Hz}$

19.32. A 5-in-diameter maple log weighing 50 lb/ft³ is 5 ft long. If, while it is floating vertically in the water, it is displaced downward from its equilibrium position, what will be the period of oscillation?
  Ans. $\tau = 2.21 \text{ s}$

19.33. Determine the natural frequency of vertical vibration of a horizontal simple beam of length l to which a mass m is fastened at the midpoint. Neglect the mass of the beam. Note that the deflection of m is $mgl^3/48EI$.   *Ans.* $f = (2/\pi)\sqrt{3EI/ml^3} \text{ Hz}$

19.34. A 100-g mass is fastened to the midpoint of a 150-mm-long vertical wire in which the tension is 15 N. What will be the period of vibration of the mass if it is displaced laterally and then released?
  Ans. $\tau = 0.1 \text{ s}$

19.35. A simple pendulum consists of a small bob of mass m tied to the end of a string of length l. Show that, for small oscillations, the natural frequency is $(1/2\pi)\sqrt{g/l} \text{ Hz}$.

19.36. Refer to Fig. 19-29. Determine the natural frequency of the system composed of a mass m suspended by a spring, whose constant is k, from the end of a massless cantilever beam of length l. [*Hint:* A unit force applied at m will cause it to deflect a total distance of $1/k + 1/(3EI/l^3)$.]

  Ans. $f = \dfrac{1}{2\pi}\sqrt{\dfrac{3EIk}{m(3EI + kl^3)}} \text{ Hz}$

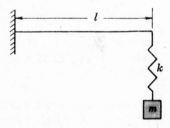

Fig. 19-29

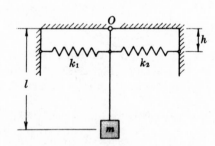

Fig. 19-30

19.37. Determine the natural frequency in Hz of the pendulum system in Fig. 19-30. The springs have constants k_1 and k_2, respectively, and are unstretched in the equilibrium position. Neglect the weight of the stiff rod. (*Hint:* The pendulum in its displaced position is acted on by a gravitational force mg and the sum of the forces in the springs.)

  Ans. $f = \dfrac{1}{2\pi}\sqrt{\dfrac{mgl + (k_1 + k_2)h^2}{ml^2}} \text{ Hz}$

19.38. In Fig. 19-31, the weight $W = 12$ lb. The spring modulus $k = 30$ lb/in. Neglecting the weight of the bell crank, determine the frequency of the system in cps. *Ans.* $f = 1.65$ cps

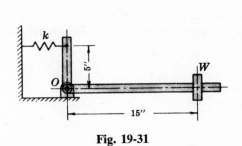

Fig. 19-31

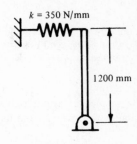

Fig. 19-32

19.39. The slender rod shown in Fig. 19-32 has a mass of 7 kg and is 1200 mm long. The spring constant is 350 N/m. Determine the frequency for small oscillations. *Ans.* 1.87 Hz

19.40. A disk with moment of inertia J_O is rigidly attached to a slender shaft (or wire) of torsional stiffness K. K is the torque necessary to twist the shaft through one radian. What will be the frequency of oscillations if the shaft is twisted through a small angle and then released? See Fig. 19-33.
Ans. $f = (1/2\pi)\sqrt{K/J_O}$ Hz

19.41. A steel disk 100 mm in diameter and 3 mm thick is rigidly attached to a steel wire 0.8 mm in diameter and 500 mm long. What is the natural frequency of this torsional pendulum? *Ans.* 0.84 Hz

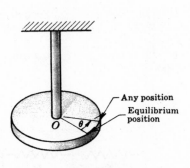

Fig. 19-33

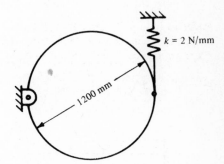

Fig. 19-34

19.42. An engine has a flywheel of 150 lb at each end of a 2-in-diameter steel shaft. Assuming that the equivalent length of shaft between the flywheels is 2 ft, determine the natural frequency of torsional oscillation in cps. The radius of gyration for each flywheel is 8.8 in and $G = 12 \times 10^6$ psi.
Ans. 36.3 cps

19.43. The homogeneous cylinder shown in Fig. 19-34 has a mass of 60 kg and a diameter of 1200 mm. The system is in equilibrium when the diameter as shown is horizontal. The spring, which is vertical in the phase shown, has a spring constant of $k = 2$ N/mm. Determine the frequency for small oscillations.
Ans. 1.41 cps

19.44. In Problem 19.43, attach a vertical dashpot to the center of mass of the homogeneous cylinder. The cylinder is initially rotated 5° clockwise and released from rest. The damping coefficient is one-tenth of the critical damping coefficient. Find the damping coefficient and the angular displacement of the cylinder when $t = 2$ s. *Ans.* $c = 17$ N · s/m, $\theta = 0.013$ rad clockwise

19.45. In Problem 19.37, replace the spring of modulus k_2 with a horizontal dashpot of damping coefficient c. Derive the expression for the critical damping coefficient.

Ans. $c_c = \dfrac{2ml^2}{h^2} \sqrt{\dfrac{kh^2}{ml^2} + \dfrac{g}{l}}$

19.46. A 3-kg mass is attached to a spring with constant $k = 2.5$ N/mm. Determine the critical damping coefficient.   Ans. $c_c = 173$ N $\cdot$ s/m

19.47. Determine the damped frequency of the system shown in Fig. 19-35.

Ans. $\omega_d = \sqrt{\dfrac{kl^2}{mb^2} - \dfrac{c^2}{4m^2}}$ rad/s

Fig. 19-35

19.48. A vibrating system consists of a 5-kg mass, a spring with constant $k = 3.5$ N/mm, and a dashpot with damping constant $c = 100$ N $\cdot$ s/m. Determine (a) the damping factor, d; (b) the damped natural frequency, ω_d; (c) the logarithmic decrement, δ; (d) the ratio of any two successive amplitudes.
Ans.   (a) $d = 0.378$, (b) $\omega_d = 24.5$ rad/s, (c) $\delta = 2.56$, (d) ratio $= 13.0$

19.49. Derive the differential equation of motion for the spring-mounted weight W shown in Fig. 19-36. The weight is subjected to a harmonic disturbing force $F = F_0 \sin \omega t$.

Ans. $\dfrac{d^2x}{dt^2} + (k_A + k_B)\dfrac{gx}{W} = \dfrac{F_0 g}{W} \sin \omega t$

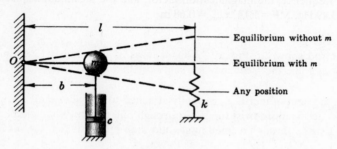

Fig. 19-36

19.50. A machine weighing 200 lb is supported on three springs, each with a constant $k = 60$ lb/in. A harmonic disturbing force of 5 lb acts on the machine. Determine (a) the resonant frequency and (b) the maximum distance the machine moves from equilibrium if the disturbing frequency is 200 cycles per minute. Assume negligible damping.   *Ans.*   (*a*) $f = 2.97$ cps, (*b*) 0.106 in

19.51. A spring-supported mass is subjected to a harmonic disturbing force of varying frequency. A resonant amplitude of 0.82 in is observed. At very high disturbing frequencies, an almost fixed amplitude of 0.07 in is observed. What is the damping ratio *d* of the system?   *Ans.* $d = 0.043$

19.52. A spring-supported mass is subjected to a harmonic disturbing force of varying frequency. A resonant amplitude of 20 mm is observed. At high disturbing frequencies, an almost fixed amplitude of 2 mm is observed. What is the damping ratio of the system?   *Ans.* $d = 0.05$

19.53. In Problem 19.49, let $k_A = k_B = 10$ lb/in, $W = 16$ lb, $F = 12$ lb and the forcing frequency $f = 1.2$ Hz. Find the natural frequency, the magnification factor, and the maximum displacement of the mass.
Ans. $f_n = 3.49$ Hz, MF = 1.13, $x_{max} = 0.68$ in

19.54. In Problem 19.38, a forcing function $8 \cos \omega t$ is applied vertically to the weight *W*. Determine the maximum frequency of the forcing function if the magnification factor is to be no more than 2.
Ans. $f = 1.17$ Hz

19.55. A motor weighing 60 lb is supported by four springs each of modulus $k = 30$ lb/in. The system is damped by a dashpot of damping coefficient *c*. Determine minimum *c* such that oscillation does not occur.
Ans. $c_{min} = 8.64$ lb-s/in

19.56. A small ship model of mass *M* is placed in a towing tank and connected to each end of the tank by springs of spring constant *k*. The model is displaced and allowed to oscillate. It is noted that the period of damped oscillation is $\frac{3}{4}$ s and the ratio of the amplitudes of two successive cycles is $\frac{3}{7}$. Determine the damping coefficient of the fluid in the tank.   *Ans.* $c = 2.27M$

19.57. A spherical mass of $\frac{1}{2}$ slug is suspended in a fluid at the end of a spring of constant $k = 4$ lb/ft. The fluid has a known damping coefficient of 5 lb-s/ft. Determine the displacement of the sphere from equilibrium after 3 s if the initial displacement is $\frac{1}{2}$ ft.   *Ans.* $x = 0.44$ in

19.58. In the preceding problem, how long will it take for the displacement to reach (*a*) $\frac{1}{10}$ of the initial displacements and (*b*) $\frac{1}{100}$ of the initial displacement.   *Ans.*   (*a*) 2.62 s, (*b*) 5.26 s

Appendix A

SI UNITS

The International System of Units (abbreviated SI) has three classes of units—base, supplementary, and derived. The seven base units and two supplementary units are listed below. Also listed are derived units with and without special names as used in mechanics.

BASE UNITS

Quantity	Unit	Symbol
length	meter	m
mass	kilogram	kg
time	second	s
electric current	ampere	A
temperature	kelvin	K
amount of substance	mole	mol
luminous intensity	candela	cd

SUPPLEMENTARY UNITS

Quantity	Unit	Symbol
plane angle	radian	rad
solid angle	steradian	sr

DERIVED UNITS WITH SPECIAL NAMES AND SYMBOLS
(Used in mechanics)

Quantity	Unit	Symbol	Formula
force	newton	N	$kg \cdot m/s^2$
frequency	hertz	Hz	$1/s$
energy, work	joule	J	$N \cdot m$
power	watt	W	J/s
stress, pressure	pascal	Pa	N/m^2

DERIVED UNITS WITHOUT SPECIAL NAMES
(Used in mechanics)

Quantity	Unit	Symbol
acceleration	meter per second squared	m/s^2
angular acceleration	radian per second squared	rad/s^2
angular velocity	radian per second	rad/s
area	square meter	m^2
density, mass	kilogram per cubic meter	kg/m^3
moment of force	newton meter	$N \cdot m$
velocity	meter per second	m/s
volume	cubic meter	m^3

SI PREFIXES
(Commonly used in mechanics)

Multiplication factor	Prefix	Symbol
$1\,000\,000\,000 = 10^{9}$	giga	G
$1\,000\,000 = 10^{6}$	mega	M
$1\,000 = 10^{3}$	kilo	k
$0.001 = 10^{-3}$	milli	m
$0.000\,001 = 10^{-6}$	micro	μ

CONVERSION FACTORS

To convert from	to	multiply by
degree (angle)	radian (rad)	1.745 329 E − 02*
foot	meter (m)	3.048 000 E − 01
ft/min	meter per second (m/s)	5.080 000 E − 03
ft/s	meter per second (m/s)	3.048 000 E − 01
ft/s²	meter per second² (m/s²)	3.048 000 E − 01
ft · lbf	joule (J)	1.355 818 E + 00
ft · lbf/s	watt (W)	1.355 818 E + 00
horsepower	watt (W)	7.456 999 E + 02
inch	meter (m)	2.540 000 E − 02
km/h	meter per second (m/s)	2.777 778 E − 01
kW · h	joule (J)	3.600 000 E + 06
kip (1000 lb)	newton (N)	4.448 222 E + 03
liter	meter³ (m³)	1.000 000 E − 03
mile (international)	meter (m)	1.609 344 E + 03
mile (U.S. survey)	meter (m)	1.609 347 E + 03
mi/h (international)	meter per second (m/s)	4.470 400 E − 01
ounce-force	newton (N)	2.780 139 E − 01
ozf · in	newton meter (N · m)	7.061 552 E − 03
pound (lb avoirdupois)	kilogram (kg)	4.535 924 E − 01
slug · ft² (moment of inertia)	kilogram meter² (kg · m²)	4.214 011 E − 02
lb/ft³	kilogram per meter³ (kg/m³)	1.601 846 E + 01
pound-force (lbf)	newton (N)	4.448 222 E + 00
lbf · ft	newton meter (N · m)	1.335 818 E + 00
lbf · in	newton meter (N · m)	1.129 848 E − 01
lbf/ft	newton per meter (N/m)	1.459 390 E + 01
lbf/ft²	pascal (Pa)	4.788 026 E + 01
lbf/in	newton per meter (N/m)	1.751 268 E + 02
lbf/in² (psi)	pascal (Pa)	6.894 757 E + 03
slug	kilogram (kg)	1.459 390 E + 01
slug/ft³	kilogram per meter³ (kg/m³)	5.153 788 E + 02
ton (2000 lb)	kilogram (kg)	9.071 847 E + 02
W · h	joule (J)	3.600 000 E + 03

---

* E − 02 means multiply by 10^{-2}

First Moments and Centroids

Lines

Kind	Figure	Length	Q_x	Q_y	$\bar{x}$	$\bar{y}$
Straight line	1	L	$\frac{1}{2}L^2 \sin\theta$	$\frac{1}{2}L^2 \cos\theta$	$\frac{1}{2}L\cos\theta$	$\frac{1}{2}L\sin\theta$
Quarter circle	2	$\frac{1}{2}\pi r$	r^2	r^2	$2r/\pi$	$2r/\pi$
Half circle	3	πr	$2r^2$	0	0	$2r/\pi$
Arc	4	$r\alpha$	0	$2r^2 \sin\frac{1}{2}\alpha$	$(r\sin\frac{1}{2}\alpha)/(\frac{1}{2}\alpha)$	0

Areas of Plane Surfaces

Kind	Figure	Area	Q_x	Q_y	$\bar{x}$	$\bar{y}$
Triangle	5	$\frac{1}{2}bh$	$\frac{1}{6}bh^2$	$\frac{1}{6}b^2h$	$\frac{1}{3}b$	$\frac{1}{3}h$
Quadrant of circle	6	$\frac{1}{4}\pi r^2$	$\frac{1}{3}r^3$	$\frac{1}{3}r^3$	$4r/3\pi$	[1]$4r/3\pi$
Quadrant of ellipse	7	$\frac{1}{4}\pi ab$	$\frac{1}{3}ab^2$	$\frac{1}{3}a^2b$	$4a/3\pi$	[2]$4b/3\pi$
Segment of circle	8	$\frac{1}{2}r^2\alpha$	0	$\frac{2}{3}r^3 \sin\frac{1}{2}\alpha$	$(2r\sin\frac{1}{2}\alpha)/(\frac{3}{2}\alpha)$	0

Volumes

Kind	Figure	Volume	Q_{xz}	$\bar{y}$
Hemisphere	9	$\frac{2}{3}\pi r^3$	$\frac{1}{4}\pi r^4$	$\frac{3}{8}r$
Cone	10	$\frac{1}{3}\pi r^2 h$	$\frac{1}{12}\pi r^2 h^2$	$\frac{1}{4}h$
Cylinder	11	$\pi r^2/h$	$\frac{1}{2}\pi r^2 h^2$	$\frac{1}{2}h$

Areas of Curved Surfaces

Kind	Figure	Area	Q_{xz}	$\bar{y}$
Cylinder, closed bottom	12	$\pi r(2h+r)$	πrh^2	$h^2/(2h+r)$
Cylinder, closed top and bottom	(not shown)	$2\pi r(h+r)$	$\pi rh(h+r)$	$\frac{1}{2}h$
Cone, open base	13	πrL	$\frac{1}{3}\pi rhL$	$\frac{1}{3}h$
Cone, closed base	14	$\pi r(L+r)$	$\frac{1}{3}\pi rhL$	$\frac{1}{3}h(1+r/L)$
Hemisphere, open base	15	$2\pi r^2$	πr^3	$\frac{1}{2}r$
Hemisphere, closed base	16	$3\pi r^2$	πr^3	$\frac{1}{3}r$

[1] True also for a semicircular area having the x axis as a base.
[2] True also for a semielliptical area having the x axis as a base.

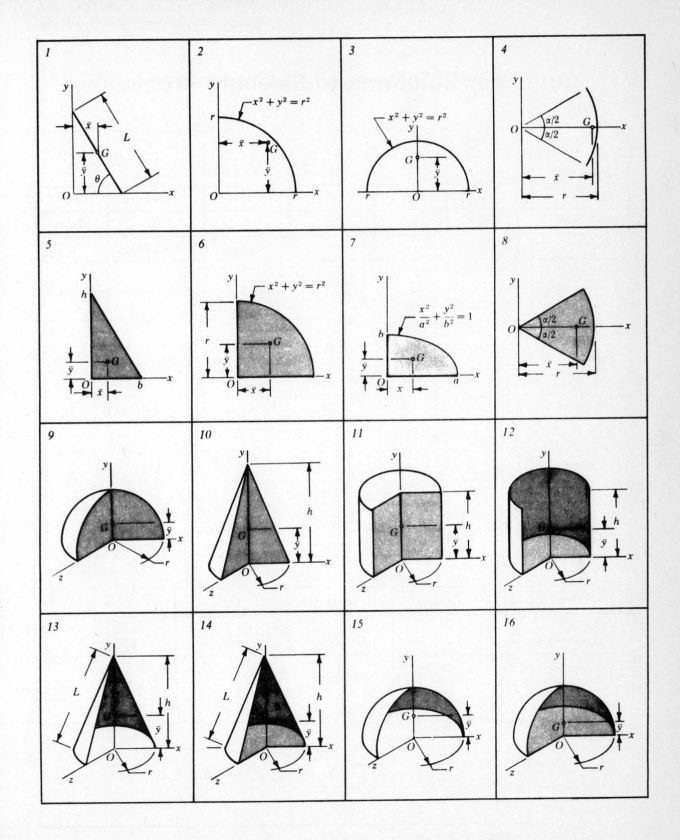

Appendix C

Computer Solutions to Selected Problems

5.18
```
 10 REM PROBLEM 5-18
 20 DIM A[6, 6], X[6], B[6]
 30 MAT READ A
 40 MAT READ B
 50 DATA 0, 0, 0, 1, 0, 1
 60 DATA 0, 0, 1, 0, 1, 0
 70 DATA 0, 0, 0, 0, 3.662, 3.662
 80 DATA 0, 0, −5.77, 0, 0, 0
 90 DATA 5.77, 0, 0, 0, 0, 0
100 DATA 0, 1, 0, −1, 0, 0
110 DATA 0, −223, 315, 1456, 852, 0
120 MAT A=INV(A)
130 MAT X=A * B
140 PRINT
150 PRINT ''B FORCES ARE:'' , X[1], X[2]
160 PRINT
170 PRINT ''D FORCES ARE:'' , X[3], X[4]
180 PRINT
190 PRINT ''C FORCES ARE:'', X[5], X[6]
200 END
```

```
B FORCES ARE:  147.66    −56.6789
D FORCES ARE: −252.34    −56.6789
C FORCES ARE:   29.3397   56.6789
```

6.8
```
 10 REM PROBLEM 6-8
 20 DIM A[3, 3], B[3], X[3]
 30 A[1, 1]=−3/SQR (34)
 40 A[1, 2]=−3/SQR (41)
 50 A[1, 3]=−3/5
 60 A[2, 1]=−4/SQR (34)
 70 A[2, 2]=−4/SQR (41)
 80 A[2, 3]=4/5
 90 A[3, 1]=3/SQR (34)
100 A[3, 2]=−4/SQR (41)
110 A[3, 3]=0
120 B[1]=−100, B[2]=0, B[3]=0
130 MAT A=INV(A)
140 MAT X=A * B
150 PRINT ''FORCES ARE:''
160 MAT PRINT X
```

```
FORCES ARE:
55.5329    45.7366    83.3333
```

7.12
```
 10 REM PROBLEM 7-12
 20 REM NEWTON-RAPHSON ITERATION FOR A ROOT OF F(X)
 30 DEF FNF (X)=X+50−X * CSH(500/(2 * X))
 40 DEF FND (X)=1+X * (SNH(500/(2 * X)) * (500/(2 * X * X)))−CSH (500/2 * X))
 50 PRINT
 60 PRINT ''ENTER APPROX ROOT'':
 70 INPUT X0
 80 PRINT
 90 PRINT ''ENTER ABSOLUTE ERROR'':
100 INPUT E
```

```
110 N=8
120 PRINT
130 FOR I=1 TO N
140    IF FND(X0) < >0 THEN 170
150    PRINT ''DERIVATIVE OF FUNCTION IS ZERO—TRY AGAIN''
160    GOTO 60
170    X=X0—FNF (X0)/FND (X0)
180 NEXT I
190 IF X< >0 THEN 220
200 PRINT ''POSSIBLE ZERO FOUND FOR ROOT''
210 GOTO 270
220 IF ABS((X—X0)/X) < E THEN 270
230 X0=X, N=2 * N
240 IF N< =1028 THEN 130
250 PRINT ''UNABLE TO CONVERGE TO DESIRED ACCURACY''
260 STOP
270 PRINT LIN(1), ''THE ROOT IS''; X; ''THE FUNCTION IS''; FNF (X)
290 END

ENTER APPROX ROOT?635
ENTER ABSOLUTE ERROR?.001
THE ROOT IS 633.163    THE FUNCTION IS 1.22070E—04
```

12.22
```
10 CLS
20 DIM TABLE (10, 2)
30 INPUT ''ENTER LENGTH OF THE CHAIN:''; L
40 INPUT ''ENTER BASE AMOUNT OF OVERHANG:''; C
50 INPUT ''ENTER BASE TIME:''; T
60 COUNT=1:
70 IF COUNT > 10 THEN GOTO 200
80     HALF = .5 * C
90     EX  =SQR((32.2/L) * T)
100    EX1=—SQR((32.2/L) * T)
110    A   = (EXP(EX))
120    B   = (EXP(EX1))
130    X   =HALF * (A+B)
140 TABLE (COUNT, 1)=X
150 TABLE (COUNT, 2)=T
160 PRINT TABLE(COUNT, 2), TABLE(COUNT, 1)
170 COUNT=COUNT+1
180 T=T+.1
190 GOTO 70
200 END
```

This program will yield the following results utilizing the input information shown below.

```
ENTER LENGTH OF THE CHAIN:? 10
ENTER BASE AMOUNT OF OVERHANG:? 1
ENTER BASE TIME:? .1
 .1              1.165367
 .2              1.339656
 .3              1.523155
 .4              1.71616
 .5              1.91897
 .6              2.131895
 .7000001        2.35525
 .8000001        2.589356
 .9000001        2.834544
 1               3.091151
Ok
```

14.9
```
10 REM PROBLEM 14-9
20 REM L=ROD LENGTH
30 REM T=ANGLE
40 REM V=VELOCITY OF A
50 REM W=ANGULAR VELOCITY
60 READ L, V
70 DATA 2.5, 4
80 PRINT
90 PRINT ''ANGLE'', ''VELOCITY''
100 FOR T=0 TO 45 STEP 5
  110 W=V/(L*(COS(T*3.14159/180)+SIN(T*3.14159/180)))
  120 PRINT T, W
130 NEXT T
140 END
```

ANGLE	VELOCITY
0	1.6
5	1.4769
10	1.38115
15	1.30639
20	1.24833
25	1.20398
30	1.17128
35	1.14882
40	1.13569
45	1.13137

Note: The computer programs for problems 7.12, 13.22 and 14.9 will run on IBM-PC or equivalent computers with Microsoft BASIC programming. The computer programs for problems 5.18 and 6.8 will run on a DEC computer or equivalent with VAX BASIC version 3.0 or higher programming.

Appendix D

SAMPLE Screens From
The Companion *Schaum's Electronic Tutor*

This book has a companion *Schaum's Electronic Tutor* which uses Mathcad® and is designed to help you learn the subject matter more readily. The *Electronic Tutor* uses the LIVE-MATH environment of Mathcad technical calculation software to give you on-screen access to approximately 100 representative solved problems from this book, together with summaries of key theoretical points and electronic cross-referencing and hyperlinking. The following pages reproduce a representative sample of screens from the *Electronic Tutor* and will help you understand the powerful capabilities of this electronic learning tool. Compare these screens with the associated solved problems from this book (the corresponding page numbers are listed at the start of each problem) to see how one complements the other.

In the companion *Schaum's Electronic Tutor,* you'll find all related text, diagrams, and equations for a particular solved problem together on your computer screen. As you can see on the following pages, all the math appears in familiar notation, including units. The format differences you may notice between the printed *Schaum's Outline* and the *Electronic Tutor* are designed to encourage your interaction with the material or show you alternate ways to solve challenging problems.

As you view the following pages, keep in mind that every number, formula, and graph shown *is completely interactive when viewed on the computer screen.* You can change the starting parameters of a problem and watch as new output graphs are calculated before your eyes; you can change any equation and immediately see the effect of the numerical calculations on the solution. Every equation, graph, and number you see is available for experimentation. Each adapted solved problem becomes a "live" worksheet you can modify to solve dozens of related problems. The companion *Electronic Tutor* thus will help you to learn and retain the material taught in this book and you can also use it as a working problem-solving tool.

The Mathcad icon shown on the right is printed throughout this *Schaum's Outline* to indicate which problems are found in the *Electronic Tutor*.

For more information about the companion *Electronic Tutor,* including system requirements, please see the back cover.

® Mathcad is a registered trademark of MathSoft, Inc.

Engineering Mechanics: Cross Product
(Schaum's Electronic Tutor Solved Problem 1.17)

Introduction   In this problem the cross product of two vectors is found. The solution is found using the cross product determinant form and by Mathcad matrices vector and matrix operators.

Statement   Find the cross product of **P** and **Q**.

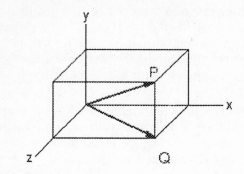

System Parameters   Vector components:

Vector P: $P_x := 2.85 \cdot ft$ $P_y := 4.67 \cdot ft$ $P_z := -8.09 \cdot ft$

Vector Q: $Q_x := 28.3 \cdot lbf$ $Q_y := 44.6 \cdot lbf$ $Q_z := 53.3 \cdot lbf$

Solution   To obtain the solution, the Cartesian unit vectors **i, j,** and **k** must be defined. Cartesian unit vectors, **i, j,** and **k**, are used to designate the x, y, and z directions, respectively.

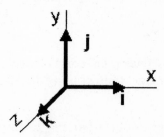

Matrices

The unit vectors are defined in 3 x 1 *matrices* .

Unit vectors: $\mathbf{i} := \begin{pmatrix} 1 \\ 0 \\ 0 \end{pmatrix}$ $\mathbf{j} := \begin{pmatrix} 0 \\ 1 \\ 0 \end{pmatrix}$ $\mathbf{k} := \begin{pmatrix} 0 \\ 0 \\ 1 \end{pmatrix}$

Determinate Form

Cross product in determinant form is calculated as shown below.

$$\mathbf{P} \times \mathbf{Q} = \begin{bmatrix} \mathbf{i} & \mathbf{j} & \mathbf{k} \\ P_x & P_y & P_z \\ Q_x & Q_y & Q_z \end{bmatrix} = (P_y \cdot Q_z - P_z \cdot Q_y) \cdot \mathbf{i} + (P_z \cdot Q_x - P_x \cdot Q_z) \cdot \mathbf{j} + (P_x \cdot Q_y - P_y \cdot Q_x) \cdot \mathbf{k}$$

$$\mathbf{PxQ} := (P_y \cdot Q_z - P_z \cdot Q_y) \cdot \mathbf{i} + (P_z \cdot Q_x - P_x \cdot Q_z) \cdot \mathbf{j} + (P_x \cdot Q_y - P_y \cdot Q_x) \cdot \mathbf{k}$$

$$\mathbf{PxQ} = \begin{pmatrix} 609.725 \\ -380.852 \\ -5.051 \end{pmatrix} \cdot \text{ft} \cdot \text{lbf}$$

Mathcad Matrices Math

Vector **P** equals each vector component, P_x, P_y, P_z, times its respective Cartesian unit vector, **i**, **j**, **k**.

$$\mathbf{P} := P_x \cdot \mathbf{i} + P_y \cdot \mathbf{j} + P_z \cdot \mathbf{k} \qquad \mathbf{P} = \begin{pmatrix} 2.85 \\ 4.67 \\ -8.09 \end{pmatrix} \cdot \text{ft}$$

Vector **Q** equals each vector component, Q_x, Q_y, Q_z, times its respective Cartesian unit vector, **i**, **j**, **k**.

$$\mathbf{Q} := Q_x \cdot \mathbf{i} + Q_y \cdot \mathbf{j} + Q_z \cdot \mathbf{k} \qquad \mathbf{Q} = \begin{pmatrix} 28.3 \\ 44.6 \\ 53.3 \end{pmatrix} \cdot \text{lbf}$$

The cross product is calculated using the cross product *vector and matrix operator*.

Vector and Matrix Operators

$$\mathbf{P} \times \mathbf{Q} = \begin{pmatrix} 609.725 \\ -380.852 \\ -5.051 \end{pmatrix} \cdot \text{ft} \cdot \text{lbf}$$

Exploration

Which method of finding the cross product did you prefer; finding the cross product using the determinant form or using the Mathcad cross product vector and matrix operator?

What is the cross product when **P** = 2.83 **i** + 4. 46 **j** + 5.33 **k** ft.

Use this worksheet to solve problems 1.47, 1.50, and 1.55.

Engineering Mechanics: Concurrent Force System
(Schaum's Electronic Tutor Solved Problem 4.1)

Introduction
In this problem the resultant of spatial concurrent forces is found. The resultant of a concurrent force system may be (a) a single force through the concurrence or (b) zero.

Statement
Force F_1, F_2, F_3, and F_4 are concurrent at the origin and are directed through points 1, 2, 3, and 4, respectively. Determine the resultant of the system.

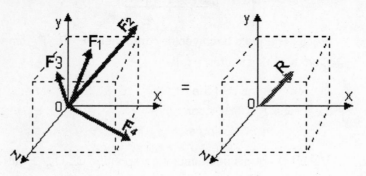

System Parameters

Number of forces: $N := 4$

Head and tail coordinates of vectors:

Subscripts

Point 0:	$x_0 := 0 \cdot ft$	$y_0 := 0 \cdot ft$	$z_0 := 0 \cdot ft$
Point 1:	$x_1 := 2 \cdot ft$	$y_1 := 1 \cdot ft$	$z_1 := 6 \cdot ft$
Point 2:	$x_2 := 4 \cdot ft$	$y_2 := -2 \cdot ft$	$z_2 := 5 \cdot ft$
Point 3:	$x_3 := -3 \cdot ft$	$y_3 := -2 \cdot ft$	$z_3 := 1 \cdot ft$
Point 4:	$x_4 := 5 \cdot ft$	$y_4 := 1 \cdot ft$	$z_4 := -2 \cdot ft$

Forces:

Force 1:	$F_1 := 20 \cdot lbf$
Force 2:	$F_2 := 15 \cdot lbf$
Force 3:	$F_3 := 30 \cdot lbf$
Force 4:	$F_4 := 50 \cdot lbf$

Solution

Range Variable

Define a *range variable,* **i,** for convenience in performing iterative calculations.

$$i := 1 .. N$$

Each spatial force, **F**, has an x component, F_x, y component, F_y, and z component F_z.

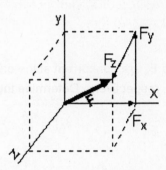

The x component of the resultant, R_x, equals the sum of the x components of the forces, F_x. The y component of the resultant, R_y, equals the sum of the y components of the forces, F_y. The z component of the resultant, R_z, equals the sum of the z components of the forces, F_z.

The x, y, and z components of each force, F_x, F_y, and F_z, are found by multiplying the force magnitude, **F**, by the direction cosines. The direction cosines equal the length of the vector, x, y, and z directions divided by the magnitude of the position vector, r.

x direction: $\cos\left(\theta_x\right) = \dfrac{F_x}{F} = \dfrac{x}{r}$ and $F_x = F \cdot \dfrac{x}{r}$

y direction $\cos\left(\theta_y\right) = \dfrac{F_y}{F} = \dfrac{y}{r}$ and $F_y = F \cdot \dfrac{y}{r}$

z direction $\cos\left(\theta_z\right) = \dfrac{F_z}{F} = \dfrac{z}{r}$ and $F_z = F \cdot \dfrac{z}{r}$

The force component calculations are performed iteratively. Because the vectors are concurrent the tail of each vector is located at Point 0 (x_0, y_0, z_0)

x components: $\quad F_{x_i} := F_i \cdot \dfrac{x_i - x_0}{\sqrt{(x_i - x_0)^2 + (y_i - y_0)^2 + (z_i - z_0)^2}}$

F_{x_i}
lbf
6.25
8.94
- 24.05
45.64

y components: $\quad F_{y_i} := F_i \cdot \dfrac{y_i - y_0}{\sqrt{(x_i - x_0)^2 + (y_i - y_0)^2 + (z_i - z_0)^2}}$

F_{y_i}
lbf
3.12
- 4.47
- 16.04
9.13

z components: $\quad F_{z_i} := F_i \cdot \dfrac{z_i - z_0}{\sqrt{(x_i - x_0)^2 + (y_i - y_0)^2 + (z_i - z_0)^2}}$

F_{z_i}
lbf
18.74
11.18
8.02
- 18.26

The x, y, and z components of the resultant, R_x, R_y, and R_z, are found by summation of the x, y, and z components, F_x, F_y, and F_z, of each force. The summation is accomplished using the *summation operator.*

Summation

x component: $\quad R_x := \displaystyle\sum_{i=1}^{N} F_{x_i} \qquad R_x = 36.78 \cdot lbf$

y component: $\quad R_y := \displaystyle\sum_{i=1}^{N} F_{y_i} \qquad R_y = -8.26 \cdot lbf$

z component: $\quad R_z := \displaystyle\sum_{i=1}^{N} F_{z_i} \qquad R_z = 19.68 \cdot lbf$

The magnitude of the resultant, **R**, is determined by the Pythagorean theorem.

$$R := \sqrt{\left(R_x\right)^2 + \left(R_y\right)^2 + \left(R_z\right)^2}$$ $R = 42.53 \cdot lbf$

The direction angles θ_x, θ_y, θ_z which the resultant **R** makes with the x, y, and z axis are found by direction cosines.

x direction: $\theta_x := acos\left(\dfrac{R_x}{R}\right)$ $\theta_x = 30.13 \cdot deg$

y direction $\theta_y := acos\left(\dfrac{R_y}{R}\right)$ $\theta_y = 101.19 \cdot deg$

z direction $\theta_z := acos\left(\dfrac{R_z}{R}\right)$ $\theta_z = 62.43 \cdot deg$

Exploration

What is the resultant and direction cosines angles when the point of concurrence, Point 0, has coordinates (10 ft, 10 ft, 10 ft)?

This worksheet can be used to find the resultant of any number of concurrent forces. Use this worksheet to solve problems 4.6, 4.7, 4.8, 4.9, and 4.10

Engineering Mechanics: Center of Pressure (Gate)
(Schaum's Electronic Tutor Solved Problem 10.28)

Introduction   In this problem the concept of center of pressure is applied to an hydraulic gate.

Statement   A rectangular gate separates fluids of two different densities. The gate is hinged at the top and rests against a stop at the bottom. Find **d**, the greatest difference in depth for which the gate will remain closed.

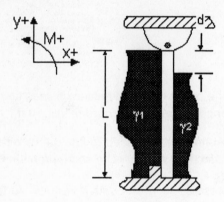

System Parameters

Gate length:	$L := 12 \cdot ft$	
Liquid 1 specific weight:	$\gamma_1 := 62.4 \cdot \dfrac{lbf}{ft^3}$	
Liquid 2 specific weight:	$\gamma_2 := 105 \cdot \dfrac{lbf}{ft^3}$	

Solution   Draw a free-body diagram of the gate as shown below. The fluid pressure on each side of the gate is zero at the fluid surface and maximum at the gate bottom. The pressure distribution is linear (triangular shape), therefore, the center of pressure is located 1/3 of the fluid depth from the bottom or 2/3 of the fluid depth from the fluid surface.

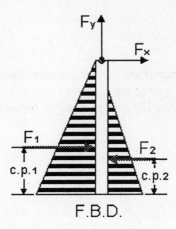

F.B.D.

There are two solution steps. The first is to write an equation of force for each side of the gate, and the second is to write an equation of summation of moments of forces about the hinge and solve for **d**.

Fluid pressure, **p**, (at a depth **h** from the surface) equals fluid specific weight, γ, times fluid depth, **h**.

$$p = \gamma \cdot h$$

Fluid force, **F**, equals fluid pressure, **p**, times area, **A**.

$$F = p \cdot A$$

A force equation in terms of fluid specific weight, depth, and area is obtained by substituting the pressure equation into the force equation.

$$F = (\gamma \cdot h) \cdot A$$

Now, a force equation for each side of the gate for a unit width of gate (notice unit width of 1) can be written. Notice, on side 1 the fluid depth at the surface is 0 and the fluid depth a the bottom is L; so the average depth is L/2 ([0+L]/2). Similarly, the average depth on side 1 is [L-d]/2.

Side 1: $F_1 = \gamma_1 \cdot \dfrac{L}{2} \cdot (L \cdot 1)$

Side 2: $F_2 = \gamma_2 \cdot \left(\dfrac{L-d}{2} \right) \cdot ((L-d) \cdot 1)$

Distance **d** is found by writing an equilibrium equation of summation of moments of forces about the hinge.

$$\Sigma M_{hinge} = F_1 \cdot \left(\frac{2}{3} \cdot L \right) - F_2 \cdot \left[d + \left[\frac{2}{3} \cdot (L-d) \right] \right] = 0$$

$$\Sigma M_{hinge} = \left[\gamma_1 \cdot \frac{L}{2} \cdot (L \cdot 1) \right] \cdot \left(\frac{2}{3} \cdot L \right) - \left[\gamma_2 \cdot \frac{L-d}{2} \cdot ((L-d) \cdot 1) \right] \cdot \left[d + \left[\frac{2}{3} \cdot (L-d) \right] \right] = 0$$

Solution of this equation for **d** is accomplished using a *solve block*.

Solve Block

Initial solution estimate: $d := 2 \cdot ft$

Given

$$\left[\gamma_1 \cdot \frac{L}{2} \cdot (L \cdot 1) \right] \cdot \left(\frac{2}{3} \cdot L \right) - \left[\gamma_2 \cdot \frac{L-d}{2} \cdot ((L-d) \cdot 1) \right] \cdot \left[d + \left[\frac{2}{3} \cdot (L-d) \right] \right] = 0 \cdot lbf \cdot \frac{ft}{ft}$$

$d := find(d)$

$d = 3.331 \cdot ft$

Exploration

If the specific weight of the fluid on side 2 is 200 lbf/ft³ what is the greatest difference in depth for which the gate will remain closed?

Engineering Mechanics: Acceleration (Terminal Velocity)
(Schaum's Electronic Tutor Solved Problem 13.53)

Introduction   In this problem the fall of an object through a resistant medium and terminal velocity is explored.

Statement   A body with mass, **M**, falls in a medium where the resistance is **k**. What is the terminal velocity?

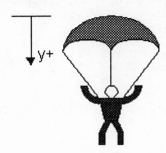

System Parameters   Body mass: $M := 1.5 \cdot kg$

Resistance: $k := 0.7 \cdot \dfrac{newton \cdot sec}{m}$

Solution   Draw a free-body diagram of the body falling as shown.

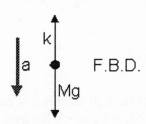

F.B.D.

Write an equation of motion for the body.

$$\Sigma F_y = M \cdot g - k \cdot v = M \cdot a \qquad \text{or} \qquad \Sigma F_y = M \cdot g - k \cdot v = M \cdot \frac{dv}{dt} \qquad \text{(Eq. 1)}$$

The result is equation 1, a differential equation that can be solved by separation of variables.

Divide by mass

$$\frac{M}{M} \cdot g - \frac{k}{M} \cdot v = \frac{M}{M} \frac{dv}{dt} \qquad\qquad g - \frac{k}{M} \cdot v = \frac{dv}{dt}$$

and let $C := \dfrac{k}{M} \qquad\qquad g - C \cdot v = \dfrac{dv}{dt}$

Separate the variables $dt = \dfrac{dv}{g - C \cdot v}$

and integrate

$$\int 1 \, dt = \int \dfrac{dv}{g - C \cdot v} \, dv$$

which yields

$$t + D = -\dfrac{1}{C} \cdot \ln(g - C \cdot v)$$

$$-C \cdot t + -C \cdot D = \ln(g - C \cdot v)$$

$$e^{-C \cdot t} \cdot e^{-C \cdot D} = g - C \cdot v \qquad\qquad \text{(Eq. 2)}$$

The constant of integration, **D**, is found by application of the initial conditions.

$$v = 0 \qquad @ \qquad t = 0$$

$$e^{-C \cdot (0)} \cdot e^{-C \cdot D} = g - C \cdot (0)$$

$$1 \cdot e^{-C \cdot D} = g \qquad\qquad \text{(Eq. 3)}$$

Function

Substitute equation 3 into equation 2, solve for velocity, **v**, and write the equation as a *function*.

$$e^{-C \cdot t} \cdot g = g - C \cdot v$$

$$C \cdot v = g - g \cdot e^{-C \cdot t}$$

$$v(t) := \dfrac{g}{C} \cdot \left(1 - e^{-C \cdot t}\right) \qquad\qquad \text{(Eq. 4)}$$

Range Variable

Now we can graph the velocity of the falling body over time. Define time, **t**, as a *range variable*.

$$t := 0 \cdot \sec .. \, 100 \cdot \sec$$

Velocity versus time is graphed on an *x-y plot*.

X-Y Plots

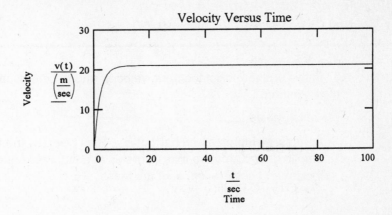

The graph shows that terminal velocity is reached after a "long" time. Mathematically, the exponential in the equation approaches zero after a "long" time. In this problem 100 seconds is a long time.

$$e^{-C \cdot (100 \cdot sec)} = 0$$

and terminal velocity is given by

$$v_{terminal} := \frac{g}{C} \cdot (1 - 0)$$

$$v_{terminal} := \frac{g}{C}$$

$$v_{terminal} = 21 \cdot \frac{m}{sec}$$

Exploration

What is the terminal velocity if the body has a mass of 4 kg and the resistance is 0.3 N s/m?

Engineering Mechanics: Harmonic Motion
(Schaum's Electronic Tutor Solved Problem 19.29)

Introduction

In this problem displacement, velocity, and acceleration of harmonic motion is explored.

Statement

A mass, **M**, vibrates with harmonic motion given by the function **x(t)**. If the amplitude is **X** and the mass vibrates at a natural frequency ω, find the maximum acceleration, **a**, of the mass.

System Parameters

Mass: $M := 5 \cdot kg$

Amplitude: $X := 100 \cdot mm$

Natural frequency: $\omega := 1750 \cdot \dfrac{2 \cdot \pi}{min}$

Function

Motion *function*: $x(t) := X \cdot \sin(\omega \cdot t)$

Solution

The velocity of the mass, **v(t)**, equals the first derivative of displacement and acceleration, **a(t)**, of the mass equals the second derivative of displacement.

Velocity: $v(t) := X \cdot \cos(\omega \cdot t) \cdot \omega$

Acceleration: $a(t) := - X \cdot \sin(\omega \cdot t) \cdot \omega^2$

Maximum acceleration will occur when the quantity sin(ωt) equals 1 (or -1).

$$a_{max} := - X \cdot 1 \cdot \omega^2$$

$$a_{max} = -3358.41 \cdot \frac{m}{sec^2}$$

Exploration

Range Variable

X-Y Plots

Let's explore the displacement, velocity, and acceleration of the mass through a complete period.

A period, τ, equals 2π divided by the natural frequency, ω.

$$\tau := \frac{2 \cdot \pi}{\omega} \qquad\qquad \tau = 0.03 \cdot \sec$$

Define time, $\mathbf{t_{range}}$, as a *range variable*.

$$t_{range} := 0 \cdot \tau, \frac{\tau}{50} \,..\, \tau$$

Graph displacement, velocity, and acceleration on an *X-Y plot*.

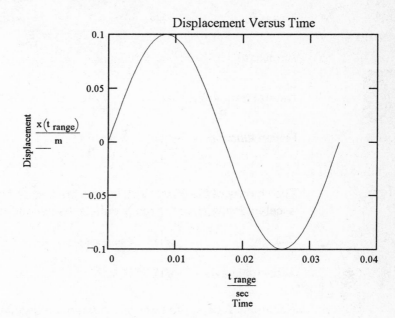

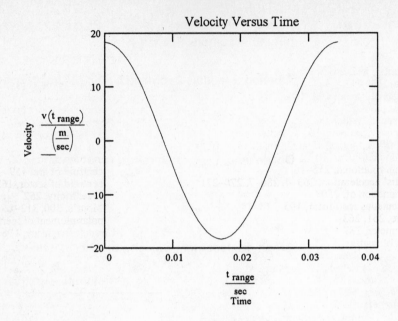

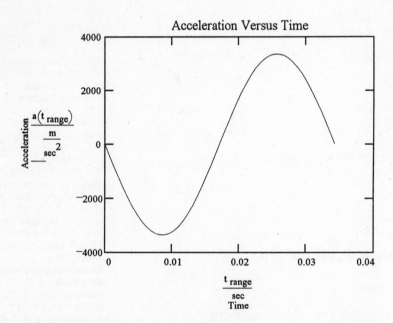

When displacement and acceleration are maximum, velocity equals _____ .

Index

The following license agreement will apply if this Schaum's Outline contains a Schaum's Electronic Tutor.

The McGraw-Hill Companies, Inc. and MathSoft, Inc. - Schaum's Electronic Tutors; MathSoft, Inc--Mathcad Engine
License Agreement and Limited Warranty.

Both the program(s) (including all data and information contained therein) and the documentation are protected under applicable copyright laws. Your right to use the program(s) and the documentation are limited to the terms and conditions described herein.

Limited Non-Exclusive License
You May: (a) use the enclosed program(s) on a single computer; (b) physically transfer the program(s) from one computer to another provided that the program(s) are used on only one computer at a time, and that you remove any copies of the program(s) from the computer from which the program(s) are being transferred; (c) make copies of the program(s) solely for backup purposes. You must reproduce and include the copyright notice on a label on any back-up copy.

You May Not: (a) distribute copies of the program(s) or the documentation to others; (b) rent, lease or grant sublicenses or other rights to the program(s); (c) provide use of the program(s) in a computer service business, network, time-sharing, multiple CPU or multiple users arrangement without the prior written consent of MathSoft; (d) translate, decompile, reverse engineer or otherwise alter the program(s) or related documentation without the prior written consent of MathSoft.

Terms
Your license to use the program(s) and documentation will automatically terminate if you fail to comply with the terms of this Agreement. If this license is terminated you agree to destroy all copies of the program(s) and documentation.

Limited Warranty
MathSoft and McGraw-Hill warrant to the original licensee that the disk(s) on which the program(s) are recorded will be free from defects in materials and workmanship under normal use for a period of ninety (90) days from the date of purchase as evidenced by a copy of your receipt. If failure of the disk(s) has resulted from accident, abuse or misapplication of the product, then McGraw-Hill, MathSoft or third party Licensors shall have no responsibility to replace the disk(s) under this Limited Warranty.

This limited warranty and right of replacement is in lieu of, and you hereby waive, any and all other warranties both express and implied, including but not limited to warranties of merchantability and fitness for a particular purpose. The liability of McGraw-Hill, MathSoft or third party Licensors pursuant to this limited warranty shall be limited to the replacement of the defective disk(s), and in no event shall McGraw-Hill, MathSoft or third party Licensors be liable for incidental or consequential damages, including but not limited to loss of use, loss of profits, loss of data or data being rendered inaccurate, or losses sustained by third parties even if McGraw-Hill, MathSoft or third party Licensors have been advised of the possibility of such damages. McGraw-Hill and MathSoft make no representation or warranty that the results obtained will be successful or satisfy licensee's requirements, and McGraw-Hill and MathSoft shall have no liability for any errors or omissions in the programs or in the data and information contain therein. This warranty gives you specific legal rights which may vary from state to state. Some states do not allow the limitation or exclusion of liability for consequential damages, so the above limitation may not apply to you.

This License Agreement shall be governed by the laws of the Commonwealth of Massachusetts and shall inure to the benefit of McGraw-Hill, MathSoft, their respective successors, administrators, heirs and assigns of third party Licensors.